Modelling Under Risk and Uncertainty

WILEY SERIES IN PROBABILITY AND STATISTICS

Established by WALTER A. SHEWHART and SAMUEL S. WILKS

A complete list of the titles in this series appears at the end of this volume.

Modelling Under Risk and Uncertainty

An Introduction to Statistical, Phenomenological and Computational Methods

Etienne de Rocquigny

Ecole Centrale Paris, Université Paris-Saclay, France

WILEY

A John Wiley & Sons, Ltd., Publication

This edition first published 2012

Registered office
John Wiley & Sons Ltd, The Atrium, Southern Gate, Chichester, West Sussex,
PO19 8SQ, United Kingdom

For details of our global editorial offices, for customer services and for information about how to apply for permission to reuse the copyright material in this book please see our website at www.wiley.com.

Library of Congress Cataloging-in-Publication Data

Rocquigny, Etienne de.
Modelling under risk and uncertainty: an introduction to statistical, phenomenological, and computational methods / Etienne de Rocquigny.
p. cm. – (Wiley series in probability and statistics)
Includes bibliographical references and index.
ISBN 978-0-470-69514-2 (hardback)
1. Industrial management–Mathematical models. 2. Uncertainty–Mathematical models. 3. Risk management–Mathematical models. I. Title.
HD30.25.R63 2012
338.501′5195–dc23

2011046726

A catalogue record for this book is available from the British Library.

ISBN: 978-0-470-69514-2

Set in 9/11pt Times Roman by Thomson Digital, Noida, India
Printed and bound in Singapore by Markono Print Media Pte Ltd

Luke 13: 4-5

Or those eighteen on whom the

tower at Siloam fell, do you

suppose they had failed in their

duty more than all the rest of the

people who live in Jerusalem?

Contents

Preface

This book is about *modelling* in general, as a support for decision-making in man-made systems and/or in relation to life and the natural environment. It is also about *uncertainty*: in the view of the author, modelling is so connected to uncertainty that the expressions of 'uncertainty modeling' or 'model uncertainty' are fraught with pleonasm. It is just as much about *risk* modelling, meaning the crafting of models to help risk assessment and support decision-making to prevent risk. Incidentally, controlling the risk generated by the use of models – a critical issue for many analysts of the world financial crises since 2008 – is a key by-product of the analysis.

Coming from a traditional – that is deterministic – scientific and engineering background, the author experienced regular discomfort in acting as a young consulting engineer in the mid-1990s: (a) struggling to calibrate models on past figures before discovering horrendous errors in the most serious datasets and previous model-based expertise, or large deviations between client-provided data and his own field surveys; (b) being challenged on the confidence that could be placed in his models by acute clients or regulators; (c) being pressured by the same clients to reduce the study budget and hence to downsize data collection and model sophistication. A memorable case was that of a small village of the French Alps where the author had to draw an official line between the floodable and non-floodable areas under intense pressure for land development. A witty-enough client, the mayor, questioned the author as to whether the *1/100-yr* return legal contours would change if account was made for missing memory about passed floods (which he knew of from the tales of an old relative), likely deviations of the river bottom over decades (which he knew of from being born there), inaccuracies in the datum of land in the village (which he knew of as being also a farm owner), or an increased study budget allowing for a more careful survey and modelling exercise (which he eventually made available to the author). The nights were unsettled after having also understood the extent of responsibility held by the consulting engineers in signing such uncertain contours.

Since that time, the author has moved quite a long way into the science and techniques of modelling in the context of risk and uncertainty, assuming responsibilities in industrial R&D, environmental management, nuclear risk assessment and the development of large academic research programmes and education on the topic. A pedagogical example – on flood protection of an industrial facility – that has long been used as a tutorial and benchmark for that work and which illustrates the entire development of this book, is directly inspired by the live case. The writing of the book was almost complete when the tragic events of March 2011 occurred at Fukushima/Sendai, recalling, if necessary, the dramatic impact of natural and technological risk and the core need for careful design, safety and maintenance procedures. The author is hopeful that the book may serve modellers to better support those decision makers facing uncertainty.

Acknowledgements

The author wishes to thank everyone who helped make this book possible. The book could not have existed without the debt owed by the author to his former employers Sogreah and *Electricité de France*, being richly inspired by his close collaborators and partners such as Serge Hugonnard and Yannick Lefebvre in particular as well as Emmanuel Remy, Christian Derquenne, Laurent Carraro and Gilles Celeux.

Nicolas Devictor, Stefano Tarantola, Philip Limbourg and the other co-authors of the Wiley book 'Uncertainty in Industrial Practice' and the European cross-industry network, as well as its first-class reviewers Terje Aven and Jon Helton, helped in crafting and challenging the generic modelling approach in early versions of the book. Others such as Nicolas Fischer, Erick Herbin, Alberto Pasanisi, Fabien Mangeant, Bertrand Iooss and Regis Farret accompanied the author in launching pioneering executive education programmes and Master's/PhD level courses at the *Ecole Centrale Paris* as well as providing key opportunities to complete the pedagogical approach and tutorial examples. Numerous Master's, PhD students and industrial attendees involved in those initiatives also questioned and helped refine the material through years of teaching practice.

Thanks are also merited by a large number of reviewers including Jérôme Collet, Bertrand Iooss, Emilia Ferrario and Enrico Zio, Yann Richet, Pierre Bertrand, Bernard Gourion, Guillaume Malingue and Eric Walter.

In the long term, moral support and encouragement from Aude Massiet du Biest proved essential.

Boscrocourt (France), October 3rd, 2011

Introduction and reading guide

Modelling – a many centuries old activity – has permeated virtually all areas of industrial, environmental, economic, bio-medical or civil engineering. The driving forces behind such a development are scientific knowledge, technological development and cheaper information technologies that deliver massive computer power and data availability. Essential to any modelling approach, the issues of proper qualification, calibration and control of associated error, as well as uncertainty or deviations from reality have arisen simultaneously as the subjectof growing interest and as areas of new expertise in themselves.

Indeed, the search for more advanced management of uncertainty in modelling, design and risk assessment has developed greatly in large-scale industrial and environmental systems with the increase in safety and environmental control. Even more so, an appetite for more accountable decision-making and for greater confidence in predictions has found a general audience in the fields of weather forecasting, health and public policy. In parallel, robust design approaches and the valuation of flexibility through option-based decision-making are more and more disseminated in various industrial sectors in the quest for an increase in customer satisfaction, design and manufacturing cost reduction and market flexibility.

Faced by such demands, a systematic approach to uncertainty linked to the modelling of complex systems proves central in establishing a *robust* and convincing approach to the *optimisation* of *safety* and *performance margins*. The applications to engineering design, environmental management and risk management are numerous; embracing also concerns for the wider issues of resource allocation involved in modelling-based activity be it in R&D, consulting, policy-making or applied research. Strategic choices are required in order to *focus* data collection, computing and analytical resources or research budgets on areas where the lack of knowledge or uncertainty proves to have the *greatest importance* from the *perspective* of the ultimate *decision-making goal* instead of the favourite or best-known field of expertise.

This book concentrates on *quantitative* models and *quantifiable* uncertainty or risk indicators. This may sound somewhat restrictive to those readers having in mind real-world situations. On the one hand, human and organisational factors play an essential role, as evidenced by the famous risk examples of the Three-Mile-Island accident (1979) in the nuclear field or the Space Shuttle Challenger disaster (1986) in aerospace. On the other, uncertainty is to a large extent poorly or not quantifiable at all. Think about quantifying the lack of knowledge about the plausibility of the September 11th terrorist attacks (2001) or another similar event. Eventually, decision-making involves much more than an interaction with analytics and quantitative figures. Those aspects are evidence of the limitations of the applicability of the book's methods, although some limited modelling contributions may still apply here and there, such as human reliability probabilistic models, or the extended deterministic-probabilistic settings that account for poorly quantifiable occurrences. The challenging frontiers of *statistical*-based *modelling* also involve some fundamental mathematical properties, introduced intuitively as follows: (a) *phenomenological stationarity*: the elements of the system – if not the system itself – need to behave similarly in a statistical and/or phenomenological sense for a model-based inference to be of help; unseen combinations or extrapolations of known elementary phenomena could be considered, but not completely-disruptive phenomenology; (b) *limited range of dependence*: strong dependence between far-flung parts of a system or distant events in time also cause trouble for statistical and computational methods, if not the credibility of the models themselves. Experts in the rising sciences of complex systems including the celebrated author of 'the Black Swan' (Taleb, 2007) would qualify 'true uncertainty' as being characterised precisely by strong deviations from those properties. Although this book introduces a number of mathematical extensions of those cases, they remain as strong scientific bottlenecks demanding care, modesty and more research.

In spite of those limitations, modelling brings with it essential tools for the existing and future regulation of industrial activity or environmental control, as well as for decision-making in corporate investment, public infrastructure or health. Consider the dramatic events that occurred in the wake of the Sendai earthquake (March 2011) and the consequences of the Fukushima nuclear disruptions. Although the natural hazards outside the nuclear plant have so far dominated in terms of the number of fatalities which occurred along the densely-populated coast, the events at the nuclear plant had major consequences for nuclear safety and acceptability. Pre-existing statistical and geological evidence on the local tsunami hazard (not to mention more sophisticated modelling of the quake and the hydrodynamics of the tsunami) could have been valued directly in the re-assessment of plant safety and in particular to remedy shortcomings in the height of the dike and the location of emergency power redundancies (see Chapter 3). To a lesser extent, probably because of the complexity of modelling world economic behaviour, the massive crises following the 2007–2008 market failures find part of their explanation with the over-estimation of risk mutualisation within extremely complex portfolios which hid a very high level of dependence upon risky assets in the American real estate sub-primes, a basic feature easily understood in standard risk modelling (see Chapter 4).

1. The scope of risk and uncertainty considered

Risk and *uncertainty* are closely linked concepts and distinguishing them is not free from controversy. Facing up to the hard task of defining these two concepts precisely, some preliminary comments will be made at this stage. *Uncertainty* is the subject of long-standing epistemological interest as it stands with a fundamental connection both to any type of *modelling* activity and the scientific consideration of *risk*. The former relates to the fact that any modelling endeavour brings with it a more or less explicit consideration of the uncertain deviation of the model from empiric evidence. The latter focuses on the detrimental (or beneficial) consequences of uncertain events in terms of some given stakes, assets or vulnerability of interest for the decision-maker. Risk and uncertainty analysis prove to be so connected in application that the precise delimitation between the practical meanings of the two terms will not appear to be central with regard to modelling: it depends essentially on the definition of the underlying system or on the terminological habits of the field under consideration. The system considered generally includes the undesired consequences or performance indicators of a given industrial facility or environmental asset in a fully-fledged risk assessment, while it may be limited to a technical or modelling sub-part seen from a narrower perspective when talking about uncertainty or sensitivity analysis. Chapter 1 reviews an entire spectrum of applications from natural risk, industrial design, safety or process optimisation to environmental impact control and metrological or numerical code validation. The underlying concepts will prove greatly similar in spite of the apparent terminological variety.

Indeed, from the perspective of the book, the practicality of studying the uncertainty or risk associated with a given system is generated by the following common features: (i) the fact that the state of the system considered, conditional upon taking some given actions, is imperfectly known at a given time; and (ii) the fact that some of the characteristics of the state of the system, incorporated in a given type of 'performance' or 'consequence', are at stake for the decision-maker. Because of (i), the best that may be looked for are *possible* or *likely* ranges for the variables of interest quantifying those characteristics. More specifically, any inference will be made under a probabilistically defined *quantity of interest* or *risk measure*,[1] such as event probabilities, coefficients of variation of the best-estimate, confidence intervals around the prediction, values-at-risk and so on. The rationale of risk or uncertainty modelling is to estimate those quantities through the aggregation of all statistical or non-statistical information available on any type of variable linked to the system, with the piece of information brought by statements depicting the system's structure and phenomenology (physical laws, accident sequence analysis, etc.). Eventually, the goal is to help in making decisions about appropriate *actions* in the context of a *relative* lack of knowledge about the state of the system and given stakes of interest: a model and the associated risk measure are only suitable for that given purpose and there is no such criterion as an *absolute* quality of modelling or recommended degree of sophistication.

Combining both statistical and phenomenological models aims to produce the best-informed inference, hopefully less uncertain than a straightforward prediction obtained through pure empiric data (observed frequencies) or expert opinion only. Note that from the book's perspective, risk modelling is much more than the derivation of self-grounded advanced mathematical tools as epitomised by the critiques of the role played by quantitative financial models in the 2008 market crises. It should be deeply grounded in statistical, phenomenological and domain-specific knowledge so that a proper validation (or invalidation) can be undertaken with a robust, uncertainty-explicit, approach. Representing honestly the *quality of information* – a dual concept of uncertainty – is about the most important recipe for building a sound model, and the book will insist on the key concepts available – such as model *identifiability*, uncertainty arising in the analysis process itself, quality of calibration and so on - that tend to be neglected by a tropism of analysts for quantitative and formal sophistication. In fact, an innovative practical measure – the *rarity index* – proportioning the true needs of decision-making to the real amount of information available (given the crucial degree of independence of both data sources and/or sub-system failures) will be introduced in order to challenge the credibility of model-building and of the risk assessment figures (End of Chapter 5, Chapter 7).

A large variety of causes or considerations gives rise to practical uncertainty in the state of a system as defined above in (i). This includes: uncertain inputs and operating conditions in the industrial processes; model inaccuracy (simplified equations, numerical errors, etc.); metrological errors; operational unknowns; unpredictable (or random) meteorological influences; natural variability of the environment; conflicting expert views and so on. A rich literature has described the variety of natures of uncertainty and discussed the key issue of whether they should or could receive the same type of quantification efforts, in particular a probabilistic representation. Such a debate may be traced back to the early theory of probability rooted in the seventeenth century while modern thinking may be inherited from *economics* and *decision theory*, closely linked to the renewed interpretation of statistics. Knight (1921) introduced the famous distinction between 'risk' (i.e. unpredictability with known probabilities) and 'uncertainty' (i.e. unpredictability with unknown, imprecise probabilities or even not subject to probabilisation). It is less often remembered how these early works already admitted the subtleties and limitations incorporated in such a simplified interpretation regarding real physical systems that greatly depart from the type of closed-form quantitative lotteries studied in the economic or decision-theory literature.

A closer look reveals evidence that various forms of uncertainty, imprecision, variability, randomness and model errors are mixed inside phenomenological data and modelling. Tackling real-scale physical

[1] The expression *quantity of interest* can be used in the context of any modelling activity (de Rocquigny, Devictor, and Tarantola, 2008) while the wording of a *risk measure* would probably better fit the context of decision-making under risk or risk assessment. Yet the two expressions refer to a strictly similar mathematical quantity (Chapter 2) and are thus completely synonymous throughout the book.

systems within the *risk assessment* community occupied to a large extent the 1980s and 1990s, closely linked to the development of US nuclear safety reviews or environmental impact assessments (Granger Morgan and Henrion, 1990; Helton, 1993). The classification of the large variety of types of uncertainty encountered in large industrial systems (Oberkampf *et al.*, 2002) has been notably debated with regard to their classification into two salient categories. The rationale for separating or not those two categories: namely the *epistemic* or reducible type which refers to uncertainty that decreases with the injection of more data, modelling or the number of runs; and the *aleatory* or irreducible type (or variability) for which there is a variation of the true characteristics across the time and space of an underlying population that may not be reduced by the increase of data or knowledge. This epistemic/aleatory distinction, which may be viewed as a refinement of the earlier uncertainty/risk distinction, will be discussed in the subsequent chapters as it gives rise to a variety of distinct probabilistic risk measures. While many recent works experimented with and discussed the use of extra-probabilistic settings such as evidence theory for the representation of epistemic uncertainty (*cf.* the review in Helton and Oberkampf, 2004), this book takes the more traditional view of double-probabilistic epistemic/aleatory modelling both because of its relatively more mature status in regulatory and industrial processes and the wealth of fundamental statistical tools that is available for error control of estimation and propagation procedures as well as more recent inverse probabilistic techniques.

2. A journey through an uncertain reality

Simple examples will help one to grasp the scope of this book and the reader is also invited to refer to de Rocquigny, Devictor, and Tarantola (2008) which gathers a large number of recent real-scale industrial case studies. Look firstly at the domain of *metrology*. Knowing the value of a quantitative variable representing the state or performance of the system requires an observational device which inevitably introduces a level of *measurement uncertainty* or *error*. Figure 1 illustrates the measurement results from a given device as compared to the reference values given by a metrological standard (i.e. a conventionally

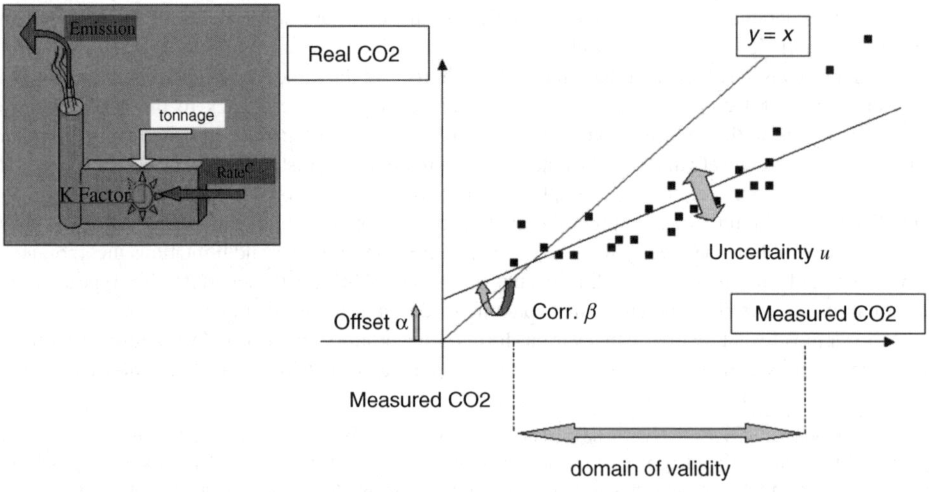

Figure 1 *Metrological uncertainty in the CO_2 emissions of an industrial process.*

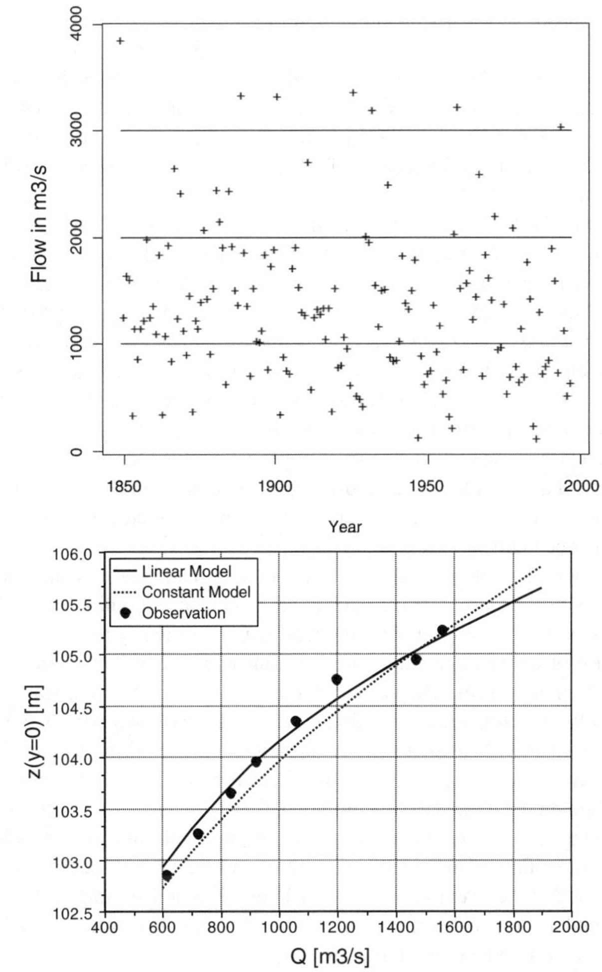

Figure 2 *(up) annual maximal flood flows from a French river – (down) observed stage-discharge relationship and competing hydraulic model parameterisations (Goutal et al., 2008).*

'finer' measurement). Uncontrollable dispersion remains even after the best calibration efforts, the first materialisation of uncertainty. Such uncertainty will need to be estimated and modelled in order to undertake, for instance, a risk assessment of an emission control system.

Within the field of natural phenomena, temporal or spatial variability is ubiquitous and has a number of impacts: natural aggressions or environmental constraints on settlements or the operation of industrial facilities. Figure 2-up illustrates the year-to-year *natural variability* of maximal flows from a French river. Because of the well-known sensitivity to initial conditions, predicting river flows through a deterministic phenomenological approach such as meteorological and hydrodynamic equations is limited to the short-term of a few days ahead for accurate trajectories. Techniques of forecasting, which will be briefly introduced in Chapter 6, have complemented such a deterministic inference by the calibration of explicit *forecasting uncertainty* allowing for probabilistic forecasts over days or weeks ahead or even for seasonal trends a few months ahead. Nevertheless, prediction of extreme events possibly happening once over decades is more the focus of risk models which are central to this book, in close relationship to the statistical extreme value theory. As the estimation of the extreme values of such flows is necessary when designing flood protection systems, it will typically be subject to a statistical model based on hydrological records: empiric records are of limited size by nature, hence generating a second type of *estimation uncertainty* linked to the statistical fluctuation of sampling (see Annex Section 10.1 for a refresher on modelling through a *random* variable).

Yet the variable of interest will generally be distinct from the variable for which records are available: a phenomenological model will be used so that the local water levels relevant to the protection system may be inferred. Calibration of such a model against available field data (see Figure 2-down) raises the issue of controlling a form of 'uncertain best fit'. Even for accurately-known flows, the prediction of water levels eventually exhibits *residual dispersion* that mixes measurement error and model and parametric uncertainty. This constitutes a third type of uncertainty that must be accounted for, quite distinct from the natural variability affecting the observed flows and more complex than direct estimation uncertainty. A number of techniques that are in the process of being developed could be used to work on those subjects. Uncertainty may stem from measurement error, improper model assumptions, imprecise spatial or temporal discretisation of the equations or unknown input parameters; in all cases, it needs to be combined to some extent with the distribution of natural variability in order to secure an overall robust design and a solid risk assessment procedure.

Uncertainty and unpredictability are also prevalent in man-made systems, motivating *reliability and failure studies*: for instance, failure dates are unknown and dispersed within a population of industrial components (Figure 3 up). Modelling such failure randomness is the key to assessing the (probabilistic) lifetime and reliability of components and hence guarantees the safety of entire systems aggregating redundant or complementary pieces. They are traditionally modelled through simple statistical models such as exponential or Weibull *lifetime* distributions. Looking at this in more detail, the phenomenology of such failures may be valuable with regard to achieving better control of its influencing factors or predicting the *structural reliability* of systems operating outside the data basis. This is done, for instance, in fracture mechanics where failure modes related to flaw initiation, flaw propagation or structural ruin are controlled both by stress and toughness. While temperature may be physically modelled as a key determinant of toughness, residual uncertainty remains in the calibration of such a model with experimental data (Figure 3-down). Predicting the overall lifetime or risk level requires turning such mixed statistical and phenomenological models for each failure mode into a combined risk measure, for instance the probability of failure over a given operation period.

The *design*, manufacturing and operation of *new* industrial *systems,* technological products and services are strong driving forces for the development of decision-support *modelling*. They also raise peculiar challenges regarding the confidence that may be placed in those tools and the associated issues of qualification, certification or optimisation: systems that do not yet exist involve by nature a type of unpredictability that resists direct statistical or evidence-based modelling. Various forms of engineering errors, manufacturing variability or deviations of operational use with respect to the initial plan are intertwined. Add the more fundamental lack of knowledge about the phenomenology of truly-innovative systems, particularly when involving human interaction, and one thus makes more complex the control of any model that can be thought of. Yet, a number of techniques of sensitivity analysis, calibration through the knowledge of sub-systems or the use of subjective probabilistic modelling will be introduced in that respect.

Eventually, complex systems at risk combine those various factors of uncertainty, variability and unpredictability. *Accident analysis* would typically require the consideration of a variety of external or internal uncertain disrupting events (hazards) that could resemble either the above-mentioned natural phenomena (e.g. flood, storm, tsunami, earthquakes, heat or coldwaves ...) and/or failure of man-made components (independently or consequently to those external hazards) leading to the failure of a given

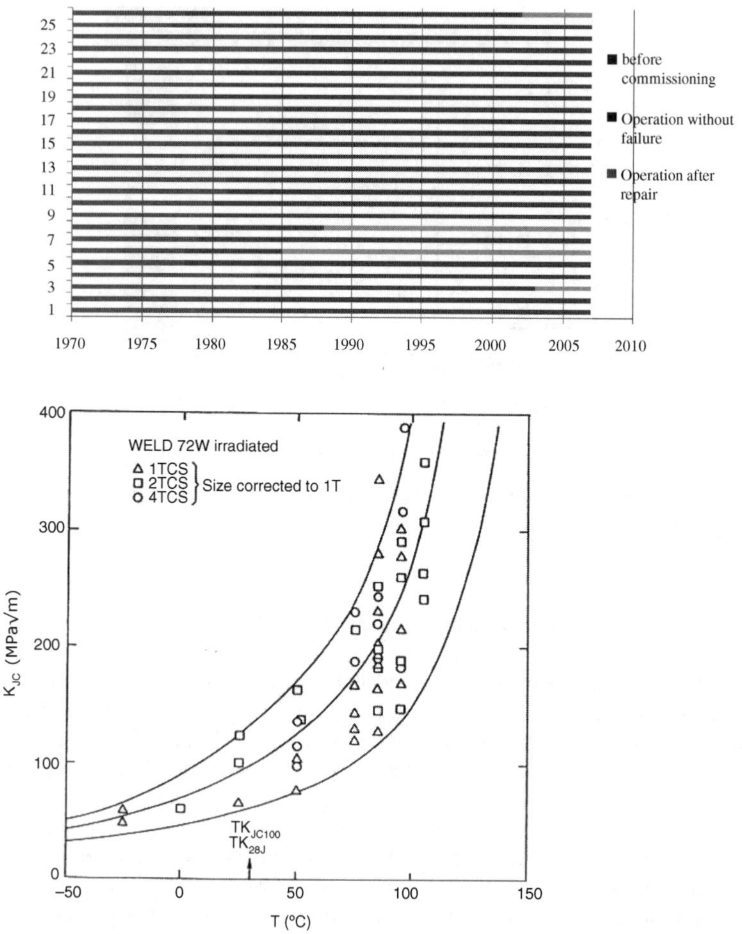

Figure 3 *Up: log of commissioning and failure dates of similar independent components of power plants (personal communication to the author) – down: dispersion of toughness or irradiated steel welds with temperature (adapted from Wallin, 1991).*

system. Such an undesired event can henceforth entail a series of detrimental consequences through subsequent explosion, spill of liquids or gases and so on ending up in material and human costs or environmental impacts (such as plant unavailability costs, large release of pollutants, number of fatalities, etc.) depending on the level of exposure and vulnerability. Again, various types of uncertainty limit the control of those downstream consequences, be it due to the uncertain efficiency of man-made protection equipments or procedures, to variability of the meteorological or environmental conditions and/or of the presence of people (hence of *exposure*) at the unknown time of the accident, and or lack of knowledge of the eventual *vulnerability* to such pollutants or aggressions. Similar features of uncertainty characterise the *environmental* or *health impact* studies of man/made systems or infrastructure, not only conditional to a failure event but also under normal operation. The bow-tie scheme (see Figure 4) illustrates a common overview of the modelling approach to such analyses through the complex chaining of a fault tree from

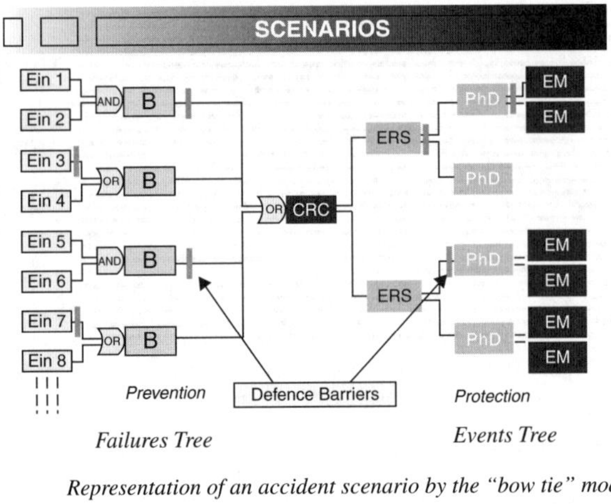

Figure 4 *Failure tree/event tree model of hazard and consequences (adapted from source: ARAMIS European project (Salvi and Debray, 2006)).*

initiators to the central undesired event and of an event tree predicting the consequences from the central undesired event.

Hazard and/or consequence prediction could link logical models, viz. fault or event trees, and phenomenological models, typically physical, structural, chemical, environmental or economic models describing quantitatively the elementary uncertain events involved. Eventually, the coupling of numerous models of all the sub-systems and events important for the understanding of a complex system end up in challenging computational programs. This is all the more so since joining their deterministic and probabilistic parts representing all kinds of phenomenology and uncertainty will require the joint use of numerical and probabilistic simulation techniques, as this book will explain in detail. Their implementation would generally require further simplification and approximation, hence giving rise to another source of *propagation uncertainty* stemming from the limitations of the computational budget rather than those of data or knowledge about the system. Although often neglected by the analyst any serious uncertainty modelling exercise ought never to introduce more uncertainty in the figures than is already in the system itself; the latter represents an objective limit for large-scale industrial studies: hopefully, some techniques discussed in this book can help in controlling it.

3. The generic methodological approach of the book

Notwithstanding such diversity of origins of uncertainty and risk phenomena within natural or man-made systems, this book will show that associated studies do involve similar *key steps* and a range of mathematical synergies. Indeed, quite distinct options remain regarding the peculiar choice of mathematical representation of uncertainty: this depends on the nature of the system and the precise specification of the variables of interest for the decision-maker. But the following generic framework will be a recurring concept throughout the book: the stakes in decision-making involve the control of the key *outputs of interest* of a system and the optimisation of associated *actions* (or designs, operation options ...) faced with uncertainty in the states of the system (Figure 5).

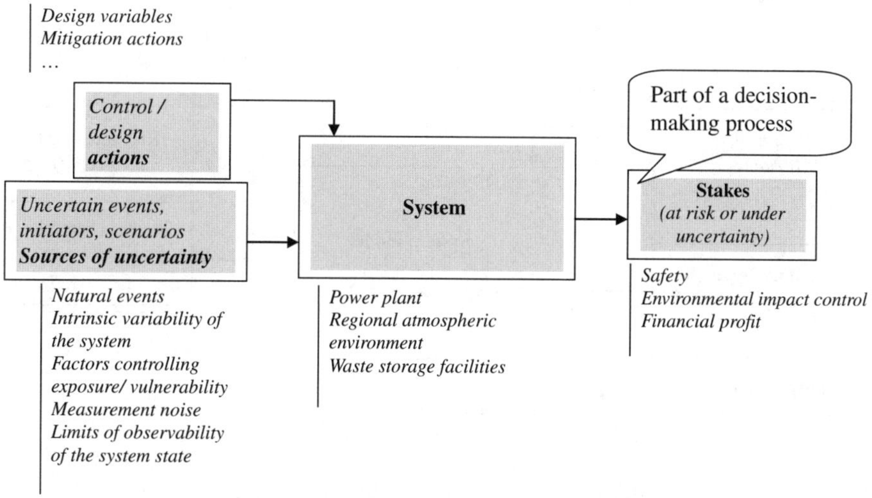

Figure 5 *The pre-existing or system model and its inputs/outputs.*

Risk/uncertainty-conscious modelling mobilises a range of statistical, physical and numerical techniques designed to encode data and expertise in a combined phenomenological and probabilistic model (*system* and *uncertainty model*). This is meant to *best* represent the behaviour of the system outputs of interest under uncertainty *given* the variables of interest for the decision-maker and the scope of possible actions; hence, it aims to support valuable or accountable decisions – formally selecting $\boldsymbol{d}$ - with respect to a certain criterion $c_Z(\boldsymbol{d})$ which will be called the *risk measure* or *quantity of interest* (Figure 6).

Eventually model-based decision-making under a uncertainty or risk assessment are all about weighing relative uncertain options and a model should help in calculating the plausible (relative) likelihoods of the outcome. The risk measure or quantity of interest, a central concept of the book, is a quantitative figure summarising – mostly in a *relative* manner – the extent of uncertainty in the stake-linked variable of interest; associated quantities are the sensitivity indices or importance factors apportioning the relative importance of the different inputs in explaining the level of risk or uncertain spread of the output. They play a central role in fulfilling the four salient goals suggested to categorise the rationale of modelling a system under uncertainty: *Understanding* the behaviour of a system, *Accrediting* or validating a model, *Selecting* or optimising a set of actions and *Complying* with a given decision-making criterion or regulatory target. The key analytical steps involved in the performance of such studies and detailed in the chapters of the book are summarised in Figure 7.

4. Book positioning and related literature

Probabilistic modelling for design, decision-support, risk assessment or uncertainty analysis has a long history. Pioneering projects in the 1980s relied on rather simplified mathematical representations of the systems, such as closed-form physics or simplified system reliability, involving a basic probabilistic treatment such as purely expert-based distributions or deterministic consequence modelling. Meanwhile, quality control enhancement and innovation programs in design and process engineering started with highly simplified statistical protocols or pure expertise-based tools. Thanks to the rapid development of

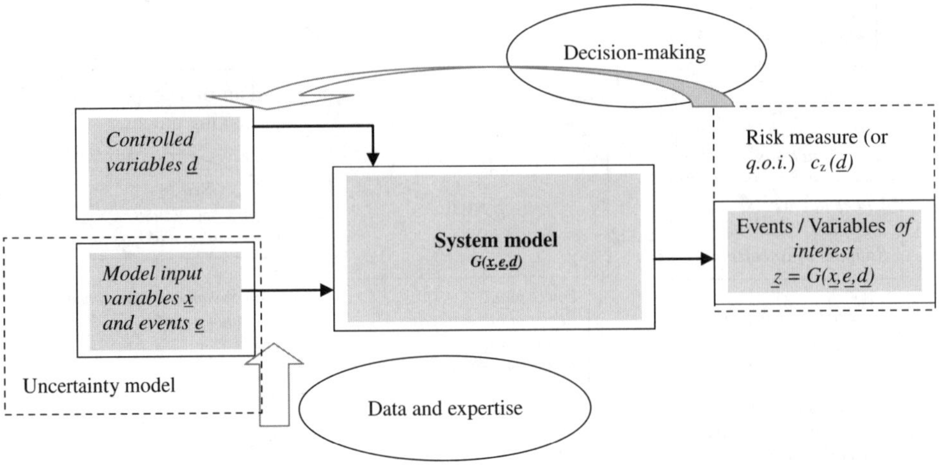

Figure 6 *The generic modelling framework.*

computing resources, probabilistic approaches have gradually included more detailed physical-numerical models. Such complex modelling implies a finer calibration process through heterogeneous data sets or expertise. The large CPU time requirement is all the more demanding since the emerging regulatory specification is intended to adequately predict rare probabilities or tail uncertainties. Conversely, the importance of critical issues of *parsimony* or control of the risk of over-parameterisation of analytics exaggerating the level of confidence that may be placed in predictions through uncontrollably-complex

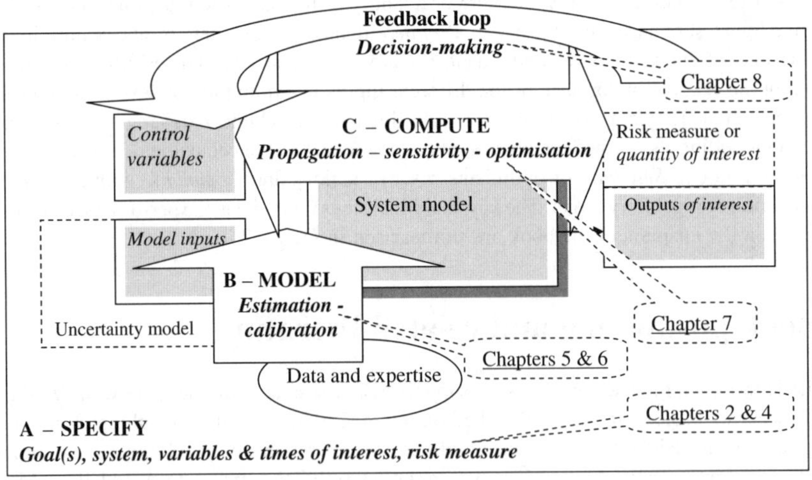

Figure 7 *Key generic steps of modelling in the context of risk and uncertainty – corresponding chapters of the book.*

models becomes all the greater. This gives rise to new challenges lying at the frontier between statistical modelling, physics, scientific computing and risk analysis.

Indeed, there seems to be insufficient co-operation between domain experts, physicists, statisticians, numerical experts and decision-support and risk analysts in applications. One of the concerns of this book is to go beyond the 'black-box' view developed in the underlying phenomenology of the environmental or industrial processes in most risk studies or asset management systems. Such an approach is thought to miss the modelling potential associated with an explicit description of their physical properties. Conversely, this book aims to lead environmental or engineering modellers facing new regulations from deterministic decision-making to more elaborate statistical and risk analysis material. This means, for instance, taking advantage of scientific computing enabling sophisticated estimation, calibration or simulation techniques, extreme event and robust estimators, dependence modelling and Bayesian algorithms. One of the central arguments of the book is that in-depth mathematical formulation of both physical properties and statistical modelling features inside a combined risk model brings scientific and regulatory value. Monotony, for instance, will be commented upon throughout the book as being both a very intuitive and rather powerful property for efficient and robust modelling, notwithstanding some limitations that can be circumvented through careful hypothesis-setting. Together with bounded behaviour or prior physical knowledge, such phenomenological knowledge proves to be helpful, for instance in tackling the challenges of estimator convergence and residual uncertainty.

An substantial literature has developed on modelling, uncertainty and risk analysis, robust design and decision-making under uncertainty. The particular focus of this book, besides concentrating on areas other than the extensively-addressed financial risks, is to provide new material at the frontier between statistics, computational and physical (or more generally phenomenological, meaning also biological, economic, etc.) modelling: this refers more specifically to the situation where information takes the form of statistical data as well as physical (or engineering) models. In such a context, the book develops a consistent mathematical approach that links the type of formulation common to risk and reliability analysts (e.g. Bedford and Cooke, 2001; Aven, 2003; Singpurwalla, 2006; Zio, 2009) with those used by structural reliability and mechanical engineering (Madsen *et al.*, 1986; Rackwitz, 2001) uncertainty and sensitivity analysts (e.g. Granger Morgan and Henrion, 1990; Cooke, 1991; Helton, 1993; Helton *et al.*, 1996; Saltelli *et al.*, 2004; de Rocquigny, Devictor, and Tarantola, 2008), environmental and natural risk modellers (e.g. Krzystofowicz, 1983; Beck, 1987; Bernier, 1987; Beven and Binley, 1992; Hamby, 1994; Duckstein and Parent, 1994; Apel *et al.*, 2004), statistical and computational research (e.g. de Groot, 1970; Kleijnen and Sargent, 2000; Coles, 2001; Kennedy and O'Hagan, 2002).

While being broadly introduced hereafter, the challenges of complex system reliability such as functional system complexity and dynamic system behaviour – that dominate the elementary physical behaviours in systems made of numerous inter-connected components and limit the significance of statistical data – will be documented in more detail in the specialised reliability literature (see the review in Zio, 2009). Correspondingly, this book is more centred on *risk* modelling than *forecast* modelling in the sense of long-term deviations due to rare events rather than short-term time correlation-based predictions that stand at the basis of weather forecasting, early-warning systems or economic forecasting (but see Chapter 6 for introductory comments). Additionally, for advanced sensitivity analysis involving statistical learning, design of computer experiments or Gaussian process modelling, the reader is referred to more specialised books such as Kleijnen, (2007) and Saltelli *et al.* (2008). The scientific material in this book refers essentially to applied mathematics, drawing on concepts from statistics, probability theory, as well as numerical analysis. Most of the proofs underlying the book's derivations are comprehensive, starting right from elementary results (see the Annexes for both a starter on statistical modelling and in-depth demonstrations or theoretical details); yet a thorough refresher might be obtained from the wealth of graduate-level mathematical textbooks or those more specific to probabilistic engineering (e.g. Ang and Tang, 1984).

Table 1 A roadmap through the book chapters.

To start with:
- Chapter 1 – review of current practices (natural risk, industry, environment, finance and economics)
- Chapter 3 – the generic pedagogical *Flood example*
- Annex Section 10.1 – a refresh on statistical modelling
- Chapter 2, Sections 2.1–2.3, 2.6–2.7 – overview of the general methodology

Central chapters:
- Chapter 2 and 4 – foundations of modelling and the rationale of probabilistic risk measures
- Chapter 5 (Sections 5.1–5.3) and 6 (Sections 6.1–6.2) – estimation methods
- Chapters 7 (Section 7.2 + Section 7.5) and 8 – computational methods and decision-making
- Chapter 9 for conclusion and further research

Advanced reading:
- Chapters 5 (Sections 5.4–5.5), 6 (Sections 6.3–6.5) and 7 (Sections 7.1, 7.3–7.4) – advanced statistical and computational methods
- Annexes Sections 10.2–10.5 for formal comments, detailed implementation and proofs
- Exercises at the end of Chapters 2–8

5. A reading guide through the chapters

The concepts and formulations in this book have an essential grounding in practice throughout the variety of domains which are reviewed in Chapter 1: natural risk, industrial risk and process optimisation, metrology, environmental and sanitary protection, and engineering economics. Common key steps and considerable mathematical synergies will be generated within a generic modelling framework described in Chapter 2. Chapter 4 discusses the practical implementation of the various risk/uncertainty measures or criteria that all come into the generic framework, but correspond to different epistemological options coming from decision-theory. This relates in particular to the classical concern of distinguishing according to the aleatory or epistemic nature of uncertainty, specifying more clearly the variability to be covered by the risk measure, mixing deterministic and probabilistic settings, or specifying temporal conventions with the building of composite risk measures (confidence intervals on top of exceedance probabilities, peak events in time, etc.).

Hence, a number of statistical and computing challenges stand as generic to a number of situations. *Estimation* issues with samples and expertise are discussed in two chapters. Firstly, in Chapter 5 with the case where direct information is available on the uncertain variables or risk components. Simultaneous estimation of both aleatory and epistemic components involves classical or Bayesian statistics (also benefiting from expressing physical properties) that allow for much richer descriptions than elementary Gaussian or uniform models, up to rare events of unlimited variability through a careful use of the extreme value theory. Chapter 6 then discusses the estimation theory in the alternative case where only indirect information is available, in the sense that physical or functional models are to be inversed in order to retrieve the useful variables. In other words, inference of uncertainty *pdf* involves hidden input variables of large physical models for which only the outputs are observable. Inverse probabilistic problems include data assimilation – regression settings for calibration and full probabilistic inversion to identify the intrinsic variability of inputs. This is a spectrum of quite different motivation-based techniques for which clear distinctions are made and advanced research is discussed. Chapter 6 also introduces the distinctive features of dynamic models in forecasting and associated data assimilation inverse techniques.

Chapter 7 covers risk *computation* or uncertainty *propagation* of the risk measures once the statistical model of the sources of uncertainty has been estimated. Statistical-numerical or structural reliability procedures are investigated from the point of view of convergence within the physical models. Indeed, viewing the algorithms from the mapping of physical spaces, with properties such as partial monotony, or known bounds, allows for greater relevance in choosing the best compromises in applied large CPU models. Regarding sensitivity and importance analysis, a short review is given of the abundant and recent literature on global sensitivity analysis (such as Sobol', meta-modelling, etc.).

Chapter 8 formulates explicitly the use of the cost function in order to introduce a starter on methods for decision-making under uncertainty, involving the classical expected utility approaches, robust design, value of information and related alternatives in economic theory, as well as the peculiar issues arising when considering decision-making over time (discounting, real option analysis). Eventually an introduction is given to the adding of a layer of stochastic *optimisation* on top of uncertainty models, an essential area for industrial decision-making and also a large source of CPU complexity when implemented in real physical models. Although the book is not centred on finance or insurance applications, links to the basic financial and actuarial financial modelling concepts are also provided.

Chapter 3 stands apart from the rest of the book: it introduces a physically-based *pedagogical example* that illustrates most of the issues and methods developed throughout the book. This example, which originated in industrial tutorials and academic courses, represents schematically the flood risk in an industrial facility: hydraulics, sedimentology, climatology, civil engineering and economics are the phenomenological components, all of them simplified into easily understandable closed-form formulae. Numerical implementation details are provided to the reader as a support for tutorial exercises. Note that many algorithms introduced by the book, and illustrated in the tutorial example, can be found within the Open TURNS open source development platform (www.openturns.org) which was launched in close connection with the dissemination course and executive education programmes led by the author since 2005.

The Annexes successively provide: a refresher on statistical modelling for non-specialists; theoretical comments on the probabilistic foundations of the models, and reflections on the origins of macroscopic uncertainty; numerical results and further details of the pedagogical example; detailed mathematical derivations of some important results and tools for the algorithms.

Last but not least, most of the chapters of the book end with a self-contained series of exercises of various levels of complexity which are intended to serve both as (i) study questions for Master/PhD level or advanced executive education programmes as well as (ii) starting ideas, reflections and open challenges for further research, often illustrated by the closed-form example with full numerical details (see Annex). As will be discussed throughout the book as well as in the concluding Chapter 9, many scientific challenges remain open in the field; incidentally, note that the pedagogical example has already served as a benchmark case for a number of recent research papers.

References

Ang, A.H.S. and Tang, W.H. (1984) *Probability Concepts in Engineering, Planning and Design*, vol. 2, John Wiley & Sons, Ltd., Chichester.

Apel, H., Thieken, A.H. *et al.* (2004) Flood risk assessment and associated uncertainty. *Natural Hazard and Earth System Sciences*, **4**, 295–308.

Aven, T. (2003) *Foundations of Risk Analysis*, John Wiley & Sons, Ltd.

Beck, M.B. (1987) Water quality modelling: A review of the analysis of uncertainty. *Water Resources Research*, **23**(8), 1393–1442.

Bedford, T.J. and Cooke, R.M. (2001) *Probabilistic Risk Analysis – Foundations and Methods*, Cambridge University Press.

Bernardara, P., de Rocquigny, E., Goutal, N. *et al.* (2010) Flood risk and uncertainty analysis: joint assessment of the hydrological & hydraulic components. *Canadian Journal of Civil Engineering*, **37**(7), 968–979.

Bernier, J. (1987) Elements of bayesian analysis of uncertainty in reliability and risk models. In Duckstein, L., Plate E.J., Engineering reliability and risk in water resources, NATO ASI Series E: Applied Sciences, **124**, 405–422.

Beven, K.J. and Binley, A.M. (1992) The future of distributed model: model calibration and uncertainty prediction. *Hydrological Processes*, **6**, 279–298.

de Groot, M.H. (1970) *Optimal Statistical Decisions*, McGraw-Hill, NewYork.

de Rocquigny, E., Devictor, N. and Tarantola, S. (eds) (2008) *Uncertainty in Industrial Practice, A Guide to Quantitative Uncertainty Management*, Wiley.

Goutal, N., Arnaud, A., Dugachard, M., de Rocquigny, E. and Bernardara, P. (2008) *Discharge and Strickler coefficient uncertainty propagation in a one-dimensional free surface hydraulic model*, Geophysical Research Abstracts, **10**, EGU2008-A-11844.

Granger Morgan, M. and Henrion, M. (1990) *Uncertainty – A Guide to Dealing with Uncertainty in Quantitative Risk and Policy Analysis*, Cambridge University Press.

Hamby, D.M. (1994) A review of techniques for parameter sensitivity analysis of environmental models. *Environmental Monitoring and Assessment*, **32**(2), 135–154.

Helton, J.C. (1993) Uncertainty and sensitivity analysis techniques for use in performance assessment for radioactive waste disposal. *Reliability Engineering & System Safety*, **42**, 327–367.

Helton, J.C., Burmaster, D.E. *et al.* (1996) Treatment of aleatory and epistemic uncertainty, *Special Issue of Rel. Eng. & Syst. Saf.*, **54**(2–3).

Helton, J.C. and Oberkampf, W.L. (2004) Alternative representations of epistemic uncertainty. *Special Issue of Reliability Engineering & System Safety*, **85**(1–3).

Kennedy, M.C. and O'Hagan, A. (2002) Bayesian calibration of computer models. *Journal of the Royal Statistical Society: Series B (Statistical Methodology)*, **63**(3), 425–464.

Kleijnen, J.P.C. (2007) *Design and Analysis of Simulation Experiments*, International Series in Operations Research & Management Science, Springer.

Knight, F.H. (1921) *Risk, Uncertainty and Profit*, Hart, Schaffner & Marx.

Krzystofowicz, R. (1983) Why should a forecaster and a decision maker use Bayesian theorem, *Water Resour Res*, **19**(2), 327–336.

Oberkampf, W.L., DeLand, S.M., Rutherford, B.M., *et al.* (2002) Error and uncertainty in modelling and simulation. *Special Issue of Reliability Engineering & System Safety*, **75**(3), 333–357.

Saltelli, A., Ratto, M., Andres, T. *et al.* (2008) *Global Sensitivity Analysis: The Primer*, Wiley.

Singpurwalla, N.D. (2006) *Risk and Reliability: A Bayesian Perspective*, John Wiley & Sons, Chichester.

Wallin, K. (1991) Irradiation damage effects on the fracture toughness transition curve shape for reactor pressure vessel steels. Joint FEFG/ICF International Conference on Fracture of Engineering Materials and Structures, Singapore.

Taleb, N.N. (2007) *The Black Swan: The Impact of the Highly Improbable*, Random House, ISBN 978-1-4000-6351-2.

Zio, E. (2009) *Computational Methods for Reliability and Risk Analysis, Series on Quality, Reliability and Engineering Statistics*, vol. **14**, World Scientific Publishing Co. Pte. Ltd., Singapore.

Notation

Mathematical quantities included in the text are always written in *italic*.

$\boldsymbol{x} = (x^i)_{i=1...p}$: Vector (of dimension p) of uncertain model inputs representing sources of uncertainties, uncertain variables or events

$\boldsymbol{e} = (e^i)_{i=1...p'}$: Vector (of dimension p') of uncertain model inputs representing uncertain events

$\boldsymbol{d}$: Vector (of dimension n_d) of fixed model inputs representing design or decision variables or conditional scenarios

$\boldsymbol{d}^{\circ}$: Vector of fixed model inputs taken at their reference value in a comparative study

$\boldsymbol{z} = (z^l)_{l=1...r}$: Vector (of dimension r) of the model output variables or events of interest

$\boldsymbol{y} = (y^k)_{k=1...q}$: Vector (of dimension q) of model observable output variables

$\boldsymbol{u}, \boldsymbol{u}_{mod}, \boldsymbol{u}_{mes}$: Vectors (of dimension q) of model error or more generally measurement-model deviations: $\boldsymbol{u}$ represents the total deviation, potentially broken down into $\boldsymbol{u}_{mes}$ a metrological error component and $\boldsymbol{u}_{mod,}$ a model error component: $\boldsymbol{u} = \boldsymbol{u}_{mod} + \boldsymbol{u}_{mes}$

$\boldsymbol{x}_t, \boldsymbol{e}_t, \boldsymbol{y}_t, \boldsymbol{z}_t, \ldots$: Vector of time-dependent variables that is vector time series. The time series nature is left implicit in cases where it does not matter much, thus omitting the subscript t in simplified notation $\boldsymbol{x}, \boldsymbol{e}, \boldsymbol{y}, \boldsymbol{z}, \ldots$

$\boldsymbol{y}_m = (y_m{}^k)_{k=1...q}$: Vector (of dimension q) of measurements on observable outputs, so that $\boldsymbol{y}_m = \boldsymbol{y} + \boldsymbol{u}$

$G(.)$ *or* $\boldsymbol{G}(.)$: Deterministic function (generally very complex) representing the system model linking uncertain or fixed input vectors to the vector of output variables of interest: $\boldsymbol{z} = \boldsymbol{G}(\boldsymbol{x}, \boldsymbol{d})$. Notation will be often simplified to $G(.)$

$\boldsymbol{H}(.)$: Deterministic function (generally very complex) representing the system model linking uncertain or fixed input vectors to the vector of observable variables: $\boldsymbol{y} = \boldsymbol{H}(\boldsymbol{x}, \boldsymbol{d})$. Notation will often be simplified to $H(.)$

$\boldsymbol{X}, \boldsymbol{E}, \boldsymbol{Y}, \boldsymbol{Z}, \ldots$: Vectors of uncertain random variables corresponding to $\boldsymbol{x}, \boldsymbol{e}, \boldsymbol{y}, \boldsymbol{z} \ldots$ In simplified notation, underlining is omitted: $X, E, Y, Z \ldots$

$\boldsymbol{X}_t, \boldsymbol{E}_t, \boldsymbol{Y}_t, \boldsymbol{Z}_t, \ldots$: Vector of time-dependent uncertain variables, that is (stochastic) vector time series corresponding to $\boldsymbol{x}_t$. The time series nature is left implicit in cases where it does not matter much, thus omitting the subscript t in simplified notation $\boldsymbol{X}$

$f_X(\boldsymbol{x} \mid \boldsymbol{\theta}_X)$: Joint density of random vector $\boldsymbol{X}$, parameterised by $\boldsymbol{\theta}_X$. In simplified notations, $\boldsymbol{X}$ stands for $(\boldsymbol{X}, \boldsymbol{E})$ the comprehensive vector of uncertain input (continuous variables or discrete events), or even $(\boldsymbol{X}, \boldsymbol{E}, \boldsymbol{U})$ the vector of uncertain inputs extended to model error. Additionally, parameter vector $\boldsymbol{\theta}_X$ can be omitted: $f_X(\boldsymbol{x})$

$F_X(x), F_Z(z)$: Cumulative distribution function of scalar random variable X or Z or more generally measure of probability

$(\boldsymbol{x}_j)_{j=1\ldots n}$: Sample (of size n) of n direct observations of $\boldsymbol{x}$, that is random vectors generally i.i.d. according to the pdf of $\boldsymbol{X}$

$(\mathbf{Z}_j)_{j=1\ldots N}$: Sample (of size N) of random outputs of interest generated by an uncertainty propagation algorithm (typically Monte-Carlo Sampling or alternative designs of experiment)

$\boldsymbol{\theta}_X$, $\boldsymbol{\theta}_U$: Vectors of parameters (of dimension n_p and n_u respectively) of the measure of uncertainty of $\boldsymbol{X}$ or $\boldsymbol{U}$: in the probabilistic setting, this comprises the parameters of the joint pdf. In simplified notations $\theta_X\,\theta_U$

$\boldsymbol{\theta_X}^{kn}$, $\boldsymbol{\theta_X}^{un}$: Vectors representing the known (kn) or unknown (un) components of vector $\boldsymbol{\theta}_X$ in an inverse probabilistic or model calibration approach

$\pi(\boldsymbol{\theta}\,|\,\zeta)$: Joint density of random vector Θ, modelling epistemic uncertainty in the parameters $\boldsymbol{\theta}_X$, $\boldsymbol{\theta}_U$ describing the distribution of (aleatory) uncertainty in $\boldsymbol{X}$, $\boldsymbol{U}$. Such level-2 distribution is parameterised by the hyper-parameters $\boldsymbol{\zeta}$

$\Xi_n = (\boldsymbol{y}_{mj}, \boldsymbol{d}_j)_{j=1\ldots n}$ [or $\Xi_{tn} = (\boldsymbol{y}_m(t_j), \boldsymbol{d}(t_j))_{j=1\ldots n}$]: Statistical information for model calibration, namely a sample of n observations $\boldsymbol{y}_{mj}$ with the n corresponding known conditions $(\boldsymbol{d}_j)_{j=1\ldots n}$ observed at past dates $(t_j)_{j=1\ldots n}$

IK: Non-statistical expert knowledge available on the sources of uncertainty, for example distribution shapes, prior estimates for parameter values or ranges and so on

$f_X^{\pi}(\boldsymbol{x}|IK, \Xi n), F_X^{\pi}(\boldsymbol{x}|IK, \Xi_n)$: Predictive posterior distribution (pdf or cdf respectively) of random vector $\boldsymbol{X}$, averaged over the epistemic uncertainty after incorporation of all background statistical (Ξ_n) or non-statistical (IK) information

unc_z: Risk measure (or quantity of interest) on scalar variable of interest z representing an (absolute) dispersion around the expected value (typically a given number of standard deviations)

$\%unc_z$: Risk measure (or quantity of interest) on scalar variable of interest z representing relative dispersion around the expected value (typically a given number of coefficient of variations)

c_V, $c_V(Z)$: Coefficient of variation of an uncertain variable of interest z, that is the following quantity of interest $c_V = EZ/\sqrt{varZ}$. Notation can be extended to $c_V(Z)$ if clearer

z_s: Real value representing a threshold not to be exceeded by a scalar variable of interest z when the quantity of interest is a threshold exceedance probability

c_z: Risk measure (or quantity of interest) upon the variables of interest z, that is a decision-making functional $c_z = \mathcal{F}[\mathbf{Z}]$ of the measure of uncertainty of $\mathbf{Z}$. For instance: the dispersion unc_z, the frequency of an undesired event, a threshold exceedance probability, the Boolean test comparing this probability to a maximal probability, the expected number of fatalities

$c_z(\boldsymbol{\theta}_X, \boldsymbol{d})$: Deterministic (however complex) function linking the parameters of the uncertainty model and fixed decision variables to the risk measure $c_z = C_z(\boldsymbol{\theta}_X, \boldsymbol{d})$

E_o, E_f,...: Events of interest

e_o, e_f...: Indicator variables corresponding to the events of interest, that is $e_c = 1$ when event E_c is realised, $=0$ otherwise, device c is working, ...

$varZ$, EZ: Variance and expectation of random variable or vector Z

p_f: Probability of threshold exceedance (also referred to as failure probability in structural reliability), that is $p_f = P(G(\boldsymbol{X},\boldsymbol{d}) < 0)$ or $p_f = P(Z = G(\boldsymbol{X},\boldsymbol{d}) > z_s)$

β: Reliability index (or Hasofer-Lind reliability index), another way to express p_f often used in structural reliability, whereby $\beta = -\Phi^{-1}(p_f)$

s_i: Sensitivity index corresponding to component X^i

s_{Ti}: Total sensitivity index corresponding to component X^i

Ω, Ω_x, Ω_{y},...: Sample space of the state of the system over which the structure of a probability space is formally developed (Ω formally containing all possible realisations for uncertain input variables $\boldsymbol{x}$, $\boldsymbol{e}$ or trajectories $\boldsymbol{x}_t$, $\boldsymbol{y}$, z...; Ω_x being restricted to the realisations of the variables indicated in subscript).

ω, ω_{tf}, $\omega_{[tf,\ tf+\Delta T]}$: State of the system, a point in sample space Ω, specifying additionally that it is considered at the (future) time of interest being a given date t_f or period of time $[t_f,\ t_f + \Delta T]$ of timespan of interest ΔT

D_x, D_y, D_z...: Domain of definition in $\boldsymbol{x}$ (or output range in $\boldsymbol{y}$ or z) of the system model or failure function

$E^+{}_x$, $E^-{}_x$, $E^+{}_N$, $E^-{}_N$,...: Hypercorners of MRM (a computational algorithm under monotony described in Chapter 7.4): respectively upper and lower hypercorners with respect to a single point $\boldsymbol{x}$ or after a set of N model runs

T_r, T_b: Respectively the return period and the size of blocks in the extreme value theory, typically $T_r = 100yr$ and $T_b = 1yr$ respectively when estimating the *1/100-yr* flood from data of past annual maximal flows

z_{Tr}: Real value representing a T_r -level for the variable of interest z, that is the $(1\text{-}1/T_r)$ quantile when considering annual maxima.

q, k_s, z_v, z_m, c_m: Input variables of the generic tutorial example of flood representing respectively: river flow, Strickler friction coefficient, riverbed elevation (downstream), riverbed elevation (upstream); cost uncertainty related to imprecise or variable vulnerability

h_d, 1_{ps}: Input controlled variables of the tutorial example representing dike height (in meters) and the choice to install a protection system (logical)

h, z_c, s, e_o, c_c: Output variables of interest the tutorial example representing respectively: flood water height (over riverbed), flood water level (elevation over datum), overflow of the dike (in meters), event of overflow (logical), and complete cost over the time period of interest

For the variables: superscript refers to the component index within a vector (Latin letters: x^i) or the order/level of a quantile (Greek letters: x^α); subscript refers to the index within a sample (x_j) except when it denotes a specific value (e.g. penalised value: x_{pn}, threshold value: z_s); a capital letter (X) denotes a *random* variable or vector (or more generally a possibly-non probabilistic *uncertain* variable) as opposed to a deterministic variable (or observations of a r.v.), noted in minuscule (x); boldening ($\boldsymbol{x}$) (respectively underlining ($\underline{x}$)) denotes a vector quantity in the text (resp. in the Figures), but this can be omitted in simplified notations. A hat ($\hat{C}_z$) classically denotes the estimator corresponding to an unknown deterministic quantity (c_z). Notation prime applied to a vector or matrix quantity (X') means the transpose; exceptionally, a prime applied to a real-valued function $H'(x)$ means the first-order derivative.

References to sections of this book are formatted as Section 1.2 while to Equations as (Equation (1.2)). Bibliographical references are cited as (Author, Year).

Acronyms and abbreviations

BBN	Bayesian Belief Network
PDMP	Piecewise deterministic Markov Process
cdf, ccdf	Cumulative distribution function, complementary cumulative distribution function
c.i.	confidence interval
CPU	Central Processing Unit – referring to computing time units
e.i.	Event of interest
dof	number of degrees of freedom, either a statistical or a mechanical/physical meaning
ES_α	Expected Shortfall at α–level, i.e. expected value of the random variable conditional to being beyond the α–quantile
EVD	Extreme Value Distribution – family of distributions associated with the Block maxima extreme value estimation method
EVT	Extreme Value Theory
FAST	Fourier Amplitude Sensitivity Test – Fourier-transform-based sensitivity analysis method
FORM/SORM	First (respectively Second) Order Reliability Method
GEV	Generalised Extreme Value Distribution – family of distributions associated with the Block Maxima extreme value estimation method
GPD	Generalised Pareto Distribution – family of distributions associated with the Peak Over Threshold extreme value estimation method
GUM	Guide for expression of uncertainty of measurement – ISO standard in metrology
HPC	High Performance Computing
i.i.d.	independent and identically distributed
IPRA	Integrated Probabilistic Risk Assessment
LHS	Latin Hypercube Sampling
MCMC	Monte-Carlo Markov Chains – an advanced sampling method
MCS	Monte-Carlo Sampling – crude Monte-Carlo sampling method
ML(E)	Maximal Likelihood (Estimator)
MSE	Mean Square Error
NCQ	Numerical Code Qualification
NRA	Natural risk analysis
POT	Peak Over Threshold, an extreme value estimation method
pdf	Probability distribution (or density) function
q.i.	Quantity of interest, i.e. risk measure
RI	Rarity index, either with respect to the sample size of data or to the number of model runs
r.v.	Random variable (or vector)
SRA	Structural Reliability Analysis

v.i.	Variable of interest
PS(or R)A	Probabilistic Safety (or Risk) Assessment
QRA	Quantitative Risk Analysis
UASA	Uncertainty and sensitivity analysis (or assessment)
VaR_α	Value at Risk, that is α–quantile of the random value of a portfolio

1

Applications and practices of modelling, risk and uncertainty

This chapter reviews classical practice in various domains involving modelling in the context of risk and uncertainty and illustrates its common and distinguishing features. In particular, the distinct model formulations, probabilistic settings and decisional treatments encountered are reviewed in association with the typical regulatory requirements in the areas of natural risk, industrial design, reliability, risk analysis, metrology, environmental emissions and economic forecasting. This will help to introduce the notation and concepts that will be assembled within the generic modelling framework developed in Chapter 2. It will also lead to a review of the associated challenges discussed in other chapters. Although unnecessary to an understanding of the rest of the book, Chapter 1 can thus be read as an overview of the areas motivating the book's applications and as an analysis of the corresponding state of the art.

In order to facilitate reading, the following sections group the review of methods and practices under subsections that refer to given classes. Obviously, some of the methods introduced in association with one field are in fact used elsewhere, but this would not be the dominant practice. *Industrial risk* denotes risks affecting industrial facilities as the consequence of *internal initiators* such as reservoir failure, pipe break and so on; on the other hand, *natural risk* covers risks triggered by *natural aggressions* (e.g. seism, flood,...) and impacting on either industrial facilities or domestic installations. At the crossroads lies the so-called *natech* risk amongst which the Fukushima/Sendai event is a recent example.

1.1 Protection against natural risk

Natural risk, an important concern for industrial or domestic facilities, has triggered an extensive field of risk research for which the ultimate goal is generally the design of protection for infrastructures or the reduction of the level of vulnerability in existing installations. Probabilistic approaches have permeated to a various extent both regulation and engineering practice, for example with regard to nuclear or hydro power facilities. Here are some notable examples of natural risk addressed:

Modelling Under Risk and Uncertainty: An Introduction to Statistical, Phenomenological and Computational Methods, First Edition. Etienne de Rocquigny.

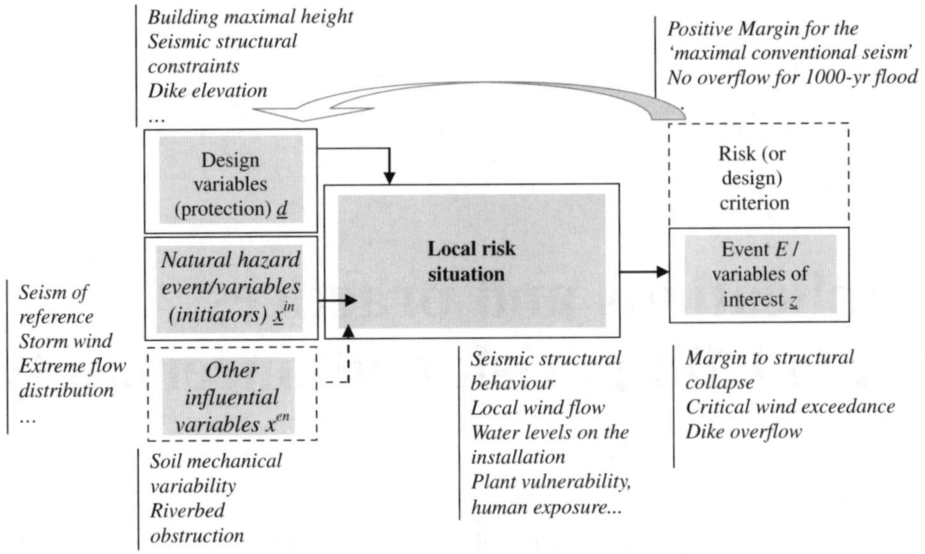

Figure 1.1 *Protection against natural risk – schematisation.*

- flood protection,
- maritime aggressions, such as waves or storm surges coupled with extreme tides,
- extreme winds,
- low flows or high temperatures (threatening the cooling of energy facilities),
- extremely cold temperatures, or associated phenomena (ice blocking, . . .),
- seism.

The typical situation is depicted in Figure 1.1. The box called '*local risk situation*' summarises all phenomena according to which a flood, seism, cold wave or any type of aggression may impact locally on the installation and generate undesired consequences. It is determined both by:

- the *natural hazard events* (flood, seism, wind series . . .) that constitute *initiators* of the risk phenomenon;
- the local configuration of the installation, that is its *vulnerability* depending on the local mechanics of the natural event and its consequences depending on the assets of all kinds that are at stake (plant operation, integrity of equipments, resulting pollution or damage to the environment, potential injuries or fatalities, . . .) and the level of protection insured by the *design choices* and *protection variables* (e.g. dike height).

Natural initiators can be generally described by a few variables such as wind speed, seismic acceleration, flood flow and so on: they will be subsequently gathered inside a vector[1] $\boldsymbol{x}^{in}$. Similarly, all protection/design variables will be formally gathered inside a vector $\boldsymbol{d}$. Additionally, official regulations or design

[1] Dimension can be very large, even infinite if the time dynamics or spatial extension of the initiator are relevant to the design (e.g. flood hydrogram, seismic accelerogram, seismic acceleration 3D field): without loss of generality the vector notation is kept at this stage, and is also understood as the result of discretisation of the associated functions.

guidelines generally specify *risk or design criteria* that drive the whole study process. The definition of such criteria combines the following elements:

- A given undesirable *event of interest* (*e.i.*) which will be denoted as *E*. Think of dike overflow caused by flood or marine surge, structural collapse, cooling system failure and so on. Such an event of interest is technically defined on the basis of critical thresholds for one or several *variables of interest* (*v.i.*) characterizing the *local risk situation*: they are represented in Figure 1.1 by vector $\mathbf{z}$. Think of the flood water level, a margin to mechanical failure, a critical local temperature and so on.
- A maximal *acceptable level of risk*: for instance, the undesired event should not occur up to the *1000-yr* flood, or for the seism of reference; or else, structural collapse should occur less than 10^{-x} per year of operation and so on.

The type of structure shown in Figure 1.1, linking variables and risk criteria, is similar to that mentioned in the book's introduction. Beyond natural risk, it will be repeated with limited variations throughout the areas reviewed in this chapter and will receive a detailed mathematical definition in Chapter 2.

1.1.1 The popular 'initiator/frequency approach'

A considerable literature has developed on the issues raised by protection against natural risk: this includes advanced probabilistic models, decision theory approaches or even socio-political considerations about the quantification of acceptable risk levels (e.g. Yen and Tung, 1993; Duckstein and Parent, 1994; Apel *et al.*, 2004; Pappenberger and Beven, 2006). The most recent discussions have focused on the cases of major vulnerability, uncertainty about the phenomena, reversibility or the precautionary principle (Dupuy, 2002). Notwithstanding all these research developments, it is useful to start with the state of the practice in *regulatory* and *engineering* matters. Most of the time, emphasis appears to be given to a form of '*initiator/frequency approach*' which consists of attaching the definition of the risk criterion to a reference level prescribed for the initiator, for instance:

- 'overspill should not occur for the 1000-yr flood',
- 'mechanical failure margin should remain positive for the reference seism'.

As will be made clear later, this consists essentially of choosing to focus on a single initiator $\mathbf{x}^{in}$ as the dominant *alea* or source of randomness controlling the hazards and the risk situation. Good examples are the extreme wind speed, the flood flow, the external seismic scenario and so on. Nevertheless, a closer look into the realisation of the undesired event *E* usually leads to identifying other potentially important sources of uncertainties or risk factors. Yet, those additional uncertain variables (which will be noted $\mathbf{x}^{en}$) may be separated and given an ancillary role if they are mentioned at all. Think of:

- the riverbed elevation which conditions the amount of overspill for a given level of flood flow;
- the soil conditions around the industrial facility that modify the seismic response;
- the vulnerability of the installations, or conversely the conditional efficiency of protection measures.

At most, the two former types of variability would be studied in the context of local sensitivity to the design, if not ignored or packed within an additional informal margin (e.g. add *20 cm* to design water level, add 20 % to seismic loading, etc.). The latter type is seldom mentioned and is even less often included in the regulatory framework.

This has a strong impact on the probabilistic formulation of the approach. Consider the undesired event *E* characterised by the variables of interest $\mathbf{z}$ (e.g. flood level, peak temperature, peak wind velocity, mechanical margin to failure). Event E is often schematised as a mathematical set stating that a certain threshold (e.g. dike level, critical temperature, critical wind, zero margin) is exceeded for a real-valued variable of interest:

$$E = \{z | z > z_s\} \tag{1.1}$$

It could more generally involve several thresholds to be exceeded by several components of interest during the same event (see Section 1.2.1 on structural reliability). Supposing that there is a model quantifying the phenomena involved in the local risk situation, the undesired event will be modelled as one of the following mathematical relationships:

$$e = 1_{z \in E} = G(\boldsymbol{x}^{in}, \boldsymbol{x}^{en}, \boldsymbol{d}) \tag{1.2}$$

$$z_s - z = G(\boldsymbol{x}^{in}, \boldsymbol{x}^{en}, \boldsymbol{d}) \tag{1.3}$$

whereby the model function $G(.)$ predicts either the event E itself, through an indicator variable e ($e = 1$ when event E is realized, $e = 0$ otherwise), or through a margin to failure $z_s - z$ that should stay positive in order to remain safe. It may be as simple as a subtraction between a scalar x^{in} (e.g. natural flood level) and the protection level d (e.g. dike elevation). It could also be a more complex model involving integral-differential equations to be solved to quantify the impact of a given natural event $\boldsymbol{x}^{in}$ (say a reference seism) on the event or variable of interest. It should depend also on the design variables $\boldsymbol{d}$ as well as on other salient environmental or system conditions $\boldsymbol{x}^{en}$.
Hence, what was called an '*initiator frequency approach*' consists formally in estimating a level x^{α} for the initiator corresponding to a given annual frequency (or probability[2]) of α. When modelling x^{in} as a random variable[3] noted X^{in}, such a level corresponds to the following quantile:

$$P(X^{in} > x^{\alpha}) = \alpha \tag{1.4}$$

where, for instance, $\alpha = 10^{-2}$ (respectively 10^{-3}) if x^{α} denotes the hundred-year (resp. the thousand-year) storm wind speed or flood flow. It may be hard to compute meaningful probabilities in some cases, such as rare seism: then x^{α} may alternatively denote a conventional level that is thought to be 'reasonably maximally plausible' without quantifying the level of risk. Either way, the approach subsequently consists in checking the following risk criterion:

$$c_Z(\boldsymbol{d}) = G(x^{\alpha}, \boldsymbol{x}^{en}, \boldsymbol{d}) \geq 0 \tag{1.5}$$

or equivalently:

$$c_Z(\boldsymbol{d}) = P[G(X^{in}, \boldsymbol{X}^{en}, \boldsymbol{d}) > 0 | \boldsymbol{X}^{en} = \boldsymbol{x}^{en}] < \alpha \tag{1.6}$$

---

[2] See Chapter 4 for comments on the theoretical differences between the two notions; differences can be neglected in practice for rare events, for which frequencies and probabilities are numerically very close.

[3] Note that while x^{α} can be defined straightforwardly as the *quantile* in dimension 1 for x^{in}, things become more complex when considering a vector initiator. This may be seen as another limit on the simplicity of the initiator-frequency approach, circumvented in practice through the consideration of an 'equivalent' event.

where $c_Z(\boldsymbol{d})$ denotes a first example of what will be called the *risk measure* throughout the book. The risk criterion is defined in comparing the risk measure to a prescribed threshold. Fulfilling such a criterion can be reformulated as:

> 'up to the α - initiator (i.e. 1000-yr flood, reference seism etc.) at least, installation is safe under design $\boldsymbol{d}$'.

Such a risk measure has more of a *conditional deterministic* nature than a fully probabilistic one. Indeed, two important remarks can be made. Note firstly that a *monotonous* relationship[4] x^α -> $G(x^\alpha,.,.)$ of the underlying physics is implicitly assumed in order to give credit to the fact that such a criterion *covers* all initiators x^{in} lower than x^α. This gives a first example of the type of *physical* consideration involved in risk assessment, to which the book will come back to in detail (see Chapters 4 and 7). Although *physically intuitive* in many natural risk situations (the stronger the windgust, the thinner the covering material, the lower the dike and so on the riskier), this hypothesis may not be certain in some cases. Secondly, the criterion may not cover satisfactorily the other sources of uncertainty affecting $\boldsymbol{x}^{en}$, for it does not formally define what values should be taken for those leftover variables. Sensible engineering practice would usually fix so-called '*penalised*' values (noted $\boldsymbol{x}^{en}_{pn}$) for those variables, meaning that some coefficients or safety margins are incorporated to multiply (or add up to) the '*best-estimate*' values noted $\boldsymbol{x}^{en}_{be}$. Note that this effort to cover the unknowns presupposes in turn the monotonicity of the alternative relationship $\boldsymbol{x}^{en} \rightarrow G(.,.,\boldsymbol{x}^{en})$. This type of approach, which will be called '*initiator-frequency + margins*', is sometimes used in the field of natural risk protection. It has inspired greater developments in a number of standards or regulations in the field of industrial safety as will be shown in Section 1.2.

1.1.2 Recent developments towards an 'extended frequency approach'

A more complete '*extended frequency approach*' has also been considered in the literature for a long time, although the names differ (cf. review by de Rocquigny, 2010). It involves an alternative risk criterion that tries to remedy the two above-mentioned shortcomings. The risk measure incorporates an extended probabilistic description of the uncertainties affecting not only the initiator but also other influential variables, as follows:

$$P\left[G(\boldsymbol{X}^{in},\ \boldsymbol{d},\ \boldsymbol{X}^{en}\right] = 0) \geq 1-\alpha \quad \text{or} \quad P\left[G(\boldsymbol{X}^{in},\ \boldsymbol{d},\ \boldsymbol{X}^{en}) > 0\right] \geq 1-\alpha \tag{1.7}$$

Fulfilling this alternative risk criterion can be reformulated as:

> 'except in less than α% of the risk situations, installation is safe under design $\boldsymbol{d}$'.

This extended approach should be seen as a rather rare and mostly recent alternative (see the Dutch regulations on flood). In natural risk, the initiator/frequency approach remains as standard practice (Table 1.1). Admittedly, this is not disconnected from some practical aspects of natural risk assessment. Indeed, concentrating on an *initiator frequency approach* enables the studies to be carried out in two distinct steps involving contrasting types of expertise:

(i) *statistics and/or physical expertise*: estimation of the initiator characteristics x^α for a target frequency α, for example the *1000-yr* flood, the *100-yr* storm or the 'reference maximal' seism, and so on;

[4] See Section 4.2.2 for more details.

Table 1.1 Protection against natural risk – examples of probabilistic levels.

Type of hazard	Initiator (yr^{-1})	Comments
Flood	1/100 – 1/500 for residential areas	Additional margins are set up on power plants
	<1/1000 for nuclear power plants	
	1/10 000 for large dams	
Storm/extreme winds	1/100 for power infrastructure network	
	1/50 to 1/100 for more standard buildings	
Earthquake	1/450 (Eurocodes)	
Sea levels	1/1000 for nuclear power plants	Variations may be found according to the conjunction of multiple hazards (tide, storm surge, estuarine flood, tsunami, . . .)

(ii) *engineering*: deterministic modelling and design of the consequences conditional to a documented target level.

Step (*i*) covers the probabilistic dimension of the risk criterion, and involves a typically extreme value statistical estimation of sample records for $\boldsymbol{x}^{in}$ (cf. Chapters 4 and 5). Samples, that will be referred to as Ξ_n with size n, may require rather elaborate statistical modelling, all the more so when $\boldsymbol{x}^{in}$ is a multivariate time series. In many cases, sample size n is small enough compared to $1/\alpha$ for the statistical fluctuations affecting the estimation of a rare quantile x^{α} not to be negligible: those *estimation uncertainties* may be quantified explicitly within the definition of the risk criterion. For instance, the French risk criterion applicable to the nuclear power criterion incorporates the upper bound of the *70%* bilateral confidence interval affecting the $x^{\alpha} =$ *1000-yr flood* estimate within an initiator-frequency approach (Equation (1.3)); on the contrary, the corresponding one for hydropower does not incorporate it in a quantitative manner. In later sections we will return frequently to the critical issue of integrating the various layers of uncertainty involved in risk assessment:

- The frequency or return period (here *1/1000yr)* measures the likelihood of the initiator, covering mostly-*irreducible* uncertainty due to the spatial-temporal variability of natural events. Some authors refer to it as encompassing the *risk*, or *aleatory* uncertainty or *variability*. Technically-speaking, it will also be called *level-1* uncertainty in subsequent chapters, as it will be modelled by random variables directly representing the unknown physical state of the event.
- The confidence level (here *70 %*) refers to the imprecision of the frequency estimation due to data limitations, and possibly also the modelling errors or imprecision regarding the description of the local phenomenae and consequences. It covers uncertainty that is – theoretically at least – *reducible*. Some authors refer to it as the *epistemic* uncertainty or simply *uncertainty* about the risk level. Technically-speaking, it will be called '*level-2*' uncertainty as it will materialise in uncertain parameters affecting the *level-1* random variables, not directly tied to the physical states.

Conversely, it may not be considered legitimate or practical to work within a probabilistic approach for *step* (*i*). This is either because samples may not be significant enough, or because of a lingering epistemological controversy about the quantification of return frequencies for very rare catastrophic events. By convention, 'reference maximal plausible events' would then replace the α-initiator level, as is the case for earthquakes in France.

The deterministic nature of *Step (ii)* has the practical advantage of simplifying the studies and reducing the computational burden. Checking that the risk criterion is fulfilled involves essentially *one run* of the model $G(.)$ representing the consequences of a given initiator for each design scenario. As will be developed in Chapter 7, managing the computational load is essential when drawing in elaborate numerical models to represent the physical phenomena involved in the risk situation. However elaborate the step (i) statistical modelling of $\boldsymbol{x}^{in}$ may be, the proposed simplification relies on the fact that the risk criterion involves a probabilistic target for just one real-valued initiator variable, excluding any probabilistic representation of the other sources of uncertainties. More elaborate probabilistic computations are required when the full probability of the undesired event (Equation (1.7)) replaces the deterministic checking of the undesired event for a given initiator quantile (Equations (1.5) and (1.6)). As a matter of fact, most engineers already carry out a few computations on the consequence models $G(x^{\alpha},.,.)$ within deterministic sensitivity studies: testing, to some limited extent, the above-mentioned informal margins in components involving the potentially influential unknowns $\boldsymbol{x}^{en}$.

Much research work has provided evidence of the utility, if not the necessity, of formally integrating other sources of uncertainty so as to get a fairer picture of the risk and hence optimise the risk management strategies in a more comprehensive risk-informed approach. This has been done in extensive Bayesian settings, for instance, which offer a natural and solid foundation for a double-level probabilistic representation of aleatory/epistemic uncertainty. Bayesian settings have been advocated for hydrological and natural risk applications for years (de Groot, 1970; Krzystofowicz, 1983; Duckstein and Parent, 1994) and in particular developed within the GLUE (generalised likelihood uncertainty estimation, Beven and Binsley, 1992) methodology and MUCM (managing uncertainty in complex models, http://mucm.group.shef.ac.uk Sheffield University program). Subsequent chapters will come back to the subject with more elaborate methods that will mostly integrate statistics and engineering steps into approaches with a higher computational cost.

It is noteworthy that, to a large extent, industrial practice retains an '*initiator/frequency + margins*' approach. Such an approach will also be referred to as mixed deterministic-probabilistic in the rest of this book. In the nuclear industry, it is traditional to call it simply '*deterministic*', since the largest part of the computational and design studies are treated conditionally with a given initiator and hence exclude any direct probabilistic treatment. More fundamentally perhaps than the computational issues, cultural limitations in *acceptability* often restrict probabilities to describing the natural hazards themselves rather than the more complete risk and uncertainty picture.

Note that the subject of natural risk assessment, particularly in its hydrological applications, is also a rich area of application for *forecasting techniques* targeting short- or mean-term predictions (e.g. for early-warning systems, flood routing or crisis management) rather than risk assessment involving extreme events. While the present section is dedicated to the latter, comments regarding the former can be found in section Section 1.5.

1.2 Engineering design, safety and structural reliability analysis (SRA)

This section concentrates on the issues associated with risks incurred by industrial facilities as the result of *internal* initiating failures or adverse events. In contrast, the preceding section involved natural initiators, hence *external* aggressions. The division between the two domains is very much a matter of tradition, and later chapters will offer generic approaches that show a common theoretical foundation.

Practices in industrial safety have nevertheless led to greater use and development of probabilistic risk assessment methods, such as structural reliability analysis (as reviewed below) or systems reliability (see Section 1.3).

1.2.1 The domain of structural reliability

Design of industrial structures, be they nuclear, aerospace, offshore, transport, and so on, has generated the need for some rules or codes to prevent failure mechanisms and secure reliability levels in the face of highly diverse sources of uncertainty affecting the operating systems (variability of material properties, of operational loads, fabrication tolerances, . . .). Beyond empirical design margins, a whole range of methods is usually referred to as the *structural reliability analysis* (SRA) (Madsen, Kenk, and Lind, 1986). This domain is not disconnected from the previous considerations regarding protection against natural risk: in fact, much emerged from the design issues related to external events such as offshore sea waves upon oil facilities or extreme winds upon buildings and so on. However the research and engineering practices are somewhat different. Beyond the fact that the communities have generally distinct backgrounds – statisticians and geosciences for natural risk, reliability and mechanics for SRA – a more fundamental difference is that 'modern' SRA readily couples the system model with a more extensive probabilistic uncertainty model.

SRA considers a structure characterised by an *event of interest* or more generally a group of events leading to failure (E_f) Figure 1.2. Failure is a matter of structural definition: it may include a number of so-called *failure modes*, physical phenomena such as brutal collapse or just crack initiation. Those failures may occur under a certain number of conditions regarding two types of variables affecting system behaviour, namely the design variables (again noted $\boldsymbol{d}$) and the physical variables (denoted as $\boldsymbol{x} = (x^i)_{i=1\cdots p}$). A failure function $G(\boldsymbol{x},\boldsymbol{d})$ – denoted as $G(\boldsymbol{x})$ in SRA whereby design options are implicitly included in $G(.)$ – encapsulates this knowledge of the different phenomena leading to failure. Be it a single mechanical model for one failure mode of a simple component, or more generally a series of sub-models describing each failure mode of each component of the system assembled through Boolean operators, it always consists of a deterministic

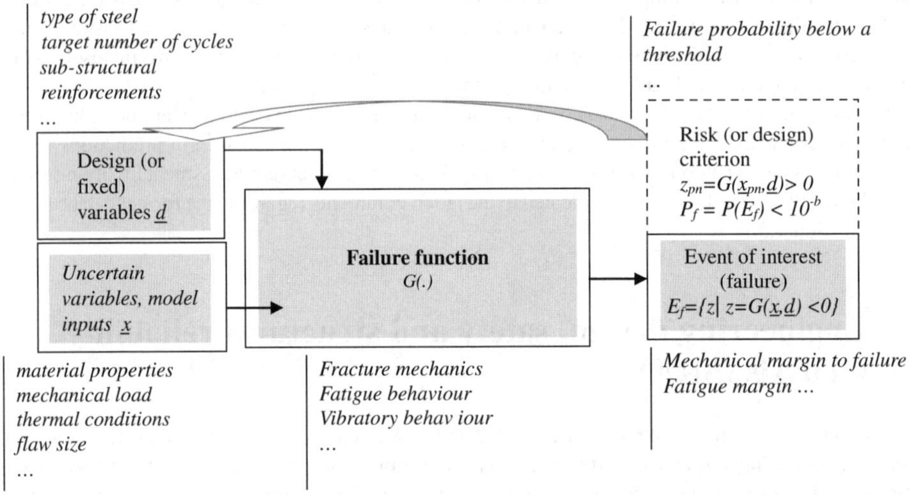

Figure 1.2 *Structural reliability analysis – schematisation.*

relationship between $\boldsymbol{x}$, $\boldsymbol{d}$ and the failure event of interest with the following convention:

$$E_f = \{\boldsymbol{x} | z = G(\boldsymbol{x}) < 0\} \tag{1.8}$$

For a given design $\boldsymbol{d}$, the *failure set* (or failure domain) thus defines a set of values of $\boldsymbol{x}$ leading to the event of interest E_f. It is a subset of the space of variation of the uncertain variables (noted Ω_x). Associated definitions are given to the *safe set* or domain $E_s = \{\boldsymbol{x} | G(\boldsymbol{x}) > 0\}$ and to the *limit state* (or failure surface) $E_l = \{\boldsymbol{x} |\ G(\boldsymbol{x}) = 0\}$. The latter, representing a zero probability in non-degenerated cases, is essentially of mathematical/algorithmic interest.

Consider, for instance, the simplified example of the reactor vessel in nuclear safety. Failure could theoretically occur as a result of the abnormal pressure-temperature of the primary fluid, itself subjected to an internal initiator such as the drop in pressure following a pipe break elsewhere in the circuit. Stress upon a pre-existing flaw inside the vessel width could then exceed the resilience margin of the material resulting in a failure event either defined as the sudden rupture or, more conservatively, as the triggering of flaw propagation. Failure modelling involves a complex finite-element thermo-mechanical model $\boldsymbol{m} = M(.)$ predicting stress and temperature fields as a function of numerous uncertain variables (properties of materials, flaw characteristics, the thermodynamics of accidental transients, the radiation received over time and the resulting fragilising, etc.) as well as design or operational conditions $\boldsymbol{d}$ (such as temperature and pressure limits, prescribed recovery times, etc.). Hence a failure margin z is computed by subtracting a stress intensity factor (noted K_I) from a toughness function (noted K_{Ic}), both functions of $\boldsymbol{m}$, defining altogether the following failure function $G(.)$:

$$\begin{aligned} \boldsymbol{m} &= M(x^1, \dots x^p, \boldsymbol{d}) = M(\boldsymbol{x}, \boldsymbol{d}) \\ z &= K_{1c}(\boldsymbol{m}, \boldsymbol{x}, \boldsymbol{d}) - K_1(\boldsymbol{m}, \boldsymbol{x}, \boldsymbol{d}) = g(\boldsymbol{m}, \boldsymbol{x}, \boldsymbol{d}) = g(M(\boldsymbol{x}, \boldsymbol{d}), \boldsymbol{x}, \boldsymbol{d}) = G(\boldsymbol{x}, \boldsymbol{d}) \end{aligned} \tag{1.9}$$

1.2.2 Deterministic safety margins and partial safety factors

As a matter of fact, the field of structural reliability has a long history of standards and regulations (Ellingwood, 1994), notably in risky industries such as the nuclear or aerospace industry. Regulatory criteria typically specify reliability requirements, namely an absence of failure over the system and/or per component. These are prescribed during a given period and for a given scenario, such as a conventional accident or stress history under hypotheses $\boldsymbol{d}$ concerning the structure. Traditional regulations resemble to a certain extent the '*initiator-frequency + margins*' approach mentioned above. Design is studied under conventional scenarios (such as pipe break, explosion, etc.) to which typical occurrence frequencies may be given, although generally on a more limited statistical analysis than for the natural initiators considered in the preceding section. A *risk criterion* then involves 'penalising' with safety margins the remaining uncertain variables $\boldsymbol{x}$, that is to say applying coefficients or safety factors f^i to their '*best-estimate*' values $\boldsymbol{x}_{be}$ and verifying reliability by deterministic calculation:

$$\begin{aligned} \boldsymbol{x}_{pn} &= \left(x^1_{be} \cdot f^1, \dots, x^p_{be} \cdot f^p\right) \\ z_{pn} &= G(\boldsymbol{x}_{pn}, \ \boldsymbol{d}) > 0 \end{aligned} \tag{1.10}$$

This approach is generally referred to in the industry – notably in nuclear design – as a *deterministic* approach or as an approach through 'penalised scenarios': it involves an elementary form of deterministic treatment of sources of uncertainty pre-supposing again that the function $\boldsymbol{x} \rightarrow G(\boldsymbol{x},.)$

considered component per component x^i is monotonous.[5] This is a mostly intuitive situation in fracture mechanics (e.g. the greater the load, the greater the risk), but is less straightforward in fluid mechanics (where competing phenomena may trigger irregular bifurcations). As mentioned in the preceding section, it has the considerable advantage of limiting the number of computations to one or a few runs in potentially complex mechanical models. It also requires the identification of an upper limit (or a lower one, according to the sign of monotony) of plausible uncertain variation for every component x^i, both physically realistic and consensual amongst stakeholders. In practice, aside from informal physical expertise, that certain limit x^i_{pn}, noting the penalised value for x^i, can correspond approximately to the quantile $x^i_{\alpha_i}$ for the component *implicitly* modelled as a random variable X^i whereby:

$$P\,(X^i > x^i_{\alpha_i}) = \alpha^i \tag{1.11}$$

A resistance property would be taken at its lower 5% value accounting for the variability of materials, while a loading variable would be taken at its upper 95% value accounting for lack of knowledge or partial randomness in the operating conditions. This is also referred to as a '*partial safety factor*' approach (Ellingwood, 1994; Melchers, 1999), whereby safety factors are defined for each partial component x^i of the vector $\boldsymbol{x}$, altogether conditioning overall safety.

Important literature continues to argue about the interest and inherent conservatism of this kind of approach which aggregates potentially heterogeneous margins, reflected by the quantiles α^i, in a barely controllable manner (cf. de Rocquigny, Devictor, and Tarantola, 2008 for a review). Remember that there are some distinctive operational advantages with this kind of approach: it enables one to reduce the often enormous number of scenarios down to a selection of hopefully conservative (if not representative) ones which are more carefully studied and given margins. It also leads to design codes which are much easier to enforce over, for example, large and heterogeneous sets of industrial products/installations than direct probabilistic targets leaving each single designer free to craft probabilistic distributions and possibly calibrate each system differently. Conversely, experience shows that it is quite difficult to translate partial safety factors or related partially deterministic approaches into an overall probabilistic risk level that could be compared to other risk situations. Chapter 4 will return to the underlying dependence issues and the interpretation of probability as a means to rank the relative likelihood of risks for decision-making.

1.2.3 Probabilistic structural reliability analysis

To circumvent those shortcomings, modern structural reliability has come to move from the partial safety to an extended randomisation of vector $\boldsymbol{x}$, in a trend comparable to the *extended frequency approach* exposed for natural risk. Formally, the risk criterion (or design limit) defined in Equations (1.10) and (1.11) could be understood as a mixed deterministic-probabilistic criterion based on the conditioning of *all* components of $\boldsymbol{X}$, as a distribution conditional to everything is simply a deterministic level:

$$z = G((x^i_{\alpha_i})_{i=1,\ldots p}, \boldsymbol{d}) > 0 \iff P\left[G(\boldsymbol{X}, \boldsymbol{d}) > 0 \middle| (X^i = x^i_{\alpha_i})_i\right] = 1 \tag{1.12}$$

[5] If $G(.)$ is not monotonous, an optimisation algorithm becomes necessary, thus multiplying considerably the number of required calculations. Such a 'strongly deterministic' approach is rarely applied in practice because of its tremendous computational challenges.

For some years now this approach has been compared with approaches known as 'probabilistic' that model explicitly all sources of uncertainty as random variables X^i. Probabilistic uncertainty modelling is developed upon the input space with the delineation of a joint probability density function f_X for vector $\boldsymbol{x}$, provided that all corresponding sources of uncertainties may be acceptably randomised. Hence, the risk measure is generally taken as the following *failure probability*:

$$p_f = P(E_f) = \int_{E_f} f_X(\boldsymbol{x})d\boldsymbol{x} = \int_{\Omega_x} 1_{G(\boldsymbol{x},\boldsymbol{d})<0} f_X(\boldsymbol{x})d\boldsymbol{x} \tag{1.13}$$

which should be kept close to zero. In fact, it is frequently found that only a part of the vector $\boldsymbol{x}$ is randomised, say the first m components $(x^i)_{i=1...,m}$ out of the total vector of p inputs. This comes down to a partial conditioning into penalised values of remaining variables such as internal initiators (e.g. reference accidents in nuclear safety). Furthermore, p_f is compared to a threshold of 10^{-b}, or at least the probability of a reference scenario $\boldsymbol{d}^o$ considered as reliable enough if the specification of an absolute probabilistic threshold is too controversial. While these approaches are known as 'probabilistic', they may again be formalised as *mixed deterministic-probabilistic* where any of the two following risk criteria replaces that of Equation (1.12):

$$p_f(\boldsymbol{d}) = P\left[G(\boldsymbol{X},\boldsymbol{d}) < 0 \middle| (X^i = x^i_{pn})_{i=m+1,...p}\right] < 10^{-b} \tag{1.14}$$

or:

$$p_f(\boldsymbol{d}) = P\left[G(\boldsymbol{X},\boldsymbol{d}) < 0 \middle| (X^i = x^i_{pn})_{i=m+1,...p}\right] < p_f(\boldsymbol{d}^o) \tag{1.15}$$

A key difficulty is related to the fact that SRA generally involves a rare failure event (now defined as the conditional event $E_f = \{\boldsymbol{x} \mid (x^i = x^i_{pn})_{i=m+1...p} \cap G(\boldsymbol{x},\boldsymbol{d}) < 0\}$, that is to say the tail-end of the Z distribution, together with rather CPU-intensive failure functions $G(.)$. The *risk measure*, a very complex multiple integral (Equation (1.13)), thus constitutes a great *computational challenge* which has generated considerable numerical research. Aside from efforts to reduce the variance of Monte-Carlo Sampling (MCS) through importance sampling techniques, specific strategies were developed to evaluate the probability of exceeding a threshold. For example, the First (resp. Second) Order Reliability Methods (FORM resp. SORM) may considerably reduce the number of $G(.)$ model runs in evaluating a very low p_f. As will be developed in Chapter 7, underlying approximations do rely crucially on the strong physical-mathematical properties of the failure function such as convexity or monotony: yet another example of the importance of combining physical and probabilistic knowledge in order to undertake accountable risk modelling, as will be discussed throughout the book. Last but not least, these methods also generate *importance factors* that is indices ranking the uncertain inputs X^i with respect to their contribution to the failure probability. More than the absolute value of the risk measure, this type of information may be contemplated in order to ameliorate the designs.

1.2.4 Links and differences with natural risk studies

Note the similarities between Figures 1.1 and 1.2: both approaches focus on a given event of interest (failure or the undesired/critical event in natural risk respectively), occurring in the system (the structure or the risk situation *resp.*) and represented by a *system model* function $G(.)$ (the failure function or the phenomenological model *resp.*). $G(.)$ depends on some uncertain variables $\boldsymbol{x}$ corresponding to the former $(\boldsymbol{x}^{in}, \boldsymbol{x}^{en})$ as well as to controllable design variables $\boldsymbol{d}$ to be adjusted according to a given risk measure (the failure probability or the frequency of the event of interest resp.).

SRA goes into the full coupling of the system model and probabilistic uncertainty modelling, enabling the computation of an integrated probabilistic risk measure. While the corresponding computational difficulties may explain why most natural risk studies are restricted to the initiator/frequency approach, they have led to more enthusiastic developments in SRA with advanced numerical thinking. Yet some challenges and limitations remain:

- Unlike the previous domain, truly statistical modelling of the sources of uncertainties is often limited; accordingly, the statistical significance of the very low probability figures computed for p_f is still of concern although it was identified early on (Ditlevsen and Bjerager, 1986).
- Randomising of all sources of uncertainties, much beyond the practices in natural risk, has often the disadvantage of staying elusive as to the proper understanding of the underlying random experiments or sample space required to theoretically settle the setting. Frequentist interpretations being generally much trickier than with natural initiators, some authors interpret the results more as subjective probabilities for relative design comparison.
- Physical hypotheses (e.g. monotony, convexity) that underlie reliability approximations or the definition of risk criterion often remain insufficiently documented.

Beyond the mere checking of an absolute risk criterion, SRA lays large emphasis on importance factors, that is the ranking of uncertain inputs regarding their contribution to reliability, as well as on the relative comparison of contrasted designs through a risk measure (the failure probability).

1.3 Industrial safety, system reliability and probabilistic risk assessment (PRA)

Reliability and risk associated with industrial facilities are highly affected by their complexity, as those *systems* comprise numerous physical components organised into a wealth of functional chains. It may be hard to identify a modern elementary object which cannot be seen as a system in itself (e.g. a temperature captor); but think about the hundreds of thousands of such elementary pieces included in industrial facilities such as power plants, nuclear waste repositories or in industrial products such as airplanes, cars, and so on. SRA covers systems to a certain extent and has developed *structural systems safety* extensions to account for multiple-failure events such as those occurring in multiple-component mechanical structures (Ditlevsen and Bjerager, 1986). Nonetheless, practices in systems reliability have traditionally involved other approaches, denominated alternatively as Quantitative Risk Analysis (QRA), Probabilistic Risk Analysis (PRA) or even Probabilistic Safety analysis (PSA) in the nuclear sector (Bedford and Cooke, 2001). Similar to what was suggested in linking formally natural risk to structural reliability studies, subsequent sections will develop the theoretical continuum relating QRA to SRA, as already advocated in the literature, for instance by Aven and Rettedal (1998).

1.3.1 The context of systems analysis

The overall context (Figure 1.3) can be analysed in a similar framework to that of Figures 1.1 and 1.2.

System reliability identifies one or several undesired event(s) of interest (or *top event*, noted again E) characterising the failure of the key functions of the system: severe reactor damage, airborne electrical shutdown, process unavailability and so on. This(ese) event(s) of interest affecting the system can be explained by a number of internal or external *initiating* elementary events (formally gathered inside the large vector $\boldsymbol{E}^{in}$), as well as *conditional* disruption events representing system components or processes that fail to contain the initiating failure (vector $\boldsymbol{E}^{sy}$). A quantitative *system model*, typically

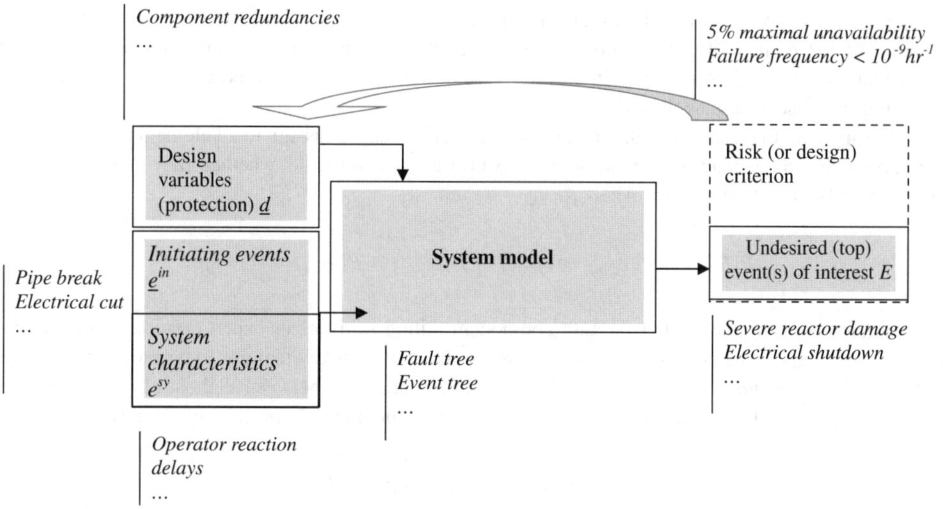

Figure 1.3 *Systems reliability analysis – schematisation.*

based on fault trees or event trees, will predict the event of interest as a function of those inputs as well as of the decision or design variables $\boldsymbol{d}$. The undesired event of interest E thus becomes a logical function of all elementary events, a basic algebraic result of those trees, such as:

$$E = \bigcup_i \left[E^{in^i} \bigcap_{j_i(\boldsymbol{d})} \left(E^{sy^{j_i}} \right) \right] = G(E^{in}) \tag{1.16}$$

where, as a non-exclusive illustration, failure of the system is assumed to occur upon at least one of a set of initiator events i associated with a series j_i of conditional disruption events that could be mitigated by design (protection) choices $\boldsymbol{d}$. More comprehensively, complex events would actually involve a recursive use of Equation (1.16) as each of those j_i disruption events would themselves be the result of a sub-chain of ancillary initiators or sub-component disruptions.

The formula has a variety of equivalent formal interpretations: (a) it can be seen as a relationship between the events defined as subsets of states of nature; (b) it can be translated straightforwardly into a relationship between the probabilities (or frequencies) associated with those events; and (c) it can also be translated into a deterministic function between the indicator variables (now noted e, in lowercase) representing each precise *elementary state* of the system components or initiators ($e = 1$ in the case of initiator realisation or component failure, $e = 0$ otherwise):

$$e = \sum_i \left[e^{in^i} . \prod_{j_i(\boldsymbol{d})} e^{sy^{j_i}} \right] = G(\boldsymbol{e}^{in}, \boldsymbol{e}^{sy}, \boldsymbol{d}) \tag{1.17}$$

In that latter case, the *system model* $G(.)$ represents the deterministic causality – or at least a presumably-strong dependence – between all sub-events or states of the system components and the final state regarding the undesired event. It constitutes the output of a *functional analysis* of the system. In real studies, Equation (1.17) may be much more complex than a simple sum of products, either because it is a conservative approximation of the exact Poincaré formula (see Annex Section 10.2), or even more so because of the complexity of sub-systems. In general, it will appear as a deterministic function of elementary states (initiators or conditional processes), however numerous those may be. Think in the thousands at least for the nuclear PSA.

Starting from that functional description, probabilistic analyses assign to all elementary events their corresponding frequencies or conditional probabilities, according to whether they are initiator or conditional failures (see Chapter 4):

$$P(E^{in\,i}) = P(1_{Ein\,i} = 1) = f^i \quad P(E^{sy\,k}) = P(1_{Esy\,k} = 1) = p^k \tag{1.18}$$

Such figures are estimated either through component failure statistics or expert opinion. Note that the dependence phenomena between elementary events are additional key features to be estimated through probabilities of *common modes*. Equation (1.17) yields a relationship between all those elementary values and the frequency of the undesired event f^e, which represents the risk measure of the QRA:

$$c_z(\boldsymbol{d}) = f^e = G[(f^i)_i, (p^k)_k, \boldsymbol{d}] \tag{1.19}$$

That risk measure can then be compared to a risk acceptance criterion such as 'system failure should be less frequent than 10^{-9} per flight-hour in the aerospace industry, or 10^{-6} per plant-year of operation in the nuclear', as illustrated in Table 1.2.

Note that the wording 'risk acceptance criterion' should be used within a limited perspective of a prescribed regulation or standard, notwithstanding the much broader and largely debated topic of risk *acceptability*. Though it largely exceeds the *modelling* scope of this book, Chapter 2 will return to some extent to the issue in the sense that a risk criterion should always be considered as a means to rank relative options, costs and benefits, never as an absolute statement. Even if very small, the residual risk of large radioactive release due to nuclear plant failure is never acceptable as such. Though risk perception and a larger societal debate have their say beyond purely technical measures, the residual risk could become less unacceptable perhaps from the perspective of the alternative residual risks of equivalent power production: think of a dam break in a hydropower plant or of the unpredictable effects of climate change due to the major CO_2 release of thermal power plants. In some cases indeed the risk measure may be formulated explicitly as a relative index so as to compare design choices or operation and maintenance modifications to a standard design or operating process. This is the case in the US nuclear field, where the Nuclear Regulatory Commission allows for changes in operational procedures provided the following risk measure does not deviate from that of the reference design.

1.3.2 Links and differences with structural reliability analysis

A simple reformulation can closely link this formula to the fully probabilistic risk measures mentioned in preceding sections, viz. the extended frequency approach of natural risk or the failure probability in SRA. Consider the system model as a deterministic function of the vector of elementary states $\boldsymbol{x} = (\boldsymbol{e}^{in}, \boldsymbol{e}^{sy})$ characterizing the overall system state. A set of elementary probabilities for each component $e^{\cdot\cdot}$, which are two-state Bernoulli random variables, plus a dependence structure consisting of, for example, common modes by pairs, is read as a *discrete joint probability density* upon the vector $\boldsymbol{x}$ of discrete random variables. The frequency of the undesired event (Equation (1.19)) appears then as the equivalent of the SRA failure probability:

$$f^e = P\,[G(\boldsymbol{e}^{in}, \boldsymbol{e}^{sy}, \boldsymbol{d}) = 1] = P\,[G(\boldsymbol{X}, \boldsymbol{d}) = 1] \tag{1.20}$$

Table 1.2 Risk acceptance criteria (in yearly frequency) in various industries.

	Targets (yearly frequency per plant)	Targeted individual risk per person per year	Observed Accident Rate (frequency of fatal accidents per year)	Comments
Nuclear plants - present (future)	Internal failure (core damage[a]) 10^{-4} (10^{-5})[i] External release of RN (LERF) 10^{-5} (10^{-6} to negligible[i])	10^{-6} (Netherlands)	$\sim 10^{-4}$ internal system failure, but $2–4.10^{-5}$ fatalities	Considering about 10 000 plant-years and a conservative figure of 2000 deaths in all civil accidents, amongst 5000–10 000 average close inhabitants to plants
Aerospace			$\sim 1.10^{-6}$	better figure expressed per flight-hour, varying from 10 to 200.10^{-8}/hr according to country, helicopter or airliner or private jets and so on
Car			1.10^{-4} [f] to 2.10^{-4} [h]	
Bridges or civil structures	10^{-4} if large exc., or 10^{-3} if normal (USA)	10^{-6} [m]	$<10^{-6}$ (UK)[m,h]	
Traffic accidents			1.10^{-4} [j] to 3.10^{-4} [h]	averaged over variable exposure of hour-person/year
All accidents			3.10^{-4} [j] (Japan) to 5.10^{-4} (USA)	
All deaths			$\sim 8.10^{-3}$ [j] to 1.10^{-2}	averaged over natural or accidental death and ages

[a] One decade less when excluding external hazard initiator.
Source: [m](Menzies, 1995); [j](Nuclear Safety Commission of Japan, 2004); [i](IAEA, 2005); [h](Madsen, Kenk, and Lind, 1986); [f]French car statistics.

where:

$$X \sim f_X = \{(f^i)_i, (p^j)_j\} \tag{1.21}$$

In that respect, QRA appears as *formally similar* to SRA. Key *practical* differences remain however:

- In QRA, the Boolean nature of the system model $G(.)$ and of model inputs $\boldsymbol{x}$ simplify considerably the computation of the risk measure, since the complex multiple-integral of SRA (Equation (1.14)) becomes here a closed-form expression (Equation (1.17)).
- Conversely, the size of $\boldsymbol{x}$ and the associated dependence structure, reflected in the logical model (fault/event tree building) and the common mode probabilities, are generally much more complex in QRA than in SRA.

Chapters 7 and 5 respectively will discuss in more detail those two essential remarks which bear an importance beyond the mere comparison of QRA and SRA. Boolean discretisation is a powerful tool to simplify the computation of risk measures such as failure probabilities even in phenomenological models departing from the logical reliability models. Functional analysis through fault trees and common mode quantification is also a powerful tool to help model complex phenomenological dependence in a tractable way.

1.3.3 The case of elaborate PRA (multi-state, dynamic)

Systems analysis involves other methods than fault trees and event trees. These basic tools do not represent the dynamic reliability phenomena, and become tricky to use if elementary events behave closer to multi-state than to simple Boolean variables. Yet, much of the observations made above remain unchanged.

Multi-state static models such as Bayesian Belief Networks (BBN) (cf. for example Kurowicka and Cooke, 2006) generalise the deterministic functional dependence coined in Equation (1.17), insofar as they link the undesired event e to multi-state indicator variables $\boldsymbol{e}^{in}$ and $\boldsymbol{e}^{sy}$ representing the state of initiator events or conditional processes:

$$e = G(\boldsymbol{e}^{in}, \boldsymbol{e}^{sy}, \boldsymbol{d}) \tag{1.22}$$

Accordingly, the computation of the risk measure f^e involves discrete probability distributions and conditional dependence structures generalising Bernoulli distributions and common modes. This can still be interpreted as a failure probability being the image of a joint pdf on the vector $\boldsymbol{x}$ of multi-state events through $G(.)$. Closed-form expressions still result in general, albeit of large dimension because of the multi-state features.

Dynamic reliability approaches (e.g. under Markov chains, Petri networks, etc.) are developed to account for the temporal variation of reliability characteristics as well as the time-dependent combinations of causalities. Think of the contrasted consequences of the failure of the i-th component occurring before or after failure of the j-th; or of emergency system actions with uncertain response delays, which modify the subsequent event tree. Then the prediction of $\boldsymbol{e}_t$ at time t draws in a time-dependent stochastic operator to the dynamic system: Equation (1.22) is replaced by Piecewise Deterministic Markov Processes (PDMP), stochastic differential equations (SDE) and so on. The probabilistic description of time-dependent elementary states of the system $\boldsymbol{e}_t^{in}$, $\boldsymbol{e}_t^{sy}$ involves time series instead of standard random variables. Risk measures, such as the expected time to failure or expected dependability of the system, are not closed-form anymore. They still appear however as elaborated *integral* functions of the system model, depending on the uncertain model inputs represented by time series $\boldsymbol{X}_t$ as well as on the possibly time-dependent design actions or maintenance strategies $\boldsymbol{d}_t$

such as the following expectation:

$$R = E[G(\mathbf{X}_t, \mathbf{D}_t)] = \int_{\Omega} g[\mathbf{X}_s, \mathbf{d}_s]\, dW_s \tag{1.23}$$

Note that more sophisticated mathematical tools may then be necessary, such as the use of stochastic integration over the entire trajectory $s \in [0,t]$ of the stochastic processes W_s representing the sources of uncertainties perturbing the time-dependent system state represented by $\mathbf{X}_t$.

Although this domain is specific to a large extent, it has, like the previous situations, the risk measure appearing as an integral of the system model along random distributions, so it may theoretically use Monte-Carlo simulation similarly to SRA. Practical approximations have to be made generally due to computational costs of the system model over a large time interval and hence a large number of discretised time steps: those may be quite different to that of SRA owing to the stochastic nature of the time-dependent integral (Equation (1.23)). This may include the complete replacement of the system model by dedicated mathematical formulations allowing for closed-form integration to avoid the excessively costly convergence of Monte-Carlo.

1.3.4 Integrated probabilistic risk assessment (IPRA)

Consider now the elaborate risk studies within which several of the previously introduced layers are being put together, such as:

- a *central (logical) system model* with some functional complexity described by QRA, denoted as: $e = G_s(e^{in}, e^{sy}, d^s)$,
- an *'upstream' natural risk* component plugged in as a more detailed description of one of the initiating events $e^{in\ i} = G_n(x^{in}, x^{en}, d^n)$ affecting the central system,
- a *'downstream' consequence phenomenological model* further describing the impact of undesired functional failure of the system e into variables of interest $z = G_c(e, x^c, d^c)$ quantifying the detrimental consequences (such as plant unavailability costs, large release of pollutants, number of fatalities...) as a function of supplementary variables x^c modifying the detriment for a given e (meteorological conditions, vulnerability, etc.).

This is the case for some of the most elaborate American nuclear risk assessments, such as the Waste Isolation Power Plant performance assessment (Helton *et al.*, 1996), or the NUREG-1150 studies (NRC, 1990). It is also the setting of the ARAMIS European project (Salvi and Debray, 2006), cf. (Figure 1.4), whereby that central logical system model is thought of as a chaining of a fault tree from initiators to the central undesired event and an event tree predicting the consequences from the central undesired event. Consequence prediction could couple a logical model, viz. an event tree, and a phenomenological model, typically a physical-chemical or environmental model predicting the extension of the explosion or the accidental pollutant spill.

A further refinement could be to plug inside the central QRA system model an SRA-type model. This would mean detailing one of the QRA conditional failure events, such as the breaking of the vessel, as a result of the initiating event plus some system failures. By considering a similar structure for each layer, the global approach can be viewed as a simple chaining of each:

- the final variables of interest are defined as those of the consequence model z,
- the vector of uncertain inputs (resp. of design actions) consists in concatenating all vectors of input states or variables of the three models $\mathbf{x} = (e^{in}, e^{sy}, x^{in}, x^{en}, x^c)$ (resp. $\mathbf{d} = (d^s, d^n, d^c)$),

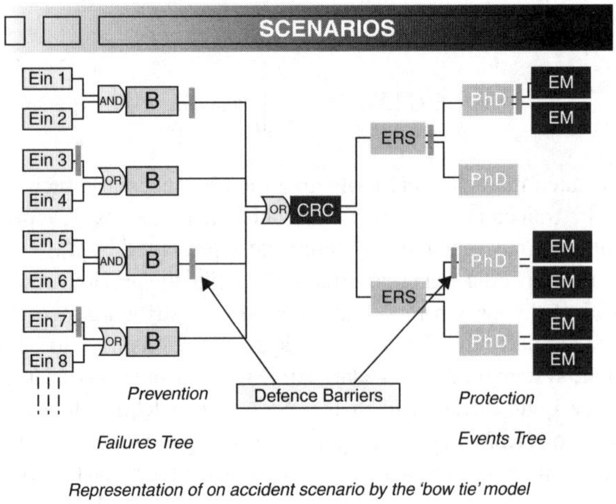

Figure 1.4 *ARAMIS Risk model for IPRA, adapted from (Salvi and Debray, 2006).*

- a compound deterministic function represents the overall system model as follows:

$$z = G_c \, o \, G_s \, o \, G_n \left(e^{in},\ e^{sy},\ x^{in},\ x^{en},\ x^{co},\ d^s,\ d^n,\ d^c\right) = G\left(\boldsymbol{x}, \boldsymbol{e}, \boldsymbol{d}\right) \tag{1.24}$$

The single (output) undesired event of interest is replaced by a whole spectrum of consequences, hence described by a *vector* of variables of interest z. The associated risk measures are multifold:

- (discrete level-based) a discrete collection of probabilities (or frequencies) corresponding to a set of selected representative levels of consequences,
- *(ccdf)* a curve of probabilities to exceed a continuous level of z for a given type of consequences.

The second option does correspond to the formal consideration of the cumulative distribution function (cdf) of the variable of interest z, or rather it's complementary which is traditionally preferred under the denomination of *ccdf* (Helton, 1993). This can be understood as equivalent to one of the options available to formalise the well-known Farmer curve (or risk profile) relating to severity and frequency (Figure 1.5-right). Note that assessing consequences for various aspects such as cost, number of fatalities, cumulated dose emitted and so on, generates a set of *ccdf* for each component z^k. Practice, inspired by regulations, tends to focus eventually on a discrete level-based risk measure. While the whole risk curve may be simulated, risk criteria generally refer to the exceedance of one or a few thresholds, such as that associated with the *Large Early Release Frequency* (*LERF*) in the nuclear field. This comes down to considering the likelihood of a discrete set of consequences (z_1, z_2, ...) or equivalently a discrete set of undesired consequence events defined as $E_1 = \{z|\ z > z_1\}$, $E_2 = \{z|\ z > z_2\}$ and so on.

Both options require a probabilistic model to be set up to describe the amount of uncertainty in each input of the chain: distribution of the initiating events or hazard variables, elementary failure probabilities of the internal system components, conditional probabilities and variability of uncertain consequences and so on. Hence, computing the *ccdf* of the consequences or the probability of exceeding a given

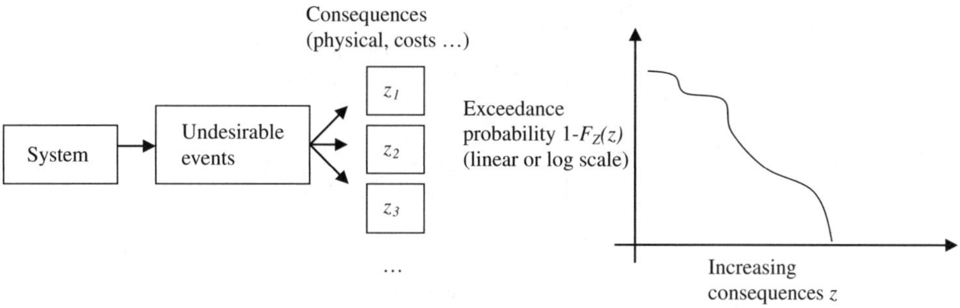

Figure 1.5 *A basic risk-consequence model (left) and ccdf (right).*

undesired threshold involves the following global integral:

$$c_z(\boldsymbol{d}) = P(z > z_s) = 1 - F_Z(z_s) = P(G(\boldsymbol{X}, \boldsymbol{E}, \boldsymbol{d}) > z_s) \tag{1.25}$$

Given the mix of continuous and discrete variables involved, computation typically resembles that of a chaining of SRA and QRA. Simplifications of the inner integral, similar to QRA involving the logical models, bring it down to closed-form conditional probabilities, leaving an SRA-type integration for the part that includes phenomenological models (see Chapter 7).

Yet a complete picture of the risk measures and computations involved requires raising the issue of the level-2 uncertainty affecting the risk components. As already mentioned in Section 1.1.2, data and expertise limitations raise the issue of the accuracy of failure probabilities of components (with respect to the potential common modes, that is non-independent behaviour of component failure), of frequencies of extreme initiator events, or of rare conditional consequences. This issue appeared rather early in the history of nuclear probabilistic risk analysis, as the famous first full-scale IPRA ever undertaken in a nuclear plant (Rasmussen *et al.*, 1975) – the *Reactor Safety Study*, also known as the WASH report – was quickly criticised on these grounds (Lewis *et al.*, 1979). Hence, models of level-2 epistemic uncertainty in the parameters of the core probabilistic model were developed. A long debate took place on the subsequent classification of the large variety of uncertainties encountered in large industrial systems into two types, namely the aleatory (or irreducible, variability) and epistemic (or reducible, lack of knowledge) categories. The rationale for explicitly separating those two categories into two levels inside the decision-making criteria has been discussed accordingly (Apostolakis, 1999; Oberkampf *et al.*, 2002; Aven, 2003). Subsequent chapters will return to that key concept. Note that computations depend critically on the choice made at that level and incorporated into compound risk measures. A level-2 uncertainty model transforms the ccdf curve illustrated in Figure 1.5 into an uncertain curve, or in other words into a set of ccdf indexed by the level-2 confidence level as illustrated in Figure 1.6.

Indeed, denoting θ_{XE} as the vector of parameters of all input distributions of the global model, the *ccdf* depends on its value:

$$1 - F_Z(z_s) = 1 - F_Z(z_s|\theta_{XE}) = P(G(\boldsymbol{X}, \boldsymbol{E}, \boldsymbol{d}) > z_s|\theta_{XE}) \tag{1.26}$$

so that the ccdf becomes a random function when a level-2 uncertainty model transforms those parameters in a random vector $\Theta_{XE} \sim \pi(\Theta_{XE} \mid \zeta)$. A variety of compound level-2 risk measures have been contemplated in IPRA practice and discussed in the literature. The expected ccdf (expected over the

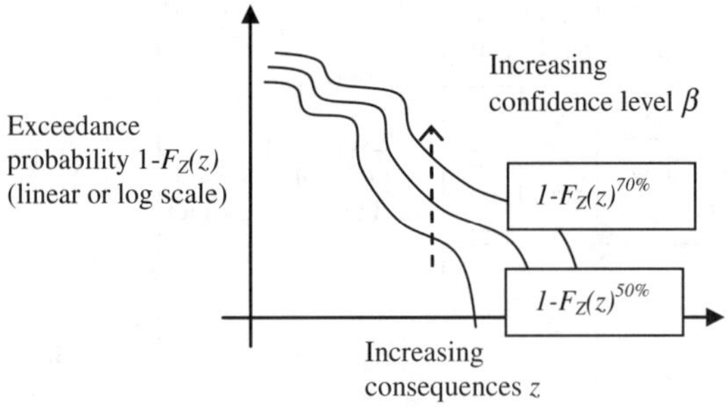

Figure 1.6 *A double-level uncertainty-risk-consequence Farmer's curve.*

level-2 distribution $\pi(\cdot)$) may be the most popular one, possibly because of its simplicity:

$$c_z(\boldsymbol{d}) = E_\pi[1-F_Z(z_s|\Theta_{XE})] = E_\pi[P(G(\boldsymbol{X}, \boldsymbol{E}, \boldsymbol{d}) > z_s|\Theta_{XE})] \tag{1.27}$$

Another common one is the (level-2) probability or confidence that the (level-1) frequency will exceed a given threshold of consequence while remaining below a risk target α (say $\alpha = 10^{-b}$):

$$c_z(\boldsymbol{d}) = P\,[1-F_Z\,(z_s|\Theta_{XE}) < \alpha] \tag{1.28}$$

Both involve elaborate integration of the chained model $G(.)$ with significant computational consequences (see Chapter 7). Those two risk measures exhibit different decision-theory properties and associated protection margins and costs (see Chapter 4). The kind of chained model involved in such elaborate risk studies and the variety of associated risk measures will also be illustrated by the pedagogical example in Chapter 3.

Although beyond the scope of the book, note also that the continued debate in the 2000s also suggested some extra-probabilistic extensions whereby the second level (or epistemic uncertainty layer) is modelled through Demspter-Shafer or evidence theory (see review in de Rocquigny, Devictor, and Tarantola, 2008), possibilitic distributions or fuzzy sets (see Helton and Oberkampf, 2004) on top of a probabilistic layer for aleatory (level-1) uncertainty.

1.4 Modelling under uncertainty in metrology, environmental/sanitary assessment and numerical analysis

The concept of uncertainty was already an essential part of the domains reviewed previously: every type of risk analysis comprises some statement about uncertainty . . . or else the risky event becomes a certainty. It is not the purpose of the present section to discuss in epistemological terms the specificities of uncertainty as differentiated from risk: Chapters 2 and 4 will return to that point. Observe so far that, beyond the domain of risk analysis as such, the study of uncertainty affecting the results of any modelling or measurement effort has involved some dedicated approaches. Those are referred to as *uncertainty and sensitivity analysis* (UASA). They stand beyond the explicit *risk analysis* as they do not model risky events and consequences in the sense of serious accidents. Consider for instance:

- Metrology;
- quality control in industrial processes;
- qualification and calibration of numerical codes;
- environmental or sanitary impact analysis;
- quantitative project management;
- oil exploration;
- climate studies;
- and so on.

1.4.1 Uncertainty and sensitivity analysis (UASA)

Notwithstanding the variable terminologies in sub-domains, several authors have pointed to a generic underlying methodological structure to the UASA studies (Helton *et al.*, 1996; de Rocquigny, Devictor, and Tarantola, 2008). Figure 1.7 sketches the conceptual framework advocated by de Rocquigny, Devictor, and Tarantola (2008) in a format comparable to the previous Figures presented in this chapter. UASA studies involve considerable similarities: a pre-existing model – anything from the simple analytical operations describing a basic measurement chain to the complex finite element model of air pollution transfer – is the central object of interest. It is studied as to how sources of uncertainties affecting its inputs affect the outputs. UASA studies tend to privilege output *variables* of interest rather than *events* of interest in the absence of *risky events*. In the UASA studies, it is also useful to distinguish between:

- full *decision criteria*, which correspond to risk acceptance criteria. For instance: the metrological chain should issue a measurement with a maximum of 3% uncertainty (generally a given multiple of the coefficient of variation); pollutant discharge should not exceed the licence threshold with 95% confidence (or 95% of the time variability) and so on,

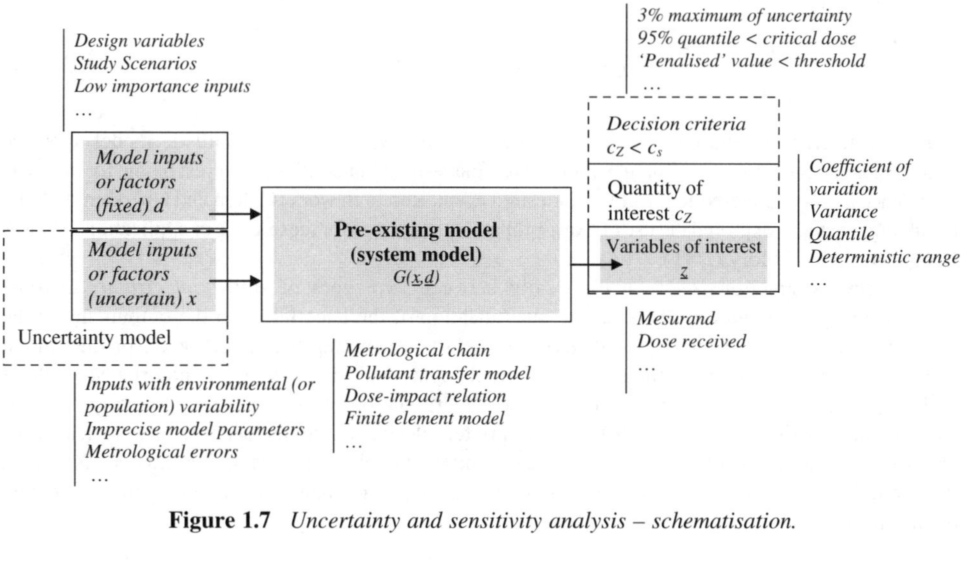

Figure 1.7 *Uncertainty and sensitivity analysis – schematisation.*

- *quantities of interest*, which correspond to risk measure. For instance: a coefficient of variation representing the output variability or inaccuracy; a quantile; or a maximal deterministic range of variation and so on.

In many studies, particularly in the numerical modelling activity or other upstream stages, there are no (or not yet clear) external regulatory or formal decision-making processes to abide by and hence no predetermined criterion for the question: 'how much uncertainty affects my predictions?'. Assessing uncertainty in terms of some given *quantity* is however inevitable, although sometimes implicit. The question 'how much uncertainty?' is often answered on the basis of the variance or coefficient of variation of the output. However, it is sometimes forgotten that the appropriate approach and answers given could be quite different if one referred to the maximal deterministic range or a one-sided confidence interval: this is where the choice of a *quantity of interest* comes up.

On the input side, UASA also distinguish uncertain model inputs (vector $\boldsymbol{x}$) from fixed ones (vector $\boldsymbol{d}$). Note that the allocation of the model inputs between 'uncertain' and 'fixed' ones is often a matter of choice rather than theoretical discussion. It can change over the course of the study and of the decision-making process. Importance ranking/sensitivity analysis and model calibration steps do play key roles with respect to that choice, as will be commented upon later.

An uncertainty model is built upon the uncertain inputs, through various probabilistic or non-probabilistic settings. Central practice consists in the classical one-level probabilistic setting: this means randomising all sources of uncertainty – be they temporal, spatial or population variability, lack of knowledge on model parameters, measurement errors, and so on – affecting vector $\boldsymbol{x}$. This is summarised into a pdf with independent or dependent inputs if appropriate, provided information is available to quantify them. The variety of settings already mentioned previously on natural risk, structural safety or integrated PRA can also be found: mixed deterministic-probabilistic settings (e.g. with some penalised fixed inputs) or even double-level probabilistic settings distinguishing so-called aleatory from epistemic components. Indeed, double-level probabilistic settings materialise naturally when undertaking an uncertainty study on the parameters of a pre-existing QRA or IPRA model. As mentioned above, the risk measure of a QRA is expressed as the following deterministic relation between the frequency of the output event and those of the elementary input events:

$$c_z(\boldsymbol{d}) = f^e = G[(f^i)_i, (p^k)_k, \boldsymbol{d}] \tag{1.29}$$

Hence, an uncertainty study of its parameters results in developing a probabilistic model upon the frequencies and probabilities $(f^i)_i$, $(p^k)_k$ themselves, the resulting quantities of interest being, for instance, confidence levels on the frequency of the undesired event. Research work (Helton and Oberkampf, 2004; Limbourg,) has also explored the use of extra-probabilistic settings for these kinds of applications, such as Dempster-Shafer Theory (DST).

A formal distinction is drawn classically between two types of study: *uncertainty analysis* (or *uncertainty propagation*) denoting the quantification of the uncertainty in the outputs inside the appropriate quantities of interest; and *sensitivity analysis* denoting the apportioning of that output uncertainty into each component of the uncertain inputs and the ranking of importance of those with respect to a given quantity of interest. Such interpretation of the terms is not always as clear-cut, so that sensitivity analysis may less appropriately designate elementary treatments such as the one-at-a-time variations of the inputs of a deterministic model or the partial derivatives; elsewhere, it may denote the entire UASA effort of understanding and managing a model in the context of uncertainty.

1.4.2 Specificities in metrology/industrial quality control

In the domain of metrology or quality control, uncertainty analysis is a basic requirement associated with the qualification of any measurement device, chain of devices or production process. Through sensitivity analysis (although the term is less classical there), it is also a forcible means to optimise costs and productivity of monitoring and quality control investments. It may sometimes be officially regulated, such as in nuclear maintenance, radiological protection or environmental control. For instance, in the application of European undertakings relating to the section of the Kyoto protocol concerning industrial emissions, a system of CO_2 emission permits has been established; it requires an explicit uncertainty declaration in order to secure fairness between real emissions and emission rights (Figure 1.8).

A metrological UASA study consists typically in aggregating the uncertainties associated with each variable x^i representing the result associated with each i-th operation or elementary sensor (e.g. flow metering of fuel, analysis of carbon content, net emission factor...) within the generally simple analytical function $z = G(\boldsymbol{x})$ expressing the final *measurand* z, for example the annual tonnage of CO_2.

In the field of metrology of industrial processes, the use of a probabilistic uncertainty model is rather common and quite standardised by the 'Guide to the expression of uncertainty in measurement' (ISO, 1995). Rather than aggregate elementary uncertainties in a deterministic manner, it assumes that uncertainty associated with each sensor or measurement operation in comparison with a reference value has an aleatory character that is modelled efficiently by probability calculations. Such uncertainty includes: noises affecting the operation of the sensor, local fluctuation of the measurement point around the average physical quantity sought, error in reading a counter or in rounding the measurement displayed, *and so on.*

The quantity of interest mostly involved is the so-called 'enlarged uncertainty', defined either as the ratio of the half-extension of a 95%-confidence interval around the expectation of z:

$$\% unc_z = \frac{1}{2.E(Z)} \left(z^{\frac{1+\alpha}{2}} - z^{\frac{1-\alpha}{2}} \right) \tag{1.30}$$

or, more approximately, as a multiple of its coefficient of variation:

$$\% unc_z \approx k.\frac{\sqrt{\mathrm{var}(Z)}}{E(Z)} = k.c_V(Z) \tag{1.31}$$

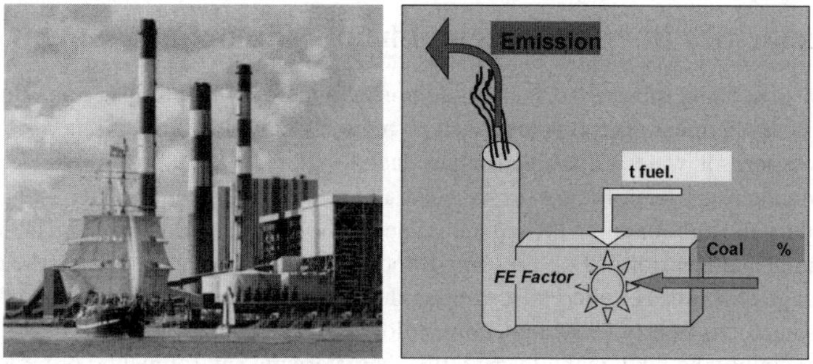

Figure 1.8 *Thermal power plant emissions (left) and metrological steps (right), (de Rocquigny, 2006).*

where $z^{\beta} = F_Z^{-1}(\beta)$ designates a quantile, F_Z standing for the cumulative distribution function of Z. The so-called enlargement factor k is being taken as 2, 3 or an intermediate value depending on the case. A typically associated decision criterion requires it should not exceed a given threshold, such as 3% maximal uncertainty in the declared value.

Uncertainty modelling is generally simple in metrology. While correlation coefficients represent dependences, fully Gaussian uncertainty models are selected: this is sometimes supported by an argument based on the symmetry of errors and the existence of multiple underlying additive physical phenomena. The fact that the uncertainties are bounded (for example because of a command and control mechanism) may nevertheless lead to the choice of a uniform or even triangular *pdf*.

Note, however, that the level-2 uncertainty associated with the estimation of the uncertainty of elementary devices is sometimes included explicitly in metrological uncertainty modelling. Think of the fluctuation of a sampling statistic $s^2_{X^i}$ in small samples around the real unknown variance $\sigma^2_{X^i}$. For instance, repeatable data-based empiric variances can be multiplied by a factor greater than 1, decreasing with the number of measures n used to estimate it, instead of simply taking the gross estimated value. This accounts for the possibility, through the effect of sampling fluctuation, of under-estimating the unknown true variance. The practice is heterogeneous, because n is not always known to the industrial end-user. This question becomes particularly important when the cost of determining sources of uncertainty is high, because of the necessary tests involving burdensome basic operations. Consider the reception of nuclear fuel assemblies before reloading in a power plant: the measurement process is very demanding, but the risk associated with too great a level of uncertainty is serious, since mistaken acceptance of a fuel assembly too large for the core grid would result in the costly unavailability of the plant.

Propagation, that is the computation of the quantities of interest, involves mostly the so-called 'uncertainty propagation law', which is a Taylor-first order approximation of the variance of the variable of interest:

$$\begin{aligned} E(Z) &\approx G(E(\mathbf{X})) \\ Var\,Z &\approx \sum_{i=1}^{p} \left(\frac{\partial G}{\partial x^i} \right)^2 VarX^i + \sum_{i_1 \neq i_2}^{p} \left(\frac{\partial G}{\partial x^{i_1}} \right) \left(\frac{\partial G}{\partial x^{i_2}} \right) Cov\,(X^{i_1}, X^{i_2}) \end{aligned} \tag{1.32}$$

Chapter 7 will discuss the fact that it constitutes indeed an *elementary* propagation method. Such an approximation relies essentially on the linearity of the system model, well suited to the purpose when the uncertainties are generally small enough with regard to the scale of potential non-linearities of $G(.)$.

1.4.3 Specificities in environmental/health impact assessment

Environmental or sanitary impact assessments comprise numerous modelling layers, such as: emission analysis, physical-chemical reactions and local or regional transport through one or multiple media (air – gaseous or aerosol, water – dissolved or suspended, soil, . . .), bio-exposition models, dose-effect, toxicological models, epidemiology, external costs and so on. UASA studies have been developed for decades to understand and contain with better confidence the uncertain impact of emissions or activities (Granger Morgan and Henrion, 1990; Frey and Rhodes, 2005). As a matter of fact, the area is fraught with a wide range of sources of uncertainty: either due to the hard task of modelling complex coupled systems and sometimes poorly known phenomenologies, discrepancies between expert opinions, the lack of data, or simply the fundamental variability characterising environmental or biological systems and populations.

In reference to the previous areas, the following features and specificities can be noted:

- Risk measures are typically quantiles: 95 % quantile of concentration of a pollutant in the discharge, 10^{-3} received dose of a potentially harmful compound and so on.
- The choice of distinguishing or not variability from uncertainty is an important point: this results in single or double-level probabilistic settings which are theoretically very similar to those encountered in the aleatory/epistemic debate in IPRA.
- Highly numerous uncertain model inputs are encountered, with a comparatively low amount of data. This implies an extended use of expert elicitation and of Bayesian settings which are much more developed than elsewhere.
- The absence of unique or well-defined pre-existing models, the widely-recognised need for calibration (Beck, 1987) and the interest for machine learning or data mining-inspired techniques (e.g. (Gauthier *et al.*, 2011) on cancer risk score).

The latter two points might be the more challenging with regard to modelling. This may be understood in the context of the intrinsic complexity and variability of the mineral or organic environment: this happens in the geological, hydrological or sedimentology fields, or with even greater complexity at the biological, ecological or health levels. The phenomenology is generally less controlled than in man-built structures or systems. The dissemination of chemical compounds in a river environment, for instance, involves many types of transfer between water-borne dissolved or suspended, sediment-borne, or living matter-borne compartments, each of which may be modelled with variable levels of geographical or phenomenological detail. Only a handful of sample measurements may be available to inform and calibrate such models. Moreover, in the absence of data, quite controversial ranges may be found for the plausible variability of the environmental coefficients which additionally depend on the type of model to which the coefficient is an input. Besides standard uncertainty treatments such as the use of Monte-Carlo simulation, the issue of model building, selection and calibration has seen greater development than elsewhere. The following section and Chapter 6 will discuss that important issue.

1.4.4 Numerical code qualification (NCQ), calibration and data assimilation

The prevalence of uncertainty considerations has grown considerably in the field of numerical analysis in the 2000s. In fact, they may be traced back to the late 1970s in nuclear thermal-hydraulic codes (Cacuci *et al.*, 1980), or indirectly to the very beginning of computer science after World War II. Monte-Carlo Sampling may be viewed as the very historical origin of computing: recall that Von Neumann's ENIAC machine in 1946 was designed essentially for Monte-Carlo neutronics. Yet, probabilistic sampling was fundamentally thought of as a solver of the neutron equations for given (fixed) input conditions owing to the underlying statistical physics; which does not represent the type of macroscopic uncertainty now prevalent in risk and uncertainty analysis.

The above-mentioned structure of UASA encompasses the typical issues raised: $G(.)$ represents a complex numerical function resulting from numerical solvers of physics-based integral-differential operators (e.g. a neutron calculation in a nuclear core). It often takes in large-dimension vectors of continuous inputs $\boldsymbol{x}$ and outputs z that discretise multi-variate functional inputs/outputs representing variable fields or time transients. Typical examples are fluid velocities or temperatures, stress fields and so on. Input parameters are affected by all sorts of uncertainties and what is looked for is, by order of subjective priority:

- ranking sensitivities of the outputs to the numerous inputs,
- rating an 'order of magnitude' of output variability, the quantities of interest being mostly variance or coefficient of variation and, more rarely, quantiles,
- calibrating a code against experimental data and testing facilities,
- updating the code prediction with new data to be assimilated.

Owing to the strong numerical background, derivatives have been playing a key role in the uncertainty and sensitivity techniques considered. They were firstly involved in 'one-at-a-time' deterministic sensitivity analysis and later through Taylor-first order approximations of the output variance. Although completely identical to metrological techniques from a probabilistic point of view, the numerical strategies deployed to obtain the least costly calculation of derivatives appearing in the formula led to the use of a variety of names: by direct derivative = *Direct Sensitivity Method* (*DSP*), or by adjunct derivative = *Adjoint Sensitivity Method* (*ASM*), (see Cacuci *et al.*, 1980), or else the 'method of perturbations' or the 'method of moments'. These appellations sometimes refer to the deterministic propagation of uncertainties or alternatively the estimation of other probabilistic moments through differential developments of an order higher than one. Since the year 2000, sensitivity ranking has incorporated more sophisticated sensitivity analysis techniques such as polynomial chaos expansions, which develop the input/output joint distributions on a much more powerful basis than Taylor series (Ghanem, 1990; Sudret, 2008). In general, these developments refer implicitly to variance as a quantity of interest. Gaussian or Uniform uncertainty models are used with comparatively less effort devoted to statistically estimating those models from real field data. Indeed, from a numerical analysis point of view, probabilistic sensitivity analysis may be seen as a powerful exploration tool over spaces of possible values for systems which may not yet exist, as is the case with upstream design stages (de Rocquigny, Devictor, and Tarantola, 2008).

Unlike the area of risk analysis, environmental or metrological studies, there is a virtually-unanimous consensus on sticking to simple probabilistic models mixing all sources of uncertainty within a single layer. Yet, in the particular domain of nuclear thermal-hydraulics which has considerable historical significance, it is traditional to account for a second probabilistic level in order to cover the numerical sampling limitations placed upon the (computational) estimation of an output quantile. The so-called Wilks method, based on (Wilks, 1941) though well beyond the scope of his initial work, implies the estimation of a '95 95' level for the output variable of interest. As will be discussed in Chapter 7, it is in fact a very robust technique to cover *propagation uncertainty* or error that is generated by a low number of Monte-Carlo runs on a costly numerical code. On the contrary, such sampling uncertainty is not accounted for in the mechanical models that were historically behind the structural reliability techniques mentioned above. Indeed, thermal-hydraulics models provide evidence of a response that is far more irregular for inputs in general and hence amenable to potentially rougher output distributions and slower resulting convergence.

An essential area of historical development has been with regard to the issue of experimental calibration (or qualification, or *Verification and Validation*) and data assimilation; the former typically in nuclear code building (Boyack, 1990) and the latter in meteorology (Talagrand, 1997). Closely related are the fields of parameter identification (Beck and Arnold, 1977; Walter and Pronzato, 1997) or model identification already mentioned in the environmental domain (Beck, 1987). All involve the comparison between model outputs and experimental data resulting in the calibration of input parameters that are more or less unknown, the estimation (and generally reduction) of model-measurement residual error, and possibly the dynamically-updated prediction of outputs (in the case of dynamic data assimilation). These are, for example, initial conditions and resulting atmospheric indicators in meteorology. It is necessary at

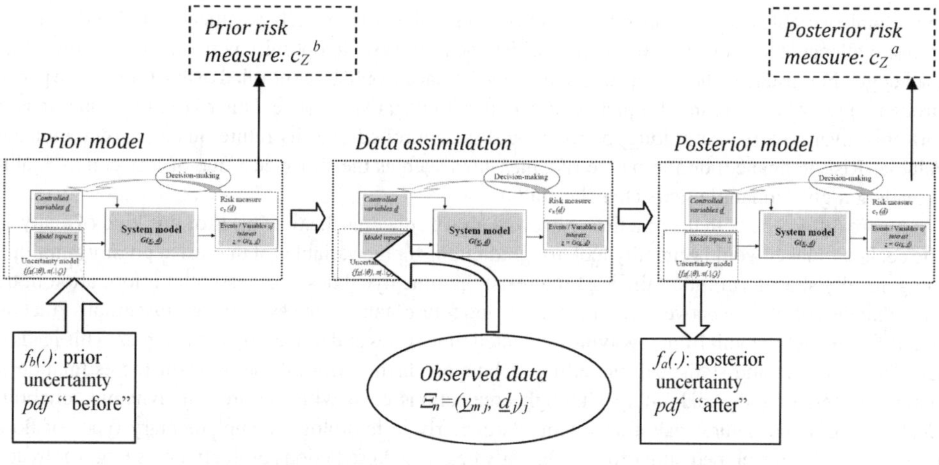

Figure 1.9 *Model calibration and assimilation – schematisation.*

this stage to complete the modelling structure by introducing a second category of model outputs that becomes essential in this case: within the high-dimensional outputs $\boldsymbol{m}$ of numerical codes, observations can be made available on output components $\boldsymbol{y}$ that are different to those on z that are later involved in predicting the quantity of interest. Hence, this family of inverse techniques generally consists in translating the information obtained on a sample of experimental measurements $(\boldsymbol{y}_j)_j$ into the calibrated value of the uncertain inputs $\boldsymbol{x}$. In a dynamic approach, generally best interpreted in a Bayesian statistical framework, there may thus be a prior uncertainty model with the associated predicted quantity of interest, followed by a posterior description and prediction after incorporation of assimilated data (Figure 1.9).

Chapter 6 will further develop this topic which proves essential for sound modelling in the context of risk and uncertainty. Substantial extensions of the algorithms developed in classical numerical analysis are needed: indeed, those algorithms mostly stick to Gaussian and/or linear contexts and are unnecessarily limited to a description of *epistemic* uncertainty in the sense that code inputs are generally considered to be unknown but physically fixed (de Rocquigny and Cambier, 2009).

1.5 Forecast and time-based modelling in weather, operations research, economics or finance

An essential area of application of modelling under uncertainty with considerable practical value and on-going research involves the prediction of variables of interest under uncertainty over future times of interest. Think of meteorological forecast, operations research and stochastic optimisation be it in econometrics (micro-economic market forecast or macro-economic growth forecast), energy or asset management, financial market forecast and portfolio analysis and so on (see, e.g. the survey of Tay and Wallis, 2000 or Elliott, Granger, and Timmermann, 2006).

This area shares essential common features with the framework developed above (in particular IPRA or UASA). Again, the decision-maker is interested in the prediction of given variables of interest characterising an underlying system, process, portfolio or market, which is itself impacted by a number of uncertain events, behaviours or mechanisms, in order to typically select the best actions to take (the controllable variables) such as operational actions best suited to the likely weather or market demand or portfolio allocations. A quantity of interest or risk measure is set to optimise performance or control the risk budget. For instance, the power supply-demand balance would be controlled over a future time period through a prescribed maximal frequency of events with negative balance. Otherwise, the value of asset portfolios over a particular holding period would be controlled over its future quantiles (denominated Value-at-Risk in finance) or alternative risk measures such as the expected value of loss conditional to exceeding a given quantile (expected shortfalls).

A core difference with the practices reviewed above concerns the time basis of this kind of analysis. The decision-maker would generally look for predictions for the variables of interest at possibly-multiple and generally short-term future times of interests; the time dynamics carry essential information in the sense that the correlation between recent, present and future states of the system are substantial so that the underlying models benefit from involving dedicated time series and probabilistic dynamics. This leads to using probabilistic and statistical modelling techniques distinct from those used for the estimation of extreme events or industrial safety although connections exist with the area of dynamic reliability (Section 1.3.3) and natural risk assessment. Particularly in hydrology, complementary types of flood models are being developed either for the day or week-ahead predictions of likely flows (e.g. for hydropower, early-warning, flood routing or crisis management) or for extreme event flood risk assessment (e.g. for the design of dam safety or protection dike). Eventually, the area of forecasting often involves *events frequent enough* to be observable.

Though not central in the scope of this book, a number of comments will be addressed to the specificities required regarding time-explicit risk measures (in Chapter 4), forecast models as distinguished from risk models (in Chapter 6) and time-based decision-making possibly involving the essential concept of options (in Chapter 8).

1.6 Conclusion: The scope for generic modelling under risk and uncertainty

1.6.1 Similar and dissimilar features in modelling, risk and uncertainty studies

Practice shows both similar and dissimilar features, as summarised in Table 1.3. Firstly, the general structure is similar in most modelling studies (Figures 1.1–1.3 and 1.7): it incorporates a system model, representing the system studied under risk and/or uncertainty, and a number of associated inputs and outputs of interest for decision-making. Furthermore, most risk or uncertainty studies incorporate a probabilistic representation of at least some of the inputs. Hence the outputs of interest are assessed from the perspective of a probabilistic quantity which is called the *risk measure* or *quantity of interest* (or design criterion, etc.). From a theoretical perspective, such measures mostly involve a generic format, namely multiple integrations of the system model over the uncertainty model.

Yet, a closer look into the various practices evidences some dissimilarity:

- The nature of system models: clearly *phenomenological* for natural risk or SRA; mostly *logical* for QRA; of a mixed nature within IPRA; and viewed as *numerical* in UASA or code qualification.
- The type of inputs/outputs: *discrete* events are considered for some (in QRA or SRA for the output side) while *continuous* variables are involved for others (such as UASA or SRA for the input side). Probability distributions thus vary from discrete laws, such as Bernoulli r.v. for simple events, to continuous laws, such as Gaussian r.v. for uncertain input variables.

Table 1.3 Overview of domain generics and specificities.

	Natural risk protection (NRA)	SRA	QRA	IPRA	UASA - metrology	UASA - environmental	NCQ, data assimilation	Forecasting/ time-based in weather, asset man^{gt}, finance
System model	Local risk situation	Failure function	Event tree, fault tree, BBN...	A combination of the NR, SRA and QRA	Pre-existing model (metrological chain)	Pre-existing model (numerical model)	Numerical code	Pre-existing model (numerical model)
Event/ variables of interest	Critical event	Failure event	Undesired event (or top event)	Continuous consequence variables or set of consequence events	Variables of interest: continuous and often scalar	Variables of interest: continuous and often vector	Variables of interest: continuous and often vector	Variables of interest: continuous and time series
Model inputs	aleatory variables (mostly continuous)	uncertain variables (mostly continuous)	Elementary events	A combination of NR, SRA, QRA	uncertain variables (measurement errors)	uncertain variables (mostly continuous)	uncertain inputs (mostly continuous)	Uncertain inputs and recent states (static and/or time-series)
Risk measure/ quantity of interest	Frequency of undesired event or quantile of variable of interest	Failure probability (undesired event)	Frequency of undesired event	Frequencies of a set of consequences or ccdf of variables of interest	Quantity of interest: C_V or c.i.	Quantity of interest: C_V or c.i.	Quantity of interest: variance or C_V	Expected value, c.i., variance or quantiles at short-term future times
Remarks	Careful statistical modelling, often involving extreme value theory	CPU challenge due to low probabilities	Very large vector of inputs and dependence	'Epistemic' and 'aleatory' components may be distinguished	Level-2 uncertainty (estimation of input uncertainty)	Large uncertain models and large use of expert opinion	CPU challenge due to large input dimension	Specificities due to the importance of temporal correlation structure and stochastic processes

- The *type of risk measures/quantities of interest* considered and associated with the decision criteria: failure probabilities or frequencies, output variance or confidence intervals, expected consequence and so on.
- The probabilistic settings and their specification: some studies clearly differentiate 'risk or vulnerability or *aleatory*' components (quantified in a frequency) from 'uncertainty or *epistemic*' components (quantified in an associated confidence interval); others aggregate overall probabilities or expectations without making an explicit distinction between the variability or levels of imprecision involved. In many cases *deterministic penalisation* of some inputs is kept, along with probability distribution for others.
- The main *objectives of the study*: to demonstrate compliance with a regulatory criterion, to optimise a design, or simply to facilitate importance ranking and deeper phenomenological understanding.
- The biggest challenges encountered: for some studies, *statistical estimation* may involve large resources and complex models while, in others, informal expert judgement is considered satisfactory; the *computation of risk measures* may be a real limiting factor in SRA, NCQ or some large UASA while it is only a minor step in QRA or natural risk; *model building* and *validation* and associated uncertainties are of key concern in environmental or biological modelling, though much less elsewhere.

1.6.2 Limitations and challenges motivating a unified framework

Previous sections have already evidenced some limitations in given domains, and the restricted permeability between them. In summary:

- Distinction of practices between NRA and SRA tends to prevent a fruitful use of the relative advances of each, such as detailed statistical modelling for the former, or performing coupled probabilistic-numerical computation for the latter.
- Careful specification of the meaning of the probabilistic models (the nature of underlying uncertainty) is not as developed in SRA or NCQ as it is in IPRA.
- The potential of high-performance computing is largely investigated in SRA or NCQ, but is insufficiently accounted for in NRA which is often limited to unnecessarily simplified risk description.
- Formalised model identification and selection techniques, largely developed in data analysis and statistical modelling, have only permeated into environmental or biological modelling, and only to a limited extent into numerical code qualification and data assimilation.
- The time basis for the variables of interest is fundamental in forecasting, time-based risk analysis in finance or data assimilation techniques owing to the crucial role of temporal correlations inside the model; yet, other domains such as NRA or IPRA also implicitly define the time basis of their measures, though not necessarily requiring a complete stochastic process as the correlations can be less tractable for, for example, extreme events.
- Rather than opposing them violently, a careful understanding of the mixed deterministic-probabilistic settings helps describe more accountably the plausible extent of lack of knowledge. This could avoid over-trusting either: deterministic design assumptions disregarding the extent of uncertainty; or conversely fully probabilistic computations based on vaguely justified pdf in the absence of data, credible expertise or simply by not specifying the meaning of the randomised quantities.

Complex industrial systems require an increasing blend of all such approaches as the search for operational margins and the regulatory control processes intensifies. For instance, the investigation of vulnerability to natural risk initiators addresses jointly: a careful specification of the meaning of the quantity of interest or risk measure regarding the underlying type of uncertainty covered, as well as the time basis; a finer modelling of the climate statistical series; a better calibration of physical-numerical models representing the local phenomenology; and a finer understanding of protection system reliability. This suffers from insufficient consistency between the reliability, statistical and physical modelling components and expertise.

In spite of their apparent diversity, there is much to share and integrate. As mentioned above, QRA may be formally presented similarly to SRA, itself linkable to the 'extended frequency approach' in NRA. The comments on advanced IPRA or UASA have demonstrated the opportunities associated with unifying frameworks explicitly: Chapter 2 will generalise them into a *generic modelling approach*. For instance, the computation of most of the risk measures or quantities of interest involves essentially the integration of a deterministic function (the system model) over the uncertainty model (a more or less elaborate density function); the nature of the inputs or outputs involved (discrete events or continuous variables) merely facilitates some integration techniques while conserving the theoretical issue (see Chapter 7). Moreover, the *estimation* needs of the various types of models, as well as those of calibration, and assimilation of identification, mostly take on similar statistical techniques, although coming with partial variants owing to the continuous, discrete or temporal nature of the variables (see Chapter 5), or to the input/output position of the variables that can be observed directly or indirectly through a numerical model (see Chapter 6). These facts, all too often hidden by domain boundaries in the literature or engineering practices, constitutes the key starting point for the use of the generic numerical tools that will be developed in the subsequent chapters.

References

Apel, H., Thieken, A.H. *et al.* (2004) Flood risk assessment and associated uncertainty. *Natural Hazard and Earth System Sciences*, **4**, 295–308.

Apostolakis, G. (1999) The distinction between aleatory and epistemic uncertainties is important; an example from the inclusion of aging effects in the PSA. Proceedings of PSA'99, Washington DC.

Aven, T. (2003) *Foundations of Risk Analysis*, John Wiley & Sons, Ltd.

Aven, T. and Rettedal, W. (1998) Bayesian Frameworks for integrating QRA and SRA methods. *Structural Safety*, **20**, 155–165.

Beck, J.V. and Arnold, K.J. (1977) *Parameter Estimation in Engineering and Science*, John Wiley & Sons, Ltd.

Beck, M.B. (1987) Water quality modelling: A review of the analysis of uncertainty. *Water Resources Research*, **23**(8), 1393–1442.

Bedford, T.J. and Cooke, R.M. (2001) *Probabilistic Risk Analysis: Foundations and Methods*, Cambridge University Press.

Beven K.J. and Binley A.M. (1992) *The future of distributed model: model calibration and uncertainty prediction,* Hydrological Processes, **6**, 279–298.

Boyack, B.E. (1990) Quantifying reactor safety margins-part I: an overview of the code scaling, applicability and uncertainty evaluation methodology. *Nuclear Engineering and Design*, **119**, 1–15.

Cacuci, D.G. *et al.* (1980) Sensitivity Theory for General Systems of Nonlinear Equations. *Nuclear Science and Engineering*, **75**.

de Groot, M.H. (1970). Optimal Statistical Decisions, McGraw-Hill, NewYork.

de Rocquigny, E. (2010) An applied framework for uncertainty treatment and key challenges in hydrological and hydraulic modeling. *Canadian Journal of Civil Engineering*, **37**(7), 941–954.

de Rocquigny E. (2006), La maîtrise des incertitudes dans un contexte industriel : 1ère partie – une approche méthodologique globale basée sur des exemples. *Journal de la Société Française de Statistique,* **147**(4), 33–71.

de Rocquigny, E. and Cambier, S. (2009) Inverse probabilistic modelling through non-parametric simulated likelihood Pre-print submitted to Inverse Problems in Science and Engineering.

de Rocquigny, E., Devictor, N. and Tarantola, S. (eds) (2008) *Uncertainty in Industrial Practice, A Guide to Quantitative Uncertainty Management*, John Wiley & Sons, Ltd.

Ditlevsen, O. and Bjerager, P. (1986) Methods of structural systems safety. *Structural Safety*, **3**, 195–229.

Duckstein, L. and Parent, E. (1994) Engineering Risk in Natural Resources Management, NATO ASI Series.

Dupuy, J.P. (2002) Pour un catastrophisme éclairé – Quand l'Impossible est certain. Ed. du Seuil. (Dutch regulations on flood risk).

Ellingwood, B.R. (1994) Probability-based codified design: past accomplishments and future challenges. *Structural Safety*, **13**, 159–176.

Elliott, G., Granger, C. and Timmermann, A. (eds) (2006) *Handbook of Economic Forecasting*, North Holland, French nuclear regulations on external aggressions (flood, seism, . . .).

Frey, H.C. and Rhodes, D.S. (2005) Quantitative analysis of variability and uncertainty in with known measurement error: Methodology and case study. *Risk Analysis*, **25**(3).

Gauthier, E., Brisson, L., Lenca, P. and Ragusa, S. (2011) Breast Cancer Risk Score: A data mining approach to improve readability. Proc. of International Conference on Data Mining (DMIN11), (Ghanem, 1990; Sudret, 2008) on polynomial chaos & SA.

Ghanem, R.-G. and Spanos, P.-D. (1991), *Stochastic finite elements – A spectral approach*, Springer Verlag.

Granger Morgan, M. and Henrion, M. (1990) *Uncertainty – A Guide to Dealing with Uncertainty in Quantitative Risk and Policy Analysis*, Cambridge University Press.

Helton, J.C. (1993) Uncertainty and sensitivity analysis techniques for use in performance assessment for radioactive waste disposal. *Reliability Engineering & System Safety*, **42**, 327–367.

Helton, J.C., Burmaster, D.E. *et al.* (1996) Treatment of aleatory and epistemic uncertainty. *Special Issue of Reliability Engineering & System Safety*, **54**(2–3).

Helton, J.C. and Oberkampf, W.L. (2004) Alternative representations of epistemic uncertainty. *Special Issue of Reliability Engineering & System Safety*, **85**(1–3).

ISO (1995) Guide to the expression of uncertainty in measurement (G.U.M.).

Krzystofowicz, R. (1983). Why should a forecaster and a decision maker use Bayesian theorem, *Water Resour Res*, **19**(2), 327–336.

(Limbourg, 2008) – Book on the use of DST on PRA.

Lewis, H.W., Budnitz, R.J., Rowe, W.D. *et al.* (1979) Risk assessment review group report to the U. S. nuclear regulatory commission. *IEEE Transactions on Nuclear Science (IEEE)*, **26**(5), 4686–4690.

Madsen, H.O., Kenk, S. and Lind, N.C. (1986) *Methods of Structural Safety*, Prentice-Hall Inc.

Melchers, R.E. (1999) *Structural Reliability Analysis and Prediction*, 2nd edn, John Wiley & Sons, Ltd.

Menzies, J.B. (1995) Hazard, risks and structural safety. *The Structural Engineer*, **73**(21), 7.

NRC (1990) Severe Accident Risk: An Assessment for Five U.S. Nuclear Power Plants. U.S. Nuclear Regulatory Commission, NUREG-1150, Washington, DC, December 1990.

Oberkampf, W.L., DeLand, S.M., Rutherford, B.M. *et al.* (2002) Error and uncertainty in modelling and simulation. *Special Issue of Reliability Engineering & System Safety*, **75**(3), 333–357.

Pappenberger, F. and Beven, K.J. (2006) Ignorance is bliss. 7 reasons not to use uncertainty analysis. *Water Resource Research*, **42** (W05302).

Rasmussen, N.C. *et al.* (1975) Reactor Safety Study: An Assessment of Accident Risks in U.S. Commercial Nuclear Power Plants. U.S. Nuclear Regulatory Commission, NUREG-75/014 (WASH-1400), Washington, DC.

Salvi, O. and Debray, B. (2006) A global view on ARAMIS, a risk assessment methodology for industries in the framework of the SEVESO II directive. *Journal of Hazardous Materials*, **130**, 187–199.

Sudret, B. (2008) Global sensitivity analysis using polynomial chaos expansions, *Reliab. Eng. Syst. Safe.*, **93**, pp. 964–979.

Talagrand, O. (1997) Assimilation of observations, an introduction. *Journal of the Meteorological Society of Japan*, **75**, 191–201.

Tay, A.S. and Wallis, K.F. (2000) Density forecasting: A survey. *Journal of Forecasting.*, **19**, 235–254.

Walter, E. and Pronzato, L. (1997) *Identification of Parametric Models from Experimental Data*, Springer Verlag, Heidelberg.

Wilks, S.S. (1941) Determination of sample sizes for setting tolerance limits. *Annals of Mathematical Statistics*, **12**.

Yen, B.C. and Tung, Y.K. (1993) *Reliability and Uncertainty Analysis in Hydraulic Design*, ASCE, New York, N.Y.

2

A generic modelling framework

This chapter introduces the generic mathematical modelling framework that underpins the entire scheme of this book. Such a framework is focused on the proper integration of statistical/physical tools within a consistent modelling and decision-making approach. It is common to most modelling practices within the context of risk or uncertainty in mathematical, analytical and computational terms, although allowing for significant differences in the probabilistic (or mixed deterministic-probabilistic) formulations according to regulations or within design-making contexts. Within this framework, Chapter 2 introduces the issues and methods covered by subsequent chapters.

2.1 The system under uncertainty

The following issues will be developed while keeping in mind the entire variety of risk or uncertainty modelling situations which were reviewed in Chapter 1. As was delineated in the corresponding illustrations, they generally involve, as summarised in Figure 2.1 below:

- a *pre-existing system* (be it industrial, environmental, metrological ...) lying at the heart of the study,
- a variety of *actions*, that is design, operation or maintenance options or any type of controlled variables or events that enable the operator to modify or control to some extent the system's performance or impact,
- a variety of uncontrolled events (external initiators, internal failures ...), operation or environmental scenarios, measurement noise, practical observation limitations or more generally any *sources of uncertainty* affecting the precise knowledge of the state of the system,
- *stakes* associated with the state of the system that are involved in the decision-making process motivating the risk/uncertainty assessment; more or less explicitly, they include: safety and security, environmental control, financial and economic optimisation and so on.

It is worth emphasising that the undertaking of a risk/uncertainty study is linked to the following underlying facts:

Modelling Under Risk and Uncertainty: An Introduction to Statistical, Phenomenological and Computational Methods, First Edition. Etienne de Rocquigny.

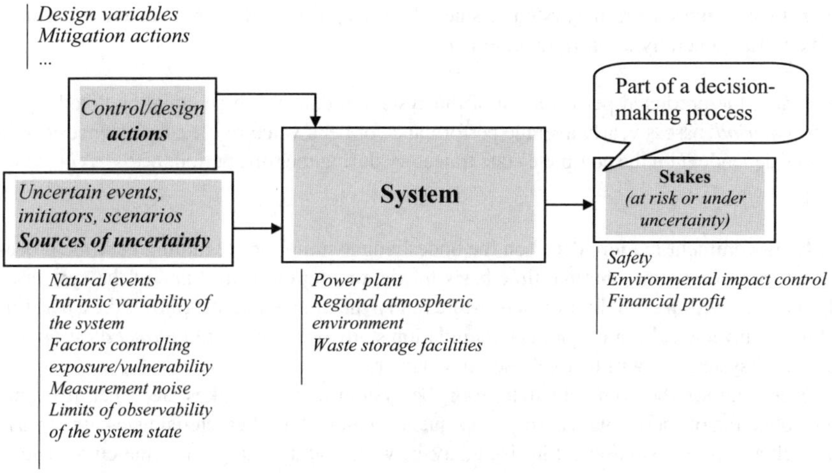

Figure 2.1 *Overall situation for risk and uncertainty assessment.*

(i) The state of the system considered, conditional to taking some given actions, is *imperfectly known* at any given time.
(ii) Some characteristics of the state of the system, incorporated in a given type of 'performance' or 'consequence', are *at stake* for the decision-maker.

Consider the following examples:

- The occurrence of flooding over a given dike height within the next year, essential for safety, cannot be predicted exactly because of meteorological unpredictability or lack of knowledge of flooding phenomena.
- Next year's net profit for a given production plan, essential for the shareholders, is uncertain due to lack of knowledge of the effective market demand, variable dependability of plants, unpredictability of the exchange rates and so on.
- The length of a device – at the time it is being measured within a given experimental setting – is imperfectly known due to internal noise or uncontrolled and influential environmental factors affecting the sensor; such length is an essential feature to secure the fit within the assembly specifications.
- . . .

The *state*[1] *of the system* may be characterised by a large number of indicators, such as discrete states (component no. 3 operational or failing) or continuous variables (water level, dollar rate, etc.). In the spirit of the theory of probability, it will be formally denoted as ω, as being a point in the large theoretical *sample space* Ω which contains virtually any possible state of the system. For the sake of completeness, the *time*

[1] Note that it is rather more a 'modelling' state than the 'true' state of the nature of the system; even if it was complex enough to represent quantitatively every characteristic, it would remain attached to a conventional definition of those characteristics from an *observable* standpoint: mechanical strains or temperature defined at a macroscopic scale, velocity or densities defined as given time-averages, etc. See also the comments on the metrological aspect in Section 4.1 whereby those observables are to some extent always *relative* values instead of absolute values.

of interest at which the state of the system is studied for the purpose of the risk/uncertainty study should be specified. It is generally a *future* time, either:

- a *given date*: the predicted performance of the system at date t_f, which could be noted ω_{tf}
- or a *given period*: assess yearly average performance or peak value of the consequences over a future year, both dependent on the complete state trajectory during the time period $[t_f, t_f + \Delta T]$, which could be denoted $\omega_{[t_f,t_f+\Delta T]}$.

Chapter 4 will comment in more detail on the underlying assumptions regarding the time processes; so far, it is sufficient to assume that the time basis for the prediction of the state of the system has been properly specified within the definition of ω. More rarely, the time of interest can also be a *past* time when undertaking an inverse calibration or data assimilation study on the basis of *indirect observations* of the past states of a system, as will be explained in Chapter 6.

However complex the complete state ω of the system may be, stakes are generally valued by a restricted collection of 'performance' (or 'consequence') indicators characterising *events or variables of interest*, such as: an undesired event in risk analysis, worse failure margin or time-cumulated environmental impact during operation period and so on. As was introduced in Chapter 1, those will be denoted in vector $z = (z^l)_{l=1\ldots r}$, the component of which is either continuous (for variables of interest) or discrete (for events of interest). Most of the time, z is a *scalar* or a *small-size vector* (e.g. $r = 1$ to 5) since the decision-making process involves essentially one or few variables or events of interest. But in some cases, z may be of a large dimension (e.g. predicted oil volumes at many potential well sites) or even a *function* denoted as $z(\boldsymbol{r})$ (e.g. flood damage over a geographical area). Note also that when the time of interest is a given *period*, the vector of interest becomes a function of time, *that is* a *time series* denoted as z_t: think of the mechanical margin as a function of the number of fatigue cycles, the net cost of oil production as a function of time, and so on. The vector notation, appropriate in most situations, will be kept hereafter unless the functional nature of the output of interest changes the derivations significantly.

Remember that the state of the system (including its indicators z) is influenced by the choice of *actions*, for example: the major operating conditions of the installation to be certified, protection design levels, controlled experimental conditions involved in the metrological process and so on. 'Actions' should be understood within an even wider meaning (cf. 'acts' in (Savage, 1954)): they stand for all the controlled characteristics of the system, the efficiency or impact of which is to be taken into account within the risk/uncertainty study. Actions will be given the formal vector notation $\boldsymbol{d}$, comprising either discrete or continuous components. Although those 'actions' refer to complex designs, operational processes, time-dependent commands and so on that would theoretically require elaborate functions (spatial fields or temporal functions . . .) to represent them, it is recognised that, most of the time, risk/uncertainty studies involve the consideration of a reduced number of choices or degrees of freedom.

As recalled in the review presented in Chapter 1, the literature and textbooks regarding risk analysis would generally distinguish the following dimensions as follows:

- *hazard*, referring to the dangerous events that can aggress or disrupt the system;
- *likelihood*, or *uncertainty* associated to the *occurrence* of those events;
- *consequences*, owing to the *exposure* and *vulnerability* of given assets or stakes (material, economic, environmental, cultural, human . . .).

Paying due attention to those separate components proves analytically helpful or perhaps even mandatory for any serious risk assessment. Hazardous material (e.g. chemical release, water pollution, . . .) or uncontrolled aggression (e.g. meteorite, tsunami, quake . . .) generate significant risk only when there is non-negligible likelihood, significant exposure and true vulnerability to the hazard. It constitutes also the basic recipe for supplying decision-making or the wider issues of risk *perception* with regard to technical material.

As regards modelling, such a threefold breakdown of risk (which was formalised notably by the (Kaplan and Garrick, 1981) triplets, see Annex Section 10.2) can be mapped schematically within the previously introduced concepts as follows: variables or events of interest z on the output side of the system model represent the consequences or stakes for the decision-maker; the uncontrolled events or sources of uncertainties on the input side of the system model represent the hazard to which a likelihood or probabilistic model of uncertainty will be associated. Yet, complex risk assessment generally requires consideration not only of the likelihood of hazards but also the uncertainty in the exposure or vulnerability that may depend on ancillary uncertain events or risk factors. Further, it becomes difficult to precisely delineate the categories of hazard and consequences when analysing the complex chain of phenomena that lead to the ultimate undesired consequences. Eventually, regulatory purposes may circumscribe the variables of interest inside risk criterion to a sub-part of the entire chain – for example regulating the release of hazardous material, or overflow of a given levee system – thus making implicit the existence of exposure and vulnerability. Thus, this book focuses on a slightly different modelling framework with the input/output structure of a system model: the inputs including all risk factors, be they hazards, uncertain events, properties controlling the response, vulnerability and consequences; the outputs being tuned to the decision-making purposes, more or less directly linked to the eventual consequences.

2.2 Decisional quantities and goals of modelling under risk and uncertainty

2.2.1 The key concept of risk measure or quantity of interest

The consideration of risk and uncertainty with regard to the system means of course that there is no intention to control the exact value of z at the time of interest, conditional on actions $\boldsymbol{d}$. Knowledge of the system may encompass a variety of past observations of some features of the system, phenomenological or logical arguments supporting the prediction of certain limits or regularities on the states of the system, all more or less directly connected to the variables or events of interest z. Nonetheless, by the above definition, there will always be knowledge limitations within the state of the system. The best that may be hoped for is to infer *possible* or *likely* ranges for z.

Many frameworks could be potentially mobilised in order to represent such uncertain values or ranges, such as probability theory, deterministic interval computation, possibility theory, fuzzy logic or the Dempster-Shafer or Evidence Theory that may be seen as theoretically encompassing all of the former. While the literature has discussed to a large extent the pros and cons of various probabilistic or non-probabilistic paradigms (see, e.g. Helton and Oberkampf, 2004), this book is centred on the use of the probabilistic framework which has a number of desirable decision-making properties, as will be briefly recalled in Chapter 4. Besides, the review in Chapter 1 has shown that most studies rely on dedicated *probabilistic* quantities to measure the likelihood of values for z.

> Such a quantity used for the inference of the outputs of interest under uncertainty will be called a *risk measure* or a *quantity of interest*, otherwise referred to as a performance measure in Aven (2003). In general, any modelling effort that is considering explicitly the associated uncertainty would involve such a *quantity of interest* (de Rocquigny, Devictor, and Tarantola, 2008) while the wording of a *risk measure* would probably better fit the context of decision-making under risk or risk assessment.

Yet the two expressions refer to a strictly similar mathematical quantity, and are thus completely synonymous throughout the book. Some examples:

- percentages of error/uncertainty of the variable(s) of interest (i.e. coefficient of variation);
- confidence intervals of the variable(s) of interest;
- quantile of the variable of interest conditional on penalised inputs;
- probabilities of exceedance of a safety threshold;
- probability (or frequency) of an event of interest;
- expected value of the consequences;
- expected benefits-costs;
- and so on.

Such a quantity should be seen as a strategic summary of the limited knowledge of the states of the system at the time of interest for any decision-making: as such, it should best represent all *data and expertise* about the system. As already shown in Chapter 1, such an inference of the risk measure in the system depends closely on the proper definition and estimation of a certain form of model of the system and associated uncertainties. A precise mathematical definition of the risk measures depends on the full specification of the type of model used, such as probabilistic distributions (*i.e.* level-1), mixed deterministic-probabilistic descriptions or possibly even level-2 distributions as in IPRA. It will be introduced in the next sections along with a generic definition of what modelling means in an uncertain context. In any case, the risk measure will subsequently be noted as $c_z(\boldsymbol{d})$ since it is computed to represent the likely values of the outputs of interest z and crucially depends on the choice of actions $\boldsymbol{d}$ that modifies the anticipated state of the system, albeit to an uncertain extent, as depicted in Figure 2.2.

2.2.2 Salient goals of risk/uncertainty studies and decision-making

The risk measure is the key quantitative tool used to make decisions under uncertainty. From a theoretical standpoint (with particular reference to (Savage, 1954)), risk and uncertainty studies should eventually help in making decisions about appropriate *actions* in the context of a *relative* lack of knowledge about the state of the system. Decision-making is always about the relative *preference* used to *select* from within the

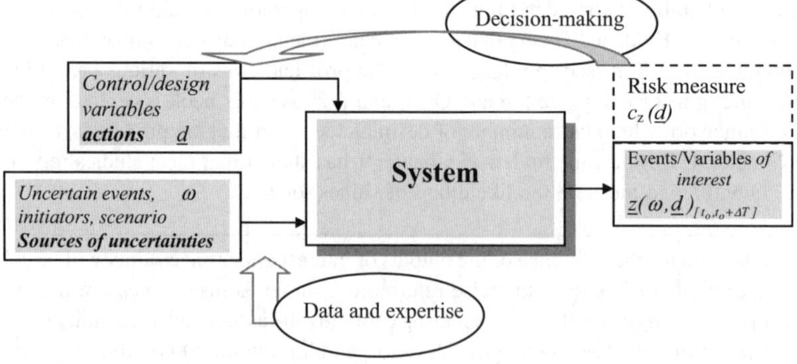

Figure 2.2 *Generic context for risk/uncertainty assessment.*

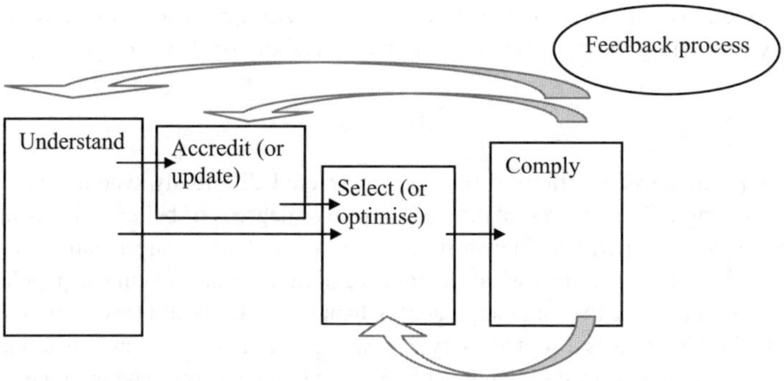

Figure 2.3 *Goals and sequencing.*

scope of *available* actions; Kaplan and Garrick (1981) expressed a similar idea in arguing that the *acceptability* of risk cannot be considered in isolation but only in the comparison of the various options that are at hand and their corresponding costs and benefits.

Thus, the risk measure $c_z(\boldsymbol{d})$ should be viewed as an appropriate means to rank the options in the face of uncertainty, as in the following examples involving options A *vs*. B:

$$c_z(\boldsymbol{d}_A) < c_z(\boldsymbol{d}_B) \tag{2.1}$$

As will be discussed in Chapter 4, that classical approach concludes with the definition of the *expected utility* as the most appropriate risk measure: the utility being a scalar function computed on the output of interest $u(z)$ representing the decision-maker's preferences (e.g. his risk aversion); 'expected' meaning the expectation over the probabilistic uncertainty model on $\boldsymbol{Z}$ that is interpreted to represent the decision-maker's subjective belief about the possible states of the system.

Yet, in more practical terms, risk and uncertainty studies are not always implemented up to the completion of that ultimate theoretical goal. Firstly, the practices reviewed in Chapter 1 in natural risk protection, industrial safety or environmental control show that regulation specifies in many cases compliance with an 'absolute' criterion. Far-reaching decision-making would weigh the relative preference between building a nuclear power plant (under flood risk) and operating a thermal power plant (under the risk induced by climate change impacts) or accepting a standard device that has been controlled with a cheap (medium-accuracy) *vs*. an expensive (high-accuracy) metrological protocol. Operational regulation or practices would often prefer to specify a fixed *decision criterion* (or risk acceptance criterion, tolerance specification, required property, etc.) to *comply* with, such as:

- 'design should ensure that the mechanical failure margin should remain positive for the reference seism in spite of uncertainty, with a probability less than 10^{-b} to be negative'
- 'system redundancy design and maintenance should guarantee that frequency of plant failure should be less than 10^{-b} per year, at a 95 % confidence level covering the uncertainties'
- 'there should be less than 3 % uncertainty on the declared measurement value for the output of interest'
- 'up to the α - initiator at least (e.g. 1000-yr flood), installation is safe under design $\boldsymbol{d}$'
- 'range of the output variable of interest should always be less than 20 %' or 'maximal value of the variable of interest should stay below a given absolute threshold'
- ...

Subsequent sections will show that such *decision criteria* appear as tests derived from the risk measure, mostly comparing the risk measure to a fixed threshold (or 'risk budget') c_s:

$$c_z(\boldsymbol{d}) < c_s \tag{2.2}$$

Note that this type of decision criterion may be interpreted differently whether one considers the probabilistic modelling to be: a representation of subjective degrees of belief (such as in the theory of expected utility) over the imperfectly known states of the system; or a representation of the observable objective frequencies of possible states within a sequence of time for the system or a population of similar systems; or even both (as in a Bayesian approach). Chapter 4 will discuss this in more depth.

To complete the picture it is also necessary to consider that many risk and uncertainty studies are undertaken in relatively upstream stages, not yet facing explicit decision-making processes or even less regulatory criterion. As reviewed in Chapter 1, this is the case with some UASA studies that are helping the development of phenomenological models; it is also the purpose of numerical code qualification. Even when these models and codes are designed to ultimately help decision-making or industrial certification, the goal of upstream uncertainty studies may be firstly to *understand* the extent of uncertainty affecting the model's prediction, particularly the relative importance of input uncertainties in order to intensify data collection and the most important research or modelling efforts. Sensitivity indices or importance factors can be computed in association with the risk measure or quantity of interest so as to apportion the relative importance of the different inputs.

Once a better understanding of the uncertainties and of the behaviour of the system model is available so that it is considered to have the potential for operational use, it may be necessary to establish formally its range of validity and to control its residual prediction uncertainty: a more or less sophisticated calibration, the validation or qualification process aims to *accredit* its validity. Although those two goals may be understood as being only indirectly linked with decision-making, it may be seen that the risk measure is still the key quantitative tool involved (see Table 2.1). Chapters 6 and 7 will illustrate the fact that the importance ranking of the uncertain inputs or the proper model calibration is different whether a variance or a failure probability in the output of interest is selected as the risk measure or quantity of interest.

To summarise, industrial practice shows that, most of the time, the goals of any quantitative risk/uncertainty assessment belong to the following categories (de Rocquigny, Devictor, and Tarantola, 2008):

- U (*Understand*): To understand the influence or rank the importance of uncertainties, thereby to guide any additional measurement, modelling or R&D efforts.
- A (*Accredit*): To give credit to a model or a method of measurement, that is to reach an acceptable quality level for its use. This may involve calibrating sensors, estimating the parameters of the model inputs, simplifying the system model's physics or structure, fixing some model inputs, and finally validating according to a context-dependent level. In a sequential process it may also refer to the *updating* of the uncertainty model, through dynamic data assimilation.
- S (*Select*): To compare relative performance and optimise the choice of maintenance policy, operation or design of the system.
- C (*Comply*): To demonstrate the compliance of the system with an explicit criterion or regulatory threshold (e.g. nuclear or environmental licensing, aeronautical certification, etc.).

There may be several goals in any given study and they may be combined over the course of a more-or-less elaborate decision-making process (see Figure 2.3). Goals S and C refer to more advanced steps in

Table 2.1 Goals and use of the risk measure.

Final goal	Formal link with the risk measure
Understand	Rank the importance of sources of uncertainty in terms of c_z
Accredit	Validate a model with respect to c_z
Select	Compare choices of actions: $c_z(\boldsymbol{d}_1) < c_z(\boldsymbol{d}_2)$
	Optimise some of the vectors $\boldsymbol{d}$ according[a] to c_Z i.e. $Max_{\boldsymbol{d}}\,[c_z(\boldsymbol{d})]$
Comply	Ensure a decision criterion is met: $c_z(\boldsymbol{d}) < c_s$

[a] More complex programs may of course imply minimising a criterion C_{z1} under constraint from another criterion C_{z2}, for example minimising the expected value of the cost of flooding under the constraint that the probability of failure remains less than 10^{-3}: see Chapter 8, Section 8.1.5.

operational decision-making, while Goals U and A concern more upstream modelling or measurement phases. The order of the list above may often be encountered in the practical sequencing of studies. Importance ranking may serve for model calibration or model simplification at an earlier stage, which becomes, after some years of research, the basis for the selection of the best designs and the final demonstration of compliance with a decision criterion. Compliance demonstration may explicitly require importance ranking as part of the process, and so on.

The *feed-back process* is an essential item in modelling studies: for instance, failure to secure compliance for a given design or changes affecting the impact of actions may lead to changing the actions in order to fulfil the criterion: this may involve changing design or operational parameters, or even investing more in information in order to reduce the sources of uncertainties. This would obviously rely on ranking the importance of uncertain inputs or events, and possibly require a new accreditation of the altered measurement or modelling chain before being able to use it in order to demonstrate compliance. Eventually, the optimisation of actions $\boldsymbol{d}$ may be distributed over various future time steps within a *real options* approach: Chapter 8 will introduce this powerful though computationally-intensive approach which, for instance, allows for the choice of a flexible design and investing in data collection in a first time period, and then finalising the design once the information is complete, all this through the maximisation of an overall expected utility over different time steps.

This process may be important in the context of the *precautionary principle* when tackling the impacts of technologies that are potentially severe, irreversible but largely uncertain due to lack of scientific knowledge. In such a case, Goal U might drive upstream uncertainty and sensitivity modelling in order to prepare the debate on setting quantitative decisional targets within risk prevention-related Goals S or C.

2.3 Modelling under uncertainty: Building separate system and uncertainty models

2.3.1 The need to go beyond direct statistics

Once the process has clarified the salient goals of the study and defined the appropriate events or variables of interest and the risk measure, it becomes necessary to estimate it effectively, in order to make a decision.

In some cases, it is possible to get *direct information* on the events or variables of interest, through measurement or expertise, so that the uncertainty in predicting z may be inferred straightforwardly through elementary statistics. Think of standardised light bulbs whose lifetime has been observed in large enough homogeneous samples so that the chance that a new bulb will remain reliable over a given period of time can be directly inferred. Yet, in most practical cases, this is not sufficient. Information such as measurement results or expert knowledge may not be directly available in the z, but rather in different variables characterizing the system state that are more amenable to observation. Or even more importantly, the choice of actions $\boldsymbol{d}$ may change the system so that past records cannot be directly relevant. Some form of modelling then becomes necessary.

Building a model in the context of risk and uncertainty is the central purpose of this book and will receive considerable attention within subsequent chapters. It should be seen fundamentally as inferring the best reflection of the extent and lack of knowledge one has about the system from the particular point of view of the quantities of interest for decision-making. Knowledge of the system may include in general:

- *Statistical observations* of the features of the system (variables or events more or less directly linked to z). They will be formally denoted as Ξ_{tn} or simply Ξ_n, the subscript number n witnessing the *limited* size of information available at the time of the analysis; the system would have typically been observed at n past dates $(t_j)_{j=1...n}$ although each observation may be multi-variate and/or heterogeneously sampled with varying sizes according to the features and degrees of freedom considered.
- *Non-statistical knowledge* of the likely extent and features of variability or *uncertainty offeatures of the system*. This may refer to the plausible bounds of variation, probabilistic distribution shapes (e.g. inferred through anologies in the literature), structure of dependence, and so on. Such knowledge will be formally denoted as *IK*, and corresponds typically to priors in a Bayesian approach.
- *Phenomenological knowledge* of the behaviour of the system as materialised in a plausible relationship between those observed or partially known features and the variables or events of interest z. Such knowledge refers to the existence of a plausible system model (or collection of plausible models) denoted as *IM* coming from a physics-based, economic or logical representation that is inherited from science, past experiments or rational thinking.

Thus, a model should *best* combine all those pieces of information *as regards the prediction of the risk measure*. While the model will eventually be a single fully-integrated predictor of the risk measure, it is helpful to view it in close association with the following two components:

- the *system model* component, representing a mostly deterministic statement on the relationships between the characteristics of the system,
- the *uncertainty model* component, representing the extent of the lack of knowledge, randomness or degrees of belief affecting the state of the system through a probabilistic representation.

2.3.2 Basic system models

To begin with, a system model may be introduced within an elementary formulation according to which a *deterministic* relationship[2] may be inferred regarding the state of the system between:

- some (input) variables and events, denoted as $\boldsymbol{x}$ and $\boldsymbol{e}$ respectively, that characterise parts of the state of the system upon which information (data or expertise) is supposed to be available,
- the controlled actions, represented by vector $\boldsymbol{d}$,

[2] Causal relationships in most cases, or more generally strong dependence between variables or events characterising the state of the system.

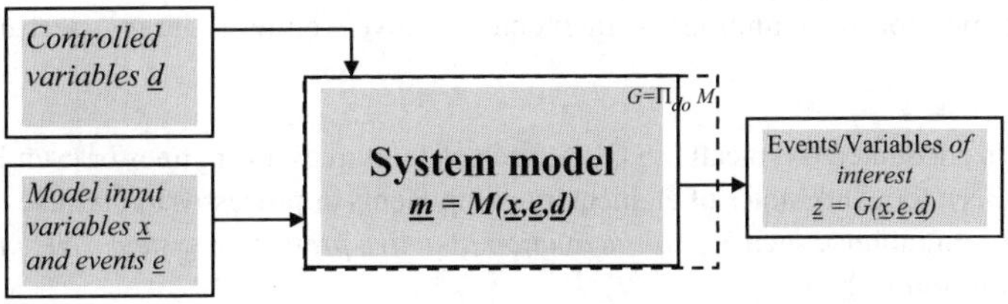

Figure 2.4 *Generic system model.*

- and a range of model outputs $\boldsymbol{m}$ including in particular the events or variables of interest z for decision-making, as expressed by:

$$\boldsymbol{x}, \boldsymbol{e}, \boldsymbol{d} \rightarrow \boldsymbol{m} = M(\boldsymbol{x}, \boldsymbol{e}, \boldsymbol{d}) \tag{2.3}$$

$$z = \Pi_d(\boldsymbol{m}) = \Pi_d o M(\boldsymbol{x}, \boldsymbol{e}, \boldsymbol{d}) = G(\boldsymbol{x}, \boldsymbol{e}, \boldsymbol{d}) \tag{2.4}$$

At this stage Π_d should be understood simply as the projection of the potentially-large vector of raw model outputs into the vector z: future discussion will provide evidence of the interest in distinguishing between a larger system model $M(.)$ and the more limited $G(.)$ tailored to the needs of final decision-making. Note that $\boldsymbol{x}, \boldsymbol{e}, \boldsymbol{d}$ may represent time series $\boldsymbol{x}_t, \boldsymbol{e}_t, \boldsymbol{d}_t$ in which case $M(.)$ and $G(.)$ could represent *operators* modelling a truly dynamic system in the sense that the outputs (a scalar z or even a time series z_t) may not be standard functions of the instantaneous inputs $\boldsymbol{x}_t$, $\boldsymbol{e}_t$, $\boldsymbol{d}_t$ but depend as a functional in the functions of time that are the past trajectories $(\boldsymbol{x}_u, \boldsymbol{e}_u, \boldsymbol{d}_u)_{u<t}$ and some associated differential quantities.[3] As was mentioned earlier for z, the notation will be kept implicit unless entailing significant differences. Figure 2.4 recalls the corresponding modelling structure, similar to that encountered throughout Chapter 1.

Equations (2.3) and (2.4) should read as follows:

> 'conditional on the knowledge of the state of the system represented by known states for input events $\boldsymbol{e}$ and values for variables $\boldsymbol{x}$, and conditional on taking actions $\boldsymbol{d}$, the system performance as quantified by the output of interest will be $z = G(\boldsymbol{x}, \boldsymbol{e}, \boldsymbol{d})$'.

Note that the system model is defined as a predictor based on *known* values for $\boldsymbol{x}$, $\boldsymbol{e}$ in order to reflect standard phenomenological knowledge such as: 'given such boundary and initial conditions, flow, topography, and so on and such aggravating events, dike breach, and so on the water level will reach that height ...'. Yet, it will later be used with *uncertain* values for those inputs, as they are a partial representation of the state of the system at the time of interest for decision-making, and are thus also suffering from lack of knowledge. Note also the crucial incorporation of the actions $\boldsymbol{d}$ within the system model inputs. The value of building a model in the present context is to help with predicting the likely values of z *conditional to* a choice of *actions*. The choice of the appropriate system model should obviously keep in mind the range of actions that ought to be covered adequately in order to be tested eventually.

[3] A re-parameterisation of the inputs through a finite number of parameters describing the input trajectories transforms the system model into a standard function of those new inputs. In the context of a risk and uncertainty study, it may often be enough to represent the variability of those new parameters instead of that of the full trajectories.

The above-mentioned definition is intended to cover quite a large spectrum of models, including:

- logical models encountered typically in QRA, such as fault trees, event trees or even Bayesian Belief Networks (BBN): they link states of elementary components or processes to the state of the undesired event (e.g. system failure), such as '*non-activation of active protection system AND failure of passive system -> plant failure*';
- physical models encountered typically in NRA or SRA, such as physical models based on integral-differential equations for example '*for 1000 m^3/s of upstream flow, and Strickler friction coefficient of 10, and ... -> river water depth is 5 m*';
- any combination of those as encountered in IPRA, including environmental or health impact models of the consequences, economic evaluations and so on;
- any numerical model considered in UASA or NCQ.

Such a generic system model could be seen as a generalisation of the general logical risk model relating uncertain initiators or reference scenarios to uncertain impacts (see Kaplan and Garrick, 1981 Aven, 2003, or Salvi and Debray, 2006), potentially including physical sub-models within the logical model. The common feature of all those models being:

- the existence of a *numerically-computable* model,
- a *deterministic* link between input and output variables of interest, both representing *features of the state* of the system.

The second point is more a matter of epistemological interpretation, but makes a difference with some modelling practices: interpreted by Aven (2003), a model is understood as linking true (observable, at least in theory) quantities characterising the state of the system. At the time of interest, a full knowledge of the true state of the system (conditional to actions $\boldsymbol{d}$) would theoretically mean that one knows perfectly the entire vector ω, which comprises vectors $\boldsymbol{x}$, $\boldsymbol{e}$ and $\boldsymbol{z}$, each of which characterises (although only partially) the true state of the system. Think, for instance, of the complete knowledge of next year's peak flood event, including that of the upstream flow and local hydraulic conditions (possible parts of $\boldsymbol{x}$ within a hydro model) that will generate the true extent of dike overflow and plant failure consequences (possible parts of vector $\boldsymbol{z}$). Conversely, the interpretation retained for a fault tree model is the logical relationship between elementary failure or safe component states and the state of the entire system (Equation (1.17) of Chapter 1). It is not the deterministic relationship between elementary failure frequencies or probabilities and system failure frequency (Equation (1.19) of Chapter 1), as these quantities do not characterise the state of the system at the time of interest but rather abstract risk measures. Although in that case the two interpretations convey the same numerical relationship, and hence the distinction has limited practical consequences, neither is it the case when assembling phenomenological and logical models where it helps to stick to the states of the system.

However, it is not suggested here that one ought to maintain this point of view as the exclusive interpretation. Model inputs will often also comprise 'modelling parameters' that correspond to non-directly observable quantities: think of the Strickler coefficient, for instance, which is a macroscopic average of complex heterogeneous friction processes (e.g. turbulent or not). The model would still be a deterministic relationship between fixed values for those model parameters and the predicted state of the system, and one could imagine an indirect experience by which to observe them. However, badly-posed problems may leave multiple choices for such parameter values making it difficult to consider them as a true characteristic of the system.

Note also that the *deterministic* nature of the relationship between features characterizing the state of the system may hide an inner *probabilistic nature* developed on a more refined physical scale or approach. Here are two important examples for industrial applications:

- a computational fluid dynamics (CFD) model including turbulence, for instance in a bi-phase 3D flow inside a reactor (e.g. Neptune_CFD, see Guelfi *et al.*, 2005) or open-air noise propagation under meteorological variability (Leroy *et al.*, 2008);
- a neutronic code predicting neutron fluxes inside a nuclear reactor resulting from random microscopic nuclear reactions and trajectories.

In both cases, the output of interest is typically a time-averaged velocity or energy flux, transforming into a deterministic quantity the underlying random variable representing temporal, spatial or microscopic (quantum mechanics) variability. Hence, viewed from such a macroscopic output of interest, the relationship is still truly deterministic as a function of high-level physical properties taken as inputs.

2.3.3 Building a direct uncertainty model on variable inputs

So far, the basic system model has been defined as a relationship between precise quantities characterising the state of the system. Nevertheless, apart from those controlled variables classified as $\boldsymbol{d}$ by the definitions above, the consideration of risk and uncertainty about the system means that $\boldsymbol{x}$ and $\boldsymbol{e}$ will be unknown to some extent at the time of interest. It is indeed the purpose of the uncertainty model to best represent the extent of lack of knowledge in those inputs in order to compound it later with the system model so as to infer probable values for $\boldsymbol{z}$.

Depending on the type of system model, the vector of model inputs $(\boldsymbol{x}, \boldsymbol{e}) = \left[(x^i)_{i=1\ldots p}, \left(e^k\right)_{k=1\ldots l}\right]$ may comprise very different types of continuous or discrete variables (particularly Boolean representing binary input events) representing all imperfectly-known characteristics of the system at the time of interest (see Figure 2.1): imperfectly known characteristics, random initiator scenarios, uncontrolled operational or environmental conditions, internal noise and so on.[4] Dimensions can be very large (a few units up to several hundreds). Remember the particular cases of uncertain inputs being time series (either in natural risk or in dynamic reliability), or spatial fields. Although it is theoretically possible sometimes to discretise those and get back to a vector model, they raise specific questions which will be dealt with to some extent in Chapter 4. Similarly with $\boldsymbol{z}$, the vector notation will be kept, for it appears appropriate in most situations.

Probability theory will be the central feature considered in this book, although mixed with deterministic settings. The uncertainty model will encode the lack of knowledge remaining after incorporation of all information available to describe the inputs, either observed statistical data Ξ_n or expertise *IK* through a combination of deterministic ranges and a joint continuous or discrete distribution describing the input uncertainty within random vectors $\boldsymbol{X}, \boldsymbol{E}$. Note that the uncertainty model considered here is a simultaneous representation of all input components, that is to say both continuous variables and Boolean events: this key feature enables one to maintain a unified perspective between QRA and SRA, for instance. In a probabilistic setting, this assumes that the input distribution is a product of cdf (or pdf) for continuous variables and/or discrete distributions.

---

[4] Model inputs may be generalised to include designators for alternative models or modelling assumptions: see (Apostolakis, 1999). Although this does not change the mathematical considerations, the interpretation in terms of states of the system may be trickier.

Defining a probabilistic structure requires a formal view of the probability space over the sample space Ω, with a number of fundamental axioms and hypotheses regarding the system model's measurability. In order to facilitate the reading, it is assumed in this chapter that such a structure is set so that a joint distribution may be made available: refer to Annex Section 10.2 for a full theoretical definition of the probabilistic structure and associated hypotheses. Once such a probabilistic setting is set, an uncertain model for $\boldsymbol{X}$ and $\boldsymbol{E}$ will consist of the following:

$$\boldsymbol{X}, \boldsymbol{E} \sim f_{XE}(\boldsymbol{X} = \boldsymbol{x}, \boldsymbol{E} = \boldsymbol{e} | \boldsymbol{\theta}_{XE}) \tag{2.5}$$

whereby f_{XE} indicates a joint mixed probability distribution (or density) function (*pdf*) associated with the vector of uncertain input variables $\boldsymbol{X}$ and events $\boldsymbol{E}$. The parameters of f_{XE} are grouped within the vector $\boldsymbol{\theta}_{XE}$: they include, for example, the parameters of position, scale and shape, or moments of marginal laws, event probability, coefficients of correlations, common mode probabilities (or parameters of the copula function), or even extended parameters of a *non-parametric* kernel model. When considering input time series $\boldsymbol{X}_t$, $\boldsymbol{E}_t$ the definition of the density f_{XE} incorporates the probabilistic model of the whole stochastic process, notably its temporal inter-correlation structure with the corresponding parameters.

Provided the availability of statistical knowledge (a sub-part of Ξ_n containing observations of the model inputs) and/or of some non-statistical knowledge of the input uncertainty (*IK* including, for instance: plausible distribution shapes, dependence structures or even approximate values for distribution parameters for some components $\boldsymbol{\theta}_{XE}^{kn}$), the encoding of the uncertainty model equates to the following statistical estimation program:

$$\left.\begin{array}{l} \left(\boldsymbol{X}_j, \boldsymbol{E}_j\right)_{j=1\ldots n} \sim f_{XE}(.|\boldsymbol{\theta}_{XE})\ i.i.d. \\ \text{estimate } \boldsymbol{\theta}_{XE} \\ \text{knowing } \Xi_n = \left(\boldsymbol{x}_j, \boldsymbol{e}_j\right)_{j=1\ldots n} \\ \text{and } IK = \left(f_{XE}, \boldsymbol{\theta}_{XE}^{kn}\right) \end{array}\right\} \tag{2.6}$$

> While involving continuous or discrete distributions respectively, input variables $\boldsymbol{x}$ and events $\boldsymbol{e}$ play the same role formally. In order to simplify notation, vector $\boldsymbol{x}$ will now stand for all uncertain inputs, be they discrete or continuous, unless the differentiation is valid. The uncertainty model $f_{XE}(\boldsymbol{X} = \boldsymbol{x}, \boldsymbol{E} = \boldsymbol{e}|\boldsymbol{\theta}_{XE})$ simplifies into $f_X(\boldsymbol{X} = \boldsymbol{x}|\boldsymbol{\theta}_X)$, and its estimation consists of fitting $\boldsymbol{\theta}_X$ to the sample $\Xi_n = \left(\boldsymbol{x}_j\right)_{j=1\ldots n}$.

Such a condensed formulation should not hide the fact that in some cases there may be only separate univariate samples for some of the input variable or event components; one may then only estimate marginal distributions instead of the joint distribution and rely exclusively on expert judgement for possible dependencies. There may even be a complete lack of statistical data for some components, so that non-statistical knowledge *IK* may be the only source for choosing the parameters $\boldsymbol{\theta}_X$.

2.3.4 Developing the underlying epistemic/aleatory structure

Beyond differences regarding the type of inputs (variables $\boldsymbol{X}$ or events $\boldsymbol{E}$) and the mixture of deterministic and probabilistic models in those inputs, the practices reviewed in Chapter 1 evidenced some differences regarding the probabilistic structure of the uncertainty model, namely the potential distinction of two levels of uncertainty encoded into simple or double probabilistic settings. It is generally referred to in the literature as the *aleatory-epistemic* distinction (Helton *et al.*, 1996), the 'level-1' modelling the *aleatory* (or irreducible) component while 'level-2' models the *epistemic* (or reducible) component, as will be discussed in more depth in Chapter 4. Any type of risk or uncertainty analysis basically develops the *first level* as a model of uncertainty about the variables or events of interest, defined previously as $\boldsymbol{X} \sim f_X(\boldsymbol{X} = \boldsymbol{x}|\boldsymbol{\theta}_X)$. More elaborate interpretations and controversies arise when considering the issue of the lack of knowledge regarding such uncertainty descriptions, as generated, for instance, by small data sets, discrepancies between experts or even uncertainty in the system model itself, in comparison with the intrinsic temporal, spatial or population variability of the system states and features. This is particularly the case in the field of risk analysis or reliability, but happens also in natural risk assessment, advanced metrology or more importantly even in environmental/health impact assessment with space or population variability across compound systems, as reviewed in Chapter 1.

Indeed, the uncertainty model $f_X(.|\boldsymbol{\theta}_X)$ needs to be estimated or inferred on the basis of all available pieces of information, such as data or expertise. As expressed by Equation (2.6), it should be viewed as a statistical estimation effort, such as that of the expected temperature from the empiric mean on a finite dataset of temperatures or of the failure rate from limited lifetime records. It inevitably generates statistical fluctuation in the parameter estimates, that is uncertainty in the appropriate value for $\boldsymbol{\theta}_X$ (see Annex Section 10.1 for an illustration). Beyond that kind of parametric uncertainty, there is even more epistemic uncertainty associated with choosing the distribution shape f_X for a given input x^i. Traditional hypothesis-testing techniques give at most only incomplete answers: is the Gaussian model appropriate for the distribution of x^i, as opposed to, for instance, a lognormal or beta model? This results in a form of 'uncertainty about the uncertainty (parameters)', although such a formulation has little value and is better replaced by 'level-2' uncertainty with respect to the basic uncertainty model. The essential characteristic of such level-2 uncertainty is that it is reducible, theoretically at least, when statistical information Ξ_n grows in size: this is a general consequence of the central-limit theorem whereby the dispersion of an empiric estimate decreases in $1/\sqrt{n}$ (see Chapter 5 for more detail)[5]. This is not necessarily the case for the states of the system represented by $\boldsymbol{x}$. Think of temperature variability: however long the sample observed in a given place, its temperature value one year ahead will remain uncertain although its probability distribution would then become very accurate (providing stationarity). Such epistemic uncertainty will also be referred to hereafter as *estimation uncertainty*, in order to mark its strong link with the step in the process where the uncertainty model is estimated.

As advocated by Helton (1994), a more complete representation of the extent of the knowledge or lack of knowledge uncertainty model has therefore to account for such an impact of the amount of information. In order to both represent uncertainty in the state of the system modelled by $\boldsymbol{x}$ and lack of knowledge of the precise value of parameters $\boldsymbol{\theta}_X$, a new probabilistic structure is defined over the extended sample space $\Omega_x \times \Omega_\theta$ which contains all of the possible combinations of their respective values. The complete (double-level) uncertainty model is defined by the following conditional structure relating to the two random vectors $\boldsymbol{X}$ and $\boldsymbol{\Theta}_X$:

$$\begin{aligned} \boldsymbol{X}|\boldsymbol{\theta}_X &\sim f_X(\boldsymbol{x};\boldsymbol{\theta}_X) \\ \boldsymbol{\Theta}_X &\sim \pi(\boldsymbol{\theta}_X|\zeta) \end{aligned} \tag{2.7}$$

[5] New information has to be consistent with that which already exists: this may be difficult with non-statistical *IK* knowledge whereby the inclusion of a larger number of experts does not always increase the consensus.

π denoting a given pdf parameterised by ζ, the (hyper-)parameters characterising epistemic uncertainty while f_X represents the aleatory part. Hence, estimation involves the following program where, once again, expert knowledge may not only include the distribution shapes but also partial knowledge of some of the component parameters noted ζ^{kn}:

$$\left.\begin{array}{l} \forall j = 1, \ldots n \quad \boldsymbol{X}_j|\boldsymbol{\theta}_X \sim f_X(.|\boldsymbol{\theta}_X)\ i.i.d. \\ \boldsymbol{\Theta}_X \sim \pi(.|\zeta) \\ \text{estimate } \pi(.|\zeta) \\ \text{knowing } \Xi_n = \left(\boldsymbol{x}_j\right)_{j=1\ldots n} \\ \text{and } IK = \left(f_X, \pi, \zeta^{kn}\right) \end{array}\right\} \tag{2.8}$$

Background information generally encountered in risk and uncertainty analysis involves small samples complemented by partial expertise. As a general statement, a Bayesian approach offers a convincing setting for the estimation of Equation (2.8) in such a situation (Aven, 2003; Pasanisi *et al.*, 2009). It includes by definition the conditional (or double-level) probabilistic structure of Equation (2.7) and offers a traceable process to mix the encoding of engineering expertise and the observations inside an updated epistemic layer that proves mathematically consistent even when dealing with very low-size samples. Indeed, expert knowledge *IK* is encoded in the following *prior* uncertainty distributions:

$$\begin{array}{l} \boldsymbol{X}_j|\boldsymbol{\theta}_X \sim f_X(.|\boldsymbol{\theta}_X) \\ \boldsymbol{\Theta}_X \sim \pi_o(.) = \pi(.|IK) \end{array} \tag{2.9}$$

whereby experts represent inside the *prior* distribution π_o their knowledge of the plausible values for $\boldsymbol{\Theta}_X$, which may be very poor through non-informative priors. Hence, the estimation process incorporates the statistical sample of Ξ_n so as to produce the *posterior* distribution π_1 for $\boldsymbol{\Theta}_X$ through the following integration, based on Bayes' law:

$$\pi_1(\boldsymbol{\theta}_X) = \pi(\boldsymbol{\theta}_X|IK, \Xi_n) = \frac{\prod_j f_X(\boldsymbol{x}_j|\boldsymbol{\theta}_X).\pi_0(\boldsymbol{\theta}_X)}{\int_{\boldsymbol{\theta}_X} \prod_j f_x(\boldsymbol{x}_j|\boldsymbol{\theta}_X)\pi_0(\boldsymbol{\theta}_X)d\boldsymbol{\theta}_X} \tag{2.10}$$

ending up with the final uncertainty model incorporating both Ξ_n and *IK*:

$$\begin{array}{l} \boldsymbol{X}|\boldsymbol{\theta}_X \sim f_X(.|\boldsymbol{\theta}_X) \\ \boldsymbol{\Theta}_X|\Xi_n \sim \pi_1(\boldsymbol{\theta}_X) = \pi(\boldsymbol{\theta}_X|IK, \Xi_n) \end{array} \tag{2.11}$$

Note that this last model generally incorporates a complex (non closed-form) function $\pi_1(\boldsymbol{\theta}_X)$ possibly requiring MCMC-like simulation methods (see Chapter 5).

A fully Bayesian approach has powerful epistemological and decision-theory features in the context of risk analysis although a traditional controversy remains to some extent with regard to classical approaches which will not be discussed hereafter (see, for instance, Robert, 2001). Yet classical statistical estimation, much more disseminated in practice, may also be used to estimate Equation (2.8) as will be discussed in Chapter 5: maximal likelihood estimation generates an asymptotic Gaussian distribution for $\pi(\boldsymbol{\theta}_X|\zeta)$, although the incorporation of partial knowledge by expertise *IK* might be formally more difficult.

Note that there is an essential difference between the temporal structure of aleatory and epistemic representations of uncertainty. The conditional distribution $f_X(\boldsymbol{x}|\boldsymbol{\theta}_X)$ should be understood as a model of aleatory uncertainty for each observation $\boldsymbol{X}_j$ of the sample $\Xi_n = (\boldsymbol{X}_j)_{j=1\ldots n}$ identically representing *past* observed states of the system, as well as of the state of the system $\boldsymbol{X}$ at the *future* time of interest. Each of those *aleatory* uncertain variables are supposed to be *independent and identically distributed, conditional to* $\boldsymbol{\theta}_X$. Meanwhile, there is a *single* – time-independent – *random variable* $\boldsymbol{\Theta}_X$ describing the *epistemic* uncertainty: it represents the common lack of knowledge in the description of their identical distributions for any past or future state of the system. When considering the prediction of the uncertain state of the system at various future dates – or equally at various locations or amongst various individuals of the population constituting the system – the variable of interest for various future dates is impacted by different realisations of the (more or less inter-correlated) aleatory process. Meanwhile, the realisation of the epistemic component takes place only once with a single value over all future dates, representing equal lack of knowledge about the characteristics of the aleatory process. More will be explained in Chapters 4–6 regarding this probabilistic structure.

Once the estimation step is over, one ends up with the following joint uncertainty model representing all available knowledge:

$$(\boldsymbol{X}, \boldsymbol{\Theta}_X) \sim f(\boldsymbol{x}, \boldsymbol{\theta}_X|IK, \Xi_n) = f_X(\boldsymbol{x}|\boldsymbol{\theta}_X).\pi(\boldsymbol{\theta}_X|IK, \Xi_n) \tag{2.12}$$

Such comprehensive notation will be kept for future reference, although simplifications may occur in practice. The true distribution π may be replaced by a single fixed value of parameters $\boldsymbol{\theta}_X$ in the following cases: for those inputs for which there is nothing but expertise, instead of a mixture of statistical data and expertise, whereby information may be too poor to encode a double-level distribution; or else when epistemic uncertainty is neglected because of large-enough samples or for the sake of simplicity. In such cases, the density π is formally a Dirac mass δ_{θ_X} centred on the fixed parameter value $\boldsymbol{\theta}_X$. Equally, it may be referred to as a *single-probabilistic* setting following the simplified probabilistic structure of Equation (2.5).

Alternatively, once estimation has been undertaken, it may not be of interest anymore to keep all of the details regarding the epistemic and aleatory types of uncertainty, as the final goal is to infer from inside a risk measure combining all of the sources of uncertainty. In those cases, one may summarise the uncertainty model keeping only the posterior (unconditional) distribution for $\boldsymbol{X}$, generated by an averaging over the epistemic component of the conditional distribution:

$$X \sim f_X^{\pi}(\boldsymbol{x}) = E_{\theta_X}[f_X(\boldsymbol{x}|\boldsymbol{\Theta}_X)|IK, \Xi_n] = \int_{\theta_X} f_X(\boldsymbol{x}|\boldsymbol{\theta}_X)\pi(\boldsymbol{\theta}_X|IK, \Xi_n)d\boldsymbol{\theta}_X \tag{2.13}$$

The subsequent discussion will address the question of for what kind of risk measures the knowledge of the summarised unconditional aleatory distribution may suffice.

2.3.5 Summary

Figure 2.5 summarises the results derived from the previous sections, introducing the basic system model as well as the building of an uncertainty model.

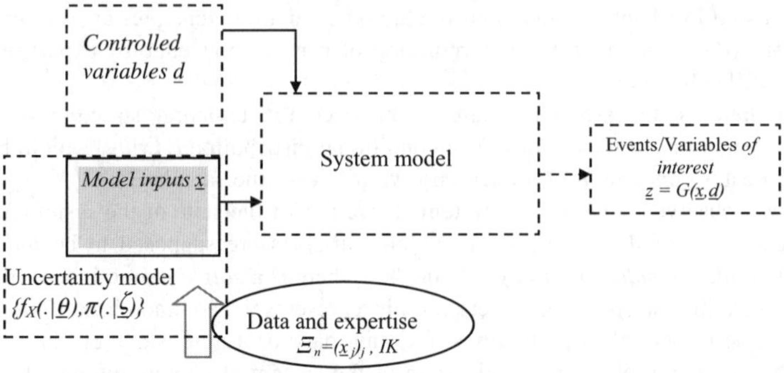

Figure 2.5 *Estimating a direct uncertainty model.*

Phenomenological knowledge was assumed to deliver a system model $G(.)$ credible enough to assist with inferring values of the output of interest z on the basis of known input values $\boldsymbol{x}$. On the other hand, statistical or non-statistical data has been mobilised to encode within an uncertainty model the extent of lack of knowledge in the model inputs. Thanks to probabilistic techniques (see section Section 2.6 or Chapter 7), the distribution of $\boldsymbol{Z}$ and associated risk measures may then be computed.

2.4 Modelling under uncertainty – the general case

2.4.1 Phenomenological models under uncertainty and residual model error

Models are always an imperfect representation of reality: beyond the basics given in Section 2.3.2, a more thorough definition of a system model should end with '... *approximately* $z = G(\boldsymbol{x}, \boldsymbol{d})$, *accurately enough for decision-making*', leaving room for residual model uncertainty or inaccuracy. The whole rationale of system modelling is that information brought by the phenomenological statement defining the system model itself (physical laws, accident sequence analysis, etc.) plus the information available on the extent of uncertainty in its inputs $\boldsymbol{x}$ will result in a *less uncertain* inference than a prediction of z based on direct observations or expertise. However, any effort to establish sound modelling in the context of risk and uncertainty should account more explicitly for such residual inaccuracies in the system model.

Indeed, a large class of modelling practice does involve an explicit representation of it to some extent. The previous section illustrated the situation where the amount of physics-based or logical knowledge resulted in a solidly-established system model M, prior to the consideration of the available statistical information. Differently, a 'phenomenological' system model, in the environmental or sanitary fields, for instance, may be regarded from the outset as the product of looser prior conceptual knowledge – a set of plausible model structures – and of statistical inference from observed data Ξ_n. Regression-like approaches will then clearly define and estimate the residual uncertainty or model-measurement discrepancy as follows:

$$observations = model\, predictions + residual$$

$$z_j = h(\boldsymbol{c}_j, \boldsymbol{\phi}) + \boldsymbol{u}_j \tag{2.14}$$

or, closer to the book notation, formulated equivalently as:

$$z_j = G\left(\boldsymbol{x}_j^{kn}, \boldsymbol{d}_j, \boldsymbol{x}^{un}\right) + u_j \tag{2.15}$$

where:

- $h(.) = G(.)$ may be a regression-based closed-form function (e.g. linear, power, logistic), a CART-based decision tree or set of ANOVA-effects (incorporating test functions such as $G\left[\left(1_{xj>xjs}\right)_j\right]$) or any combination of those,
- z_j represents the available observations made on the output of interest at given past times t_j and under known experimental conditions $\boldsymbol{c}_j = \left(\boldsymbol{x}_j^{kn}, \boldsymbol{d}_j\right)$; they possibly include some that are measured but later uncontrollable, denoted as $\boldsymbol{x}_j^{kn}$, and/or others that are well controlled, denoted as $\boldsymbol{d}_j$,
- $\boldsymbol{\phi} = \boldsymbol{x}^{un}$ represent model inputs (or parameters) that are *unknown* prior to model calibration,
- $\boldsymbol{u}_j$ represents the inescapable residual left unexplained in comparing observations with model outputs.

In that context, standard model building generally involves testing different models $G(.)$, in particular regarding the minimisation of residuals $\boldsymbol{u}_j$ remaining in comparison with observed data. Those competing models differ either through the value of calibration parameters or through their fundamental structure. Indeed, the comparison between observations and model predictions generally provides one with some informational basis to estimate residual model inaccuracy $\boldsymbol{u}$ although it obviously represents an uncertain variable: by nature, such a residual represents the variability *resisting any deterministic explanation* through the model-building process, thus displaying 'erratic' behaviour. Similar to the uncertain input variables or events $\boldsymbol{x}$, experience shows that a probabilistic representation of $\boldsymbol{u}$ is often appropriate, providing a number of important checks (see Chapter 6). This means modelling residual uncertainty as a random variable $\boldsymbol{U}$ distributed according to a pdf f_U. Such a description of those model-measurement residual values does amount to a *new component of the uncertainty model* described earlier. Think simply of taking the empiric standard deviation of the residuals and subsequently inferring that the error is normally distributed with the same deviation. Such an approach enables the refinement of the initial definition of the system model introduced above, as follows:

> 'conditional to the knowledge of the partial state of the system represented by known states and values for input events or variables x, and conditional on taking actions d, the system performance as quantified by the output of interest will be $z = G(\boldsymbol{x}, \boldsymbol{d})$, modulo a residual uncertainty distributed as **u**'

2.4.2 The model building process

A system model intended to infer probable values for the output of interest z under uncertainty should be built carefully into a process mobilising all available information, be it phenomenological, or *direct or indirect* statistical observations of the system. The available observations denoted as $\left(y_{mj}\right)_{j=1\ldots n}$ refer to any system features, variables or degrees of freedom that can be easily instrumented or observed, and thus may differ from those model output variables z which will eventually be involved in decision-making (see Figure 2.6). It is necessary to introduce another category of model outputs that better corresponds to those observations: the *observable model outputs*, denoted as $\boldsymbol{y} = \left(y^k\right)_{k=1\ldots q}$. These may be defined

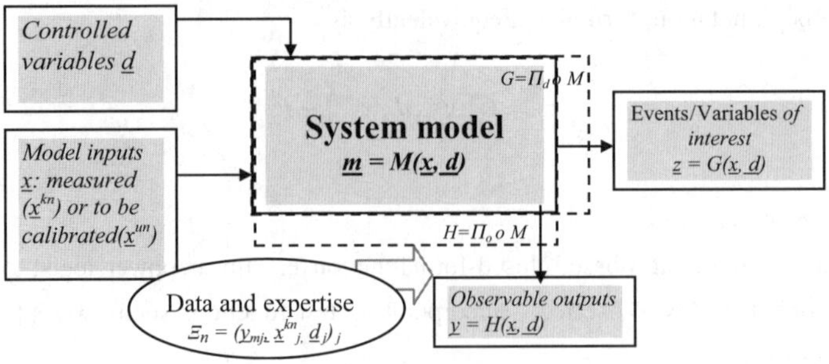

Figure 2.6 *System model estimation.*

formally as the result of another projection for the system model $M(.)$, as distinct from that which defined the predictive system model $G(.)$ in Equation (2.4):

$$y = \Pi_o(m) = \Pi_o \, o \, M(x, d) = H(x, d) \tag{2.16}$$

$$x = \left(x^{kn}, x^{un}\right) \tag{2.17}$$

Fundamentally, model building consists in fitting a variety of plausible models to the observed data as defined in Equations (2.14) and (2.15). Such a process either involves competing model structures (formally indexed by s in M_s) or input variables, such events or model parameters to be calibrated (represented by x^{un}). Calibrating a system model under uncertainty requires the following elements:

- available statistical data (a part of Ξ_n), namely a sample relating observations, that is measured observables $\left(y_{mj}\right)_{j=1\ldots n}$, to known experimental conditions, designs or measurement locations represented by $\left(d_j\right)_{j=1\ldots n}$ and/or $\left(x_j^{kn}\right)_{j=1\ldots n}$ depending on whether they correspond to deliberate *actions* that vary *controllably* or to measured inputs (variables or events) at past dates that become *uncertain* at the future time of interest;
- unknown model inputs x^{un} that need calibration.

Thus, select the $M_s(.)$ and x^{un} that 'best' explain the observed sample through the observational model:

$$\begin{aligned} \Xi_n &= \left(y_{mj}, x_j^{kn}, d_j\right)_{j=1\ldots n} \\ y_j &= \Pi_o \, o \, M_s\left(x_j^{kn}, x^{un}, d_j\right) \end{aligned} \tag{2.18}$$

Note that this vector formulation should be understood as reflecting somewhat heterogeneous datasets in practice: throughout the measurement unit $j = 1\ldots n$, not all degrees of freedom of the state of the system may be jointly observed, so that vector function $\Pi_o \, o \, M_s\left(x_j^{kn}, ., d_j\right)$ may have zero components for some values of j.

Chapter 6 will discuss in more detail how 'best' may be understood in a probabilistic sense. As explained in the previous section, system model building and calibration is undertaken while estimating *at the same time* an uncertainty model representing the associated observation-measurement residuals $(\boldsymbol{u}_j)_{j=1\ldots n}$. These may be observed as additive residuals:

$$\begin{aligned} \boldsymbol{u}_j = \boldsymbol{y}_{mj} - \boldsymbol{y}_j = \Pi_o \, o \, M_s\left(\boldsymbol{x}_j^{kn}, \boldsymbol{x}_j^{un}, \boldsymbol{d}_j\right) \\ \Leftrightarrow \boldsymbol{y}_{m\,j} = \Pi_o \, o \, M_s\left(\boldsymbol{x}_j^{kn}, \boldsymbol{x}_j^{un}, \boldsymbol{d}_j\right) + \boldsymbol{u}_j \end{aligned} \tag{2.19}$$

or more complex functional combinations that best standardise model-data misfits (see Chapter 6, Section 6.1.5) which may formally be included in the definition of an extended observational system model involving residuals:

$$\boldsymbol{y}_{m\,j} = \Pi_o \, o \, M_s\left(\boldsymbol{x}_j^{kn}, \boldsymbol{x}_j^{un}, \boldsymbol{d}_j, \boldsymbol{u}_j\right) \tag{2.20}$$

In practice, $\boldsymbol{u}_j$ mixes a variety of uncertainty sources that often cannot be discerned: not only model inadequacy (conceptual limitations or possibly numerical solving errors) but also metrological noise and artefacts, ambiguous or differing definitions of the measured observable and the modelled observable (e.g. space or time location and averaging) and so on. Although the concept of modelling unexplained variability is paradoxical to some extent and controversial in the literature (Aven, 2003; Droguett and Mosleh, 2008), its probabilistic control through such a calibration approach accounts to some extent for the model prediction inaccuracies. A *Popperian* view on *scientific* model building as advocated by Beck (1987) would require one to specify the observational conditions by which the system model may be rejected. Chapter 6 will describe the checks required to handle residuals legitimately with a probabilistic calibration approach, including the distinction between independent *calibration* and *validation sub-samples*. Modelling the residuals $\boldsymbol{u}$ could at its simplest take the form of a Gaussian random vector *i.i.d.* throughout the sample but more refined techniques can account for non-Gaussian or heteroscedastic behaviour (e.g. average measurement-model deviations increase or decrease for large or small values for $\boldsymbol{d}_j$) or even model the complex dependence $\boldsymbol{u}(\boldsymbol{x}, \boldsymbol{d})$ as a stochastic process such as the popular Kriging or Gaussian process techniques developed in the field of geostatistics or machine learning (cf., e.g. Krige, 1952; Welch *et al.*, 1992).

Provided the model is not rejected, it is thought to be worth estimating – that is model in a statistical approach – the model inaccuracy and include it later within the predictive model in order not to underestimate prediction uncertainty (Kennedy and O'Hagan, 2002). Estimating the observational model involving both $\boldsymbol{x}^{un}$ and $\boldsymbol{u}$ is undertaken through a range of techniques discussed in Chapter 6, mixing *regression*-like estimation and *inverse probabilistic techniques* since data is made available on variables which do not coincide straightforwardly with the model parameters $\boldsymbol{x}^{un}$. Associated discussion and choices are necessary in order to specify clearly, for instance, the fixed or aleatory model applied to input parameters while keeping them identifiable, that is a reasonable amount of parameters to be estimated. The basic approach, as in (non-linear) regression, is to estimate them as a *physically fixed albeit unknown* vector of parameters throughout the $j = 1 \ldots n$ sample. However, phenomenological knowledge or expertise may suggest that the state of the system represented in $\boldsymbol{x}^{un}$ does significantly vary from one piece of observation to another, while not being capable of measurement (or else it would be included in $\boldsymbol{x}_j^{kn}$). An alternative model allows for the *unknown input* variables, events or model parameters to be fully random, that is to *vary according* to the *j-th observation* as i.i.d. random realisations $\boldsymbol{X}_j^{un}$. This means that it becomes an additional component of the uncertainty model being represented by an additional pdf f_X^{un} that needs to be estimated. That latter formulation also covers the basic case – unknown but fixed parameters $\boldsymbol{x}^{un}$ throughout the sample – assuming formally that the density f_X^{un} is a Dirac mass $\delta_{\boldsymbol{x}\,un}$ centred on the fixed albeit unknown model parameter value. Eventually, the joint estimation of the appropriate system model, unknown input variable and events and residual uncertainty consists in the following statistical estimation program:

$$\left.\begin{aligned}
&\boldsymbol{Y}_{mj}|_{\boldsymbol{x}_j^{kn},\boldsymbol{d}_j} = \Pi_o \circ M_s\left(\boldsymbol{X}_j^{un}, \boldsymbol{x}_j^{kn}, \boldsymbol{d}_j, \boldsymbol{U}_j\right) \\
&\boldsymbol{X}_j^{un}|\boldsymbol{\theta}_x^{un} \sim f_X\left(.|\boldsymbol{\theta}_x^{un}\right)\ \boldsymbol{U}_j|\boldsymbol{\theta}_u \sim f_U(.|\boldsymbol{\theta}_u) \\
&\boldsymbol{\Theta}_x^{un}, \boldsymbol{\Theta}_u \sim \pi(.|\zeta) \\
&\text{estimate } \{M_s, \pi(.|\zeta)\} \\
&\text{knowing } \Xi_n = \left(\boldsymbol{y}_{mj}, \boldsymbol{x}_j^{kn}, \boldsymbol{d}_j\right)_{j=1..n} \\
&\text{and } IK = \left(f_X, f_U, \pi, \zeta^{kn}\right)\ IM = (M_s)_s
\end{aligned}\right\} \tag{2.21}$$

Similar to the estimation of the direct uncertainty model in Section 2.3, epistemic uncertainty is left in the value of the model parameters so that a double-level conditional structure has been given to the random variables ($\boldsymbol{X}^{un}$, $\boldsymbol{U}$, $\boldsymbol{Q}_X^{un}$, $\boldsymbol{\Theta}_U$). Such epistemic uncertainty is a formal generalisation of the standard errors routinely given in the estimation of regression coefficients in a phenomenological model, for example dose-effect curve in health control. While classical maximal likelihood estimation is instrumental in estimating such a model and associated asymptotic variances, a *Bayesian* model calibration approach (Kennedy and O'Hagan, 2002) is again a solid framework for encapsulating the three types of knowledge listed in Section 2.3.1. Inside a *predictive risk model* there are statistical observations that are incorporated through Ξ_n, non-statistical knowledge *IK* incorporated in the distribution shapes as well as phenomenological knowledge *IM* attached to the class of plausible models. Outside of simplified Gaussian contexts, full Bayesian calibration over physical models is costly both on the computational side (see Chapter 5 for the powerful Monte-Carlo Markov Chains MCMC methods) and on the specification side, as it requires a careful delineation of prior distributions, models of residuals and associated likelihood functions and so on. The GLUE approach is very popular in the hydrological community – the generalised likelihood uncertainty estimation introduced by Beven and Binley (1992) and much developed since, including by the MUCM Sheffield University program (http://mucm.group.shef.ac.ik). It can be introduced as a simplified Bayesian alternative involving a so-called 'informal likelihood' function (interpreted partially as the likelihood of a statistical model of residuals) as well as a sampling process simplifying the Bayesian integration into an averaging over acceptable values for the calibrated uncertain inputs (so-called 'behavioural solutions').

Note that all the developments undertaken in this section assumed the availability of statistical information appropriate for the estimation of the system model. In many cases, there may not be any statistical sample at all so that model uncertainty will not be observable as such: expertise may still provide an estimate, drawing for instance on analogies with similar models calibrated on past observations. Instead of being estimated statistically, distributions $f_X{}^{un}$, f_U may be specified on that basis: however, the concept of assessing the residual inaccuracy of a model without having relevant experimental data to refer to is quite controversial from an epistemological point of view (Aven, 2003).

2.4.3 Combining system and uncertainty models into an integrated statistical estimation problem

The extended definition of the system model as introduced in the previous section incorporated a partial uncertainty model through a probabilistic representation of the residual model uncertainty $\boldsymbol{u}$ as well as the unknown model input components $\boldsymbol{x}^{un}$. Yet, the remaining model inputs $\boldsymbol{x}^{kn}$ were assumed to be well-known within the model estimation process, which is not the case when moving on to prediction. Indeed, the system model has been defined as a relationship between precise quantities characterising the state of the system; an essential ingredient for its calibration being the availability of past observations, whereby at least parts of the model inputs $\boldsymbol{x}^{kn}$ representing the true state of the system were known. Apart from those controlled variables classified as $\boldsymbol{d}$ by the definitions above, the consideration of risk and uncertainty in the system means that *all* components of $\boldsymbol{x}$ will be unknown to some extent at the time of interest.

An additional uncertainty model must therefore be built upon those components $\boldsymbol{x}^{kn}$ that were known at the stage of system model calibration. This means building a *direct* uncertainty model, as explained in Section 2.3, to encode the knowledge (statistical or expert-based) of the variability of those degrees of freedom at the future time of interest. For instance, if the future variability is thought to be similar to that reflected in the set of measurements used for model calibration, the uncertainty model may be estimated on the basis of the sample of observations $\boldsymbol{x}_j^{kn}$.

> The *flood* model
> The pedagogical flood model in Chapter 3 provides an illustration: past water levels (the $\boldsymbol{y}_{mj}$) were measured jointly with maximal flood flows (the $\boldsymbol{x}_j^{kn}$) occurring at random dates the calibration was fed into a flow-height hydraulic model. The same flood flow sample will also assist with the estimation of a pdf for the uncertain future maximal flood flow at the time of interest ($\boldsymbol{X}^{kn}$).

Besides, those for which there are no measured model outputs $\boldsymbol{y}_{m\,j}$ may be used to estimate the pdf for $\boldsymbol{X}^{kn}$ either jointly with the former sample or by replacing it if the calibration sample corresponded to a non-random or biased experimental design misrepresenting future variability.

Thus, from the point of view of estimation, the $\boldsymbol{x}^{kn}$ components play a double role according to the part of the process that is being considered. It is helpful to denote them differently for the sake of statistical consistency. When there are observations $\boldsymbol{x}_j^{kn}$ associated with measured model outputs $\boldsymbol{y}_{m\,j}$, they play the same role as $\boldsymbol{d}_j$ as explicative (or conditioning) variables $\boldsymbol{c}_j$ with known values in the system model calibration. They may be incorporated formally within a larger vector $\boldsymbol{d}_j$. Besides, those values $\boldsymbol{x}_j^{kn}$ and/or additional ones for which there are no measured model outputs, are used to estimate a pdf for the uncertainty model of $\boldsymbol{X}^{kn}$, in a similar statistical approach to the basic Section 2.3 case of a direct uncertainty model on $(\boldsymbol{x}, \boldsymbol{e})$. From that point of view, they may simply be incorporated within vector $\boldsymbol{x}_j$.[6]

> To simplify the notations, the superscripts '*kn*' will be omitted here below, $\boldsymbol{x}_j^{kn}$ being incorporated either into $\boldsymbol{d}_j$ or into $\boldsymbol{x}_j$ as is relevant.

[6] It is essential to note that this does not imply any undue doubling of the statistical sample as they are used as $\boldsymbol{d}_j$ exclusively in order to estimate the conditional law of Y_{mj}, not that of X^{kn}.

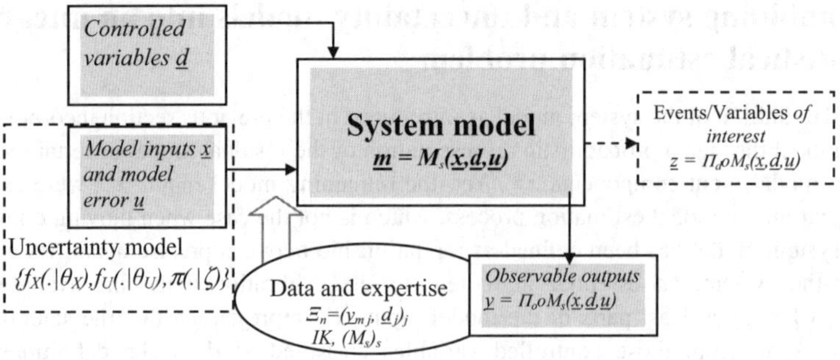

Figure 2.7 *Combined estimation program (observational system model).*

Eventually, the final model should be viewed as a fully-integrated predictor of the likely outputs z. As already shown with the estimation of model residuals, the system and uncertainty models prove to be closely-associated when it comes to estimating and validating them on the basis of the available information. Provided with a few formal conventions, the estimation of the system model and of the uncertainty model shares a common statistical structure. Indeed, for the input components for which information (statistical observations or expert knowledge) is directly available, denote $Y_m^k = X^i$ which means formally considering $\Pi_o o\ M_s$ to be the identity operator, and the residual component U^k as constantly zero. Altogether, with simplified notation, the estimation consists of the following summarised program:

$$\left.\begin{aligned}
&\boldsymbol{Y}_{mj} = \Pi_o \circ M_s\left(\boldsymbol{X}_j, \boldsymbol{d}_j, \boldsymbol{U}_j\right)\\
&\boldsymbol{X}_j|\boldsymbol{\theta}_x \sim f_X(.|\boldsymbol{\theta}_x),\ \boldsymbol{U}_j|\boldsymbol{\theta}_u \sim f_U(.|\boldsymbol{\theta}_u)\\
&\boldsymbol{\Theta}_x, \boldsymbol{\Theta}_u \sim \pi(.|\boldsymbol{\zeta})\\
&\text{estimate } \{M_s, \pi(.|\boldsymbol{\zeta})\}\\
&\text{knowing } \Xi_n = \left(\boldsymbol{y}_{mj}, \boldsymbol{d}_j\right)_{j=1\ldots n}\\
&\text{and } IK = \left(f_x, f_U, \pi, \zeta^{kn}\right)\ IM = (M_s)_s
\end{aligned}\right\} \tag{2.22}$$

resulting in the final estimation of the combined system and uncertainty model (Figure 2.7):

$$\begin{aligned}
&\{M_s, f_X, f_U, \pi\}\\
&(\boldsymbol{X}, \boldsymbol{U}, \boldsymbol{\Theta}_X, \boldsymbol{\Theta}_U) \sim f(\boldsymbol{x}, \boldsymbol{u}, \boldsymbol{\theta}_X, \boldsymbol{\theta}_U|IK, \Xi_n) = f_X(.|\boldsymbol{\theta}_X).f_U(.|\boldsymbol{\theta}_U).\pi(.|IK, \Xi_n)
\end{aligned} \tag{2.23}$$

Keeping in mind that this encompasses a large number of distinct situations according to the components of the vectors $\boldsymbol{Y}_m$ and $\boldsymbol{X}$ that are considered, a general statement regarding most risk and uncertainty studies is that *the choice and building of a suitable combination of the system and uncertainty model results fundamentally in a statistical estimation problem.* Such a generalised formulation has the following key consequences:

- The statistical techniques appropriate for modelling (parametric or non-parametric kernels) estimation (e.g. maximal likelihood, Bayesian estimation), goodness-of-fit (e.g. Kolmogorov or Anderson-

Darling tests) or model selection (e.g. AIC or BIC, i.e. Akaike or Bayesian Information Criteria) may serve potentially throughout the process be it directly with traditional data or indirectly through a system model. This may require overcoming the complications associated with the mixed discrete-continuous nature of variables and more importantly with the inversion of the system model.
- There is a general issue of identifiability to be studied when confronting model parameterisation with the *limited amount of information available*.

Table 2.2 provides a synoptic view of the various types of system models according to the definitions of the components of the observational model and uncertainty distributions f_X, f_U and π. Chapters 5 and 6 will illustrate in depth the statistical tools available for estimating the combined model, based generally either on classical maximal likelihood or Bayesian estimation.

2.4.4 The combination of system and uncertainty models: A key information choice

Phenomenological knowledge about a given system usually allows for a variety of potential system models M_s of greatly-varying complexity. When considering flood risk, think of the graduation between coarse stage-discharge closed-form curves and elaborate 3-dimensional non-stationary hydrodynamic equations, possibly including advanced turbulence description. On the other hand, when considering the safety of a large industrial facility, system reliability models may describe the global state of a score of key sub-systems (combustion, cooling, effluent treatment, ...) or detail every the state of every macro-component (turbines, circuits, ...) without mentioning unrealistic descriptions of all micro-components (valves, switches, pipes ...), and all of this may be modelled in simplified static relationships or elaborate time dynamics. While a more sophisticated system model may appear to be more accurate from a design or formal phenomenological point of view, it may change considerably when attaching to it an uncertainty model in the context of a risk or uncertainty analysis. Facing all types of sources of uncertainty, error or variability in the future operation over the time-scale of the analysis, the key question becomes how to best circumscribe information (or lack of knowledge) about the possible states of the system with the most appropriate level of detail.

Putting it more mathematically, the key question is to control the degrees of freedom of the system at the best level according to the information available. For a given amount of data, that is the number of independent scalar observations contained in the vector sample $\left(\boldsymbol{Y}_{mj}\right)_{j=1\ldots n}$, the number of model parameters ($\boldsymbol{\theta}_x$, $\boldsymbol{\theta}_u$) that may be identified with confidence is limited as the epistemic uncertainty $\pi(.)$ remaining after such an estimation rises with their number. From a statistical point of view, the increase in complexity of the model beyond a certain point results in a loss of accuracy, that is in a wider uncertainty model As encountered classically in regression, the decrease in the residual model error $\boldsymbol{u}$ experienced through an increasing parameterisation has to be paid for in the statistical estimation error of those parameters.

Of course, phenomenological knowledge should equally be valued within the balance of information, particularly in cases of scarce data. Adding a phenomenological layer inside the system model, when there is a good deal of confidence in the deterministic relationships it embodies, is an efficient means to go beyond the data unavailability and often allows for the recovery of other pieces of data on different variables. It is also essential to integrate phenomenological knowledge and expertise into the uncertainty model itself, particularly on the possible bounds of the uncertainty distributions or the dependence between variables. It reduces the excessive amount of uncertainty that would result from relying exclusively on statistical data, a feature made explicit in a Bayesian approach. In fact, in the absence of data beyond this type of phenomenological knowledge, the principle of maximal entropy is often taken as a basis for specifying the uncertainty model (see Chapter 4, Section 4.2.3).

Table 2.2 Main system models to be estimated.

Type	Observational model	Uncertain model parameterisation 1st level «risk»	2nd level «uncertainty»	
Direct observations – statistics (e.g. environmental, lifetime, ...)	$\Pi_o \circ M_s(.) = \boldsymbol{X}_j \textit{ or } \boldsymbol{E}_j$	$\boldsymbol{X}_j \sim f_s(.\mid\boldsymbol{\theta}_x)$ or $E_j = 1_{t=T_j}\ T_j \sim f_s(.\mid\boldsymbol{\theta}_T)$ s competing models	$\boldsymbol{\theta} \sim \pi(.\mid IK, \Xi)$	See Chapter 5
Extreme value modelling/ renewal process	$\Pi_o \circ M_s(.) = \sum_j 1_{E_j}(t) X_j$	$\boldsymbol{X}_j \sim f_s(.\mid\boldsymbol{\theta}_x)$ and $E_j = 1_{t=T_j}\ T_j \sim f_s(.\mid\boldsymbol{\theta}_T)$ s competing models	Idem	
Data assimilation/parameter identification	$\Pi_o \circ M_s(.) = H(\underline{x}, d_j) + U_j$	$\boldsymbol{x} \sim \delta_{\underline{x}=\boldsymbol{\theta}_x}$ to be updated $\boldsymbol{U}_j \sim f_U(.\mid R)$ with R known	$\boldsymbol{x} = \boldsymbol{\theta} \sim \pi(.\mid IK, \Xi)$ often Gaussian noted as $N(x_a, A)$	See Chapter 6
Phenomenological modelling (e.g. biostats)	$\Pi_o \circ M_s(.) = H_s(\boldsymbol{\theta}, d_j) + U_j$	$\boldsymbol{\theta}$ and $\boldsymbol{U}_j \sim f_U(.\mid\theta_U)$ to be calibrated s competing models	$\boldsymbol{\theta} \sim \pi(.\mid IK, \Xi)$	
Full variability identification/ probabilistic inversion	$\Pi_o \circ M_s(.) = H(\boldsymbol{X}_j, d_j) + U_j$	$\boldsymbol{X}_j \sim f_X(\boldsymbol{X}_j = \boldsymbol{x}\mid\boldsymbol{\theta}_x)$ $U_j \sim f_U(\boldsymbol{U}_j = \boldsymbol{u}\mid\boldsymbol{\theta}_u)$	Idem	

Thus, the combination of the system and uncertainty models should provide the best representation of the extent of uncertainty or lack of knowledge, or conversely of the information available. A general principle of *parsimony* applies here in the choice of the level of detail of both the system and uncertainty models, well in line with good practice in statistical *learning approaches*.

2.4.5 The predictive model combining system and uncertainty components

Once a suitable combination of models $\{\boldsymbol{M}_s, \boldsymbol{\theta}_X, \boldsymbol{\theta}_U, \pi\}$ has been estimated, the focus shifts to the prediction of the outputs of interest z, and particularly that of the risk measure. The outputs of interest inherit the double random variable structure of the input uncertainty model $(\boldsymbol{X}, \boldsymbol{U})$ as being its image by the system model, which shifts from its observational formulation:

$$\begin{aligned} &\boldsymbol{Y}_m = \Pi_o \, o \, M_s(\boldsymbol{X}, \boldsymbol{d}, \boldsymbol{U}) \\ &(\boldsymbol{X}, \boldsymbol{U}, \boldsymbol{\Theta}_X, \boldsymbol{\Theta}_U) \sim f_X(.|\boldsymbol{\theta}_X).f_U(.|\boldsymbol{\theta}_U).\pi(.|IK, \Xi_n) \end{aligned} \tag{2.24}$$

into its predictive formulation:

$$\begin{aligned} &\boldsymbol{Z} = \Pi_d \, o \, M_s(\boldsymbol{X}, \boldsymbol{d}, \boldsymbol{U}) \\ &(\boldsymbol{X}, \boldsymbol{U}, \boldsymbol{\Theta}_X, \boldsymbol{\Theta}_U) \sim f_X(.|\boldsymbol{\theta}_X).f_U(.|\boldsymbol{\theta}_U).\pi(.|IK, \Xi_n) \end{aligned} \tag{2.25}$$

It is useful to suggest that *from the point of view of prediction*, there is not much difference between the uncertain model inputs $\boldsymbol{X}$ and the model-measurement residuals $\boldsymbol{U}$. While being of a slightly different nature regarding the definition of the states of the system and model results, both vectors encompass a description of the likely extent of uncertainty affecting a feature of the modelled system, with an impact on the likely values of the output of interest. Both will require identical sampling in the prediction phase in order to fully represent the degree of output uncertainty.

To simplify the notation, $\boldsymbol{U}$ will be omitted here below whenever considering the prediction phase, as being incorporated within a larger vector $\boldsymbol{X}$.

Altogether, the predictive model will be formulated as follows (see Figure 2.8):

$$\begin{aligned} &\boldsymbol{Z} = \Pi_d \, o \, M_s(\boldsymbol{X}, \boldsymbol{d}) = G(\boldsymbol{X}, \boldsymbol{d}) \\ &(\boldsymbol{X}, \boldsymbol{\Theta}_X) \sim f_X(.|\boldsymbol{\theta}_X).\pi(.|IK, \Xi_n) \end{aligned} \tag{2.26}$$

where $\boldsymbol{X}$ includes all sources of uncertainty deemed important for decision-making: uncertain model inputs, events as well as model error. $\boldsymbol{Z}$ being a function of $\boldsymbol{X}$, the output vector inherits the image structure of the uncertainty model, as follows:

$$(\boldsymbol{Z}, \boldsymbol{\Theta}_X) \sim f_Z(.|\boldsymbol{\theta}_X).\pi(.|IK, \Xi_n) \tag{2.27}$$

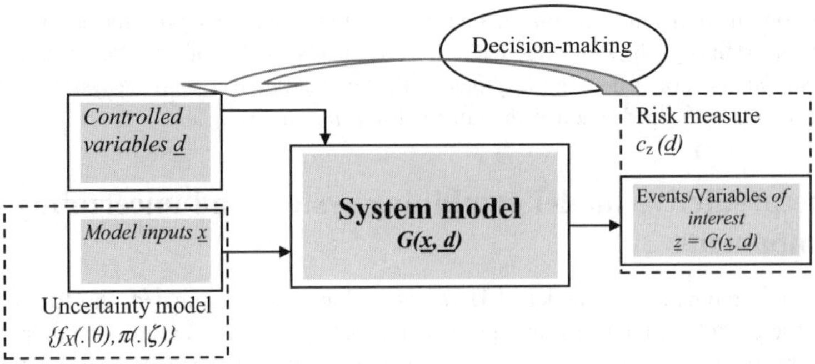

Figure 2.8 *Predictive system model.*

In spite of this formal notation, remember that the resulting distribution of $\boldsymbol{Z}$ is generally unknown and costly to compute, since M_s may be a complex system model: this is the purpose of *uncertainty propagation* methods which will receive more attention later.

2.5 Combining probabilistic and deterministic settings

Probability is central to the approach taken in this book, although it proves necessary to mix it with deterministic settings. This section will first comment on the meaning of probabilistic uncertainty models and introduce the mixed probabilistic-deterministic settings.

2.5.1 Preliminary comments about the interpretations of probabilistic uncertainty models

Different epistemological interpretations correspond to the various probabilistic structures available for uncertainty modelling. This choice is closely linked to that of the various natures of uncertainty about the state of the system which may be represented, such as natural time or space variability, lack of knowledge and so on: Chapter 4 will return to that discussion. At this stage, only preliminary observations will be made.

The first interpretation, *frequentist* or *classical*, would consider $\boldsymbol{x}$, $\boldsymbol{e}$ and z as observable realisations of uncertain (or variable) events (or properties of the system) occurring several times independently so that, at least in theory, the frequency records of both variables would allow for the inference and validation of the probability distribution functions (for both inputs and output). In that context, modelling probabilistic distribution functions in the inputs may be seen as a basis for the inference of some output quantities of interest, such as a probability to exceed a regulatory threshold or an expected cost. Taking such a frequentist basis for decision-making benefits from straightforward risk control interpretations and has the advantage of a potential validation through long-term observations, as practised daily in environmental control, natural risk regulations or insurance records.

Other views, involving totally different interpretations, can be developed in the same mathematical setting (say standard probabilistic) to quantify uncertainty. A classical *subjective* interpretation of the same setting might lead to a consideration of the probability distributions as a model of the decision-maker's subjective preferences following a 'rational preference' set of axioms (such as in Savage, 1954), or degrees of belief without the necessary reference to frequency observations of physical variables. As will be explained below, dealing with the expected value of a function of the output random vector

Z called the *utility* may then be used in a decision-making process, benefiting from solid decisional properties. This may not necessarily need validation by long-term observations, which, in fact, may often be impractical in industrial practice: think about such cases as the design choices of an industrial product that does not yet exist.

Alternatively, when considering the global sensitivity analysis of a complex physical or environmental system model in upstream model development stages, one may rely on a sort of *functional analysis* interpretation. Using probabilistic distributions in the inputs $\boldsymbol{x}$ and $\boldsymbol{e}$ of a system model and considering any variance in the model output, z has some desirable numerical space-exploration or global averaging properties that allow for well-defined sensitivity ranking procedures. However, the inputs and outputs of such a model may not be observable at all, on the one hand being rather abstract model parameters in upstream research processes or, on the other, not corresponding to reproducible random experiments.

While not being centred on a discussion of the epistemological foundations of risk and uncertainty modelling and supporting in detail a given interpretation framework, this book advocates in general:

- a clear specification of the variable of interest (and the associated future time(s) of interest, as well as the underlying system or population of systems) that should be observable;
- a review of the natures of uncertainty to be covered in the probabilistic targets, distinguishing, at least in the early stages of the analysis, aleatory or variability from the distinct sub-types of epistemic uncertainty (data estimation, model imprecision, propagation uncertainty);
- a double-probabilistic structure (interpreted in a Bayesian approach) for handling distinctively the multiple natures of uncertainty and helping to identify the reducible part;
- when both observations and formalised background expertise are available, a Bayesian estimation framework; though a classical estimation framework will often be used in its place for the sake of simplicity or with 'pure observational' data;

Practice will often lead to the use of *simplified* (or summarised) *single probabilistic structure* in the *later stages* of the analysis, or even working *conditionally* to partially-deterministic hypotheses, as discussed in the following section.

2.5.2 Mixed deterministic-probabilistic contexts

Risk measures may differ as to the extent to which the quantity computed on the output of interest z corresponds to probabilistic or deterministic uncertainty models in the various uncertain inputs $\boldsymbol{x}$. Indeed, it should be reckoned that in most studies, besides describing some of the uncertain inputs by probabilistic distributions, at least some other *uncertain inputs* (variables or events) are *fixed*. This is the case because:

- for some model inputs, the decision-process will *conventionally fix the values* despite the acknowledgement of uncertainties: for any comparative purpose, by a conventional 'penalisation' that is the choice of a fixed 'pessimistic' scenario and so on.
- uncertainties affecting some model inputs are considered to be negligible or of *secondary importance with respect to* the *output* variables of interest

Following the conventions introduced in Chapter 1, these inputs will be denoted as $\boldsymbol{x}_{pn}^{en}$ (meaning $\boldsymbol{x}_{pn}^{en}$, $\boldsymbol{e}_{pn}^{en}$ in full notation). Note that a more precise specification of the risk measure should involve the explicit *conditioning* of those fixed inputs:

$$c_z(\boldsymbol{d}) = c_z\left(\boldsymbol{d}|\boldsymbol{x}_{pn}\right) \tag{2.28}$$

As an essential point with regard to building a generic description of modelling in the context of risk and uncertainty, such a notation appears to be a *convenient means of unifying the deterministic or probabilistic descriptions of uncertainty*. Indeed, as noted in Chapter 1 and the comments on SRA, one may view a purely deterministic approach as a mixed deterministic-probabilistic criterion based on the conditioning of all components of $\boldsymbol{X}$:

$$z_{pn} = G(\boldsymbol{x}_{pn}\,\boldsymbol{d}) > 0 \Leftrightarrow z = G\left(\left(x^i_{\alpha_i}\right)_{i=1,\ldots p}, \boldsymbol{d}\right) > 0 \Leftrightarrow P\left[G(\boldsymbol{X}, \boldsymbol{d}) > 0 \middle| \left(X^i = x^i_{\alpha_i}\right)_{i=1,\ldots p}\right] = 1 \quad (2.29)$$

Hence the criterion could be written equivalently with the face value of a deterministic or probabilistic risk measure:

$$c_z(\boldsymbol{d}|\boldsymbol{x}_{pn}) > 0 \text{ } where \text{ } c_z(\boldsymbol{d}|\boldsymbol{x}_{pn}) = G(\boldsymbol{x}_{pn}, \boldsymbol{d}) \quad (2.30)$$

$$\Leftrightarrow c_z(\boldsymbol{d}|\boldsymbol{x}_{pn}) = 1 \text{ } where \text{ } c_z(\boldsymbol{d}|\boldsymbol{x}_{pn}) = P[G(\boldsymbol{X}, \boldsymbol{d}|\boldsymbol{x}_{pn}) > 0] \quad (2.31)$$

Conversely, even a so-called 'fully probabilistic' approach that develops a probabilistic distribution in $\boldsymbol{X}$ should be viewed as *conditional* on some basic implicit hypotheses. Think of the appropriateness of the physical model simplifications; the lack of knowledge of local 3D-effects; the completeness of the list of accidental sequences in a fault-tree and so on. Thus, the following notation would still be appropriate:

$$c_z(\boldsymbol{d}) = P[Z = G(\boldsymbol{X}, \boldsymbol{d}) > 0] = P[Z = G(\boldsymbol{X}, \boldsymbol{d}) > 0|\boldsymbol{x}_{pn}] = c_z(\boldsymbol{d}|\boldsymbol{x}_{pn}) \quad (2.32)$$

Although the conditioning in some input hypotheses is pervasive, the notation will often be simplified into $c_z(\boldsymbol{d})$. Within the uncertainty model, it would be equivalent in practice to assume either two options:

- $\boldsymbol{x}_{pn}$ receives a point distribution, that is a Dirac-density centred on $\boldsymbol{x}_{pn}$ for the continuous random vectors and a probability-one Bernoulli distribution on event $\boldsymbol{e}_{pn}$,
- whatever their underlying distribution, it works conditional on it being equal to $\boldsymbol{x}_{pn}$.

The second option is more flexible in a study process where one would consider sequentially distinct risk measures conditioned on a variable set of input hypotheses. Of course, beyond their formal consistency, one should be reminded that two risk measures based on differing probabilistic-deterministic mixtures do have very different meanings. In several areas, debates have taken place between the defenders of probabilistic and deterministic settings, as mentioned in the case of partial safety factors in SRA, or in that of the initiator-frequency approach in NRA. Indeed, these various risk measures correspond to rather different epistemological interpretations of the nature of uncertainty and possess distinct decision theory properties. In industrial practice, the allocation of the model inputs between 'uncertain' and 'fixed' appears a matter of choice rather than a theoretical discussion. Besides the fact that it may considerably change the study, particularly the data collection and computational costs, as mentioned in the review of NRA, they do correspond to different *design margins*. This will be illustrated in Chapter 4 in the very important case of monotonous system models, where deterministic *vs.* probabilistic settings for the different uncertain inputs have a direct visible impact on the design margins. This allocation can change in the course of the study and during the decision-making process. The unified notation proposed above facilitates this dynamic change, as the moving of one input from the probabilistic uncertainty model to a fixed deterministic value, for instance (or the reverse), does not change the theoretical framework. The risk measure:

$$c_z^1(\boldsymbol{d}) = P\left[Z = G(\boldsymbol{X}, \boldsymbol{d}) > 0 | \left(X^i\right)_{i=q,q+1...p} = \boldsymbol{x}_{pn}\right] \tag{2.33}$$

is simply replaced by the following alternative:

$$c_z^2(\boldsymbol{d}) = P\left[Z = G(\boldsymbol{X}, \boldsymbol{d}) > 0 | \left(X^i\right)_{i=q-1,q,q+1...p} = \boldsymbol{x}_{pn}\right] \tag{2.34}$$

Importance ranking/sensitivity analysis and model calibration steps do play key roles with respect to that choice of allocation between fixed and probabilised inputs, as will be commented upon in Chapters 6 and 7.

To be complete, an extended formulation of mixed deterministic-probabilistic risk measures should also include the cases where a fully-deterministic treatment of uncertainty in some inputs is contemplated, although the system model is *not monotonous* with respect to those. Suppose a case for components $\left(x^i\right)_{i=q,...q+1,...p}$ for which the available information leads one to consider uncertain intervals $\left[x^i_{min}, x^i_{max}\right]$. Then the risk measure could be:

$$c_z(\boldsymbol{d}) = Max_{[ximin,ximax]}(p-q)P\left[Z = G(\boldsymbol{X}, \boldsymbol{d}) > 0 | \left(X^i\right)_{i=q,q+1...p}\right] \tag{2.35}$$

which includes formally also the exclusively deterministic version:

$$c_Z(\boldsymbol{d}) = \max_{\boldsymbol{x} \in X} G(\boldsymbol{x}, \boldsymbol{d}) \tag{2.36}$$

Those risk measures are different to any of the previously mentioned risk measures (conditioned on 'penalised' lower or upper bounds) in the non-monotonous case where 'penalisation' is meaningless. Such risk measures, which could be labelled 'strongly deterministic' to mark the difference with the penalised ones, lead generally to serious computational challenges, as will be discussed in Chapter 6. In the context of the deterministic-probabilistic debates most would indeed refer to penalised risk measures when talking about deterministic alternatives, or possibly to a much simplified version of the strongly deterministic, defined as the maxima over a *finite* set of scenarios for the non-probabilised inputs.

Note that from the point of view of prediction (or uncertainty propagation), $\boldsymbol{x}_{pn}$ plays the same role as the components of vector $\boldsymbol{d}$ of controlled actions as the conditioning of the probabilistic risk measures. They are obviously different with respect to the scope of actions that may be undertaken; yet, be it a deliberate scenario attached to a choice of design or operation or a conventional scenario to cover sources of uncertainties that are not required to be modelled as random variables, both lead to the same *mathematical* analysis. This is why the notation may be simplified.

To simplify notation, $\boldsymbol{x}_{pn}$ will be omitted here below whenever considering the prediction phase, as being incorporated within a larger vector $\boldsymbol{d}$.

Deterministic penalisation may equally be derived on the second probabilistic level describing epistemic uncertainty in $\boldsymbol{\theta}_X$. Chapter 1 showed it to be standard practice in some areas of natural risk assessment or metrology: take an inflated value for the variance of metrological uncertainty or the parameters of an extreme value distribution modelling a natural hazard in order to 'safely' cover the inaccuracies associated with low sample sizes. In that case, probabilistic distribution $\pi(.)$ is replaced by a penalised value $\boldsymbol{\theta}_X^{pn}$ for the corresponding components, which is similar formally to conditioning the risk measure to a fixed value for the random variable $\boldsymbol{\Theta}_X^{pn}$ or introducing a Dirac distribution for $\pi(.)$. A monotonous hypothesis is also required in order to state that such a practice remains conservative, although it is not always as simple when considering the second probabilistic level, as will be discussed in Chapter 7.

Although beyond the scope of this book, recent interest in the Dempster-Shafer Theory (DST) settings can be also understood from the perspective of such mixed deterministic-probabilistic contexts, as it is possible to interpret this non-probabilistic paradigm as an extension of probability mixed with some form of deterministic (interval) representation (de Rocquigny, Devictor, and Tarantola, 2008).

2.6 Computing an appropriate risk measure or quantity of interest and associated sensitivity indices

A large variety of risk measures may therefore be considered: formulating a different probabilistic property such as exceedance probability or variance; distinguishing or not the aleatory and epistemic components of uncertainty in several possible ways; and incorporating more or less mixed deterministic-probabilistic settings. Leaving to Chapter 4 the discussion of their implications regarding the nature of the underlying systems and the purpose of decision-making, let it be noted that the regulatory or practical formulation of the risk or uncertainty assessment may sometimes specify the appropriate risk measure clearly, as illustrated in Table 2.3.

Table 2.3 Formulation of decision criteria and corresponding paradigms.

Formulation of the risk measure or decision criterion	Domain of typical application	Type of risk measure
'there should be less than 3% uncertainty on the declared value for the output of interest'	Uncertainty analysis or calibration	Variance
'the physical margin should remain positive in spite of uncertainty, with a probability less than 10^{-b} to be negative'	Structural safety	Exceedance probability
Large Early Release Fraction should keep below a given dose for an expected frequency of 1/10 000 (i.e. averaged over the epistemic uncertainty affecting the frequency value)	Nuclear IPRA	Quantile
'frequency of failure should be less than 10^{-b} per year, at a 95 % confidence level covering the uncertainties'	Systems reliability, PSA ...	Double-level exceedance probability

2.6.1 Standard risk measures or q.i. (single-probabilistic)

Remember that the estimation phase results in the following predictive model:

$$\begin{aligned} &\boldsymbol{Z} = G(\boldsymbol{X}, \boldsymbol{d}) \\ &(\boldsymbol{X}, \boldsymbol{\Theta}_X) \sim f_X(.|\boldsymbol{\theta}_X).\pi(.|IK, \Xi_n) \\ &(\boldsymbol{Z}, \boldsymbol{\Theta}_X) \sim f_Z(.|\boldsymbol{\theta}_Z).\pi(.|IK, \Xi_n) \end{aligned} \tag{2.37}$$

whereby the joint (input or output) distributions represent both aleatory and epistemic (input or output) uncertainty components. The simplest case is to work out the basis of the following *combined aleatory and epistemic* uncertainty distribution. It is expressed as follows on the input side:

$$\boldsymbol{X} \sim f_X^\pi(\boldsymbol{x}) = E_{\boldsymbol{\theta}_X}[f_X(\boldsymbol{x}|\boldsymbol{\Theta}_X)|IK, \Xi_n] = \int_{\boldsymbol{\theta}_X} f_X(\boldsymbol{x}|\boldsymbol{\theta}_X)\pi(\boldsymbol{\theta}_X|IK, \Xi_n)d\boldsymbol{\theta}_X \tag{2.38}$$

which is the *posterior* (unconditional) or *predictive distribution* in a Bayesian approach. Indeed, the distribution f_X^π represents the best view of the likely variability of the whole vector of sources of uncertainty (including the model inaccuracy $\boldsymbol{U}$) associated with model $G(.)$, averaged over the possible values for parameters $\boldsymbol{\theta}_X$ due to lack of knowledge remaining after incorporation of all information (IK, Ξ_n) available *at the time of analysis*. Hence its image distribution by $G(.)$ defined by the following *cdf* also represents the combined view of the uncertainty in $\boldsymbol{Z}$:

$$\begin{aligned} F_Z^\pi(z) = P(Z < z|IK, \Xi_n) &= \int_\theta P(Z < z|\boldsymbol{\theta}_X)\pi(\boldsymbol{\theta}_X|IK, \Xi_n)d\boldsymbol{\theta}_X \\ &= \int_\theta \left[\int_x 1_{G(X,d)<z} f_X(\boldsymbol{x}|\boldsymbol{\theta}_X)d\boldsymbol{x} \right] \pi(\boldsymbol{\theta}_X|IK, \Xi_n)d\boldsymbol{\theta}_X \\ &= \int_x 1_{G(\boldsymbol{x},\boldsymbol{d})<z} f_X^\pi(\boldsymbol{x})d\boldsymbol{x} \end{aligned} \tag{2.39}$$

Such a distribution and its associated density f_Z^π are instrumental in determining risk measures that summarise the probabilistic description of uncertainty into a *single-probabilistic* distribution or, in other words, 'averaging over the uncertainties of the risk components'. Such a formulation also covers the *essential practical case* in which epistemic uncertainty is neglected in a single-level setting so that the combined pdf f_X^π of $\boldsymbol{X}$ equates simply to its conditional distribution $f_X(.|\boldsymbol{\theta}_X)$: the subsequent *conditional* risk measure becomes a standard function of $\boldsymbol{\theta}_X$.

A variety of such risk measures or quantities of interest may be associated as listed in Table 2.4, possibly also derived under mixed deterministic-probabilistic versions.

Chapter 1 reviewed their use in various contexts. Probabilities of threshold exceedance or of undesired events are the typical risk analysis or structural reliability risk measures, if not simply penalised values in traditional structural engineering. Quantiles (or confidence intervals, their bilateral equivalent) are pervasive in the areas of environmental or uncertainty analysis. Multiples of standard deviation or coefficients of variation are the preferred choice in metrology, with dedicated names. Expectations are used instead in the context of decision analysis, through the consideration of a utility as the variable of interest, as discussed below.

Table 2.4 Risk measures or q.i. in simple probabilistic setting.

Type of risk measure	Risk measure or quantity of interest	Decision Criterion to Comply (absolute threshold)	Decision Criterion to Select (relative threshold)
Penalised value	$G(x_{pn}, \mathbf{d})$ or $Max_X\, G(x_{pn}, \mathbf{d})$	$z_{pn} < z_s$	$z_{pn}(\mathbf{d}) < z_{pn}(\mathbf{d}^{\circ})$
Probability of threshold exceedance	$P[Z > z_s \mid (X^i = x^i_{pn})_{i=q+1\ldots p}]$	$P[Z > z_s] < p_s$	$P[Z > z_s \mid \mathbf{d}] < P[Z > z_s \mid \mathbf{d}^{\circ}]$
Probability (or frequency) of undesired event	$P(E_f) = P\left[Z = e_f \middle| \left(X^i = x^i_{pn}\right)_{i=q+1\cdots p}\right]$	$P[E_f] < p_s$	$P[E_f \mid \mathbf{d}] < P[E_f \mid \mathbf{d}^{\circ}]$
quantile (conditional)	$z^{\alpha} = F_Z^{-1}(\alpha)$ where $F_Z(z) = P\left[Z < z \middle| \left(X^i = x^i_{pn}\right)_{i=q+1\ldots p}\right]$	$z^{\alpha} < z_s$	$z^{\alpha}(\mathbf{d}) < z^{\alpha}(\mathbf{d}^{\circ})$
Absolute uncertainty	$unc_z = k.\sqrt{\mathrm{var}(Z)}$ or $unc_z = \dfrac{z^{\frac{1+\alpha}{2}} - z^{\frac{1-\alpha}{2}}}{2}$	$unc_z < i_s$	$unc_z(\mathbf{d}) < unc_z(\mathbf{d}^{\circ})$
Relative uncertainty	$\%unc_z = k.\dfrac{\sqrt{\mathrm{var}(Z)}}{E(Z)}$ or $\%unc_z = \dfrac{z^{\frac{1+\alpha}{2}} - z^{\frac{1-\alpha}{2}}}{2.E(Z)}$	$\%unc_z < \%i_s$	$\%unc_z(\mathbf{d}) < \%unc_z(\mathbf{d}^{\circ})$
Expectation (usually expected utility)	$E(Z)$	(seldom used)	$E(Z \mid \mathbf{d}) < E(Z \mid \mathbf{d}^{\circ})$

The table also contains the decision criteria that may be associated with such risk measures, which generally compare the risk measure to a threshold. Such a threshold may be prescribed independently of the system, '*absolute*' such as 10^{-b}/yr, or '*relative*' as a comparison between options involving the same system, such as the impact of potential actions or a reference design. In practice, there is often no explicit decision criterion. In other words, the absolute or comparative threshold is often implicit, so that one has a tendency to speak about the 'criterion' while referring simply to the risk measure itself. The phrase 'show the relative uncertainty' suggests, for example, 'show that it is weak', or 'less than elsewhere'.

In practice, the risk measure is often defined in a single variable of interest representing the overall performance or safety indicator. More rarely it may happen that the criterion bears on *vector* $\boldsymbol{z}$. The event of interest can then be defined as the logical association of the tests above for each output of interest such as $E_f = \{z|z^1 > z_s^1 \; and/or \; z^2 > z_s^2\}$ with the risk measure $P(E_f)$, a situation referred to in the reliability domain as 'a multiple failure event', the failure event being the intersection of $\{z_s^l - Z^l < 0\}$ for $l = 1, \ldots r$.

2.6.2 A fundamental case: The conditional expected utility

In spite of such apparent variety, most of these risk measures may be interpreted formally as an *expected utility conditional on penalised assumptions*, which proves to be the fundamental risk measure or quantity of interest:

$$c_z(\boldsymbol{d}) = E\left[U(Z)|\left(\boldsymbol{X}^i\right)_{i=q+1,\ldots p} = \boldsymbol{x}_{pn}, \boldsymbol{d}\right] \tag{2.40}$$

The expectation is integrated over the predictive output distribution f_Z^π. Such a definition covers a large variety of risk measures. In general, the expected utility is intended to gather together the eventual outcomes (including the vulnerability of the final assets that is important to the decision-maker and represented in z) and the associated probabilities of occurrence of the potentially multi-dimensional performance represented in variables of interest $\boldsymbol{z}$. Yet, the expected utility also covers degenerate cases corresponding to design (or risk level) compliance: the exceedance probability corresponds to an *extreme* utility which would value only the threshold exceedance (or the intersection of threshold exceedance in the case of multiple outputs of interest):

$$\begin{aligned} U_s(z) &= 1 \; \textit{if} \; z < z_s \\ U_s(z) &= 0 \; \textit{if} \; z > z_s \end{aligned} \tag{2.41}$$

Assume that one is interested in controlling z defined as the maximal value of Z_t over a period of time $[t_o, t_o + \Delta T]$. The (degenerate) expected utility then comes as:

$$c_Z(\boldsymbol{d}) = EU_s(z) = E1_{Max_{[t_o,t_o+\Delta T]}Z_t(\omega) \leq z_s} = P\left[Z = Max_{[t_o,t_o+\Delta T]}Z_t \leq z_s|\boldsymbol{x}_{pn}, \boldsymbol{d}\right] \tag{2.42}$$

Similarly, the probability of an undesired event is equivalent to a 'pure-reward' expected utility, equalling 0 for the undesired event and 1 otherwise. Thus, also for the deterministic 'point value' which was proved to be formally equivalent to an exceedance probability.

Quantiles are sometimes selected as a risk measure. However, a closer look at the associated decision criteria shows generally that an implicit exceedance probability is targeted. Indeed, when an industrial standard or regulation requires the declaration of the 95 % quantile for a chemical discharge, it is meant to be compared to a pre-existing environmental or sanitary threshold. So that one could equally require the computation of the threshold (non-) exceedance probability and check whether that figure is higher than 95 %. Besides, a metrological 'standard (or extended) uncertainty' is generally makeshift for a quantile. Indeed

G.U.M. tolerates the use of a given number of standard deviations as the practical approximate of a confidence interval under an implicit Gaussian hypothesis. Hence, it may also be linked formally to an expected utility.

However, not all risk measures may be viewed as expected utilities: Chapter 8 will show the limitations associated with such a concept, which were made popular by the Allais paradoxes on rational decision-making (Allais, 1953) and which generated alternative proposals designed to best represent a decision-maker's attitude towards risk and uncertainty.

2.6.3 Relationship between risk measures, uncertainty model and actions

For the sake of computation, it is important to note that the single-probabilistic risk measures take the general form of a *functional* of the joint distribution of the uncertain variables and events of interest **Z** (and are thus not a straightforward *function* of *z*):

$$c_z = F\left[f_Z^\pi(.)\right] \tag{2.43}$$

Chapter 1 showed that most of the time, such a functional means an *integral* over the predictive distribution f_Z^π of **Z**. Indeed, as mentioned above, most risk measures appear as expected utilities for a certain type of utility function:

$$\begin{aligned} c_Z(\boldsymbol{d}) &= E\left[U(\boldsymbol{Z})|IK, \Xi_n, \boldsymbol{x}_{pn}, \boldsymbol{d}\right] = \int_{\boldsymbol{Z}} U(z) f_Z^\pi\left(z|\boldsymbol{x}_{pn}, \boldsymbol{d}\right).dz \\ &= \int_{\boldsymbol{x}} UoG(\boldsymbol{x}, \boldsymbol{d}) f_X^\pi\left(\boldsymbol{x}|IK, \Xi_n, x_{pn}\right).d\boldsymbol{x} \end{aligned} \tag{2.44}$$

Other quantities of interest that are sometimes considered in metrology, such as a multiple of the standard deviation, again involve integrals over *Z* although not as an expected utility:

$$\begin{aligned} unc_z &= k.\sqrt{\mathrm{var}(Z)} = k.\sqrt{\int_z Z^2 f_X^\pi(z).dz - \left(\int_z Z f_X^\pi(z).dz\right)^2} \\ &= k.\sqrt{\int_{\boldsymbol{x}} G(\boldsymbol{x}, \boldsymbol{d})^2 f_X^\pi(\boldsymbol{x})d\boldsymbol{x} - \left(\int_{\boldsymbol{x}} G(\boldsymbol{x}, \boldsymbol{d}) f_X^\pi(\boldsymbol{x})d\boldsymbol{x}\right)^2} \end{aligned} \tag{2.45}$$

Hence the risk measure c_z is a *functional* of the pdf f_Z^π of Z, instead of a deterministic function of *z*. Note firstly that its computation is in great generality achievable through a blend of simulation and optimisation techniques (see Chapter 7). Thanks to the propagation of the predictive distribution of **X** through the system model *G*(.) (cf. Equation (2.38)) note also that it appears as a standard *function of the control variables **d*** as well as of the information available on the system (IK, Ξ_n):

$$c_z = F\left[f_Z^\pi(.)\right] = G\left[f_X^\pi(.|IK, \Xi_n), G(., \boldsymbol{d})\right] = C_z(\boldsymbol{d}, IK, \Xi_n) \tag{2.46}$$

This is mathematically obvious in the formulation of the integrals above: while this function is correlatively very complex, far from closed-form in general, it is theoretically a deterministic function. Important consequences will result for computing efficiently the impact of actions within the decision-making process (see Chapters 7 and 8). Additionally, this shows that any update in the information available to the analyst would enable the corresponding updating of the risk measure (see Chapter 6). Indeed, the risk measure captures by definition the lack of knowledge in the variables of interest *z* at the future time of interest remaining after incorporation of all information (IK, Ξ_n) available *at the time of analysis*.

Note that the above-mentioned integrals are standard (Lebesgue-integrals) in the static case of simple random vectors $\boldsymbol{X}$, $\boldsymbol{Z}$... while the introduction of stochastic integration may be necessary when dealing with time series $\boldsymbol{X}_t$, $\boldsymbol{Z}_t$... Broadly, this leaves unchanged the previous comments: formally, the resulting risk measure remains a standard function of $\boldsymbol{d}$; practically, Monte-Carlo simulation can still be undertaken through time discretisation, although computationally costlier.

2.6.4 Double probabilistic risk measures

In the general case, it is desirable to control the impact of the double epistemic-aleatory structure of uncertainty on the output $\mathbf{Z}$: think of the natural risk assessment where T-level maximal flood flow or wind speed may be inflated by a α-level of confidence to cover the risks associated with sample limitation, or of a frequency of failure (or disease) over a population of sub-systems or individuals inflated by a level of confidence. They become formally a little more complex, such as the range of possible values for the coefficient of variation or the exceedance probability. Instead of being functionals of the predictive output distribution f_Z^{π} only, such risk measures are functionals built on the full joint output distribution:

$$c_z = F[f_Z(.), \pi(.|IK, \Xi_n)] \tag{2.47}$$

A proper understanding of the decisional meaning of such double-probabilistic risk measures involves decision-theory refinement and the associated pros and cons, discussed in Chapter 4. This is described in Table 2.5.

A common double probabilistic risk measure is the (level-2 or epistemic) confidence that the (level-1 or aleatory) frequency of an event of interest e_p – say plant failure – remains below a risk target α (say $\alpha = 10^{-b}$):

$$\begin{aligned} c_z &= P_{\boldsymbol{\Theta}_X|IK,\Xi_n}\left[P_{X|\theta_X}\left(E_p = e_p|\boldsymbol{\Theta}_X, \boldsymbol{d}\right) < \alpha\right] \\ &= \int_{\theta_X} 1_{\left[\int_{x} 1_{G(\boldsymbol{x},\boldsymbol{d})=e_p} f_X(\boldsymbol{x}|\theta_X) d\boldsymbol{x} < \alpha\right]} \pi(\boldsymbol{\theta}_X|IK, \Xi_n) d\boldsymbol{\theta}_X \end{aligned} \tag{2.48}$$

Such a risk measure shows a double level of integration in a costlier formulation than the basic expected utilities (see Chapter 7). It is typically integrated in a decision criterion that requires 'frequency of the undesired event of interest is less than 1/1000-yr at a α-level of confidence', encountered in protection against natural risk or integrated probabilistic risk assessment.

Alternatively, a *deterministic level-2* risk measure (above a probabilistic level-1) introduces *theoretically* a maximisation step above the integration step:

$$c_z(\boldsymbol{d}) = Max_{\boldsymbol{\theta}_X \in D_\theta}\left[P\left(E_p = e_p|\boldsymbol{\theta}_X, \boldsymbol{d}\right)\right] = Max_{\boldsymbol{\theta}_X \in D_\theta} \int_{\boldsymbol{x}} 1_{G(\boldsymbol{x},\boldsymbol{d})=e_p} f_X(\boldsymbol{x}|\boldsymbol{\theta}_X) d\boldsymbol{x} \tag{2.49}$$

However, remember that the most common understanding of a level-2 deterministic setting would simply be to *fix* penalised values for some parameters of the uncertainty model. Think of an upper bound of the measurement variance in metrology, or of the penalised parameters of the extreme value distribution in natural risk:

$$c_z(\boldsymbol{d}) = P\left(E_p|\boldsymbol{\theta}_X^{pn}, \boldsymbol{d}\right) = \int_{\boldsymbol{x}} 1_{G(\boldsymbol{x},\boldsymbol{d})=e_p} f_X(\boldsymbol{x}|\boldsymbol{\theta}_X^{pn}).d\boldsymbol{x} \tag{2.50}$$

Table 2.5 Double-probabilistic risk measures.

Setting	Uncertainty model	Risk measure or quantity of interest	(Potential) decision criterion
Probabilistic with level-2 deterministic	Joint pdf (on $\boldsymbol{X}$), the parameters of which, $\boldsymbol{\theta_X}$, are given intervals of variation, or more generally, subsets.	Max. of probabilistic q.i. (variance *var*(Z), standard deviation, coef. of variation, probability of exceeding a threshold $P(Z > Z_s)$ or quantiles $z^{\alpha\,\%}$ ...) when $\boldsymbol{\theta_X}$ varies within its interval or subset.	Max. $c_V(Z) < 3\,\%$ Max. $P(Z > z_s) < p_s$ Max. $z^{95\,\%} < z_s$
Double probabilistic	Joint pdf (on $\boldsymbol{X}$), the parameters of which, $\boldsymbol{\theta_X}$, themselves follow a given joint pdf π.	Level-2 quantile of a level-1 quantile, level-2 expectation of a level-1 quantile, *and so on*	$P[P(Z > z_s) < p_s] > 95\,\%$ $P[E(Z) < z_s] > 95\,\%$

2.6.5 The delicate issue of propagation/numerical uncertainty

Whatever the risk measure selected, it needs to be computed within a practical study. Generally, it involves a complex system model embedded in a large multiple-integral. Although there is a wealth of dedicated techniques to compute efficiently any of those (the subject of Chapter 7), numerical limitations may arise at this stage. This is principally due to the limitations in the number of system model runs that are computationally affordable. These limitations rise steeply with the rarity of the exceedance probability or the number of input dimensions involved in the definition of the risk measure. Other types of numerical errors may also occur in solving numerically the system model equations because of a limited underlying numerical precision (of the coding of real-valued numbers) of the low-layer variability due to computer operating systems.

Thus, a new type of uncertainty appears at this stage, although the literature is somewhat silent on this delicate subject. It is sometimes referred to as sampling uncertainty or variability when sampling techniques are involved in the propagation step. Or else as 'confidence' in the results, numerical error and so on. It will be discussed in detail in Chapter 7, as it is a key feature to be mastered when undertaking reliable risk or uncertainty studies in large industrial applications. It will be referred to principally as *propagation uncertainty* in subsequent chapters to mark clearly its link to the propagation step, in a clear distinction with the two levels of uncertainty already encountered, the *fundamental* (aleatory or level-1) uncertainty, and the *estimation* (epistemic or level-2) uncertainty. Propagation uncertainty is also of an epistemic or reducible nature, as it is, theoretically at least, amenable to be reduced by an increasing computational budget or by better numerical schemes.

Although the practice is seldom used, there happen to be level-2 risk measures within which the second-level corresponds to propagation uncertainty instead of estimation uncertainty. It is particularly the case in nuclear thermal hydraulics, where an extensive use of the so-called *Wilks method* – in fact Monte-Carlo sampling with a robust confidence interval in the quantile estimator – accompanies risk measures such as 'peak cladding temperature 95 % quantile computed with 95 % upper confidence' (de Rocquigny, 2005).

Triple-level (or higher) risk measures have never been encountered by the author in practice. Although they are a theoretical possibility, as mentioned in the specialised literature on sampling techniques (Lemmer and Kanal, 1986), they would become too complex to be tractable in operational decision-making or regulation.

2.6.6 Importance ranking and sensitivity analysis

Beyond the computation of a given risk measure for a given uncertainty model and set of actions, an essential output of most risk or uncertainty studies is a statement about the relative importance of the various uncertain inputs. Whether the process is regulatory-driven or performance-driven, shared opinion has it that the sole probabilistic figure is not enough, nor is it the most important output of a study. The uncertainty model is never complete enough, nor a stable feature in the long term: more measurement, data or expertise may emerge in the later steps, and the system's operation may depart from its initial configuration over time. Thus, it is all too important to understand also the relative impact of the various inputs on the risk measure in order to answer typical questions such as: 'where should I invest more resources in data, modelling detail or process control in order to reduce uncertainty or mitigate risk?'

Sensitivity analysis (or *importance ranking*) refers formally to the computation and analysis of so-called *sensitivity* or *importance indices* of the uncertain input variables or events $\boldsymbol{x}$ with respect to a *given quantity of interest* in the output z. In fact, such a computation involves a propagation step, for example with sampling techniques, as well as a post-treatment specific to the sensitivity indices considered. Statistical treatment has to be made of the input/output relations controlling the quantities of interest or risk measure as a function of the uncertainty model. Sensitivity indices have essentially been developed with respect to *variance* (or variance-based) risk measures in z. They include, for instance, an index derived from the Taylor quadratic approximation of variance:

$$s_i^{Taylor2} = \frac{\left(\frac{\partial G}{\partial x^i}\right)^2 \mathrm{var}(X^i)}{\sum_j \left(\frac{\partial G}{\partial x^j}\right)^2 \mathrm{var}(X^j)} \tag{2.51}$$

Such an index is widely used in practice because it requires few calculations of $G(.)$ and the formula is quite intuitive: an uncertain input is only important insofar as its own uncertainty $\sigma^2_{X^i}$ is strong and/or the physical model is sensitive to it (i.e. $\partial G/\partial x^i$ is high). Yet such index is limited to system models that are linear in $\boldsymbol{x}$ or acceptably linearised at the scale of the uncertainties. At the other end of the complexity spectrum come the 'Sobol' indices, very general but computationally challenging, such as the following conditional variance:

$$S_i = \frac{\mathrm{var}(E(Z|X^i))}{\mathrm{var}Z} \tag{2.52}$$

Chapter 7 will briefly discuss the associated issues and a more in-depth review may be found in books such as Saltelli *et al.* (2004).

Consideration of the sensitivity indices is essential in the analysis process for several reasons:

- As a representation of the proportion of the risk measure apportioned to the i-th input, the sensitivity indices are essential guides to rank the usefulness of further modelling or data acquisition so as to better control or even reduce the output risk and uncertainty.
- It can be proved that the same indices computed in an observational model also help to assess the quality of the uncertainty modelling or estimation process of the input uncertainty through indirect observations (see Chapter 6 for the linear case).
- Sensitivity indices help also in accelerating the propagation methods through dedicated sampling (see Chapter 7 and the conditional Monte-Carlo method).

Similar to the uncertainty model itself, two levels of sensitivity analysis are to be found:

- with respect to the components of level-1 uncertainty, that is the variables or events $\boldsymbol{x}$ and $\boldsymbol{e}$ standing as the inputs of the system model;
- with respect to the components of level-2 uncertainty, that is the parameters describing the lack of knowledge of the relative amounts of uncertainty.

Although distinct, they are obviously mathematically linked: a significant level-2 importance index requires generally that the parameter involved is attached to a model input which has significant importance in level-1. In nuclear probabilistic safety analysis, level-1 may be considered as important factors in the risk (frequency) of reactor damage, helping to target risk mitigation efforts, while level-2 refers to uncertainty sensitivity indices that are helping to target modelling or operating return efforts. Chapter 8 will also introduce ancillary useful quantities of interest that can be computed when the output variable is a cost, namely the Expected Value of Perfect Information (EVPI) and Expected Value of Ignoring Uncertainty (EVIU): they can also be viewed as a means to rank the importance of the input uncertainties.

2.7 Summary: Main steps of the studies and later issues

Quantitative risk and uncertainty studies involve the prediction of the key characteristics of a system (vector $\mathbf{Z}$ of variables/events of interest) at a given time, the state of which depends on actions (vector $\boldsymbol{d}$) taken in an imperfectly known manner. Whatever the diversity of the contexts reviewed in Chapter 1, two closely-associated models are mobilised:

- the system model $M_s(\boldsymbol{x}, \boldsymbol{d}, \boldsymbol{u})$ and associated observational and predictive formulations,
- the uncertainty model $\boldsymbol{X} \sim f_X(.|\boldsymbol{\theta}_X).\pi_\theta(.|IK, \Xi_n)$ in its input variables, events, parameters and residuals.

Together they help estimate the appropriate *risk measure* (or quantity of interest): a risk measure $c_z(\boldsymbol{d})$ is a functional that summarises the level of risk/uncertainty affecting $\mathbf{Z}$ conditioned by the given actions in the system. It plays a central role in fulfilling any of the four salient goals of studying the system *under uncertainty*: *understanding* the behaviour of the system; *accrediting* or validating a model; *selecting* or optimising a set of actions; and *complying* with a given decision-making criterion or regulatory target.

Phenomenological knowledge and statistical or non-statistical information are all necessary to estimate the two associated models in a joint effort to best represent the information available. Selecting the adequate risk measure requires a definition of the uncertainty setting that mixes deterministic and probabilistic features. This delicate choice involves not only decision theory but also the consideration of regulatory standards and pragmatic pros and cons. Probabilistic-numerical techniques are needed to compute the risk measure, mostly through multiple integration of the system and uncertainty models, as well as statistical techniques for the associated sensitivity indices.

At this stage a risk/uncertainty study can be summarised formally as involving a number of generic tasks that are summarised within the following main steps (Figure 2.9):

(a) *Specify* the dominant goal (U, A, S or C), the system model, the variables and time of interest as well as the risk measure or quantity of interest:
- choose variables z according to decision-making, as well as a system model $G(.)$ predicting z at the time of interest as a function of decision variables ($\boldsymbol{d}$) and all input characteristics $\boldsymbol{x}$ (uncertain variables $\boldsymbol{x}$ and events $\boldsymbol{e}$) defined so as to retrieve information and represent uncertainty affecting the system;
- choose a representation of uncertainty, associating deterministic conditioning and a single or double-level probabilistic setting;
- choose a risk measure or q.i. in accordance with the decision-making process.

(b) *Model* (i.e. estimate, identify or calibrate) $\{M_s, f_X(.|\boldsymbol{\theta}_X), \pi_\theta(.|IK, \Xi_n)\}$ the combined system and uncertainty models on the basis of all statistical and non-statistical information available in the system at the time of analysis.

(c) *Compute* the risk measure $c_Z(\boldsymbol{d})$, that is propagate the input uncertainty model through $G(.)$ to estimate the output uncertainty $f_Z^\pi(z|IK, \Xi_n, \boldsymbol{d})$ and/or the associated quantities of interest and sensitivity indices.

(*Feedback loop*) Return to step B (or maybe A) to further refine the models and/or inject more data in the most important parts and/or change design $\boldsymbol{d}$, *for example* to better comply with a decision criterion $c_Z(\boldsymbol{d}) < c_s$ or possibly to optimise $c_Z(\boldsymbol{d})$.

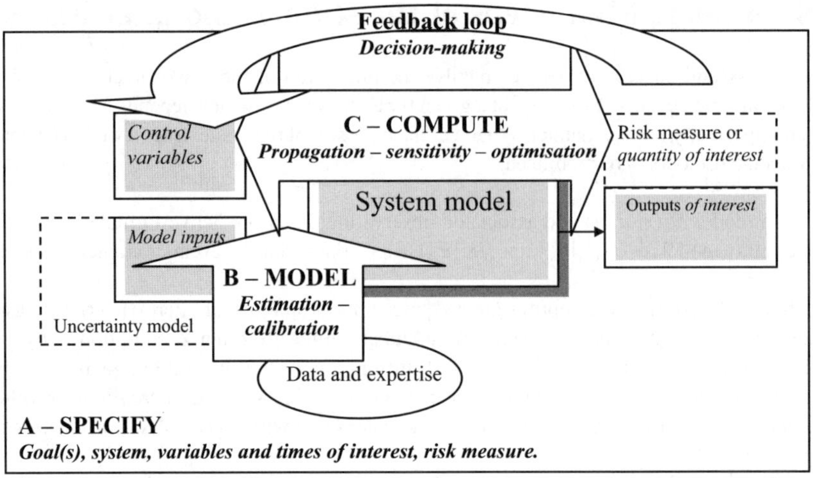

Figure 2.9 *Key generic steps of modelling in the context of risk and uncertainty.*

These key steps embody in a generic process the majority of statistical and numerical modelling techniques associated with modelling in the context of risk and uncertainty. They will be described subsequently in the book as follows:

- Chapter 4 will discuss *step A*, in the context of decision making under risk and associated theories.
- Chapter 5 will discuss *step B*, completed later on by Chapter 6 for indirect data involving inverse techniques.
- Chapter 7 will then discuss the statistical and numerical challenges and methods used in *step C* for the computation of risk measures and sensitivity analysis, and the use of the latter sensitivity indices for the *Feedback loop*: many of those methods can be found within the Open TURNS open source development platform (www.openturns.org).
- Chapter 8 will enlarge the models into an economic perspective, and thus discuss the challenges associated with the selection or optimisation of actions under uncertainty in the *Feedback loop*.

Eventually, the optimisation of actions $\boldsymbol{d}$ under risk/uncertainty inside the *Feedback loop* – be it through the acquisition of more information, data or model refinement or through the choice of more appropriate designs, operational choices or investments – may be the most important process. This is particularly the case when making the time basis more explicit and allowing for the breakdown of actions $\boldsymbol{d}$ over various future time steps within a *real options* approach. It also stands as the most challenging from a computational perspective as it involves the cycling of the study process a number of times and is a general requirement for stochastic optimisation (see Chapter 8). It remains one of the remaining scientific challenges.

Exercises

There follows a set of exercises referring to the concepts developed in this chapter:

2.1 Suggest a simple closed-form system model accompanied by a real-valued uniformly-distributed uncertainty model; write the formulation of a threshold-exceedance risk measure as function of the following quantities: actions, model parameters and the observed sample. Is there a closed-form expression to compute the resulting risk measure?

2.2 Discuss triple-level risk measures making propagation uncertainty explicit as well as their aleatory/epistemic components, either through a probabilistic or mixed deterministic-probabilistic structure: how does it allow handling the degree of reducibility? How can it be communicated? What are the computational implications?

References

Allais, M. (1953) Le comportement de l'homme rationnel devant le risque: critique des postulats et axiomes de l'école américaine. *Economica*, **21**(4), 503–546.

Apostolakis, G. (1999) The distinction between aleatory and epistemic uncertainties is important; an example from the inclusion of aging effects in the PSA. *Proceedings of PSA'99*, Washington DC.

Aven, T. (2003) *Foundations of Risk Analysis*, John Wiley & Sons, Ltd.

Aven, T. and Rettedal, W. (1998) Bayesian Frameworks for integrating QRA and SRA methods. *Structural Safety*, **20**, 155–165.

Beck, M.B. (1987) Water quality modelling: A review of the analysis of uncertainty. *Water Resources Research*, **23**(8).

Bedford, T.J. and Cooke, R.M. (2001) *Probabilistic Risk Analysis – Foundations and Methods*, Cambridge University Press.

Beven, K.J. and Binley, A.M. (1992) The future of distributed model: model calibration and uncertainty prediction. *Hydrological Processes*, **6**, 279–298.

de Finetti, B. (1974) *Theory of Probability*, vols. **I and II**, John Wiley & Sons, Ltd., New York.

de Rocquigny, E. (2005) A statistical approach to control conservatism of robust uncertainty propagation methods; application to accidental thermal hydraulics calculations. Proceedings of ESREL-05, Tri City, Poland.

de Rocquigny, E. (2006) La maîtrise des incertitudes dans un contexte industriel: 1ère partie – une approche méthodologique globale basée sur des exemples; 2nd partie – revue des méthodes de modélisation statistique, physique et numérique. *Journal de la Société Française de Statistique*, **147** (4), 33–106.

de Rocquigny, E., Devictor, N. and Tarantola, S. (eds) (2008) *Uncertainty in Industrial Practice, A guide to Quantitative Uncertainty Management*, under publication by John Wiley & Sons, Ltd.

Ditlevsen, O. and Bjerager, P. (1986) Methods of Structural Systems Safety. *Structural Safety*, **3**, 195–229.

Droguett, E.L. and Mosleh, A. (2008) Bayesian methodology for model uncertainty using model performance data. *Risk Analysis*, **28**(5), 1457–1476.

Granger Morgan, M. and Henrion, M. (1990) *Uncertainty – A Guide to Dealing with Uncertainty in Quantitative Risk and Policy Analysis*, Cambridge University Press.

Guelfi, A., Boucker, M., Hérard, J.-M. *et al.* (2005) A new multi-scale platform for advanced nuclear thermal-hydraulics status and prospects of the NEPTUNE project. 11th International Topical Meeting on Nuclear Reactor Thermal-Hydraulics Avignon, France, October 2–6 2005.

Helton, J.C. (1994) Treatment of uncertainty in performance assessments for complex systems. *Risk Analysis*, **14**, 483–511.

Helton, J.C. and Oberkampf, W.L. (2004) Alternative representations of epistemic uncertainty. *Special Issue of Reliability Engineering & System Safety*, **85**(1–3).

Helton, J.C., Burmaster, D.E. *et al.* (1996) Treatment of aleatory and epistemic uncertainty. *Special Issue of Reliability Engineering & System Safety*, **54**(2–3).

ISO (1995) Guide to the expression of uncertainty in measurement (G.U.M.).

Kaplan, S. and Garrick, B.J. (1981) On the quantitative definition of risk. *Risk Analysis*, **1**(1), 11–27.

Kennedy, M.C. and O'Hagan, A. (2002) Bayesian calibration of computer models. *Journal of the Royal Statistical Society: Series B (Statistical Methodology)*, **63**(3), 425–464.

Knight, F.H. (1921) *Risk, Uncertainty and Profit*, Hart, Schaffner & Marx.

Krige, S.G. (1952) A statistical approach to some basic valuations problems on the witwatersrand. *Journal Chemical, Metallurgical and Mining Society*, **52**(6), 119–139.

Law, A.M. and Kelton, W.D. (2000) *Simulation Modelling and Analysis*, 3rd edn, McGraw Hill.

Lemmer, J.F. and Kanal, L.N. (eds) (1986) *Uncertainty in Artificial Intelligence*, Elsevier.

Leroy, O., Junker, F., Gavreau, B. *et al.* (2008) Calibration under Uncertainty – a coupled physical-statistical approach and first application to ground effects on acoustic propagation. 13th Long-Range Sound Propagation (LRSP) Symposium, Lyon.

Madsen, H.O., Kenk, S. and Lind, N.C. (1986) *Methods of Structural Safety*, Prentice-Hall Inc.

Nilsen, T. and Aven, T. (2003) Models and model uncertainty in the context of risk analysis. *Reliability Engineering & System Safety*, **79**, 309–317.

Oberkampf, W.L., DeLand, S.M., Rutherford, B.M. *et al.* (2002) Error and uncertainty in modelling and simulation. *Special Issue of Reliability Engineering & System Safety*, **75**(3), 333–357.

Pasanisi, A., de Rocquigny, E., Bousquet, N. and Parent, E. (2009) Some useful features of the Bayesian setting while dealing with uncertainties in industrial practice. Proceedings of the ESREL 2009 Conference, 3, pp. 1795–1802.

Paté-Cornell, M.E. (1996) Uncertainties in risk analysis; Six levels of treatment. *Reliability Engineering & System Safety*, **54**(2–3), 95–111.

Quiggin, J. (1982) A theory of anticipated utility. *Journal of Economic Behaviour and Organization*, **3**, 323–343.

Robert, C.P. (2001) *The Bayesian Choice*, Springer.

Saltelli, A., Tarantola, S., Campolongo, F. and Ratto, M. (2004) *Sensitivity Analysis in Practice: A Guide to Assessing Scientific Models*, John Wiley & Sons, Ltd.

Salvi, O. and Debray, B. (2006) A global view on ARAMIS, a risk assessment methodology for industries in the framework of the SEVESO II directive. *Journal of Hazardous Materials*, **130**, 187–199.

Savage, L.J. (1954) *The Foundations of Statistics*, Dover Publication, Inc., N.Y.

Welch, W.J., Buck, R.J., Sacks, J. *et al.* (1992) Screening, predicting, and computer experiments. *Technometrics*, **34**, 15–25.

3

A generic tutorial example: Natural risk in an industrial installation

This chapter introduces an example of a tutorial that will be used later to illustrate most of the mathematical derivations in the book, benefiting from a comprehensible physical interpretation and easily derivable numerical figures. It sets out the model-based design of a simple asset, an industrial installation protected by a dike, in the context of an uncertain (and thus risky) external environment, a floodable river. It includes hydraulics, sedimentology, climatology, civil engineering and economics as its phenomenological components, all of them simplified into easily understandable closed-form formulae. This chapter introduces the gradual modelling layers regarding risk and uncertainty: effective computations and results will be provided in subsequent chapters as illustrations of most of the methods discussed.

3.1 Phenomenology and motivation of the example

Consider a vulnerable installation located on a river bank: housing, industrial facilities or power plant benefiting from the cooling resource and so on. The risk situation can be characterised schematically by the following components:

- **hydro**: – flood flows q may entail high water levels, potentially flooding the installation if the maximal level z_c occurring during a flood event overflows the crest of the dike (supposedly at level z_d), (see Figure 3.1)
- **systems reliability**: – conditional on dike overflow (an event defined by $s = z_c - z_d > 0$), an elementary protection system, comprising plant isolation active or passive devices, exhaust pumps and so on, which may or may not contain the consequences of failure,
- **economics**: – the final detrimental consequences depend upon the effective vulnerability of the facilities, largely unknown at the time of the hypothetical event. They are encapsulated in an overall

Modelling Under Risk and Uncertainty: An Introduction to Statistical, Phenomenological and Computational Methods, First Edition. Etienne de Rocquigny.

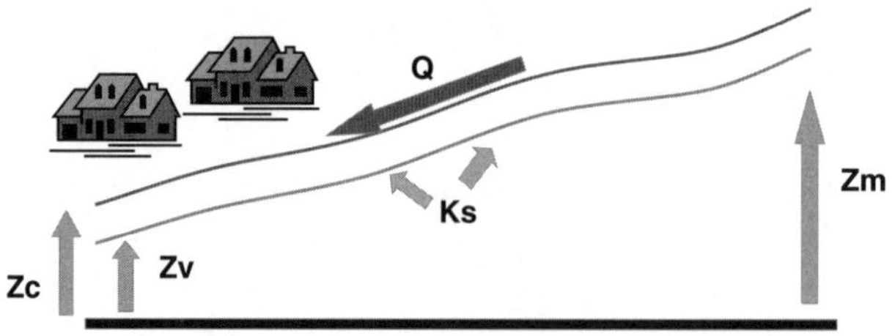

Figure 3.1 *Flood flow schematisation (taken from Limbourg and de Rocquigny, 2010).*

cost of damage c_d. It adds up to the more-manageable investment costs c_i resulting from design decisions so as to form the complete cost c_c to be optimised.

Referring back to the generic concepts in Chapter 2, the outputs of interest (vector z) may here include various variables or events according to the decision-making framework and components included in the system model:

- the overflow s,
- the overflow event $E_o = \{s \mid s > 0\}$,
- the plant failure event E_p,
- or the total cost c_c.

In fact, the decision-making framework can involve schematically either (or successively):

(1) *Demonstrating compliance with a design criterion*, ensuring that the installation is well protected up to a conventional risk level, say a *thousand-year return period*, in other words $c_Z < 10^{-3}$; the risk measure may include the system's reliability component or represent simply the overflow over the dike.
(2) *Optimising the design in a cost benefit (or expected utility) approach* to the dike and the protection system in the face of potential accidental costs and adjustable investment costs.

In fact, two points of views may be taken into account in order to address option (1): undertake a *risk assessment* exercise for a given design, for instance of an existing facility; or undertake a *design* exercise for a given risk criterion, for instance for future facilities or the refurbishment/up-grading of existing ones. The latter may be the first step before responding to issue (2). Although very similar in practice, examples given below will illustrate the computation differences.

Simple analytical models will be derived for each component in subsequent sections, followed by a review of the sources of uncertainty, data and expertise available. Refer to *Annex Section 10.4* to obtain the complete numerical detail.

3.1.1 The hydro component

The hydro component comprises all phenomena leading to the potential dike overflow:

$$s = z_c - z_d = z_c - z_b - h_d \tag{3.1}$$

Dike elevation z_d being the sum of the *known* decision variable (the dike height h_d) and parameter (the elevation of natural riverbank z_b), predicting the overflow means predicting the flood maximal water level in front of the installation z_c. This involves hydrology, for the determination of the flood flows generated by heavy rains upstream of the river reaches; hydro-sedimentology, for the determination of the riverbed conditions (bed topography, obstructions, ...), that are often modified by historic floods, and that in turn largely modify the flood flow; and finally hydraulics, for the determination of the resulting flood water level.

Most detailed studies would potentially involve elaborate models on all three layers: hydrological studies could model the upstream flow hydrogram Q_t as the result of a mixture of rain or flow statistical estimation and rain-flow conversion models; hydro-sedimentology would model the 2-D time-varying topography, mapping riverbed *elevation*($\mathbf{r}$, t); thus the hydrodynamic model would predict 2-D flood water levels $z(\mathbf{r}, t)$ through partial differential equations (typically Navier-Stokes) depending on the results of the two first layers.

These models would of course require a great deal of data, expertise and study time. A much simpler model can schematise these phenomena as follows:

- Maximal flood flow q is considered to be accessible through hydrological statistics (working only on the peak of the hydrogram).
- Riverbed conditions are represented by the Strickler's coefficient k_s of hydraulic friction[1] and the riverbed's changing elevations (z_m upstream and z_v at the level of the floodable site).
- Slope i over a fixed-length reach l varies accordingly: $i = \frac{z_m - z_v}{l}$.
- Water depth h over the riverbed varies as a power-function of flood flow, friction coefficient, slope and supposedly-fixed river width b.

Hence the basic hydraulic model:

$$h = \left(\frac{q}{k_s * \sqrt{i} * b} \right)^{3/5} \tag{3.2}$$

$$s = z_c - z_d = h + z_v - z_b - h_d \tag{3.3}$$

The output variable of interest of the *hydro* system model is basically the overflow s; which may also be expressed in the form of the event of interest 'dike overflow' defined simply by $e_o = 1_{0<s}$. Thus, the complete system model of the hydro component is as follows:

$$z = (s,\ e_o) = (G_h(q, k_s, z_m, z_v, h_d),\ 1_{0<s}) \tag{3.4}$$

Although being simple enough to be used later on in this book as an illustrative example, it should be noted that it is an accurate physical formulation for the mean water depth. Provided there are time-stationary, space-uniform and large width-over-depth ratio hypotheses, Equation (3.2) turns out to be able to solve St-Venant one-dimensional hydrodynamic equations where $S(h)$ and $R_H(h)$ represent the wet section and hydraulic radius and $r \in [0, l]$ denotes the abscissa along the river (Goutal et al, 2008; Lencastre, 1999):

$$\begin{aligned} &\frac{\partial q}{\partial r} + b\frac{\partial h}{\partial t} = 0 \\ &\frac{\partial q}{\partial t} + \frac{\partial}{\partial r}\left[\frac{q^2}{S(h)}\right] = bgh\left(i - \frac{\partial h}{\partial r}\right) - g\frac{q^2}{bhR_H(h)^{4/3}K_s{}^2} \end{aligned} \tag{3.5}$$

---

[1] The friction coefficient follows an inverse convention, a classic of hydraulics, according to which the more k_s decreases, the higher the friction.

Note the monotonous behaviour of overspill $s = G_h(q, k_s, z_m, z_v, h_d)$ as a function of all the input variables. Overflow increases when flow q increases, downstream riverbed elevation z_v increases, friction increases (smaller k_s), or slope reduces because of lower upstream elevation z_m. Note also the reverse signs of monotonous dependence for downstream and upstream riverved elevations. Overflow decreases clearly when the dike is designed to be higher. And we may understand non-linear dependence by way of the fact that water depth increases less quickly than flow because flow speed and section both increase at the same time (power 3/5 for h as a function of q), and through kinetic energy there are losses of gravity's potential energy (square-root over the slope).

3.1.2 The system's reliability component

Suppose now that a limited protection system – comprising plant isolation devices, exhaust pumps and so on – can be designed to further protect the installation from failure even in the case of dike overflow. Think, for instance, of protecting redundant power systems that are essential for a nuclear plant, as shown dramatically by the Sendai/Fukushima event of 2011. For this component of the study, note that the installation failure becomes the event of interest, while dike overflow becomes a mere initiating event.

The protection system would probably not be able to contain equally all types of overflow. Detailed models could possibly describe the varying efficiency of the protection according to the level of flood initiator. To start with, let us simply say that the protection system is only worthwhile for a limited range of overspill (say event $E_{os} = \{s \mid 0 < s < 2m\}$) beyond which overspill is completely and surely uncontained (the large overspill event being denoted as E_{ol} so that $E_o = E_{os} \cup E_{ol}$). Suppose then that two redundant isolation systems or processes (flood gates, exhaust pumps, evacuation procedures, etc.) are in place, of which only one is necessary to block the consequences of a flood. Each isolation system is comprised of two components, an active one with total efficiency once activated, and a passive one, with only limited efficiency but not needing activation. Then, conditional upon the event of limited overspill E_{os}:

- if either of the two isolation systems is activated, the plant is safe, probability of failure is zero. There is, however, a non-negligible probability of non-activation of any of the two systems (corresponding events are denoted as V_1 (resp. V_2)): assume also a common mode for those two failure events (denoted as V_{12}) representing, for instance, the likelihood of being unable to activate system 2 when activation of 1 has failed for example due to common electrical power shut-down, access difficulties and so on, besides reasons specific to any of the two systems (failure events independent of each other are denoted as V_{1i} and V_{2i} respectively),
- conditional upon non-activation of system 1 (resp. 2), then the passive component may contain plant failure with only a limited probability. Events of failure of passive component 1 (resp. 2) are denoted as L_1 (resp. L_2) and are assumed to be independent and equally probable.

This system can be modelled by an elementary event tree or equivalently a fault tree (see Figure 3.2, which does not include the large overspill event).

The event of plant failure E_p results as a logical function of all elementary events, a basic algebraic result of those trees complemented by the event of a large overspill E_{ol}:

$$E_p = [E_{os}\ AND\ L_1\ AND\ L_2\ AND\ (V_{12}\ OR\ (V_{1i}\ AND\ V_{2i}))]\ OR\ E_{ol} \tag{3.6}$$

As mentioned in Chapter 1, note that this formula has a variety of interpretations: (a) it can be seen as a relationship between the subset of the sample space representing the events; (b) it can be translated straightforwardly into a relationship between the probabilities (or frequencies) associated with those

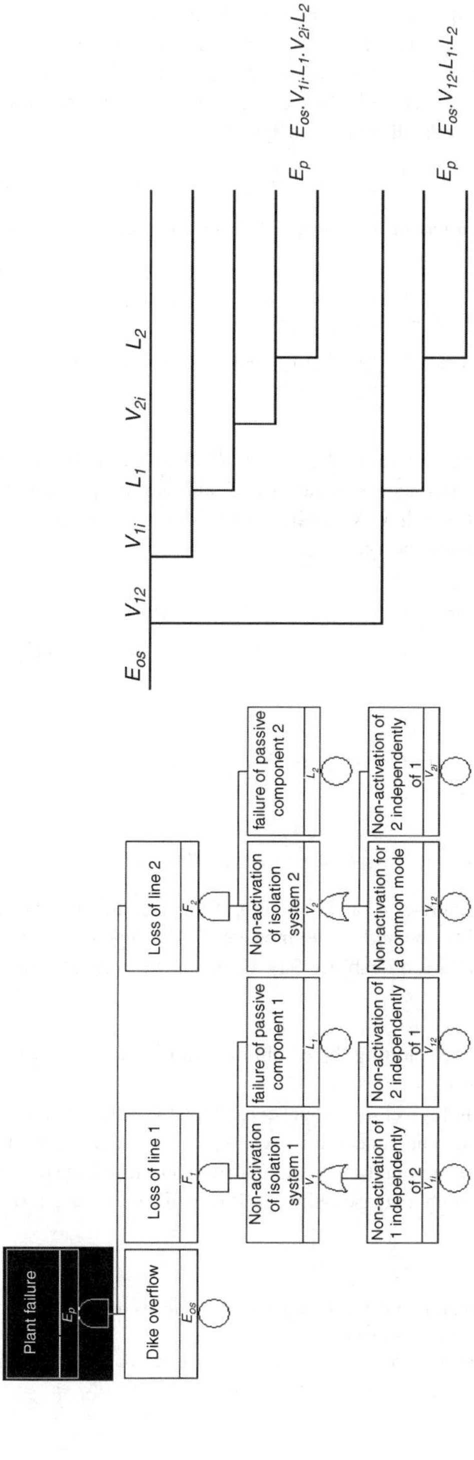

Figure 3.2 *Protection system fault tree (left) and event tree (right) - event of large overspill is ignored for the sake of simplicity.*

events; and (c) it can be seen as a deterministic function between the indicator variables that do represent the deterministic causality between (supposedly known) states of nature, that is points in the sample space.[2] Retaining this last interpretation, a new *system model* can be generated, relating the final event of interest to new model inputs. A new decision variable $\boldsymbol{d}$ is included: the indicator variable 1_{ps} representing formally the choice to install or not the overall protection system.

$$e_p = e_{os}.[1_{ps}.l_1.l_2.(v_{12} + v_{1i}.v_{2i}) + (1 - 1_{ps})] + e_{ol} = G_s(e_{os}, l_1, l_2, v_{12}, v_{1i}, v_{2i}, e_{ol}, 1_{ps}) \tag{3.7}$$

Hence, plug in the previous hydro component to generate a more complete system model presented formally as $G = G_s o\ G_h$:

$$\begin{aligned} e_p &= 1_{0<s<2}\,(s)\,.[1_{ps}.l_1.l_2.\,(v_{12} + v_{1i}.v_{2i}) + (1 - 1_{ps})] + 1_{2<s}\,(s) \\ &= 1_{0<s<2}\,(G_h(q, k_s, z_m, z_v, h_d)).[1_{ps}.l_1.l_2.(v_{12} + v_{1i}.v_{2i}) + (1 - 1_{ps})] + 1_{2<s}\,(G_h(q, k_s, z_m, z_v, h_d)) \end{aligned} \tag{3.8}$$

whereby input vector $\boldsymbol{x}$ becomes the larger vector $(q, k_s, z_m, z_v, l_1, l_2, v_{12}, v_{1i}, v_{2i})$. The vector of variables and event of interest z may keep both e_p and s, since in the case of failure of the protection system it may still be useful to know the amplitude of overflow. Assuming that overflow is contained if the protection system works, the completed system model becomes:

$$\begin{aligned} e_p &= G_s(1_{0<s<2}\,(G_h(q, k_s, z_m, z_v, h_d)), l_1, l_2, v_{12}, v_{1i}, v_{2i}, 1_{2<s}(G_h(q, k_s, z_m, z_v, h_d)), 1_{ps}) \\ s &= G_h(q, k_s, z_m, z_v, h_d).[1_{s<0} + 1_{s>2} + 1_{0<s<2}.(1_{ps}.l_1.l_2.\,(v_{12} + v_{1i}.v_{2i}) + (1 - 1_{ps}))] \end{aligned} \tag{3.9}$$

hence

$$z = G(q, k_s, z_m, z_v, l_1, l_2, v_{12}, v_{1i}, v_{2i}, h_d, 1_{ps}) = G(\boldsymbol{x}, \boldsymbol{e}, \boldsymbol{d}) \tag{3.10}$$

where

$$\boldsymbol{x} = (q, k_s, z_m, z_v) \quad \boldsymbol{e} = (l_1, l_2, v_{12}, v_{1i}, v_{2i}) \quad \boldsymbol{d} = (h_d, 1_{ps})$$

Note that monotonous behaviour characterises e_p as a function of all uncertain inputs.

As mentioned earlier, extensions of this would model more finely the dependence between the level of overspill and a component of the failure probability, say V_1 being the non-activation of protection system 1. This may be done either:

- globally, through a 'vulnerability curve' plotting the conditional failure probability as a function[3] of overspill $P(V_1 = 1 \mid S = s) = vuln(s)$,
- with more detail, through a phenomenological sub-model explaining the status of activation as a causal function of both overspill and some other uncertain event $v_1 = G_a(s, x_p)$. Think of activation as requiring a minimal time between the time when the operator arrives on site and the time when the site is completely flooded. The speed of flooding increases with overflow while the time to arrival is uncertain.

---

[2] Note that this deterministic function is defined in the *real* sample space which is only a restricted part of $\{0,1\}^7$, excluding some impossible couples such as $e_{os} = 1$ and $e_{ol} = 1$.

[3] Such a function is the direct correspondant of the seismic fragility curve, a popular concept in seismic PRA expressing a component failure probability as a function of peak ground acceleration (PGA).

3.1.3 The economic component

In many cases, the decision to design a dike of a given height and with a complementary protection system will be driven essentially by regulation, such as: protect up to a conventional risk level, for example a 1000-year return period level. Then the two components of the study already introduced may be sufficient, in a design-to-criterion approach. In some cases, however, it may be useful to go into a cost analysis in order to optimise the protection investment: typically beyond the regulatory minimal level of level, where a further 'offsetting' of the potential damage costs may be profitable. The complete cost c_c to be optimised over given period of operation would include:

- the investment and maintenance costs c_i, that depend deterministically on the decision variable $\boldsymbol{d}$ (dike height and installation or not of the protection system);
- the cost of damage c_d if there is an overflow, comprising:
  - the damage to the installation, that increases partly with the overflow level (material damages) but also includes a large base cost particularly for an industrial facility whose production losses due to unavailability would quickly overtake material damage costs;
  - possibly also the damage to the dike itself, the cost of which would increase with the dike's height, but quickly become independent of the overflow level.

Given a single flood event, the complete cost then arises as follows:

$$c_c = c_i(z_d, 1_{ps}) + c_d(s, c_m, h_d) \tag{3.11}$$

More complete studies involve a summation of the uncertain damage costs over multiple events potentially occurring during the period of operation (see *Chapter 8*, Section 8.2). Simple cost functions have been derived to model the complete cost c_c, including a new uncertain input, the coefficient of the damage cost function c_m representing the intrinsic vulnerability of the installation. As will be seen subsequently, the damage costs are uncertain for two reasons at least. Firstly, a large variability may result from the current state of activity of the site (night/day, week/weekend ...). Secondly, extensive lack of knowledge of the costs of damage before any such accident is likely to happen, even supposing that a precise level of overflow is anticipated.

Figure 3.3 (left) introduces the dike component of cost function c_i: it increases quicker than linearly with a height at low values because of the likely increase of the length of the dike length required, but then

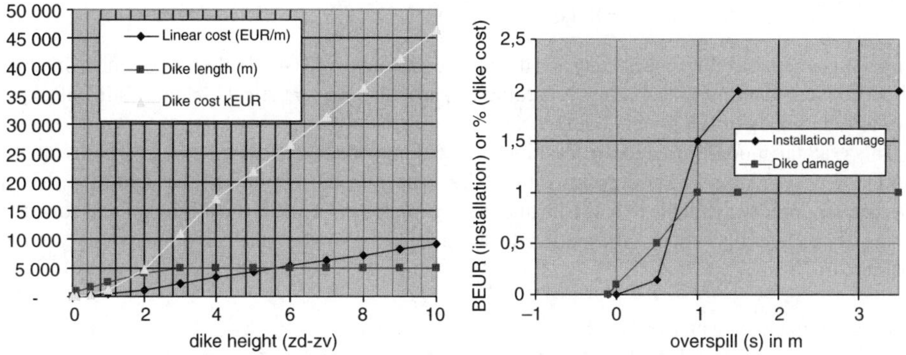

Figure 3.3 *Cost functions for dike investment (left), for installation damage or dike damage (right).*

the full perimeter of the installation is covered. Figure 3.3 (right) introduces the installation and dike components of damage cost function c_d as a function of overspill, for given values of vulnerability coefficient and dike height. Refer to Annex Section 10.4 to find the numerical details.

Thus, cost optimisation would concern the choice of dike height and of protection installation with regard to the distribution of the complete cost c_c. As will be discussed in Chapter 8, several economic risk measures can be considered for minimising such risk/uncertainty optimisation:

- the expected value of the total cost: this simple measure could be rather risky,
- the probability of catastrophic damage, an event e_c that could be formulated as when c_c exceeds a solvency threshold ($E_c = \{c_c \mid c_c > c_s\}$): this would assume a modelling of greater risk aversion,
- the expected value of a utility or other transformed functions of the cost $u_c = U(c_c)$. A more-advanced criterion including a decision theory model of the decision-maker's preference (including potential risk-aversion) would involve a utility function of the cost or even a Quiggins-non-linear transformation of the cost distribution (see Chapter 8).

In all cases, the most complete system model needs to be worked on, plugging in the total cost functions of the Section 3.1.2 model combined with the hydro and systems reliability component. Hence the new output vector of interest is predicted (where, for the first regulatory decision-making step, the overflow and event of plant failure are kept) as a function of a larger vector $\boldsymbol{x}$ (including the vulnerability coefficient) and the same vectors $\boldsymbol{e}$ and $\boldsymbol{d}$ as before:

$$z = (s,\ e_p,\ c_c,\ e_c,\ u_c) = G(\boldsymbol{x}, \boldsymbol{e}, \boldsymbol{d}) \tag{3.12}$$

where:

$$\boldsymbol{x} = (q, k_s, z_m, z_v, c_m) \quad \boldsymbol{e} = (l_1, l_2, v_{12}, v_{1i}, v_{2i}) \quad \boldsymbol{d} = (h_d, 1_{ps})$$

Refer to Annex Section 10.4 to obtain the complete formula for the largest system model. It also implies the monotonous behaviour of e_c or c_c as a function of all inputs, $\boldsymbol{x}$, $\boldsymbol{e}$ or $\boldsymbol{d}$, because the cost functions are themselves monotonous as a function of the overspill s as well as of the cost uncertainty factor c_m.

3.1.4 Uncertain inputs, data and expertise available

Previous sections have defined the following system model, linking variables or events of interest to the input (see Table 3.1).

$$z = (s,\ e_p,\ c_c,\ e_c,\ u_c) = G(\boldsymbol{x}, \boldsymbol{e}, \boldsymbol{d}) \tag{3.13}$$

Many sources of variability, uncertainty and imprecision are involved (see Table 3.2), to which one should add epistemic *estimation* uncertainty, regarding in particular the statistical characteristics of extreme flows or reliabilities of the system components.

The gradual modelling steps in Section 3.2 will introduce different ways to represent those sources of uncertainties. This depends on the risk criterion, but obviously also upon the data and expertise supposedly available to build the uncertainty model. As a rough illustration of real conditions, varying (but generally poor) information levels according to the sources of uncertainty will be considered:

- limited samples for water levels;
- large samples for flow (149 observations, see Figure 3.4 and Annex Section 10.4);
- mostly expertise for friction conditions, but limited bivariate samples on riverbed elevations;
- expertise on reliability characteristics and economic variabilities.

Table 3.1 Summary of the various system model inputs/outputs.

	Hydro	H + Systems reliability	H + SR + Economic
uncertain input variables (x)	q, k_s, z_m, z_v	...	$\ldots, c_m$
uncertain input events (e)	none	$l_1, l_2, v_{12}, v_{1i}, v_{2i}$	...
decision variables (d)	h_d	$\ldots, l_{ps}$	...
system model	G_h	$G_s o\ G_h$	$G_e\ o\ G_s\ o\ G_h$
variables/events of interest (z)	$s,\ e_o$	$\ldots,\ e_p$	$\ldots,\ c_c,\ e_c,\ u_c$
usual risk measures	$P(s > 0) = P(e_o)$	$\ldots, P(e_p)$	$\ldots,\ E(c_c), P(e_c), E(u_c)$

Table 3.2 Nature of uncertainty and type of data available.

Model inputs	Nature of uncertainty	Type of data available
q	Natural (meteorological) randomness Flowmeter imprecision	Time series records (e.g. daily over ~100 yr)
k_s	Natural (sedimentological) variability; (hydraulic) model uncertainty	Expertise and indirect measurements (model calibration)
z_m, z_v	Natural (sedimentological) variability; lack of knowledge (loose spatial sampling)	Limited records (e.g. 10–30 point measurements)
$l_1, l_2, v_{12}, v_{1i}, v_{2i}$	Intrinsic randomness (reliability components); lack of knowledge (behaviour under severe overflow)	Expertise or limited reliability data in analogous conditions
c_m	Time variability (industrial activity); lack of knowledge (damage processes and production recovery time)	Expertise

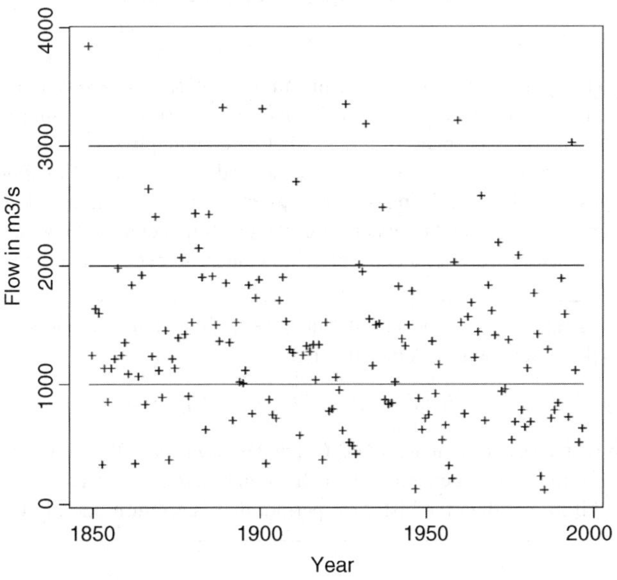

Figure 3.4 *The data sample for annual maximal flood flows.*

The overall picture of the example is now ready for analysis. Note that this pedagogical example, developed for some years in the context of academic and executive education (de Rocquigny, 2004; de Rocquigny, 2006), is obviously a rough schematisation of real systems for the sole purpose of illustration. Meant to be generic, such straightforward generalisations could involve the replacement of flood by any natural risk affecting a installation (sea surges, storm winds, seism, ...) or even any internal initiating events in industrial risks (failure of a component such as a chemical tank, boiler, etc.).

3.2 A short introduction to gradual illustrative modelling steps

Several approaches to uncertainty assessment are thus possible for one or other of these criteria: (1) *demonstrating compliance with a design criterion*, securing the fact that the installation is well protected up to a conventional specification or risk level; (2) *optimising the design in a cost benefit (or expected utility) approach* to the dike and the protection system in the face of potential accidental costs and adjustable investment costs. The gradual modelling steps in Section 3.2 will introduce different ways to solve those two decision-making problems on a mixed probabilistic-deterministic basis, gradually moving on to the more probabilistic. Two very elementary approaches will firstly be discarded, as being too poor to be retained.

Firstly, elementary engineering thinking might refer to a *basic deterministic study*. Consider the use of so-called 'penalised' values for each of the inputs. In fact, when the system model is monotonous, it is simple to understand that the highest value for the output over intervals of variation for each input component corresponds to the value computed at the lowest or highest input bound depending on the sense of monotony (increasing or decreasing):

$$Max_{\boldsymbol{x}\in D_x} G(\boldsymbol{x}, \boldsymbol{d}) = G(\boldsymbol{x}_{pn}, \boldsymbol{d}) \tag{3.14}$$

where, for the hydro component:

$$\begin{aligned} D_x &= [q_{\min}, q_{\max}] \, x \, [k_{s\,\min}, k_{s\,\max}] \, x \, [z_{m\,\min}, z_{m\,\max}] \, x \, [z_{v\,\min}, z_{v\,\max}] \\ x_{pn} &= (q_{\max}, k_{s\,\min}, z_{m\,\min}, z_{v\,\max}) \end{aligned} \tag{3.15}$$

Penalised values would typically be taken at the maximum historically observed, or an expert opinion. This elementary method generates quite a poor binary risk measure, expressing simply the fact that protection is complete or non-existent according to whether dike height is below or above a unique 'safe level'. Confidence in the protection level cannot be measured by any level of likelihood. Design to cost optimisation is impossible. Despite being standard practice in many engineering contexts, such an approach is very rare for flood studies where the aleatory character of flood levels has long been recognised: one would always refer to a concept of flow quantile even if there is very little data.

> As a numerical illustration, taking penalised values as the maxima of the observed sample for Q or according to expert judgement for other inputs:
>
> $$\boldsymbol{x}_{pn} = (3854m^3/s,\ 15m^{1/3}/s,\ 54mNGF,\ 51mNGF)$$
>
> leads to a 'pseudo-worst case' design of *59.6mNGF*, that is *4.1m* above the natural riverbank. It is hard to assess the protection level associated with such a combination; most engineers would probably reckon that it is a bit higher than the historical period, that is a return period of $\sim$*1/150-yr*.

Secondly, the estimation of a risk measure such as plant failure probability might be a straightforward count of the historical frequency of plant failure. Such an approach, a *direct reliability estimation* as practiced in massively-spread components (such as light bulbs), is generally not practical in complex risk systems because of the lack of relevant failure records, arising from several factors: (i) plant failure records are too short in relative terms (since plants are generally very reliable compared to their operation duration since commissioning[4]) if not completely blank when considering design stages; (ii) plant failure comes from multiple factors mixing external aggressions with internal failure causes, so that a direct count would considerably blur the investigation of the regulated risk source (e.g. flood risk); (iii) plant operation undergoes rapid evolution, and considerable local peculiarities so that direct counts of time or population records would be biased by massive non-stationarity or heterogeneity.

3.2.1 Step one: Natural risk standard statistics

The simplest way to answer question (1) is to estimate the simplified risk measure:

$$c_Z = P(E_o) = P(S = Z_c - z_b - h_d > 0) \tag{3.16}$$

through the closest consistent statistical records available. Take the water depth records $(h_{mj})_{j=1\ldots n}$ that may be simply added to a fixed level for Z_v, for instance a penalised value $z_{v\,pn}$ defined by an expert. Two options may be considered for the water level model:

- Adjust a binomial model for the indicator variable 'exceedance of a given threshold z_d-$z_{v\,pn}$'. This option is the most intuitive one. Yet, it is not practical since the targeted reliability generally exceeds the potential of the available data sample. For instance, only a few tens of yearly observations would be available for a targeted 100-yr or even 1000-yr protection.
- Adjust a complete distribution for the random variable H_m, such as a Gumbel, lognormal or normal distribution: this would relieve some of the limitations of the sample size, enabling extrapolation to a certain extent

Figure 3.5 and Table 3.3 recapitulates this first option. The *system model* is reduced to its simplest form: the comparison of the dike level to the water depth added to a fixed riverbed elevation generates the *variable of interest* which is overflow (or its opposite, the margin to overflow, in a convention that is closer to that in the structural reliability literature). The probabilistic *risk measure* is simply the probability that overflow becomes positive; note indeed that the probability is *conditional on* the penalised assumption for riverbed elevation. A data sample is used directly through statistical estimation to build the uncertainty model, which is simply the *cdf* of the only source of uncertainty being probabilised, that is flood water depth, so that $X = H$. No epistemic uncertainty is modelled at this stage, taking a fixed value for the parameters of the *cdf*.

The computation of the risk measure is straightforward here. It does not require any Monte-Carlo or other propagation method, since:

$$c_Z = P(E_o) = P(z_d - H - z_{v\,pn} < 0) = P(H > z_d - z_{v\,pn}) = 1 - F_H(h_d + z_b - z_{v\,pn}) \tag{3.17}$$

The numerical results are summarised in Table 3.4: cf. Chapter 5 for more detail on the statistical best fit for water heights. Note the importance of the choice of a statistical model fitting the data: replacing

4 $t_{op} \ll 1/p_f$ where t_{op} denotes the operation time (in yrs) since commissioning and p_f the (yearly) failure probability.

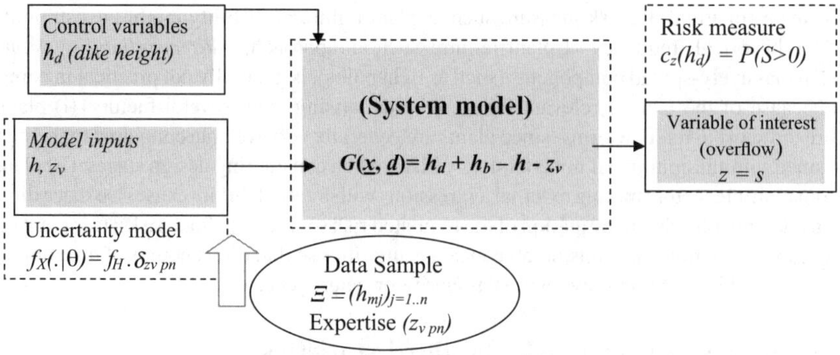

Figure 3.5 *Step One modelling process.*

Table 3.3 Step One – summary characteristics.

	Step One	Comments
system model	*Elementary* $G(.) = h_d + z_b - h - z_v$	Elementary natural risk estimation
uncertain inputs ($\boldsymbol{x}$, $\boldsymbol{e}$)	variables ($\boldsymbol{x}$): h, z_v	
decision variables ($\boldsymbol{d}$)	h_d	
variables/events of interest (z)	s, e_o	
risk measure c_z	$P(S > 0) = P(E_o)$	
Uncertainty model	A pdf is estimated for H. A fixed value ($z_{v\ pn}$) is taken for Z_v. No epistemic uncertainty ($\pi = \delta_\theta$): $\boldsymbol{\theta}_X$ has a single value ($\boldsymbol{\theta}_H$, $z_{v\ pn}$)	Estimation involves simple statistics on H and expertise on Z_v
Propagation (computation of c_z)	Simple addition	

Table 3.4 Step One – Main results.

Risk assessment

	No dike	$h_d = 1m$	$h_d = 2m$	$h_d = 3m$
Empiric model	$5\ 10^{-2}$	$2\ 10^{-2}$	$8\ 10^{-3}$	??
Statistical best fit (lognormal)	$6\ 10^{-2}$	$2\ 10^{-2}$	$8\ 10^{-3}$	$3\ 10^{-3}$
Alternative statistical fit (normal)	$3\ 10^{-2}$	$3\ 10^{-3}$	$1\ 10^{-4}$	$2\ 10^{-6}$

Design (*i.e. dike heights in meters*)

	1/50	1/100	1/1000
Empiric model	1.0	1.5	?? (>2.9)
Statistical best fit (lognormal)	1.1	1.8	4.2
Alternative statistical fit (normal)	0.2	0.5	1.3

the statistical best fit (lognormal) by a Gaussian model, largely irrelevant because of the asymmetry of the phenomenon considered here (see Chapter 5), would reduce the assessed risk level by 1 to 3 decades and the dike design by a few metres.

As will become clearer in Step Three, the main shortcoming of this first step lies in the fact that it needs a substantial record for the variable directly attached to the risk being considered, here the water level facing the industrial installation. Often a large sample of direct measurements is not available as the measurements are rare or suffer from calibration error (as evidenced in the data, see Chapter 10). Note here that the records that are used quantify the water depth instead of water level: the addition of a fixed riverbed elevation instead of representing its past variability distorts the proper inference of the water level. Using a 'penalised' value will obviously increase the water levels inferred. Using past water depths may also simply be irrelevant, since water depth encapsulates many underlying phenomena, some of which may have changed between the period of observation and the present status of the installation environment. For instance, it is generally assumed that water flow being generated by upstream catchments is insensitive to the building of the installation so that its modelling may rely on direct past records (assuming climatic stationarity). Nevertheless, flow conditions such as topography or friction coefficients often change due to the installation so that it may be physically inappropriate to use past records to build a statistic.

3.2.2 Step two: Mixing statistics and a QRA model

Suppose now that a protection system has been set up so that the installation does not always fail in the case of dike overspill. Provided the level of protection is significant enough and acceptable to regulation for it to be accounted for inside the risk assessment target, one would consider replacing the previous risk measure by the following:

$$c_Z = P(E_p) \tag{3.18}$$

The simplest way to estimate such a risk measure involves:

- setting up the systems reliability model as mentioned earlier,
- estimating $p_{os} = P(E_{os})$ and $p_{ol} = P(E_{ol})$ through techniques similar to that of Step One, thus simplifying the hydro component for quasi-direct estimation,
- estimating through expert judgement the other uncertain input events involved in the fault tree:

$$(P(E^i))_{i=1,\ldots p'} = (P(V_1), P(V_2), P(V_{12}), P(L_1), P(L_2)) = (p_1, p_2, p_{12}, p_{1i}, p_{2i}) \tag{3.19}$$

The elementary event probabilities (conditional upon the event of limited overspill) could be assessed typically as:

- '90 % chance that installation fails conditional on protection system 1 failure' ($p_I = 0{,}9$);
- or '25 % to 60 % chance that activation of protection system 1 fails conditional on limited overspill ' and so on.

Note that this involves implicitly a choice of time basis (e.g. per hour or year of operation) depending on the time representation of the conditional random events: see Chapter 4 for more on that point.

To put it simply, it is not necessary to describe epistemic uncertainty in such probabilities, so that the probabilistic model remains single-levelled as in Step One. Because of the nature of the systems reliability model – a simple fault tree – computation of the risk measure is simplified into the following closed-form

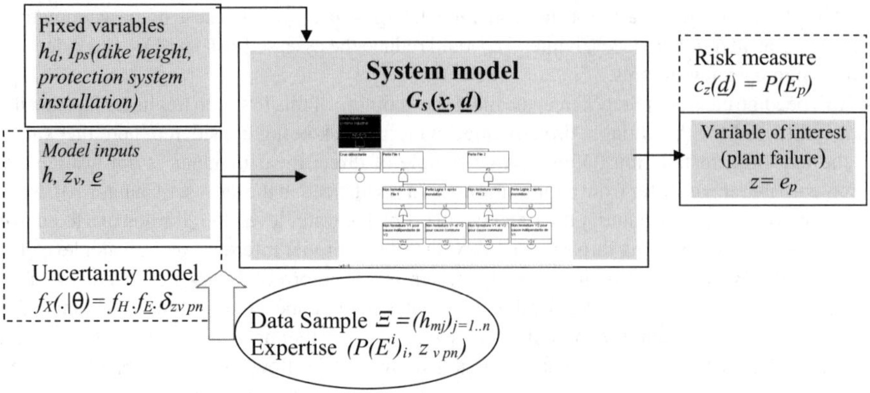

Figure 3.6 *Step Two modelling process.*

solution:

$$
\begin{aligned}
c_Z &= P(E_p) = P([E_{os}\,AND\,L_1\,AND\,L_2.AND\,(V_{12}\,OR\,V_{1i}\,AND\,V_{2i})]\,OR\,E_{ol}) \\
&= p_{os}.[1_{ps}.p_1.p_2.(p_{12}+p_{1i}.p_{2i})+(1-1_{ps})]+p_{ol}
\end{aligned}
\tag{3.20}
$$

Step Two (see Figure 3.6 and Table 3.5) undergoes the same limitations as Step One regarding the description of the initiator. It only refines the risk analysis by a more comprehensive account of the protection system's reliability through a simple fault tree analysis. Component failures are estimated only through expert judgement without any phenomenological model of the 'mechanics' of failure, nor any real statistical estimation of the failure rates. Although the example is very simple, Step Two may be understood as standard practice in the field of QRA (see Chapter 1).

Table 3.5 Step Two – summary characteristics.

	Step Two	Comments
system model	*Systems reliability* $G(.) = G_s(e_{os}, e_{ol}, \boldsymbol{e}, 1_{ps})$ $(e_{os}(s), e_{ol}(s)) = (1_{0<s<2}, 1_{2<s})$ $s = z_d - h - z_v$	Classical QRA model: mixes a fault tree and a simple initiator description
uncertain inputs $(\boldsymbol{x}, \boldsymbol{e})$	variables $(\boldsymbol{x})$: h, z_v events $(\boldsymbol{e})$: $l_1, l_2, v_{12}, v_{1i}, v_{2i}$	
decision variables $(\boldsymbol{d})$	h_d, 1_{ps}	
variables/events of interest (z)	e_p	
risk measure c_z	$P(E_p)$	
Uncertainty model	A pdf is estimated for H. A fixed value $(z_{v\,pn})$ is taken for Z_v. Bernoulli variables represent $\boldsymbol{e}$. No epistemic uncertainty $(\pi = \delta_\theta)$: θ_X has a single value $(\boldsymbol{\theta}_H, z_{v\,pn}, p_1, p_2, p_{12}, p_{1i}, p_{2i})$	Estimation involves simple statistics on H and expertise on Z_v and the input events $(\boldsymbol{e})$ probabilities and dependences (common modes).
Propagation (computation of c_z)	Closed-form solution	

Table 3.6 Step Two – Main results.

Risk assessment (*including protection system*)				
	No dike	$h_d = 1m$	$h_d = 2m$	$h_d = 3m$
Empiric model	2 10–2	?? > 6 10^{-3}	?? > 2 10^{-3}	???
Best fit statistical (logn)	2 10^{-2}	8 10^{-3}	3 10^{-3}	1 10^{-3}
Design (*i.e. dike heights*)				
	1/50	1/100	1/1000	
Empiric model	no dike	?? 0.5	???	
Best fit statistical (logn)	0.1	0.8	3.2	

Note that the adjunction of the protection system reduces the risk by a factor 3 or one metre less on the height of the required dike to meet a 1/1000-yr protection (see Table 3.6).

3.2.3 Step three: Uncertainty treatment of a physical/engineering model (SRA)

Step Three improves the description of the initiator in order to remedy the limitations mentioned for the earlier steps. Instead of estimating the water level through an imperfect procedure that oversimplifies the variability (through a fixed riverbed elevation), more reliable records are used for the estimation of other uncertain inputs (notably flow) from which the water level may be derived through the hydro model. To start simply, Step Three is limited to the prediction of the simplified risk measure already used in Step One: $c_Z = P(E_o) = P(S = Z_c - z_d > 0)$.

Although Step Three enables the building of a full probabilistic uncertainty model for the uncertain inputs, practice sometimes relies on mixed deterministic-probabilistic models, such as the association of a pdf for Q and of deterministically-penalised values for the other inputs. Chapter 4 will discuss the resulting interpretation of the various risk measures (more or less *conditional* probabilities) and the impact on the design margins. It is essential to note also that the computation of the risk measure is impacted on by that choice. As mentioned in Chapter 1, if only one input is probabilised (initiator-frequency approach), a single computation of the hydro component is generally enough. When more than one input is probabilised (extended frequency approach) a full-scale probabilistic propagation such as Monte-Carlo sampling becomes necessary (see Figure 3.7 and Table 3.7). Unlike Step Two, the system model is no longer a fault tree-style Boolean combination and the integration of c_z no longer benefits from a closed-form (see Chapter 7 for details). The detailed result shown in Section 10.4 show the key impact of a relevant choice for input distribution shapes as in Step One. Moving from the statistical best fit retained in Table 3.8 (a Gumbel model for the flow) to alternative choices (a normal or lognormal model) would change the design by a few metres; modelling dependent inputs instead of independent random variables also changes the design, though to a lesser extent in the present case (less than 0.5m), see Section 10.4.

3.2.4 Step four: Mixing SRA and QRA

Step Four is a simple combination of Step Three and Step Two enabling the prediction of the complete risk measure $c_Z = P(E_p)$ while benefiting from the improved description of the initiator through the hydro model (see Figure 3.8 and Table 3.9).

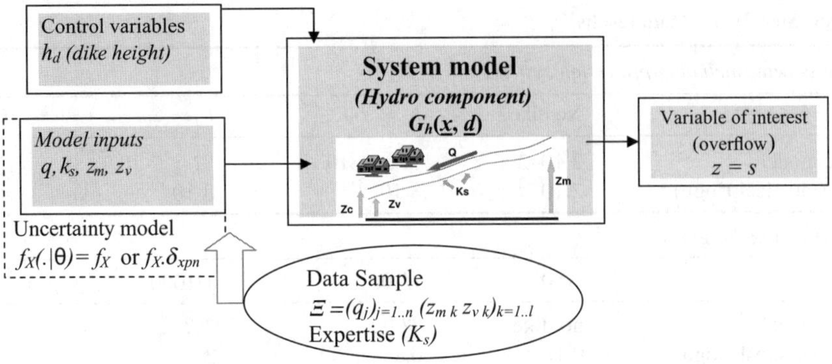

Figure 3.7 *Step Three modelling process.*

Technically, Step Four may be seen as a combination of SRA and QRA (Aven and Rettedal, 1998), sometimes referred to as IPRA. The computation of the risk measure could be undertaken theoretically by MCS over the entire input of variables and events. As will be discussed in Chapter 7, it is usefully replaced by a combined use of the closed-form integration for the QRA part and the standard probabilistic computation for the SRA part.

Note that, similar to comparing Step Two to Step One, the adjunction of the protection system reduces the risk by a factor of 2 to 3 or one metre less on the height of the required dike to meet a 1/1000-yr protection (see Table 3.10). The initiator/frequency approach combined with a deterministic model again proves very conservative with respect to full probabilistic uncertainty modelling.

Table 3.7 Step Three – summary characteristics.

	Step Three	Comments
system model	*Hydro* $G(.) = G_h(q,k_s,z_m,z_v,z_d)$	Equivalent in complexity to a classical SRA model
uncertain inputs ($\boldsymbol{x}$, $\boldsymbol{e}$)	variables ($\boldsymbol{x}$): q,k_s,z_m,z_v	according to choices, one to four of those inputs may be probabilised or deterministically penalised
decision variables ($\boldsymbol{d}$)	h_d	
variables/events of interest (z)	s, e_o	
risk measure c_z	$P(s > 0) = P(E_o)$	Risk measure becomes a conditional probability on the deterministically-fixed inputs if there are any.
Uncertainty model	A pdf is estimated for Q. For the other inputs, either fixed values are taken or pdf are modelled. Correlation coefficients can be included. $\boldsymbol{\theta}_X = (\boldsymbol{\theta}_q, \ldots)$	Estimation involves statistics and expertise on the uncertain inputs, including potential correlations.
Propagation (computation of c_z)	Standard probabilistic computation (MCS, etc.)	Involves multiple runs of the hydro model G_h

Table 3.8 Step Three – Main results.

Risk assessment				
	No dike	$h_d = 1m$	$h_d = 2m$	$h_d = 3m$
Proba. on flow + deterministic	42%	19%	6%	2%
Full probabilistic	1.2%	0.35%	0.13%	0.06%

Design (*i.e. dike heights*)			
	1/50	1/100	1/1000
Proba. on flow + deterministic	3m	*3.6*	*5.2*
Full probabilistic	*no dike*	*0.1*	*2.3*

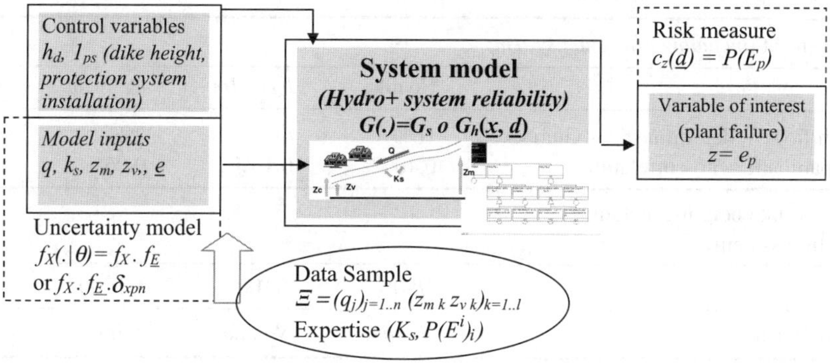

Figure 3.8 *Step Four modelling process.*

Table 3.9 Step Four – summary characteristics.

	Step Four	Comments
system model	*Hydro + Systems Reliability* $G(.) = G_s o G_h$	Equivalent in complexity to a classical SRA model
uncertain inputs ($\boldsymbol{x}$, $\boldsymbol{e}$)	variables ($\boldsymbol{x}$): q, k_s, z_m, z_v events ($\boldsymbol{e}$): $l_1, l_2, v_{12}, v_{1i}, v_{2i}$	according to choices, one to four of those input variables may be probabilised or deterministically penalised
decision variables ($\boldsymbol{d}$)	h_d, l_{ps}	
variables/events of interest (z)	e_p	
risk measure c_z	$P(E_p)$	Risk measure becomes a conditional probability on the deterministically-fixed inputs if there are any.

(*continued*)

Table 3.9 (*Continued*)

	Step Four	Comments
Uncertainty model	A pdf is estimated for Q. For the other input variables, either fixed values are taken or pdf are modelled. Bernoulli variables represent input events $\boldsymbol{e}$. $\boldsymbol{\theta}_X = (\boldsymbol{\theta}_q, \ldots, p_1, p_2, p_{12}, p_{1i}, p_{2i})$	Estimation involves statistics and expertise on the uncertain inputs, including potential correlations between variables and common mode failures between events.
Propagation (computation of c_z)	Mixes Standard probabilistic computation (MCS, etc.) for S and closed-form for $E_p \mid s$	Involves multiple runs of the hydro model G_h

Table 3.10 Step Four – Main results.

Risk assessment (*including protection system*)				
	No dike	$h_d = 1m$	$h_d = 2m$	$h_d = 3m$
Proba. on flow + deterministic – Gumbel	16%	6%	2%	0.6%
Full probabilistic – no correlation	0.40%	0.13%	0.06%	0.03%
Design (i.e. dike heights, including protection system)				
	1/50	1/100	1/1000	
Full probabilistic	*No dike*	*No dike*	*1.3m*	

3.2.5 Step five: Level-2 uncertainty study on mixed SRA-QRA model

Step Four has already developed a rather rich description of the risk situation, mixing a phenomenological model of the flood process (SRA-style) with an analytical description of the protection system's reliability (QRA-style). However, owing to the limitations of expertise and data samples, a great deal of uncertainty may affect the estimation of the input model, such as expert disagreement on the reliability of elementary components or large confidence intervals in the estimation of the pdf parameters of extreme flows. Fixed values were assumed for the previous steps which lacked any description of epistemic uncertainty.

Step Five consists of adding a description of those estimation uncertainties within a level-2 uncertainty model (see Figure 3.9 and Table 3.11). The system model is unchanged, as is the case with its inputs and outputs. The principal change affects the uncertainty model, and hence the risk measure. A level-2 uncertainty model is set to quantify the uncertainty in $\boldsymbol{\theta}_X$ either through a deterministic range or through a pdf. Both are estimated on the basis of expertise or estimation confidence intervals. The previous risk measure $c_Z = P(E_p)$ now gives multiple possible values out of which one may select a new risk measure being either the expectation over level-2 or a range of values: there will be more discussion about that delicate choice in Chapter 4. Accordingly, multiple computations of the Step Four level-1 risk measure with varying values for $\boldsymbol{\theta}_X$ become necessary, although some shortcuts may be found, as will be discussed in Chapter 7.

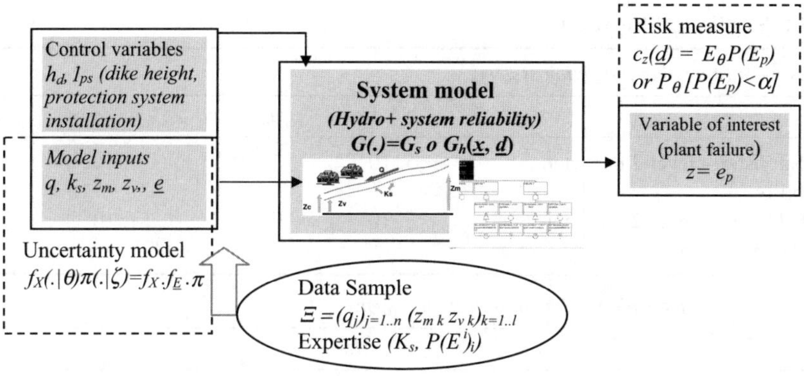

Figure 3.9 *Step Five modelling process.*

Summary results are given in Table 3.12 while Section 10.4 details the computations. Input distributions are taken with the best fit for flow (Gumbel), neglecting input dependence; design choices are compared with or without the protection system while the risk assessment always includes it. Results broadly evidence that the inclusion of epistemic uncertainty changes the design only slightly in respect of

Table 3.11 Step Five – summary characteristics.

	Step Five	Comments
system model	*Hydro + Systems Reliability* $G(.) = G_s o G_h$	Identical to Step Four
uncertain inputs $(\boldsymbol{x}, \boldsymbol{e})$	variables $(\boldsymbol{x})$: q, k_s, z_m, z_v events $(\boldsymbol{e})$: l_1, $l_2, v_{12}, v_{1i}, v_{2i}$	
decision variables $(\boldsymbol{d})$	h_d, 1_{ps}	
variables/events of interest (z)	e_p	
risk measure c_z	$E_{\boldsymbol{\theta}}(P(E_p \mid \boldsymbol{\theta}))$ *or* $[P(E_p \mid \boldsymbol{\theta}_{min}), P(E_p \mid \boldsymbol{\theta}_{max})]$ *or c.i. on* $P(E_p \mid \Theta)$	Multiple risk measures according to choices on level-2 uncertainty model
Uncertainty model (level-1)	Identical to Step Four $\boldsymbol{\theta}_X = (\boldsymbol{\theta}_Q, \ldots p_1, p_2, p_{12}, p_{1i}, p_{2i})$	Identical to Step Four
Uncertainty model (level-2)	Deterministic: ranges are set for each component of $\boldsymbol{\theta}_X$. Probabilistic: pdf (Gaussian) are modelled for each component of $\boldsymbol{\theta}_X$ within the epistemic distribution $\pi(\cdot \mid \zeta)$.	Ranges are taken from 5%–95% confidence bounds on estimation uncertainty when there is data, or otherwise from expert bounds. Gaussian pdf are modelled through equalling 5–95% quantiles to the ranges.
Propagation (computation of c_z)	Incorporates level-1 computation (mixed standard probabilistic computation and closed-form) within deterministic bound computation or probabilistic simulation for level-2	Involves multiple runs of the hydro model G_h

Table 3.12 Step Five – Main results.

Risk assessment (*including protection system*)

	No dike	$h_d = 1m$	$h_d = 2m$	$h_d = 3m$
Full probabilistic – level-2 expected probability	$5.\ 10^{-3}$	$2.\ 10^{-3}$	$7.\ 10^{-4}$	$3.\ 10^{-4}$
Full probabilistic – level-2 85% epistemic quantile on probability	$8.\ 10^{-3}$	$3.\ 10^{-3}$	$1.\ 10^{-3}$	$5.\ 10^{-4}$

Design (*i.e. dike heights, with or without protection system*)

	1/50	1/100	1/1000
Full probabilistic – level-2 expected probability (*resp. with protection system*)	No dike (*No dike*)	0.3m (*No dike*)	2.5m (*~1.6m*)
Full probabilistic – level-2 85% epistemic quantile on probability (*resp. with protection system*)	No dike (*No dike*)	0.6m (*No dike*)	3.0m (*2.2m*)

expected values ($+0.1 - 0.2m$ of dike, failure probability increase by factor ~ 1.2); a level-2 risk measure involving the upper bound of a *70 %*-confidence interval does add an additional margin of $+0.4 - 0.7m$ of dike height or increases by a factor ~ 1.5 the probability of failure for a given dike level. By far, this is less than the margin brought by a simplified deterministic-probabilistic 'initiator-frequency approach' (see Step Three).

Similar to what happened with Step Four, installing a protection system is equivalent to $\sim 0.8 - 1m$ of additional dike level.

3.2.6 Step six: Calibration of the hydro component and updating of risk measure

Step Six stands apart from the others: it consists of calibrating the hydro component in order to later refine the risk measure inference. As will be discussed in detail in Chapter 6, it may be interpreted in several ways. One may consider it as a necessary initial step of model calibration and validation before using it within a risk analysis. This may alternatively be viewed as a means of refining an initial study in an advanced 'data assimilation' step: incorporate more data within the model and refine (ideally reduce) the uncertainty model that was initially based purely on possibly-vague expertise.

Step Six involves two sub-steps. The first sub-step consists of the proper *calibration* or *identification* phase. A new data sample is incorporated so as to modify the uncertainty model. However, unlike the previous steps, the estimation is not performed directly through a straightforward statistical distribution fitting. If data is regarded as describing model outputs rather than uncertain inputs, its incorporation involves the inversion of the hydro model (see Figure 3.10 and Table 3.13). For instance, simultaneous records $(q_j, h_{mj})_{j=1\ldots n}$ of yearly maxima for flow and corresponding water depth will be used for the estimation of the distribution of K_s, the friction coefficient of uncertainty (and/or of riverbed elevation uncertainty) through the part of the hydro model predicting $h = H(q, k_s, z_m, z_v)$. The technical details may be found in Chapter 6, requiring typically a special type of regression model and ending up with an updated uncertainty model (see Table 3.14).

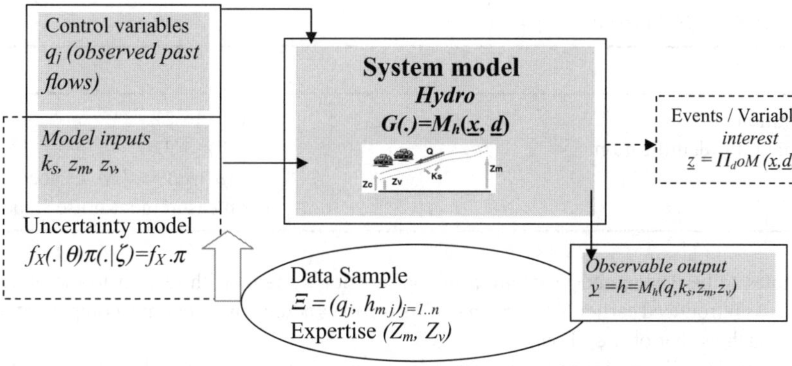

Figure 3.10 *Step Six modelling process – estimation.*

The second sub-step consists of *updating* the *risk measure* on the basis of this new uncertainty model. It is identical to that of Steps Three, Four or Five depending on whether one looks for the overflow or plant failure risk measure, and whether a level-2 model is included within the calibration.

3.2.7 Step seven: Economic assessment and optimisation under risk and/or uncertainty

Step Seven is designed to answer the cost benefit optimisation question. Beyond fulfilling a regulatory criterion for plant failure, the operator may want to investigate the optimal level of investment in the

Table 3.13 Step Six – summary characteristics.

	Step Six	Comments
system model	(for sub-step identification) Hydro sub-model $h = H(q,k_s,z_m,z_v)$ (for sub-step updating) – similar to steps 3, 4 or 5	Only a part of the hydro model is used, in order to predict the *observable* variables of interest for which data is available.
uncertain inputs ($\boldsymbol{x}$, $\boldsymbol{e}$) decision variables ($\boldsymbol{d}$) variables/events of interest (z) risk measure c_z	Similar to Step 3, 4 or 5	
Uncertainty model	(for sub-step identification) On the basis of an existing uncertainty model, new values are estimated for some of the components of $\boldsymbol{\theta}_X$ (for sub-step updating) – similar to steps 3, 4 or 5	Estimation involves an elaborate mix of statistical and numerical (inverse) algorithms
Propagation (computation of c_z)	Similar to Step 3, 4 or 5	

Table 3.14 Step Six – Main results

	K_s
Model calibration	$E(K_s) = 27$
Intrinsic variability identification	$E(K_s) = 27$ $var(K_s) = 7$ *to* 13 according to measurement noise hypotheses

control variables (dike design and installation of a protection system) with respect to the anticipated damage costs. As introduced earlier in this chapter, this involves adding the economic component into the other models, such as that of Step Four.

The new variables of interest are now the total cost c_c, an associated catastrophic event e_c or utility u_c. Various associated risk measures are possible, and will be investigated in Chapter 8. Estimation of the uncertainty model is similar to that of Step Four except that additional expertise is needed to work out the damage cost uncertainty for which data is unlikely to be available. That last uncertain input is likely to be independent of the other sources of uncertainty. Computation resembles that of Step Four insofar as one is only interested in computing the cost consequences for one given design (see Figure 3.11 and Table 3.15). Switching to design optimisation, such as minimising expected total cost or maximising expected utility, greatly complicates the computation tasks (see Chapter 8).

The table below provides some key results (without the protection system for the sake of simplicity). Minimising the total cost over 30 years of operation under risk in *expected value* ends up with a *3.5m* dike, but the residual cost as evidenced either by the *99 %*-value at risk VaR_α or the expected shortfall ES_α remain high. Other criteria (e.g. expected utility maximisation for a risk-averse profile) could lead to a different compromise along the Pareto front between expectation and standard deviation minimisation, resulting in a higher dike of *5–6m*. Chapter 8 will also show that the inclusion of discounting over time could possibly lower the recommended optimum by *0,5–1m* for increasing rates, ending up with an optimum between *2.5m* and *5m*. This is higher than the *1/1000*-return period designs (without a comparable protection system), illustrating the fact that a cost-benefit analysis could lead the operator to build a higher/safer design than the basic safety regulatory requirement.

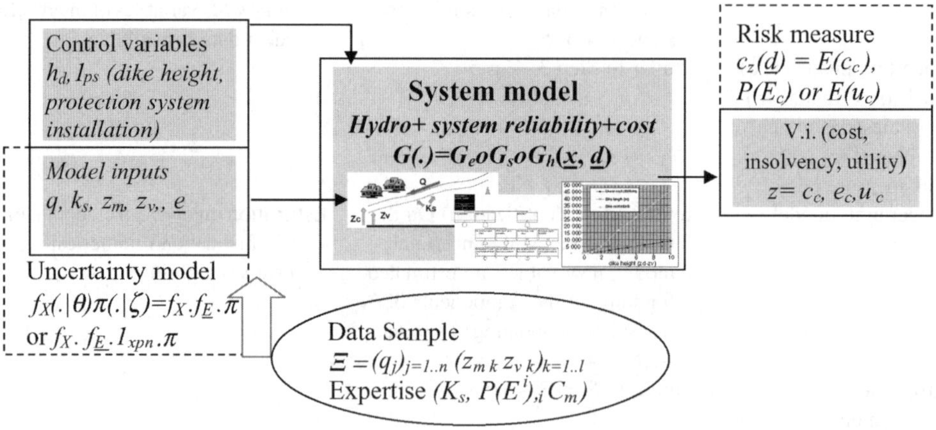

Figure 3.11 *Step Seven modelling process.*

Table 3.15 Step Seven – summary characteristics.

	Step Seven	Comments
system model	*Hydro + Systems Reliability + Economic G* (.) $= G_e o G_s o G_h$	Fully-chained risk analysis, from initiator phenomenology to utility consequences
uncertain inputs ($\boldsymbol{x}$, $\boldsymbol{e}$)	variables ($\boldsymbol{x}$): q, k_s, z_m, z_v, c_m events ($\boldsymbol{e}$): $l_1, l_2, v_{12}, v_{1i}, v_{2i}$	
decision variables ($\boldsymbol{d}$)	h_d, 1_{ps}	
variables/events of interest (z)	c_c, e_c, u_c	
risk measure c_z	$E(c_c)$, $P(e_c)$, $E(u_c)$	
Uncertainty model	Similar to Step Four + pdf on $c_m \boldsymbol{\theta}_X = (\boldsymbol{\theta}_q, \ldots, p_1, p_2, p_{12}, p_{1i}, p_{2i}, \boldsymbol{\theta}_C)$	Similar to Step Four + expertise on damage cost uncertainty (c_m)
Propagation (computation of c_z)	Incorporates level-1 computation (mixed standard probabilistic computation and closed-form) within	Optimisation renders it much more computationally-intensive

Table 3.16 Step Seven – Main results.

C_c/T *in M€/yr* (no discounting)	Dike height			
	1m	3,5m	5m	8m
Expectation	*3.4*	*1.2*	*1.2*	*1.6*
Standard Deviation	14	6	4	2
$P(C_c >$ *fixed cost*$)$	*12%*	*1.6%*	*0.5%*	*0.1%*
VaR_α (ES_α) with $\alpha = 90\%$	*0.8 (34)*	*0.6 (5)*	*0.9 (3)*	*1.6 (2)*
VaR_α (ES_α) with $\alpha = 99\%$	*81 (94)*	*3 (48)*	*0.9 (22)*	*1.6 (8)*

Similar to Step Five, a level-2 uncertainty study of the economic consequences is theoretically possible. Yet it has not been included in the results remembering that, in complex systems beyond the pedagogical example exhibited, a Step Seven based on a single probabilistic setting may already involve rather expensive studies (see Table 3.16).

3.3 Summary of the example

A sequence of steps has been proposed in order to help with understanding the embedding of various statistical and numerical modelling methods within consistent studies though real studies would not necessarily follow such a schematic process. The table below identifies the chapters that provide the main illustrations of the various steps (see Table 3.17).

Table 3.17 Summary characteristics of the various steps.

	Main features	Links to classical risk/uncertainty practices	Main methods illustrated	Chapter providing numerical details
Step One	Simple natural risk estimation	Initiator-frequency approach in NRA	Direct Estimation	Chapter 5
Step Two	Mixing statistics and a QRA model	QRA		Chapter 5
Step Three	Uncertainty on a phenomenological model (SRA)	Extended frequency approach in NRA, or SRA	Direct estimation (for flow) and Risk computation	Chapter 5 (estimation) and Chapter 7 (computation)
Step Four	Mixing QRA and SRA	IPRA	Risk computation (case of mixed continuous/discrete uncertain inputs)	Chapter 7
Step Five	Level-2 uncertainty study on a mixed QRA-SRA	IPRA distinguishing aleatory/epistemic	Risk computation (case of double-level uncertainty model)	Chapter 7
Step Six	Calibration of hydro component and updating or risk measure	NCQ	Inverse calibration and updating	Chapter 6
Step Seven	Economic assessment and optimisation under risk/uncertainty	Project analysis, finance and engineering economics	Risk computation and optimisation	Chapter 8

Exercises

The following is a list of exercises referring to the concepts developed in this chapter:

3.1 Study the monotony of a hydraulic component involving steady-flow non-uniform St-Venant equations. Discuss the appropriateness of a worst-case approach to the associated hydrologic and hydraulic parameters.

3.2 Discuss the model structure of other natural risk phenomena, such as storm damage, hail, waves and tsunami, earthquake: does it resemble that of the flood example? How would that impact upon the corresponding Steps One to Seven?

3.3 Suggest a finer model predicting the non-activation of the protection system as a function of the level of overspill and subsequent physical mechanisms: how would it change the process of Step Four? What would such a finer model bring to the study process?

References

Aven, T. and Rettedal, W. (1998) Bayesian frameworks for integrating QRA and SRA methods. *Structural Safety*, **20**, 155–165.

de Rocquigny, E. (2004) Tutorial 'Incertitudes'. 14th Lambda-Mu Conference, Bourges, France.

de Rocquigny, E. (2006) La maîtrise des incertitudes dans un contexte industriel: 1ère partie – une approche méthodologique globale basée sur des exemples; 2nd partie – revue des méthodes de modélisation statistique, physique et numérique. *Journal de la Société Française de Statistique*, **147** (4), 33–106.

Goutal, N., Arnaud, A., Dugachard, M. *et al.* (2008) Discharge and strickler coefficient uncertainty propagation in a one-dimensional free surface hydraulic model. *Geophysical Research Abstracts*, **10**, EGU2008-A-11844.

Lencastre, A. (1999) *Hydraulique Générale*, Eyrolles.

Limbourg, P., and de Rocquigny, E. (2010) Uncertainty analysis using Evidence theory – confronting level-1 and level-2 approaches with data availability and computational constraints, submitted to R.E.S.S.

4

Understanding natures of uncertainty, risk margins and time bases for probabilistic decision-making

In order to guide the key choice of the quantity of interest or risk measure, which is the first step in a risk/uncertainty-conscious modelling approach, this chapter discusses in more depth the large variety of choices which may be encountered in practice. These are *probabilities* (or frequencies) of significant deviations or undesired events, quantiles; more complex level-2 measures (e.g. confidence intervals on failure probabilities); utility expectations; variance of a variable of interest; and any of those probabilistic quantities *conditional on deterministic* values (or scenarios) for some inputs. The chapter discusses their meaning and properties in more detail, as they:

- may be used to cover differing natures and sources of uncertainty such as intrinsic variability or aleatory randomness, lack of knowledge, imprecision, model inaccuracy, and disagreement between experts (Section 4.1)
- involve differing forms of accounting for the corresponding *margins*, in particular given the pervasiveness of the monotonous behaviour of the systems at risk (Section 4.2)
- provide different interpretations regarding temporal aspects, being based on what are often implicit stochastic time series; extreme value theory is introduced in that respect (Section 4.3)
- display different decision theory properties that prove more or less helpful according to the salient goals of the study, as defined in Chapter 2 (Understand, Accredit, Select or Comply) (Section 4.4)

Modelling Under Risk and Uncertainty: An Introduction to Statistical, Phenomenological and Computational Methods, First Edition. Etienne de Rocquigny.

4.1 Natures of uncertainty: Theoretical debates and practical implementation

4.1.1 Defining uncertainty – ambiguity about the reference

Uncertainty is such a pervasive concept that any attempt to give a clear-cut and complete definition may be presumptuous. Indeed, the subject spans a wide range of scientific and practical knowledge. From the narrower *quantitative* perspective of the book, Chapter 2 delineated that uncertainty is viewed as the extent of imperfect knowledge of the state $(\boldsymbol{x}, \boldsymbol{z})$ of a system conditional on taking given actions $(\boldsymbol{d})$ at a given time (or period of time). It is quantified mathematically through the risk measure or quantity of interest over the possible or likely values of the output(s) of interest.

In essence, modelling under uncertainty will try to predict likely deviations from a predicted state or likely subsets of the space of possible states. A tempting interpretation of such a broad definition could be: 'when the appropriate time (or period of time) comes up, the *true realisation* of the state of the system should fall within the extent predicted by the risk measure'. Note that such a definition presupposes implicitly the existence and observability of a *true* value for the state of the system. This is why a first condition, when undertaking uncertainty assessment, is to specify the system and output of interest considered carefully so as to avoid confusing it with *ambiguity*.

However, this is *problematic* to some extent. On the one hand, the *specification* or definition of any physical or economic quantity to be measured is *never infinitely precise*. At a micro-physical scale, external artefacts (such as background radiation, physical-chemical surface exchanges, ...) or even thermal Brownian motion of the molecular sub-structure constantly impact upon the length of a given rod in a small time scale or even blur the notion of frontier as a basis for the definition of length. Much the same could be said about economic variables characterising system performance which are sensitive to short-term micro-variations such as hour-to-hour rate fluctuations or micro-accounting movements. On the other hand, should the quantity be defined very precisely, the true value is never observable but only approached by sensors based on metrological benchmarks that are themselves uncertain. Zooming further down into quantum mechanics, the Heisenberg uncertainty principle establishes that any physical system is intrinsically uncertain, and can be disturbed by observational devices (see Annex Section 10.3).

In spite of such difficulties in handling a true value, the notion of a reference value for the state of the system remains a necessary definition, particularly in the metrological domain where quantities of interest show deviations between measurement results and reference values. The Guide for the Expression of Uncertainty of Measurement (GUM) recommends referring by preference to 'the best value available' (as obtained by reference benchmarks with very stable results) rather than any 'true value' of a reference value. Accordingly, the guide advocates eschewing the so-called '*error*' defined as the absolute deviation between measurement result and (unobservable) true value; it is better to focus on '*uncertainty*' defined as the fluctuation of the result of a measurement that is supposed to have reasonably corrected all the systematic effects[1] as compared to a reference benchmark. Such an interpretation will be retained throughout this book beyond the metrological domain, assuming that, on the basis of an adequate *specification step* in the modelling process, there exists a *conventional reference* to which the state of the system may be related.

---

[1] Note that the random variable representing such uncertainty may include a 'bias' (which therefore remains relative) if its expectation value is not zero; however, most metrologists would reserve the term 'uncertainty' for the random variable of expectation zero, viz. restricted to fluctuation after deduction of any bias.

A large variety of causes or considerations give rise to uncertainty in the state of a system as defined above, such as:

- natural variability of the environment, notably that of weather and climate,
- natural variability of life, including that of human sensitivity,
- variability of manufacturing and operating conditions in the industrial processes,
- variability of human behaviour (be it as a customer, operator, stakeholder, civilian . . .),
- subsequent uncertain inputs and initial conditions of natural or man-made systems,
- metrological errors,
- model inaccuracy (simplified equations, numerical errors, etc.),
- conflicting expert views,
- and so on.

Thoughout the book the focus is on acknowledging and modelling the extent of uncertainty for the purpose of decision-making, Annex Section 10.3 provides some introductory reflections about the microscopic *sources* and phenomena leading to macroscopic uncertainty.

4.1.2 Risk vs. uncertainty – an impractical distinction

A rich literature has described the variety of natures of uncertainty and discussed the key issue of whether they should or could receive the same type of quantification efforts, in particular as a probabilistic representation. Such a debate may be traced back to the early theory of probability rooted in the 17th century while modern thinking may be inherited originally from *economics* and *decision theory*. Both Knight (1921) and Keynes (1921) introduced the famous distinction between:

- '*risk*', defined as unpredictability with known probabilities,
- and '*uncertainty*', defined as unpredictability with unknown, imprecise probabilities or not even subject to probabilisation.

From an economic perspective, the *former* category is typically amenable to insurance. In other words, it may be reduced through portfolio mutualisation, the premium price being linked directly to the (known) probability of the event requiring coverage. Conversely, the *latter* is theoretically un-insurable as the premium cannot be computed clearly. Uncertainty is also related to the possibility of innovation.

It is often forgotten how these early works admitted the subtleties and limitations incorporated in such simplified interpretations regarding real physical systems. Indeed, the economic and decision theory literature has generally been restricted to simple decision settings, such as closed-form quantitative lotteries, without tackling in detail the physical bases of industrial systems. A closer look shows that various forms of uncertainty, imprecision, variability, randomness and model errors are mixed up with phenomenological data and modelling. Think, for instance, about riverbed roughness for which the topography at any time is generally incomplete. It varies significantly in time during or after flood events, and is modelled incompletely within simplified hydrodynamics and the associated numerical riverbed mesh. Knowledge of such systems is seldom as binary as when incorporated in such statements as 'probabilities are known' vs. 'unknown'. In fact, it is quite rare to handle physical systems for which probabilities are 'well-known'. Even for phenomena that are classically considered to be 'random-like' such as weather variables, datasets are too small to estimate unambiguous probabilistic models, or else stationarity becomes questionable when they become large enough. Unlike idealised lotteries, the frequencies of given states of physical systems are only partially known and it is never even certain, for lack of observations, that one is sampling repetitively and independently from an identical population.

The extent to which climate mechanics (which some liken to deterministic chaos) satisfy Kolmogorov's axioms probably remains an unanswerable question.

Hence the word *uncertainty* will be understood to a wider extent than that of the famous economic distinction, covering any situation with *unpredictability* of the outputs of interest. Section 4.4 will discuss in more detail the fact that a *probabilistic risk measure* may be seen more as a *modelling choice* or representation of beliefs suited to a particular decision-making purpose than a reflection of the properties of nature. This book therefore avoids speaking about 'stochastic uncertainty' as differentiated *in nature* from other types of unpredictability.

4.1.3 The aleatory/epistemic distinction and the issue of reducibility

An important debate When tackling real-scale physical systems, the variety of natures of uncertainty has been greatly discussed within the *risk assessment* community in the 1980s–90s in close connection to the development of US nuclear safety reviews or environmental impact assessments (Granger Morgan and Henrion, 1990; Helton, 1994). The debate has notably concerned the classification of the large variety of uncertainty encountered in large industrial systems into two salient categories, and the rationale for separating or not those two categories: namely the *epistemic* (or *reducible*, lack of knowledge, ignorance, imprecision, subjective...) type referring to uncertainty that decreases with the injection of more data, modelling efforts, expertise or number of runs; and the *aleatory* (or *irreducible*, intrinsic, random, objective...) type for which there is a variation of the true characteristics that may not be reduced by the increase of data or knowledge.

Accordingly, some authors link the choice of particular mathematical settings to the different natures of uncertainty involved in the system studied. The framework of a double-layer (aleatory + epistemic) probabilistic uncertainty model, as introduced in Chapter 2, was one of the early popular tools emerging from the aleatory/epistemic debate (Helton and Burmaster, 1996). Other authors have denied the use of probabilistic treatment for epistemic uncertainty and reserved it exclusively for aleatory uncertainty that is supposed to be 'naturally' random, thus renewing the early distinctions made by the economic literature. Non-probabilistic paradigms such as the Dempster-Shafer theory may be advocated instead. More simply even, deterministic coverage of some epistemic sources (coupled with a probabilistic description of aleatory uncertainty) may be preferred in some cases, as shown in Chapter 1. This is the case with natural risk where the data limitations to estimate rare quantiles may be covered by, for example, a 15 % margin over extreme flows, or in metrology where variances of critical sensors may be deliberately inflated (penalised) so as to cover insufficient calibration.

Conversely, the aleatory/epistemic distinction is considered to be impractical or irrelevant for other practical studies. Measurement uncertainty is an example of a practical source within which it is hard to discern epistemic and aleatory components. The ISO (1995) international standard has in fact removed the reference to the epistemic/aleatory structure as a mandatory distinction in analyses and treatments, which therefore receive a common probabilistic sampling approach. Similarly, major developments have applied the impact of uncertainty to complex numerical models with the help of functional and numerical analysis. Statistical computer science (design of experiment) under the name of *sensitivity analysis* (Cacuci *et al.*, 1980) initiated the exploration of that impact, and major developments have occurred since with an extension to global analysis (Saltelli *et al.*, 2004) within which single probabilistic settings are used systematically whatever the nature of uncertainty involved.

The book approach – distinguishing carefully the various types of aleatory/variability and epistemic uncertainty in a mixed setting It is important to keep in mind from the debate summarised above that there are reasonable pros and cons for statements which question essentially the way available expertise and data may be encoded in a probabilistic approach and the way in which decision-making can at a practical level handle risk measures (Helton and Oberkampf, 2004; de Rocquigny, Devictor, and Tarantola, 2008; Limbourg and de Rocquigny, 2010). However, this book favours in general the use of a double probabilistic uncertainty model including a dedicated description of the aleatory and epistemic layers, mixed possibly with partial deterministic conditioning, as introduced in Chapter 2.

Adopting a probabilistic approach calls for a careful definition of the aleatory/epistemic distinction. Remember that, as already discussed, there is always *epistemic uncertainty* whenever *aleatory phenomena* are modelled in physical systems. It is important to distinguish between two main types:

- **Estimation uncertainty** is the first type of epistemic uncertainty that arises from the limitations in the information made available to estimate the probabilistic models, be it the limited size of a data sample, expert discrepancies, or metrological errors in the observations. This impacts upon the plausible values for the statistical *parameters* but more importantly even upon the selection of the appropriate *model type* (distribution shapes, or the type of system model more appropriate).
- **Propagation uncertainty** is the second type of epistemic uncertainty that arises from the limitations in the computational process required to find the final risk measures once the system and uncertainty model have been estimated. This includes not only the limited size of runs used by sampling algorithms, but also the numerical errors of the non-probabilistic propagation methods or the limitations of solvers and of the mesh sophistication of the system model itself.

There are essential differences regarding firstly the type of actions that can be taken to reduce them: the acquisition of more data, finer observations or expertise for the former; a larger computational budget for the latter. Consequently, the second probabilistic layer representing epistemic uncertainty would *not* mix the two types of uncertainty. It is essential for regulatory studies to specify clearly what should be covered by the level of confidence required by the target on top of the risk measure (e.g. 1/1000-yr flood with 70 % confidence): either estimation uncertainty, in the most frequent case; or propagation uncertainty, a much rarer case seen, for instance, for the control of loss-of-coolant accidents in the nuclear industry (see Section 7.3). Though theoretically imaginable, three-level risk measures gathering two levels of confidence for the two kinds of epistemic uncertainty have never been encountered by the author. A comprehensive review should also mention the trickier case of advanced model building. The two kinds of uncertainty might become intertwined through an inverse algorithm requiring uncertainty propagation for the purpose of estimation (see Chapter 6).

Reducibility and the study process For decision-making, an essential view of the distinction is to think about the level of *reducibility* of different types of uncertainty: view the *epistemic* type as the *reducible* part. When considering flow uncertainty, for instance, the complexity of long-term meteorological forecasts would prevent for a long time (if not forever) the ability to predict the level and the future date of an extreme flood, leaving *irreducible* uncertainty in flood flow, a classic example of the aleatory phenomenon. Assessing it through probabilistic distribution then leaves an amount of *estimation* epistemic uncertainty which is reducible in theory, although its *practical reducibility* is often questionable. Acquiring new data on large floods may be an attractive option for reducing margins, though in some cases it may prove incompatible with the planning of a project. One could contemplate the recovery of regional hydrological data, on the premise that some phenomenological similarity could allow for data completion with neighbouring rivers. This could involve costly modelling research. Uncertainty in other contributing variables such as friction might prove easier to reduce to some extent, through improved model calibration, for instance, or by elicitating expertise. *Propagation uncertainty* resulting from computational limitations should generally be viewed as practically reducible as the computational costs have generally become less than the costs of getting more data and observations. Thus, comes understanding of the epistemic nature as reducibility brings essential cost-benefit considerations into the process, which may be facilitated by a clear separation of the types of uncertainty at some point in the investigation.

At the end of a study process (which possibly included the acquisition of more data), both *epistemic and aleatory components mingle within the overall unpredictability* – in the eyes of the analyst – of the variable of interest *at a given time* of interest. Moving then on to the realisation of a well-specified system at a given future time of interest, and assuming that the variable of interest can be observed, one might consider that 'eventually, all uncertainty is epistemic'. This requires putting aside the cases where there are multiple future times of interest, multiple individuals or sub-systems described within the outputs of interest, over which target frequencies are defined. The next section discusses in more depth the issue of *variability* in the specification of the system model and decision-making.

4.1.4 Variability or uncertainty – the need for careful system specification

Distinguishing between 'variability' and 'uncertainty' is popular in environmental or health impact analysis (Hoffmann and Hammonds, 1994; Zheng and Frey, 2005) and has also been discussed in risk analysis. Such a distinction can be viewed as an extension of the previous one as there are more than one type of variability to be considered for some systems (mostly aleatory/irreducible but possibly also epistemic):

- Variability over *time* of a given variable of interest characterising the system (e.g. the river flow). This is the most classical example of aleatory uncertainty; it is sometimes necessary to further distinguish a temporal variability which is known or fairly predictable, such as a seasonal or periodic pattern, from the uncertain part of the time variation.
- Variability of some distributed characteristics of the system. This refers either to a *spatial* variation of some its properties (e.g. the friction properties and river bottom elevation along the length or across the width of the river) or to a variation of those properties throughout a *population* of individuals or sub-systems constituting the system (e.g. human vulnerability to a given pollutant inside a local population, mechanical properties within a fleet of parallel sub-systems, ...). Though being theoretically different from the lack of knowledge/epistemic uncertainty in the value of the property at a given point (or for a given individual) arising from measurement shortcomings, the variability of that property over the system could also stand as *epistemic* in some cases. A comprehensive data collection campaign could indeed turn the unknown variability into a known variation, thus raising again the *cost-benefit* issue of reducibility.

The proper handling of variability in decision-making is closely related to the *clear definition* of the *system* and *time of interest*, and *output variable of interest*, and is proved to depend essentially on the proper delineation of the risk measure. Consider, for instance, a system such as a single fuel rod in a nuclear power plant, the output of interest being its reference length z. Such a physical feature is affected by irradiation and thermal processes and consequently, an excessively-modified length may block the reloading of nuclear assemblies. It is therefore measured before reload and compared to an acceptable threshold, a process for which the metrological quantity of interest may be a 95 % quantile of length. The characteristics of the metrological device and its prior calibration generally secure knowledge of the sensor variance ${\sigma_m}^2$ and of its Gaussian behaviour. Thus, a simple computation of such a quantile or confidence interval can be done in order to cover measurement uncertainty around the raw result z_m. Therefore, when basing the comparison on the measurement result, the unknown reference rod length can be guaranteed to remain below the threshold z_s with 95 % likelihood.

Table 4.1 Mixing uncertainty and variability in a metrological process.

Output of interest	95 % compliance level for the maximal critical length	Comments
length of the measured fuel rod	$z_s - 1.6\sigma_m$	Variability is not taken into account – a loose criterion
length of 'one' fuel rod (i.e. randomly selected)	$z_s - 1.6\sqrt{(\sigma_m^2 + 2\sigma_v^2)}$	Variability is mixed with measurement uncertainty – a moderate criterion
length not exceeded by 95 fuel rods in average	$z_s - 1.6\left[\sigma_v + \sqrt{(\sigma_m^2 + \sigma_v^2)}\right]$	Variability is separately added onto measurement uncertainty – a tough criterion

Consider now a population of say 100 fuel rods of similar design. Inevitable manufacturing inconsistencies or dispersion of irradiation and thermal processes will surely mean that they end up with 100 different lengths. This means that if one could afford the separate and sufficiently accurate measurement of each of the 100 fuel rods, there would be an observable dispersion of the results well beyond the measurement uncertainty itself. Or, more realistically, there may be expert knowledge or modelling results helping to infer a reasonable order of magnitude of its variance ${\sigma_v}^2$. Such variability is understandably distinct from measurement uncertainty *in nature*. Yet, assuming that only one measurement result of a sampled fuel rod is available to assess the risk for the system of 100 rods, they become mixed together within the overall unpredictability of a given state of the system albeit very differently according to how it is defined. Table 4.1 illustrates three possible choices and the associated results for a quantity of interest that is always defined as a 95 % quantile. U, V_i all denote standard normal and independent variates representing respectively the metrological error for the single measurement and the rod length deviations, meaning the deviation of the 'true' lengths of each rod from the length of the $i = 1$ rod that has been measured. Additionally, all rods are assumed to have a length that is distributed with a common (unknown) mean and (known) variance, independently both from the other rods and from the measurement error; the latter is also assumed to be unbiased and of known variance.

Obviously, the extent of coverage associated with each risk measure differs notably. The last guarantees a comfortable likelihood that the whole population of rods would pass the acceptance threshold, while the first is obviously mis-representing the potential variability and thus there remains a significant possibility for some rods to be truly beyond the threshold and thus blocking reload of assemblies. The second mixes together the chances for measurement error or a potential deviating individual, ensuring a 95 % overall confidence level. Note that the experimental designs required to check such a performance of the criteria involve very different conditions. For the second case, the experimental device and the choice of rod that is measured should be sampled together repeatedly. In the third case, a compound experiment should test N_1 different experimental devices and, for each of those sample N_2 rods to be measured.

Comparable choices in aggregating uncertainty and variability could be used in environmental or health impact assessment (see Owen Hoffmann and Hammonds, 1994), given the biological variability within a population plus the uncertain dose-effect or bioaccumulation as inferred from tests undertaken on a given individual (the epistemic uncertainty corresponding to the measurement error of one rod). This could also be a component in a system with *spatially* distributed properties, for which one infers the likelihood of outputs of interest such as porosity for a soil. It could even refer to a *temporal* distribution such as the monthly effluent discharge into a river. Thus, it is very important to *specify the system and output variable of interest clearly* when there is an underlying 'population' involved.

4.1.5 Other distinctions

Finer distinctions include (Oberkampf *et al.*, 2002; de Rocquigny, 2006):

- *epistemic* uncertainty *vs. error*,[2] depending on whether the ignorance uncertainty is 'inevitable' or 'deliberate' in spite of the availability of knowledge;
- uncertainty that is *parametric*, associated with model inputs according to the level of information available on those inputs, *vs. model uncertainty*, affecting the adequacy in relation to reality of the model itself: structure, equations, discretisation, numerical resolution and so on.

The former distinction is again essentially a cost-benefit issue in practice. The best models, the most accurate numerical algorithms or highest-experienced experts, the richest datasets or the finest measurement tools may not be economical or practical for a given study. This depends on the acceptability of a certain limited degree of error, defined by the practitioner, to allow for the use of a simpler model. In this case, the degree of deliberate error will somehow be controlled by an additional margin; this may also include, more insidiously, the occurrence of involuntary study errors in spite of quality assurance. Of course, the meaning of the risk measure is not exactly the same when covering 'deliberate' ignorance or inaccessible knowledge. Correspondingly, it might be preferable for error to be covered in some cases by an overall deterministic margin without being sampled together with the other sources of uncertainty.

The latter distinction is quite instrumental in practice, once the system model is clearly defined; though the difference is not so clear-cut in theory, because it depends on the conventional definition of a 'model' and the status of model calibration or selection (Chapter 6). According to the physical-industrial domain considered, traditions differ as to the inclusion or not of model uncertainty alongside the uncertainty in 'parametric' inputs. It appears frequently in physical fields where calibration of meso-scale models is common, such as in fluid mechanics. To consider modelling parameters as uncertain inputs is commonly taken to cover both epistemic uncertainty regarding the un-modelled details of the phenomena involved (such as wavelets, micro-variability of riverbed conditions, etc.) and aleatory uncertainty, such as intrinsic turbulence structures, not predicted by average models. In other fields, such as risk analysis, there may be a great deal of theoretical controversy regarding the acceptability of including quantified model uncertainty (Nilsen and Aven, 2003).

Conclusions – handling various natures of uncertainty In concluding this section, there is always *a need for precise specification of the system and the output of interest* so as to clearly delineate the *natures of uncertainty covered by the risk measure*, in order to provide a patent interpretation of its content. It has been seen that in some cases one would use compound risk measures in order to separate those natures explicitly. It can be a double probabilistic separating the aleatory from the epistemic or variability across a population of sub-system characteristics from uncertainty in the individual properties of each; or mixed deterministic-probabilistic variants to cover separately numerical errors or epistemic sources of uncertainty within deterministic margins. Section 4.4 will return to the pros and cons of single versus double probabilistic settings in the light of the dominant goals and decision-making features of a model study. Nevertheless, it should always be useful in the course of a study to think about the types of uncertainty involved and their level of reducibility as a preliminary 'check-list'.

---

[2] According to the area considered, remember that 'error' may carry many other meanings than the one mentioned.

4.2 Understanding the impact on margins of deterministic vs. probabilistic formulations

Chapters 1 and 2 have shown the abundant use of mixed deterministic-probabilistic settings in natural or industrial risk, whereby some sources of uncertainty (e.g. epistemic, error,...) are 'covered' by the use of 'penalized' values, conditional on which probabilistic risk measures are built with the remaining uncertain inputs. This section will discuss in more depth the key properties that motivate such treatment implicitly, in order to clarify the associated assumptions and impacts on decision-making.

4.2.1 Understanding probabilistic averaging, dependence issues and deterministic maximisation and in the linear case

An essential feature of *probabilistic uncertainty modelling* is related to its fundamental *averaging effect*, that is error compensation, or risk mutualisation. To start with, a linear system model will clarify its meaning and help understand to what extent it is preferable (or not) to a deterministic alternative.

Error compensation in the linear model Consider the following linear system model predicting a scalar output of interest z from a given uncertain input vector $\boldsymbol{x}$ and set of actions $\boldsymbol{d}$:

$$z = G(\boldsymbol{x}, \boldsymbol{d}) = \sum_{i=1\ldots p} G^i(\boldsymbol{d})x^i + G^0(\boldsymbol{d}) \tag{4.1}$$

Although simple enough, note that such a model already covers a large domain: it does not need to be linear with respect to actions $\boldsymbol{d}$, and may even provide a fair approximation of a non-linear model with respect to uncertain inputs through linearisation insofar as the uncertainties are small enough (cf. Section 7.3.2). In order to simplify later comments, assume additionally that each factor $G^i(\boldsymbol{d})$ is positive, possibly requiring a prior change of sign in the definition of the components x^i. Assuming that uncertainty sources are *independent*, elementary probabilistic results ensure that:

$$\operatorname{var} Z = \sum_{i=1\ldots p} G^i(\boldsymbol{d})^2 \operatorname{var} X^i \tag{4.2}$$

In other words, output variance comes as the sum of all input variances multiplied by the squared input-output sensitivities (derivatives). This generates the simplified propagation method called the Taylor quadratic approximation (Section 7.3.2). Because factors are all positive, elementary algebra further gives:

$$\sigma_z = \sqrt{\operatorname{var} Z} = \sqrt{\sum_{i=1\ldots p} G^i(\boldsymbol{d})^2 \sigma_{X^i}^2} \leq \sqrt{\left(\sum_{i=1\ldots p} G^i(\boldsymbol{d})\sigma_{X^i}\right)^2} = \sum_{i=1\ldots p} G^i(\boldsymbol{d})\sigma_{X^i} \tag{4.3}$$

This simple equation is quite meaningful: it evidences the *reduction* of uncertainty or error compensation from input to output. This is true for a risk measure based on variance or standard deviation; it is also the case for a α-quantile risk measure, assuming, for instance, a Gaussian distribution (u_α denoting the standard normal α-quantile that is $u9_{5\%} = 1.6$; x_α^i denoting the α-quantile of the i-th input):

$$c_z(\boldsymbol{d}) = z_\alpha = G^o(\boldsymbol{d}) + u_\alpha.\sigma_z \leq G^o(\boldsymbol{d}) + \sum_{i=1\ldots p} G^i(\boldsymbol{d})u_\alpha.\sigma_{X^i} = G\left(\left(x_\alpha^i\right)_{i=1\ldots p}, \boldsymbol{d}\right) \tag{4.4}$$

This elementary property illustrates the heavy impact of choosing a probabilistic setting. In essence, it guarantees that a probabilistic risk measure (say a 95 % loss of performance) will result in a smaller margin than the risk measure obtained through a deterministic aggregation of similar inputs (loss of performance for all input factors taken at 95 % values). This may be seen either as:

- a forceful argument stating that deterministic input-based safety factors are 'over-conservative', all the more so when a large number of inputs are penalised altogether;
- or a forceful argument stating that a probabilistic-based study is dangerous as it automatically reduces the margins, and all the more so when the number of uncertain inputs increases.

Impact of the structure of input uncertainty Two essential comments need to be made. First, the amount of uncertainty reduction *depends on the relative importance factors* of the inputs (see Section 7 for extended discussion on sensitivity and importance factors). Assuming that input X_1 explains almost all the variance, that is:

$$\operatorname{var} Z = \sum_{i=1...p} G^i(\boldsymbol{d})^2 \operatorname{var} X^i = (1+\varepsilon) G^1(\boldsymbol{d})^2 \operatorname{var} X^1 \tag{4.5}$$

where $\varepsilon \ll 1$, either because model sensitivity to the first input is large or because input variance is dominant. Reduction of standard deviation due to error compensation then becomes negligible:

$$\sigma_z = \sqrt{\operatorname{var} Z} = G^1(\boldsymbol{d})\sigma_{X^1} \cdot \sqrt{1+\varepsilon} \approx G^1(\boldsymbol{d})\sigma_{X^1}(1+\varepsilon/2) \approx G^1(\boldsymbol{d})\sigma_{X^1} \tag{4.6}$$

Similarly, reduction is negligible when a $p+1$-th uncertain input with low importance $G^{p+1}(\boldsymbol{d})^2 \operatorname{var} X^{p+1} \ll \operatorname{var} Z$ is added to the system model. Conversely, maximal reduction is achieved when all inputs have comparable importance. Suppose that:

$$\forall i,\ G^i(\boldsymbol{d})\sigma_{X^i} \approx G^1(\boldsymbol{d})\sigma_{X^1} \tag{4.7}$$

Then:

$$\left(\sum_{i=1...p} G^i(\boldsymbol{d})\sigma_{X^i}\right)^2 \approx p^2.G^1(\boldsymbol{d})^2 \operatorname{var} X^1 \approx p \sum_{i=1..p} G^i(\boldsymbol{d})^2 \operatorname{var} X^i = p \operatorname{var} Z \tag{4.8}$$

In such a case, error compensation results in a (statistically unsurprising) reduction speed of the inverse square root of the input dimension:

$$\sigma_z \approx \frac{1}{\sqrt{p}} \sum_{i=1...p} G^i(\boldsymbol{d})\sigma_{X^i} \tag{4.9}$$

Such uncertainty reduction is obviously linked to the independence hypothesis. Intuitively, a random and independent behaviour ensures a certain 'compensation of error' as it is unlikely that all inputs deviate on the same side at the same time. This is easily supported by the more complete formula below including linear covariance or correlation coefficients:

$$\operatorname{var} Z = \sum_{i=1}^{p} G^i(\boldsymbol{d})^2 \operatorname{var} X^i + \sum_{i_1 \neq i_2} G^{i_1}(\boldsymbol{d}) G^{i_2}(\boldsymbol{d}) \operatorname{cov}\left(X^{i_1}, X^{i_2}\right) \tag{4.10}$$

Assuming that all inputs are fully and positively correlated leads to:

$$\forall i_1, i_2,\ \mathrm{cov}(X^{i_1}, X^{i_2}) = corr(X^{i_1}, X^{i_2})\ \sigma_{X^{i_1}} \sigma_{X^{i_2}} = 1.\sigma_{X^{i_1}} \sigma_{X^{i_2}}$$

$$\Rightarrow \mathrm{var}\, Z = \sum_{i=1}^{p} G^i(\boldsymbol{d})^2 \sigma_{X^i}^2 + \sum_{i_1 \neq i_2}^{p} G^{i_1}(\boldsymbol{d}) G^{i_2}(\boldsymbol{d}).\sigma_{X^{i_1}} \sigma_{X^{i_2}} = \left(\sum_{i=1}^{p} G^i(\boldsymbol{d}) \sigma_{X^i} \right)^2$$

$$\Rightarrow \sigma_Z = \sum_{i=1}^{p} G^i(\boldsymbol{d}) \sigma_{X^i} \tag{4.11}$$

This means that the whole extent of *uncertainty reduction is suppressed* by the full positive *correlation* between inputs. The output α-quantile is then the exact image of the system model taken at input α-quantiles, similar to a deterministic approach. Conversely, it may be seen that negative correlation between some factors does actually further reduce output uncertainty. This intuitively increases the chance that a deviation on one side by a given input will be compensated by a deviation on the opposite side. In other words, the *real difference* between the deterministic and probabilistic aggregation of uncertainty lies in the *dependence assumptions*. Dependence issues are a key to controlling the value of a risk measure: Chapter 5 will introduce more elaborate concepts for modelling dependence beyond the linear correlations.

Mutualisation of risk factors and parametrisation of system models Such observations are also common sense in the field of insurance or banking. The whole rationale of insurance is based on mutualising elementary risk (policies) so that the overall risk measure is greatly reduced; similarly the constitution of diversified portfolios of financial assets reduces the VaR_α that is the α-quantile. The simple case for both involves straightforward additions of elementary inputs, that is a linear system. This obviously depends on the true *diversified nature* of policies of assets as measured by their inter-correlations, or in other words the fact that the underlying risk factors are effectively separate and independent. Such a feature may prove increasingly difficult to control in an interconnected world, although still robust with policies covering, for instance, natural perils in some parts of the globe and car accidents in other places. Part of the major world crises following the 2007/2008 financial collapse can be sourced to an overestimation of the diversification of financial assets which proved to depend – at a world level – on a very limited number of highly risky assets in the US real estate subprimes, though substantially shielded from the analysis by complex derivative settings. Eventually, the popular expression of *systemic risk* could be understood as a level of *global* dependence generated by the failure of a sub-system so significant that it ruins any effort of risk mutualisation.

There are limited generalisations of these properties beyond linear models (see Section 4.2.2) or Gaussian distributions. Note that the formula stating variance reduction does not rely on any distribution assumption for the inputs. It is still valid for linear models with non-Gaussian inputs, such as *non-Gaussian* metrological systems with small-enough input uncertainties (so as to be able to linearise) for which a common quantity of interest or risk measure is based on variance: a multiple of the standard deviation or coefficient of variation. Alas, this does not cover quantiles as a risk measure. It is well known in the field of insurance and banking that a portfolio VaR_α (i.e. the risk measure taken on the output of interest) is not guaranteed to be lower than the sum (i.e. the image through system model) of VaR_α of the inputs when non-Gaussian, even when the model is linear. This is seen as a classic limitation of value-at-risk as a risk measure (Embrechts, McNeil, and Straumann, 1999).

Another important conclusion concerns the *parametrisation of a system model* under uncertainty. A natural analytical trend leads to sophisticated system models, in particular thanks to the increasingly available computational power. Ever-finer meshes including the replacement of 1D by 3D descriptions, and more complex physical laws, chemical reactions or ecological layers lead to a constant increase in the dimensions of input variables or parameters. Most of those are uncertain, not least because datasets or expert knowledge generally lags behind such numerical complexity. The equations above (albeit limited to a linear model) suggest that a simplistic consequence of such increasing parametrisation is the risk of *uncontrolled reduction of output variance*. Chapter 6 will discuss the statistical mechanisms by which genuine *uncertainty* actually *increases* if the associated amount of data and expertise is constant.

The case of multiplicative models Extensions can be made to linearised multiplicative (or power-product) system models. Consider the following:

$$z = G(\boldsymbol{x}, \boldsymbol{d}) = g^0(\boldsymbol{d}) \prod_{i=1...p} (x^i)^{g^i(\boldsymbol{d})} \tag{4.12}$$

Such system models can be found in the field of environmental or health impact analysis (Slob, 1994). Think of the dissemination of pollutants passing through each successive layer of the ecological chain and leading to final exposure according to imprecise or environmentally-variable fractions, or of dose-response uncertainty. It is also pervasive in phenomenological models found in material science, mechanics or other engineering sciences, in economics and public policy analysis, or in risk analysis involving the logical models of systems in series (see Chapter 1).

As will be developed in Chapter 7 under the name of the Taylor quadratic approach for uncertainty propagation, distinguishing this non-linear model yields an approximate linear model:

$$z = EZ + (z - EZ) = G(E\boldsymbol{X}, \boldsymbol{d}) + \sum_{i=1...p} \frac{\partial G}{\partial x^i}\Big|_{E\boldsymbol{X},\boldsymbol{d}} (x^i - EX^i) + o(\boldsymbol{x} - E(\boldsymbol{X})) \tag{4.13}$$

The previous comments on variance decomposition can thus be extended to an approximate quadratic development:

$$\operatorname{var} Z \approx \sum_{i=1...p} \left(\frac{\partial G}{\partial x^i}\Big|_{E\boldsymbol{X},\boldsymbol{d}} \right)^2 \operatorname{var} X^i \tag{4.14}$$

The peculiar multiplicative structure of the model now leads to a simple relationship between the input and output coefficients of variation, that is a quantity of interest quantifying the *relative uncertainty* of the inputs and output:

$$c_Z = c_V(Z) = \sqrt{\frac{\operatorname{var} Z}{(EZ)^2}} \approx \sqrt{\sum_{i=1...p} \left(\frac{1}{EZ} \frac{\partial G}{\partial x^i}\Big|_{E\boldsymbol{X},\boldsymbol{d}} \right)^2 \operatorname{var} X^i}$$

$$= \sqrt{\sum_{i=1...p} \left(\frac{\partial \log G}{\partial x^i}\Big|_{E\boldsymbol{X},\boldsymbol{d}} \right)^2 \operatorname{var} X^i} = \sqrt{\sum_{i=1...p} \left(\frac{g^i(\boldsymbol{d})}{E_x^i} \right)^2 \operatorname{var} X^i} = \sqrt{\sum_{i=1...p} (g^i(\boldsymbol{d}))^2 c_V(X^i)^2} \tag{4.15}$$

In other words, it means that the % of uncertainty in the output prediction turns out to be approximately the quadratic mean of the % of uncertainty in the inputs (weighed by the square exponents of the inputs). Consider, for instance, a system model being the product of p factors with approximately $q\%$ of uncertainty in each: it means that the output uncertainty will be around $\sqrt{p} \times q\%$. A product of risk factors thus inflates the risk (instead of reducing it as for an additive compensation of errors) though at a slower speed than the number of factors. In fact, the consideration of correlations leads to an analogous result as in the previous sections. Fully and positively-correlated inputs in a multiplicative model would lead to a % of output uncertainty equal to the sum of input % of uncertainty instead of the quadratic mean, and thus $p \times q\,\%$ for p factors with approximately $q\,\%$ of uncertainty in each.

The same comments could be derived through the consideration of the log-model which stands as a linear combination of the log-inputs; the variance of a log is approximately the square coefficient of variation of Z:

$$\mathrm{var}(\log Z) \approx \left(\frac{\partial \log Z}{\partial Z}|_{EZ}\right)^2 \mathrm{var}\, Z = \frac{\mathrm{var}\, Z}{(EZ)^2} = c_V(Z)^2 \tag{4.16}$$

Considering Gaussian distribution for the additive log-inputs is therefore equivalent to considering multiplicative lognormal inputs. Because of the pervasiveness of power-product system models in environmental and health impact and associated economics (Slob, 1994), a lognormal distribution for the associated sources of uncertainty may appear more appropriate than a normal distribution in those contexts, with the additional advantage of modelling strictly positively valued factors. It is also commonly found in PRA, for instance to represent epistemic uncertainty in component failure probabilites, or even in seismic PRA where both aleatory and epistemic uncertainty are taken lognormal. Incidentally, a power-product of a large number of independent inputs tends to the lognormal shape by virtue of the Central-Limit Theorem: applied on the log of the model, it is equivalent to the sum of a large number of independent inputs becoming Gaussian-shaped.

4.2.2 Understanding safety factors and quantiles in the monotonous case

A larger class of system models, the monotonous ones, is of essential importance in the field of risk analysis. This section will further discuss the fact, already shown in Chapter 1, that they are implicitly in the mind of many risk analysts. Monotony also justifies to a large extent the use of mixed deterministic-probabilistic risk measures.

Recall first what is meant by monotony (cf. Section 7.4 for more detail). It refers to the following intuitive physical property: increase one of the physical inputs x^i such as stress, failure frequency of a given component, flood flow, effluent toxicity or claim intensity (or more generally 'risk' or 'load' variables); or decrease toughness, dike level, degradation rate or cover capital (or more generally 'safety' or 'resistance' variables) while all other components of $\boldsymbol{x}$ are fixed, and a risk-type of output of interest (plant failure rate, dike overflow, death due to toxicity, insolvency) will increase or decrease when considering a safety-type of output (mechanical margin), wherever the variation starts from.

Mathematically, with a system model mapping D_x, a subset of $\mathbf{R}^p$ into either $\mathbf{R}$ or [0,1], when considering an event of interest, *global monotony* of $G(\boldsymbol{x},\boldsymbol{d})$ over D_x (or simply *monotony* when non-ambiguous) is defined as follows:

$$\begin{aligned}&\forall i, \exists s^i \in \{1,-1\}, \forall a \geq 0, \boldsymbol{d}, \boldsymbol{x} = (x^1, \ldots x^i, \ldots x^p) \in D_x \\ &G(x^1, \ldots, x^i + s^i a, \ldots x^p, \boldsymbol{d}) \leq G(x^1, \ldots, x^i, \ldots x^p, \boldsymbol{d})\end{aligned} \tag{4.17}$$

where s^i represents the sign of monotonous dependence: $s^i = 1$ when $G(.)$ is decreasing with i-th component x^i, and $s^i = -1$ when it is increasing. Similar to that discussed in the previous paragraph, this will be assumed, having made the necessary changes at the outset of the study. Partial monotony refers to the situation where Equation (4.17) is only valid for a given subset of indices i. The definition is extended trivially to uncertain input events, by simply ranking the binary state by 1 greater than 0, or more generally to those multimodal discrete input events that, for instance, represent an increasing degradation of the component or an increasingly severe initiator.

Linear models are monotonous but this much wider class of model also includes many non-linear models: the flood example, many structural mechanical or civil engineering models, environmental transfer models (based on reservoirs and fluxes), system reliability models, and so on. Intuitively, any *consistent* fault tree, for instance, would always keep the output event $z = 1$ (failing overall system), whenever $e^i = 0$ (i-th sub-system is safe) is shifted to 1 (i-th sub-system fails) in the input vector. As shown in the review in Chapter 1, any time a design engineer, a risk analyst or an economist handles the concept of 'pessimistic' or 'penalised' values or scenarios, implicit reference is made to the monotony of the system model. Consider indeed the following deterministic risk criterion ensuring that the risky output of interest z remains under threshold z_s for a set of input quantiles ($\alpha_1, \ldots \alpha_p$):

$$G\left(x^1_{\alpha_1}, \ldots, x^i_{\alpha_i}, \ldots x^p_{\alpha_p}, \boldsymbol{d}\right) \leq z_s \tag{4.18}$$

Assuming (implicitly) that $G(\cdot)$ is monotonous means compliance to the threshold for any situation for which each of the inputs remains below a (supposedly high) quantile, i.e. α_i% of the cases for any i-th input considered separately. As mentioned in Chapter 1, the traditional partial safety factor approach in structural reliability is based on such a property, involving a margin to failure as the output of interest of a system model combining the resistance and stress functions required to be positive by the decision criterion:

$$G(\gamma.\boldsymbol{x}) = g_R\left(x^1_k/\gamma^1_R, \ldots x^r_k/\gamma^r_R\right) - g_S\left(x^{r+1}_k\gamma^{r+1}_S, \ldots x^p_k\gamma^p_S\right) \geq 0 \tag{4.19}$$

where γ denotes partial safety factors and $\left(x^i_k\right)_{k=1\ldots p}$ denotes the so-called *characteristic values* for the uncertain inputs, generally taken as quantiles of the resistance variables ($x^1, \ldots x^r$) or load variables ($x^{r+1}, \ldots x^p$) respectively. The application of partial safety factors (all taken greater than 1) provides additional coverage beyond that ensured by the use of quantiles associated with the characteristic values.

Moreover, such a coverage property may be extended to mixed deterministic-probabilistic risk measures. Assume that G is at least partially monotonous with respect to $(x^i)_{i = q,\, q+1, \ldots p}$. The mixed risk measure based on the conditional exceedance probability also proves monotonous with respect to the 'penalised' components:

$$\begin{aligned}
&\boldsymbol{x}_{pn} < \boldsymbol{x}'_{pn} \\
&\Rightarrow G\left(\left(x^i\right)_{i=1,\ldots q-1}, \boldsymbol{x}_{pn}, \boldsymbol{d}\right) < G\left((x^i)i = 1, \ldots q-1, \boldsymbol{x}'_{pn}, \boldsymbol{d}\right) \\
&\Rightarrow \left\{\omega | G\left(\left(x^i\right)_{i=1,\ldots q-1}(\omega), \boldsymbol{x}_{pn}, \boldsymbol{d}\right) > z_s\right\} \subset \left\{\omega | G\left(\left(x^i\right)_{i=1,\ldots q-1}(\omega), \boldsymbol{x}'_{pn}, \boldsymbol{d}\right) > z_s\right\} \\
&\Rightarrow c_z\left(\boldsymbol{x}_{pn}, \boldsymbol{d}\right) = P\left[Z = G(\boldsymbol{X}, \boldsymbol{d}) > z_s | \left(X^i\right)_{i=q,q+1,\ldots p} = \boldsymbol{x}_{pn}\right] < c_z\left(\boldsymbol{x}'_{pn}, \boldsymbol{d}\right)
\end{aligned} \tag{4.20}$$

Quite intuitively, *whatever* the deterministic or mixed deterministic-probabilistic *risk measure* chosen, it is always *conservative* to *increase the value of a fixed input* for which the system model is *monotonous*. Engineering practice rightfully uses this property quite often. To prove the structure will hold up to the 10^{-3} (i.e. a conditional failure probability less than 10^{-3}) over a given range of operational factors, it is enough to prove it will do so with the most pessimistic combination. This property is unchanged when correlated inputs are considered.

Alas, as was already mentioned for the linear model with non-Gaussian inputs, this does not guarantee in general that the output α-quantile is always lower than the image of the input α-quantiles. When the system model is monotonous, and when a deterministic criterion based on input quantiles is fulfilled:

$$G\left(x_{\alpha_1}^1, \ldots, x_{\alpha_i}{}^{i}, \ldots x_{\alpha_p}^p, \boldsymbol{d}\right) \leq z_s \tag{4.21}$$

the property that always holds is the more restrictive following lower bound (see Chapter 7 for a demonstration):

$$\prod_i \alpha_i = \prod_i P\left(X^i \leq x_{\alpha_i}^i\right) \leq P\left[G(X^1, \ldots, X^p, \boldsymbol{d}) \leq z_s\right] \tag{4.22}$$

In other words, a deterministic design based on input quantiles is guaranteed to be safe for a confidence level equal at least to the product of input confidence levels: taking 95% confidence for $p = 3$ inputs guarantees output confidence of $(95\%)^3 = 86\%$; it falls to only 60% when $p = 10$ such uncertain inputs are combined. Note that such a lower bound for the output confidence level virtually requires all inputs but one being set close to its maximal value (100%-quantile) in order that the α-quantile of the last one (the weakest input) secures a confidence level of α for the output.

Conversely to what was guaranteed in the linear Gaussian case, the increase of input dimensions unfortunately reduces control over the output risk measure. Yet this is the least that may be directly guaranteed when simply assuming monotony Chapter 7 will show that monotony enables the accelerated computation of an output quantile. The flood example below also proves that in fact a much safer design may result than the mere lower bound.

Flood example
Under a 1/1 000-*yr* return initiator-frequency approach, flow is taken at a 99.9 % yearly quantile, while the other uncertain variables are given penalised values, typically -2σ for friction (i.e. ~97.5 % under Gaussian distribution) and close to maximal values for riverbeds (~100 % quantiles). The deterministic criterion leads to:

$$z_s = G\left(q^{1000}, k_{s\,pn}, z_{m\,pn}, z_{v\,pn}\right)$$

Under a Gumbel model, the design level was found to be 60.7m, that is a 5.2m dike. The associated lower bound for safety gives $P_{f-} = 0.999{*}0.98{*}1{*}1 \cong 98\,\%$ (or a 1/50-*yr* return). A full probabilistic computation assessed the corresponding safety to be around 99.98 % (or a 1/5 000-*yr* return), see Annex Section 10.4.

Getting back to the natural risk assessment practices reviewed in Chapter 1 (Section 1.1), recall the alternative 'initiator-frequency' *vs.* 'extended frequency' approaches. The former is essentially comparable to the above-mentioned safety factor approach. The return period bases the choice of a quantile on the supposedly dominant random source, the external initiator such as flood flow or extreme wind; the other inputs are given penalised values, those being more or less implicitly based on tentative quantiles or expert figures. A new understanding is now possible: the system model embodied by the 'local risk situation' (cf. Section 1.1) is mostly monotonous as one handles a series of aggravating factors (risk variables) and protection structures (safety variables), which justifies the use of a safety factor approach. However, the previous comments show the limitations of such an initiator frequency approach taken in the light of an overall probabilistic target. While it is intended to prove that:

'conditional to an initiator of a T-return period, the installation is safe (under design d)'

it cannot be guaranteed in general to mean that:

'global safety of the installation is higher than a T-return period (under design d)'

when taken for the initiator only (except when taking the absolute maximal values for all other inputs). Conversely, a fully probabilistic approach may indeed prove in some cases that the safety level is much higher than simply the return period.

In spite of those limitations, monotony proves to be very natural and rather efficient, at least to *start a simple risk analysis*: assume 'worst case' or pessimistic values so as to quickly assess whether a more detailed study is necessary. The monotony of the system model will also prove to have ancillary consequences for the impact on the safety level of including or not the input dependence, as well as on the level-2 risk measures (see Chapter 5 and 7).

4.2.3 Probability limitations, paradoxes of the maximal entropy principle

This section will further illustrate that the choice of a probabilistic model comes with strong assumptions that must be kept in mind, particularly in the absence of statistical data. The previous section has already stressed the underlying dependence assumptions. Additionally, the issue of choosing a distribution shape has a major impact on a probabilistic measure and raises classical paradoxes that will be reviewed briefly.

Consider the situation where the only piece of information available on a given uncertain input X is a plausible range between a lower x_l and upper bound x_u. To put it simply, take the following system model and risk measure based on a single input:

$$z = G(x, d) = d^* x^a \tag{4.23}$$

$$c_z(d) = P(Z < z_s | d) > \alpha \tag{4.24}$$

where a is a fixed coefficient representing, for instance, the environmental attenuation of a pollutant discharge depending on d; the concentration z is required to remain below a given threshold z_s representing a maximal acceptable level of pollution; this has to be guaranteed for sufficient confidence α. A popular probabilistic model for X would be a uniform distribution bounded by the known range. This is generally supported by the *principle of maximal entropy* (Shannon, 1948; see review in Cover and Thomas, 1990, 2006), stating that a distribution choice should be that which is no more informative than what is known, in the sense of maximising the appropriate entropy conditional on all available knowledge. Entropy is required here to be defined for a continuous distribution:

$$H(f_X) = -\int f_X(x) \log(f_X(x)) dx = E[-\log(f_X(X))] \tag{4.25}$$

Table 4.2 The impact of choosing the variable for the entropy principle.

Numerical examples $[x_l, x_u] = [1,2]$; $d = 1$; $z_s = 4$	Uniform model on X	Uniform model on Y
$a = 2$	$c_z = 100\,\%$	$c_z = 100\,\%$
$a = 4$	$c_z = 41\,\%$	$c_z = 20\,\%$
$a = 10$	$c_z = 15\,\%$	$c_z = 0{,}3\,\%$
$a = 20$	$c_z = 7\,\%$	$c_z = 3.10^{-6}$

Maximisation should be undertaken under the constraint that the distribution is appropriately bounded, as follows:

$$\underset{f_X}{Max}\left[H(f_X)\middle|\int_{[x_l,x_u]} f_X = 1\right] \tag{4.26}$$

This results classically in the choice of a uniform distribution. Other useful derivations would lead to the choice of an exponential distribution – conditional on knowing only the expected value and the random variable being positively-valued; or that of a Gaussian distribution – conditional on knowing only the expectation and variance.

However, assuming that X is modelled as uniform, the resulting distribution is not uniform in Z anymore because of the power transformation. Why not apply then the principle of maximal entropy to the following alternative parametrisation of the input, closer to the output of interest, for which equivalent knowledge states that this new variable is bounded (by $(x_l)^a$ and $(x_u)^a$)?

$$z = G(y, d) = d^*y \quad where \quad y = x^a \tag{4.27}$$

The result of such a distinct uncertainty model is quite different regarding the risk measure, as demonstrated in Table 4.2 and Figure 4.1.

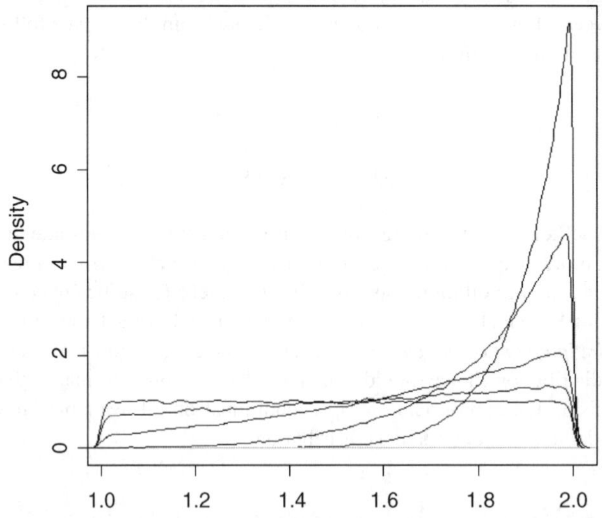

Figure 4.1 *Comparison of X distributions for which the Maximal Entropy principle has been applied resp. to X (flat curve) or to Y with a = 2, 4, 10, 20 (increasingly-peaked curves).*

Clearly, the stronger the non-linearity (the power index in the example), the deeper the impact of choosing where to apply the principle of maximal entropy: the relative value of the risk measure is impacted up to six orders of magnitude. This absence of invariance as a function of the selection of variables is known as one of Bertrand's paradoxes of maximal entropy. From a decision-making perspective, it would be tempting to privilege the second choice for which the principle is applied to the variable *closest to the output of interest*. Pursue further this intuitive idea in a multidimensional case, assuming knowledge of all inputs is restricted to the ranges of each and that the system model remains monotonous:

$$\begin{aligned} z &= G(\boldsymbol{x}, \boldsymbol{d}) \\ \forall i, \quad & x^i \in \left[x_l^i, x_u^i\right] \end{aligned} \tag{4.28}$$

One ends up with the fact that the output of interest Z should simply follow a uniform distribution between the absolute minimum and maximum values, that is the images of such combinations in the inputs (assuming the proper prior sign transformations ensure increasing monotony for each input):

$$\begin{aligned} & Z \sim U[z_l, z_u] \\ & [z_l, z_u] = [G(\boldsymbol{x}_l, \boldsymbol{d}), G(\boldsymbol{x}_u, \boldsymbol{d})] \end{aligned} \tag{4.29}$$

which means that a risk measure such as a quantile is simply computed by a ratio of the distance to the absolute bounds:

$$c_z, (d) = P(z < z_s | d) = (z_s - z_l)/(z_u - z_l) \tag{4.30}$$

This is intuitively a poor result for probabilistic modelling: there is not much to support such a quantitative risk figure given the actual level of information. In such a situation, it may be at least as questionable to model inputs by uniform distributions, compute a seemingly more informative (non uniform) distribution for the output of interest and draw an alternative figure for a threshold exceedance risk measure.

This extreme illustration stresses that it is far from equivalent to encode the following statements regarding a bounded uncertain input:

- 'all that is known is that *possible* values for X lie between x_l and x_u'
- 'the *probability* distribution of X is bounded by x_l and x_u'

This adds a further warning on the use of probabilistic setting for risk and uncertainty analysis, beyond the dependence issues evidenced earlier. It illustrates the difficulty of choosing between distribution shapes in the absence of sufficient statistical data. Again, in such a case with *monotonous* system models, it may prove just as knowledgeable to stick with a fully deterministic uncertainty model, stating simply that the output of interest is known to remain between lower and upper bounds.

4.2.4 Deterministic settings and interval computation – uses and limitations

Previous sections have made apparent a number of reasons for using partial or complete deterministic uncertainty models for the inputs. In fact, a large amount of literature has been developed on the interval computation techniques enabling standard propagation of such uncertainty models. Indeed, beyond uncertainty and risk analysis, a number of other applications motivate deterministic modelling by intervals (Jaulin *et al.*, 2001):

- optimal control and automatics;
- command/control;
- robust design;
- and so on.

While they provide relief for the possibly delicate statements on dependence, or distribution shapes, deterministic settings display strong limitations in practice for our purpose:

- For non-monotonous system models, uncertainty propagation means a multi-dimensional non-linear optimisation process, which is quite expensive computationally.
- Choosing relevant boundary values is always a delicate issue, and has particularly explicit impacts in monotonous cases.

The *former* refers to the following large optimisation under constraint, where D_x denotes the set of possible values for $\boldsymbol{x}$, corresponding to the domain of its joint pdf in a probabilistic model:

$$c_z(\boldsymbol{d}) = Max_{\boldsymbol{x} \in D_x}\{G(\boldsymbol{x}, \boldsymbol{d})\} \tag{4.31}$$

D_x may represent something more sophisticated than a simple p-dimensional product of intervals for each input, incorporating known constraints limiting the possible interactions between inputs. Outside of monotonous system models, it is always a great numerical challenge, because $G(.)$ is generally neither linear nor convex where there is a large number p of variables. The accuracy of the result is often indiscernible because of the local minima. Its simplified version is often used, consisting of maximising the output over an experimental design constructed on the limits of the components of $\boldsymbol{X}$. There is then no guarantee of accuracy in the non monotonic general case.

For the *latter*, except when there is enough knowledge to identify the absolute maximal values for the inputs, it is very hard to control the level of coverage secured by a deterministic penalisation. It is precisely the role of a mathematical *measure* such as probability to quantify subsets in the context of set theory. Conversely, taking all worst-case combinations generally proves to be 'over-conservative' in the sense that its plausibility is very limited, and brings very little information to decision-making. The frequent choice is therefore to base the input bounds more or less on input probabilistic quantiles, with the above-mentioned limitations illustrated in the flood example in Section 4.2.2.

4.2.5 Conclusive comments on the use of probabilistic and deterministic risk measures

Chapter 1 showed the pervasive use of mixed deterministic-probabilistic risk measures to handle uncertainty, and Section 4.1 discussed the fact that some authors link the use of deterministic or probabilistic representations to the nature of uncertainty sources. This section illustrates further that:

- Moving for some inputs from a deterministic to a probabilistic model involves a strong statement about the dependence between them. The linear case pointed up the risk of unduly reducing uncertainty, that is underestimating risk, by misrepresenting dependence.
- Additionally, the risk measure is strongly impacted by the choice of input pdfs, which proves problematic in the absence of statistical information as outlined by the Bertrand paradoxes with the entropy principle.
- Thus, one better understands the use of deterministic 'penalisation' for those inputs lacking enough information to support any credible choice of pdf (or statement on dependence). It proves to be justified

in the case where the system model is monotonous at least with respect to the above and where the risk measure relates to threshold exceedance: upper quantiles, exceedance probabilities or deterministic exceedance. Such cases are indeed very common in risk analysis.

- However, such penalisation only covers up to the value chosen, and monotony does not guarantee a satisfactory bounding of the corresponding fully probabilistic risk measure. The basis of α-quantile inputs does not ensure the output α-quantile as exemplified in finance by the limitations of value-at-risk. Except when there is enough knowledge to identify the absolute maximal values for the inputs, a deterministic penalisation may never be proved to be conservative; and conversely, it may also prove to be 'over-conservative' in the sense that the fully probabilistic performance is much higher.

Understandably, the use of mixed deterministic-probabilistic – in other words *conditional* probabilistic – risk measures results in a generic trade-off required to reflect the true level of information available. Eventually however, the use of conditional probabilities entails some decision-making difficulties not least because two risky designs or options may be compared only as long as their deterministic conditioning is similar. This proves quite challenging for the resulting heterogeneity of risk assessment for various industrial projects or *a fortiori* sectors within the projects.

4.3 Handling time-cumulated risk measures through frequencies and probabilities

This section explores the *time dimension* underlying any consideration of modelling in the context of risk and uncertainty. As was mentioned in Chapter 2, the state ω of the system considered, conditional upon taking some given actions $\boldsymbol{d}$, is imperfectly known at a given time of interest: either at a given date t_f or during a given period $[t_f, t_f + \Delta T]$. Abandoning the ambition of accurate prediction, modelling will consider a risk measure or quantity of interest as a means to control the lack of knowledge in ω at that time, as viewed by the analyst to the best of his knowledge. A thorough specification of the risk measure should therefore specify the time basis upon which the state of the system is studied. Except in the case of inverse calibration or data assimilation studies (see Chapter 6) where a prior analysis of the system involves a past time of interest to model indirect observations of the past states of the system, it generally refers to the future. Multiple definitions may be considered time-wise, such as frequencies or probabilities of average or maximal events or performance levels. As a number of consequences concern the associated risk measures, it is worth considering them carefully before possibly reformulating with easier-to-handle time-implicit formulations.

4.3.1 The underlying time basis of the state of the system

The states of the system considered generally undergo a dynamic variation in time, albeit often implicit: think of the river flow, the plant operation characteristics, the running or failure status of components, the daily oil price and so on. Each characteristic of the system may be considered as an (uncertain) time series: this applies to any of the uncertain inputs X^i or to the output variable of interest Z. Hence, the system state should involve theoretically a large-dimension vector of time-continuous variables (X_t, Z_t), that is real-valued continuous time series, as represented in Figure 4.2 (top left).

In order to allow for modelling and computation, the practical specification of the system inputs and outputs mostly involves some simplification regarding the time basis:

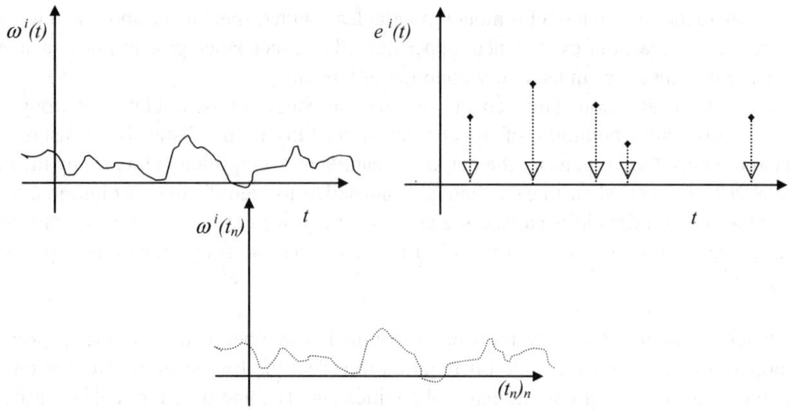

Figure 4.2 *Underlying time processes for the system state: (up left) time-continuous variables; (up right) time-continuous events; (down) time-discrete variables.*

(i) Consider events $e^i(t)$ that happen only at discrete dates, that is mostly zero but taking non-zero states (either discrete or real-value modalities) at random dates $(T_j)_j$: component failures in reliability and maintenance, large flood flows (i.e. above a certain threshold), significant seismic events (i.e. soil acceleration above a certain threshold), insurance events or claims (with claim arrival times) and so on (Figure 4.2 (top right)).

(ii) Consider a discretised time basis with one value of the inputs/outputs $\omega^i(t_n)$ per time step: daily oil price (at market closure), daily temperature as measured at 12 a.m. and so on (Figure 4.2 (below)).

Physically, one may generally view these variables of interest as derived from the sampling of some underlying time-continuous characteristics. This is obvious in alternative (ii), but is also true in (i) when considering the following definitions:

$$E^i(t) = 1_{Xi>zs}(t) \qquad (T_j)_j = \{t | 1_{Xi>zs}(t) = 1\} \tag{4.32}$$

For instance, think of a series of components, the failure of which E^i occurs as the result of increasing crack lengths X^i that eventually exceed a given structural threshold x_s. Despite the existence of such an underlying time component, many models involve explicitly only static variables. In fact, recall the generic methodological notation developed in Chapter 2 which did not display time explicitly:

$$c_z = F[\boldsymbol{Z}] = F[G(\boldsymbol{X}, \boldsymbol{d})] \tag{4.33}$$

How should such a risk measure be understood regarding the time basis?

A fixed date for the time of interest The easiest interpretation is that the uncertainty is expressed with respect to the unknown in the system state *at a fixed date* t_f. Take, for instance, the following quantities of interest:

- standard deviation of Z_t, the future temperature in a given location, taken at $t = t_f$, *for example* midday on Monday of next week,
- quantile of the daily return (tomorrow) or cumulated return after one month (starting from today),

- probability of encountering an undesired event on the next day (or the next time step) in a short-term environmental or sanitary warning system,
- probability of a cold wave (or heat wave) occurring on the planned date of a rocket launch,
- and so on.

The system model should then be understood as the inference of the relationship between the values of the variable of interest and those of uncertain and known inputs for given actions or operational conditions *at a given date*:

$$z = z(t_f) = G(\boldsymbol{x}(t_f), \boldsymbol{d}(t_f)) = G(\boldsymbol{x}, \boldsymbol{d}) \tag{4.34}$$

Multiple interpretations are possible regarding this prediction over time, in line with the multiple interpretations already mentioned for the probabilistic settings (cf. Section 2.1.4 and Section 4.1.2).

A *subjectivist* interpretation might consider the inference of likely values for the variable of interest in future z_{tf} as a representation of the decision-maker's beliefs. In fact, one may consider a single event, the future realisation of which would simply issue a single value, without repeated occurrences. Think of the return rate at the end of an investment project, the lifetime (before failure) of a given system or the coldness of next year's winter and so on. From that perspective, the underlying time process Z_t need not be stationary or independently distributed. Yet, the subjectivist view does not formulate how uncertainty modelling is undertaken on the basis of available information. Regarding the time basis of some of the uncertain inputs or factors driving the system performance, information does not generally take the exclusive form of *pure expertise* or *subjective beliefs* but relies also on time series with at least partially observed historical series.

Within the simplest *frequentist* interpretation, uncertainty modelling could assume that input data involves *i.i.d.* observations $(\boldsymbol{X}_j)_{j=1,\ldots n}$ allowing for the direct estimation of the marginal law of the underlying time series $\boldsymbol{X}_t$. Such time series being assumed to be stationary and independent,[3] the estimated law would be sufficient to infer future realisations at any time. Likewise, the distribution of future instantaneous realisations of the output Z_t may be inferred through propagation of the input distribution. In such as an *i.i.d.* situation, it is not useful anymore to include t inside the system model notation so that the risk measure will be formulated with a time-implicit approach. Think, for instance, of a simple direct model of say a share price or a river flow, for which the risk measure is a α–quantile that is VaR_α for the next day, or a given day in the future:

$$z_{tf} = G(\boldsymbol{x}) = x_{tf} \tag{4.35}$$

$$c_z = z_\alpha \tag{4.36}$$

Suppose that a historical daily time series is available:

$$(x_j = x(t_j))_{j=1\ldots n} \tag{4.37}$$

Such an *i.i.d.* interpretation would transform the time-based output of interest z_{tf} into the prediction of a standard random variable $Z = X$ on the basis of direct data; subsequently, direct estimation of the quantile may be undertaken either empirically or through the fit of a pdf (see Chapter 5).

Similar to the prediction at a fixed date of interest, another important case of application concerns the *conditional risk assessments*, as encountered for example in structural reliability analysis, or *scenario-based* quantitative risk or vulnerability analysis, where the system model helps in controlling the risk of failure or

[3] At least at a sufficiently-large time scale to exclude short-term time correlations, and a sufficiently-limited time scale to prevent non-stationarity. Refer to Section 4.3.3 for further comment.

of important damage *conditional* upon a given initiator event or scenario. The time of interest becomes then the uncertain *date* of occurrence of the event or scenario, generally rare enough to occur in a future distant enough to be independent from observed past states of the system. The conditional probabilities of consequences are studied at that time, involving uncertain realisations of system model inputs at that date which could also have been estimated through assumptions of *i.i.d.* realisations in the past.

Such a direct, *i.i.d.* frequentist interpretation of the time basis needs some extension as it firstly relies to a certain extent on the independence of time realisations of the discrete time series. This is not immediately obvious in risk analysis. The signal is often highly auto-correlated in time, at least in the short term: hour-to-hour river flows, second-to-second seismic accelerations or wind speeds, day-to-day currency rates and so on. Careful definition of discretised variables of interest may be a prerequisite to cancelling such auto-correlation of the time series, typically retrieving monthly (or even yearly) *maxima* or mean river flows, or values *exceeding* a high enough *threshold*. One eventually comes up with a probabilistic definition of the output of interest that is a standard random variable, time becoming implicit. Additional extensions are needed as the uncertain inputs or actions $(\boldsymbol{x},\boldsymbol{d})$ affecting the variable of interest at time t_f might not only refer to instantaneous states at time t_f, but also to time-fixed actions or model parameters; it might even include time-dependent states modelled over a dynamic risk model: see Section 4.3.3 and Section 6.1.4 in Chapter 6 for further comment.

A period of interest More is needed beyond the case of single time performance. Modelling for decision-support is generally concerned with richer descriptions of system behaviour over a given *period of time* $[t_f, t_f + \Delta T]$, as follows:

- in safety control, probabilistic measures on the *peak value* of the consequences over a future period of operation (one year, 30 years of plant lifetime . . .), or on the *occurrence* of an undesired event *at any time* within the same period, or else its average *frequency*;
- in performance optimisation, the *yearly average* performance value, the number of failure (or *unavailability*) events and so on;
- in insurance regulation, the ruin probability over a given period of time, that is the probability that the reserve variable over a given portfolio becomes negative at any date during the period.

In those cases where the output of interest refers to a whole time period, various alternative specifications of the risk measure become possible.

4.3.2 Understanding frequency vs. probability

Suppose the decision-maker is interested in controlling an output of the system over a given period of time $[t_f, t_f + \Delta T]$, with respect to the likelihood of event(s) of interest, such as exceedance of a given safety, performance or environmental threshold or any undesired event. Two basic probabilistic measures may be defined, namely:

- the *probability* of occurrence (at any time over the period $[t_f, t_f + \Delta T]$),
- or the *frequency* of occurrence (per time unit).

It is useful to describe the two measures in more detail as they are often referred to ambiguously in practical risk measures. Indeed, while their theoretical definition is different, they prove to be numerically close in many cases.

The key difference between the two concerns the consideration of multiple occurrences over the time period. Consider an 'eventised' random characteristic of the system, that is either a naturally-discrete event (such as component or system failure) or a threshold-exceedance of an underlying continuous characteristic, and its associated occurrence times:

$$E(t) = 1_{Z>zs}(t) \qquad (T_k)_{k=1...} = \{t \in [t_f, t_f + \Delta T] | E(t) = 1\} \tag{4.38}$$

Think, for instance, of initiating events in the context of QRA, be it internal (component failures or process temperature exceeding a safety threshold) or external (such as floods, storms exceeding a given threshold, etc.) or insurance claims generated by events (firebreak, storm . . .). In theory at least, there could be more than one event over the given period of interest; the number of events is a discrete random variable:

$$N_{E(t)}(\Delta T) = Max(k | T_k \leq t_f + \Delta T) = 0, 1, 2, \ldots n, \ldots . \tag{4.39}$$

The notation will be simplified into $N(\Delta T)$ when the underlying event of interest is unambiguous. Frequency of occurrence is then defined as the expected rate of occurrences:

$$F_q(\Delta T) = \frac{E[N(\Delta T)]}{\Delta T} = \frac{1}{\Delta T} \sum_{n \geq 0} nP[N(\Delta T) = n] \tag{4.40}$$

The probability of occurrence is simply the probability that (at least) one event occurs over the period $[t_f, t_f + \Delta T]$:

$$P(\Delta T) = P[N(\Delta T) \geq 1] = 1 - P[N(\Delta T) = 0] \tag{4.41}$$

In reliability engineering, such a probability is referred to as the failure probability, being complementary to the reliability function $R(\Delta T)$ over time ΔT.

$$p_f = 1 - R(\Delta T) \tag{4.42}$$

Note the useful elementary property:

$$\begin{aligned} P(\Delta T) &= 1 - P[N(\Delta T) = 0] = \sum_{n \geq 0} P[N(\Delta T) = n] - P[N(\Delta T) = 0] \\ &= \sum_{n \geq 1} P[N(\Delta T) = n] \leq \sum_{n \geq 1} n.P[N(\Delta T) = n] = \Delta T.F_q(\Delta T) \end{aligned} \tag{4.43}$$

Straightforward differences may be deducted:

- the probability of occurrence is bounded by one, which is not the case with the frequency of occurrence;
- using a unit period of time (e.g. $\Delta T = 1$ year):
  - probability of occurrence is always numerically *smaller or equal* to the frequency of occurrence;
  - probability and frequency of occurrence are *strictly equal* when *multiple events are impossible* over the time period;
  - provided that the probability of occurrence of any *event is rare* and at most only *weakly correlated to another*, probability and frequency of occurrence are *approximately equal*.

The latter property relies on the fact that the terms which rank superior to one in the series defining the frequency become negligible with respect to the first one, the probability that exactly one event occurs. Informally, this is the case when the occurrence of multiple events is not much more than the product of the probability of each one, each of which being very low. Suppose indeed that the time series allows for an indefinite number of unit events to occur during the time period, with complete independence: this is the case with a standard Poisson process of rate λ where all events are independent and benefit from a similar occurrence probability $\lambda\Delta T$ over the time period. Then, classically:

$$P[N(\Delta T) = n] = \exp(-\lambda\Delta T).\frac{(\lambda\Delta T)^n}{n!} \tag{4.44}$$

$$F_q(\Delta T) = \frac{1}{\Delta T}\sum_{n\geq 1} n.\exp(-\lambda\Delta T).\frac{(\lambda\Delta T)^n}{n!} = \exp(-\lambda\Delta T).\lambda.\sum_{n\geq 1}\frac{(\lambda\Delta T)^{n-1}}{(n-1)!} = \lambda \tag{4.45}$$

Hence:

$$P(\Delta T) = 1 - P[N(\Delta T) = 0] = 1 - \exp(-\lambda\Delta T) \approx \lambda\Delta T \quad \textit{if} \quad \lambda\Delta T \ll 1 \tag{4.46}$$

Note eventually another key difference between a frequency and a probability when cumulating distinct (supposedly independent) time periods or equivalently multiplying by n_p the time period. Frequency is additive or multiplicative, while this is not the case with the probability of occurrence:

$$P(n_p\Delta T) = P[N(n_p\Delta T) \geq 1] = 1 - P[N(n_p\Delta T) = 0] = 1 - \prod_{n_p} P[N(\Delta T) = 0] = 1 - (1 - P(\Delta T))^{n_p} \tag{4.47}$$

Intuitively, there is a constraint against the fact that the probability of having a yearly 1 % probable event (1/100-yr return period) cumulates up to a probability of 1 over 100 years: in fact, it is only 1-(1-0.01)100 = 63 %. However, when dealing with small occurrence probabilities with respect to the time period, that is $n_pP(\Delta T) \ll 1$, probability again becomes approximately multiplicative, similarly to frequencies:

$$P(n_p\Delta T) \approx 1 - (1 - n_pP(\Delta T)) = n_pP(\Delta T) \tag{4.48}$$

Such properties are essential in risk analysis where one tackles *rare* undesired events. It explains why the two definitions are often more or less ambiguously mixed: in probabilistic safety assessment (PSA), it is common to speak about a frequency for the initiators while (conditional) probabilities may be preferred for the chain of possible subsequent failure events. Frequencies have the advantage of being 'conservative' (always higher than the probability) and rigorously additive or multiplicative; however, probabilities of occurrence may be closer to the risk measure wanted by the decision-maker when, for instance, a single event is already unacceptable.

In fact, these properties generalise for the cases where *variability* over a population of sub-systems is involved in the output of interest. Think, for instance, of the performance or safety control of a fleet of planes or of power plants. One may then consider the probability of occurrence of at least one undesired event in at least one sub-system (or on the 'least reliable') over the time period. Alternatively, consider the frequency of 'event x sub-system' over the time period. The previous formulae and approximations generalise with n_p denoting the number of sub-systems (assumed to be independent and distributed identically), thus the usual units for failure frequencies in plant x year or flight x hour.

4.3.3 Fundamental risk measures defined over a period of interest

Two fundamental risk measures have therefore been defined as the probability of an undesired event $E(t)$ over the period $[t_f, t_f + \Delta T]$, either as a discrete event or more generally a threshold-exceedance:

$$c_Z(\boldsymbol{d}) = P\left[N_{E(t)}(\Delta T) \geq 1|\boldsymbol{d}\right] \; or \; c_Z(\boldsymbol{d}) = P[N_{Z_t \geq z_s}(\Delta T) \geq 1|\boldsymbol{d}] \tag{4.49}$$

or as the corresponding frequency:

$$c_Z(\boldsymbol{d}) = F_q[E(t)|\boldsymbol{d}] \; or \; c_Z(\boldsymbol{d}) = F_q[Z_t \geq z_s|\boldsymbol{d}] \tag{4.50}$$

In the case of the threshold-exceedance of a continuous process, the first risk measure can be re-expressed into a time-independent formulation involving a new variable of interest, the maximum of $\boldsymbol{Z}_t$ over the period of interest:

$$\begin{aligned} c_Z(\boldsymbol{d}) &= P[N_{Z_t \geq z_s}(\Delta T) \geq 1|\boldsymbol{d}] = 1-P[N_{Z_t \geq z_s}(\Delta T) = 0|\boldsymbol{d}] \\ &= 1-P\left[Max_{[t_f, t_f+\Delta T]} Z_t \leq z_s|\boldsymbol{d}\right] \\ &= P\left[Z = Max_{[t_f, t_f+\Delta T]} Z_t \geq z_s|\boldsymbol{d}\right] = 1-F_M(z_s) \end{aligned}$$

More complex risk measures may be derived on the basis of the distribution of the maximum denoted as F_M. Quantiles z_α (equivalently called *VaR* (Value-at-Risk) in the financial field for a cost or profit variable, cf. Section 4.2.1) can be defined conversely to the probability as threshold z_s for which there is a probability of α of being exceeded by the maximum over the period ΔT, i.e. at least one event of exceedance $E(t) = 1_{Z>zs}(t)$ at any time over the period:

$$c_z = z_\alpha = F_M^{-1}(1-\alpha) \tag{4.51}$$

Closely-related is the T_r-level z_{Tr}, that is the maximal level of the variable of interest Z_t for a *return period* T_r, a concept which has multiple definitions. The first definition relates to the *quantile of a yearly maximum*, a $T_r = 100$-year return level being the $(1 - 1/T_r) = 99\,\%$ upper *quantile* of the maximum variable over a period ΔT of 1 year, $[t_f, t_f + 1]$:

$$c_z = z_{Tr} = F_M^{-1}(1-1/T_r) \tag{4.52}$$

Alternatively, it is the level that is exceeded by the yearly maximum for the first year after an average number T_r years (often much longer than the period of interest ΔT):

$$T_r(z_T) = E\left[Min\left\{n \middle| Max_{[t_f+n, t_f+n+1]} Z_t \geq z_T\right\}\right] = T_r \tag{4.53}$$

or else the level for which the frequency of annual exceedances over a period $[t_f, t_f + T_r]$ lasting T_r years is one:

$$F_q(T_r)[Z_t \geq z_{Tr}] = 1 \tag{4.54}$$

As noted in the previous section, the probability that the maximal level over a period of interest $[t_f, t_f + T_r]$ lasting T_r years exceeds the T_r-level z_{Tr} equates as 1-$(1 - 1/T_r)^{Tr}$. It is approximately 1-$e^{-1} = 63\,\%$ assuming $T_r \gg 1$, a good approximation whenever $T_r > 20$. Note also that the definitions would be slightly different when considering the frequencies of exceedance of Z_t without involving prior annual maximisation.

The *expected shortfall* or tail Value-at-Risk, a popular extension of the VaR that possesses fruitful properties for decision-making (see Sections 4.4.3 and 7.1.4), is defined as a conditional expectation for exceeding a quantile of the maximal event over $[t_f, t_f + \Delta T]$:

$$TVaR_\alpha = E(Z|Z > VaR_\alpha) = E(Z|Z > z^\alpha) \tag{4.55}$$

Indeed, conditionally-defined quantities can be defined more generally with variables of interest taking non-zero values that are conditional only upon threshold exceedance. This is the case with damage costs, for instance, for which the variable of interest could be the cumulated value over the trajectory of Z_t:

$$\begin{aligned} (T_k)_{k=1\ldots} &= \{t \in [t_f, t_f + \Delta T] | Z_t > z_s\} \\ C_c &= \sum_{k=1,\ldots} c_i(Z_{T_k}, T_k, d | Z_{T_k} > z_s) \end{aligned} \tag{4.56}$$

This new variable of interest becomes formulated as a time-implicit scalar random variable, and hence the expected value or quantiles or else exceedance probabilities of the cumulated damage costs can be studied as the risk measure: see Chapter 7 for the flood example. Similarly, a ratio of unavailability cumulating the successive durations of system unavailability following failure events could be defined as a new variable of interest in a conditional way.

In the most general case, risk measures could be defined as a functional of the whole trajectory of $\mathbf{Z}_t$ (and hence of those of the inputs $\mathbf{X}_t$) not just depending on the exceedances of rare events. Underlying stochastic assumptions are required to give appropriate mathematical definitions to the complex integrals involved, such as the expectation of an additive integrand depending on $\mathbf{Z}_t$:

$$R = E[U(Z_t, D_t)] = \int_{s,\Omega} u[\mathbf{Z}_s, \mathbf{d}_s] dP(Z_s) ds \tag{4.57}$$

Eventually, a particular issue may be raised by the *operational control* of compliance *during the period of interest* and a potentially time-decreasing 'risk budget'. A risk measure may, for instance, compute the duration of exceedance of a critical threshold probability (e.g. allowed outage times in nuclear power), and the case may be that the prescribed maximal duration has been overcome before the end of the period of interest though no detrimental event occurred. Specific decision-making challenges arise and may require anticipating the so-called *time-consistent* risk measures (Boda and Filar, 2005).

4.3.4 Handling a time process and associated simplifications

Once the proper risk measure has been defined over the period of interest, the time basis of the input variables or events has particular consequences for modelling and computation (or uncertain propagation) that may justify some simplifications. Indeed, the time-dependent system model typically looks as follows:

$$\mathbf{Z}_t = \mathcal{G}(\mathbf{X}_t, \mathbf{d}) \tag{4.58}$$

with notation $\mathcal{G}$ stressing the fact that the system model is possibly no longer *a simple function* of the instantaneous inputs but more generally an *operator* over the stochastic process describing the uncertain trajectory of $\mathbf{X}_t$ and hence of the variables of interest $\mathbf{Z}_t$ over the period of interest $[t_f, t_f + \Delta T]$ for decision-making. In the 'richest' cases, the study process involves:

- modelling directly the *time series* of all uncertain inputs $\boldsymbol{X}_t$ or $\boldsymbol{E}_t$, for example through discretised autoregressive processes (AR, ARMA, GARCH, etc.) or Brownian motion-derived time-continuous stochastic processes for continuous variables or other Markovian processes for discrete events;
- computing the (dynamic) system model operator through time-based dynamic Monte-Carlo sampling, representing trajectories of the system state over the entire time period and hence deriving the risk measures.

Those are standard practices in power production financial risk analysis. However, they are generally limited to rather simple system models and simplified uncertainty models (mainly Gaussian, binomial or derived processes) for patent computational and statistical estimation reasons. A very different strategy proves to be more adequate for the analysis of rare risk or peak uncertainty whereby:

- the rarity of the uncertain events that weigh most in the risk measure makes it unacceptable to simplify probabilistic assumptions into Gaussian-derived dynamics that fail to represent the distribution tails well;
- those uncertain events involve separate time dynamics that happen during a time scale short enough to be neglected with respect to the overall time period (e.g. a few hours or days for a flood event or reactor accident, compared to a regulatory period of interest of 30 years of operation);
- outside those events, time dynamics do not bring any valuable piece of information for the variable of interest (e.g. the cost of catastrophic damage, the cash-flow of everyday operations; the windspeed during storm, the average wind speed before the event);
- an adequate understanding of their consequences involves system models that require a minimal level of sophistication to provide accuracy that excludes a full simulation over the period of interest.

This involves simplifying the time dynamics into a quasi-static formulation allowing for more complex probabilistic assumptions and sophisticated system models centred on those events. It proceeds as follows:

- The underlying time processes are modelled only for the main input variable or event responsible for the important events. The key modelled items are the time occurrence of such events and their associated magnitude.
- Conditional on those occurrences, the system model may be formulated as a quasi-static causal chain, involving only standard random variables to represent its uncertain inputs.

Typically, this means representing the main initiator events $\{E^i_t\}_{i=1...p}$ through Poisson processes of rate λ_i (or compound Poisson processes associating continuous magnitudes of initiator-associated phenomena $\boldsymbol{X}^i$ conditional on the occurrences of the process) and then defining the system model as the static relation between the output performance of interest Z_{ti} and the uncertain inputs conditional upon the occurrence of any of those initiator events. A series of conditional risk measures is computed accordingly, either $c_{Z_{ti}}|e_i = P(Z_{ti}|e_i)$ or $c_{Z_{ti}}|e_i = E(Z_{ti}|e_i)$, through normal uncertainty propagation, for example through static Monte-Carlo sampling of the conditional system model. Eventually, cumulative formulae summarise the results for all of those conditional scenarios with respect to the scalar (time-independent) variable of interest in order to compute the final time-based risk measure. For instance, the probability of occurrence of the maximum Z for each outcome Z_{ti} conditional upon the initiator events of failure rate λ_i (or equivalently as a failure tree with associated OR gates between the undesired event Z and the failure of any the component Z_{ti} in series) is shown as follows:

$$c_Z = P(Z|[t_f, t_f + \Delta T]) = 1 - \exp\left[-\sum_i \lambda_i \Delta T . P(Z_{ti}|e_i)\right] \approx \sum_i \lambda_i \Delta T . P(Z_{ti}|e_i) \qquad (4.59)$$

The latter approximation is acceptable when the elementary products are small compared to one and hence probability and frequency become approximately equal. In other words, the approximate formula popular in nuclear safety and design states that:

$$\textit{overall risk} = \sum (\textit{initiator frequencies} * \textit{conditional probabilities})$$

A cumulative sum of the sub-event consequences would alternatively be considered when dealing, for instance, with a cumulated *damage cost* or a *ratio of unavailability* over the period. In that approach, the full simulation of time dynamics is replaced by that of rare sub-events, complemented by closed-form final balance.

4.3.5 Modelling rare events through extreme value theory

In fact, there is a close link between those comments and the statistical framework of *extreme value theory* (EVT) (Reiss and Thomas, 2001; Coles, 2001). With the purpose of estimating rare quantiles (or exceedance probabilities) of a random variable of interest with limited datasets, EVT has been developed on the premise that, in many cases, the asymptotic behaviour of extreme realisations of a random variable falls within a limited number of known distribution shapes.

Two basic formulations lead to the foundational theorems. Consider first the *law of maximal value* of an increasing set of random variables of interest Z_j all *i.i.d.* according to a cdf denoted as $F_Z(z|\theta)$. Elementary probability calculus shows that the law of maxima results as follows:

$$F_n(z) = P\left[Z = Max_{j=1\ldots n} Z_j \le z\right] = F_Z(z|\theta)^n \tag{4.60}$$

Under given non-degeneracy and regularity conditions, Fisher and Tippett (1928) and Gnedenko (1943) proved that its asymptotic behaviour, provided there is appropriate rescaling when the set gets larger, resembles that of a given set of limit distributions *GEV*(.) in the sense of a convergence in law:

$$\begin{aligned} & P\left[Max_{j=1\ldots n} Z_j \le \sigma_n x + \mu_n\right] \xrightarrow{L} GEV(x) \\ & \Leftrightarrow \left| F_n(z) - GEV\left(\frac{z-\mu_n}{\sigma_n}\right)\right| \underset{n\to\infty}{\rightarrow} 0 \end{aligned} \tag{4.61}$$

where *GEV*(.) is one of Gumbel, Weibull or Fréchet, together known under the title of Generalised Extreme Value distribution (GEV) as follows:

$$\forall z \left| 1 + \xi\left(\frac{z-\mu}{\sigma}\right) > 0,\ GEV(z|\mu,\sigma,\xi) = \exp\left\{-\left[1+\xi\left(\frac{z-\mu}{\sigma}\right)\right]^{-\frac{1}{\xi}}\right\}\right. \tag{4.62}$$

where μ, σ and ξ are location, dispersion and shape parameters respectively, the latter denoting the light-tail upper-bounded behaviour ($\xi < 0$. *Weibull*) or heavy (or fat, long ...)-tail lower-bounded behaviour ($\xi > 0$, *Fréchet*). The *Gumbel* model ($\xi = 0$) is very popular, for instance, in flood risk analysis and may be rewritten as follows:

$$\text{Gumbel}(z|\mu,\sigma) = \exp\left\{-\exp\left[-\left(\frac{z-\mu}{\sigma}\right)\right]\right\} \tag{4.63}$$

Standing between Weibull and Fréchet as a limit model when $\xi \to 0$, the Gumbel model is also limitless either on its lower or upper side. Values of the variable of interest can be theoretically larger than any threshold for a very small probability (or lower but the decrease is very fast on the lower part, thus negligible in practice). Though uncertain *physical* variables will be bounded by nature, such a mathematical model still proves interesting as their upper bound is always difficult to assess *a priori* as surprisingly severe catastrophic events happened in history (see the comments in Section 5.2.6). To put it simply, note that those three distribution shapes prove 'max-stable', that is the shape of the distribution of maxima is identical to that of the underlying random variable, since:

$$\begin{aligned} F_{nM}(z) &= P\left[Z = Max_{j=1\ldots M}\left(Max_{i_j=1\ldots n} Z_{i_j}\right) \le z\right] = F_n(z|\theta)^M = GEV(z|\mu_n, \sigma_n, \xi)^M \\ &= \exp\left\{-M\left[1+\xi\left(\frac{z-\mu_n}{\sigma_n}\right)\right]^{-\frac{1}{\xi}}\right\} = \exp\left\{-\left[1+\xi\left(\frac{z-\left[\mu_n + \frac{\sigma_n M^\xi}{\xi}\left(1-M^\xi\right)\right]}{\sigma_n M^\xi}\right)\right]^{-\frac{1}{\xi}}\right\} \\ &= GEV\left(z\left|\mu_n + \frac{\sigma_n M^\xi}{\xi}\left(1-M^\xi\right), \sigma_n M^\xi, \xi\right.\right) \end{aligned} \tag{4.64}$$

while for the Gumbel model, only μ_n is replaced by $\mu_n + \sigma_n logM$ when considering the maximum over M block maxima. They are in fact the only continuous cdf to behave in such a max-stable way.

The alternative formulation involves the consideration of the *law of peaks over threshold* (*POT*), that is the distribution of a random variable of interest Z conditional on its exceeding an increasing threshold (see Figure 4.3). Elementary probability calculus shows that the law results as follows:

$$F_u(z) = P[Z \le z | Z > u] = \frac{F_Z(z|\theta) - F_Z(u|\theta)}{1 - F_Z(u|\theta)} \tag{4.65}$$

Under given non-degeneracy and regularity conditions, Picklands (1975) proved that its asymptotic behaviour, provided that appropriate rescaling is undertaken when the threshold gets higher, resembles that of a given set of limit distributions $G(.)$:

$$\left| F_u(z) - G_u\left(\frac{z-\mu_u}{\sigma_u}\right) \right| \underset{u \to u_\infty}{\to} 0 \tag{4.66}$$

where $G(.)$ is one of Pareto distributions, together known under the Generalised Pareto Distribution (*GPD*) as follows:

$$GPD_u(z|\mu_u, \sigma_u, \xi_u) = 1 - \left[1 + \xi_u\left(\frac{z-\mu_u}{\sigma_u}\right)\right]^{-\frac{1}{\xi_u}} \tag{4.67}$$

and where u_∞ denotes the endpoint of the distribution either $u_\infty = +\infty$ or a finite upper bound. Similar to *GEV* maxima distributions, μ_u, σ_u and ξ_u are location, dispersion and shape parameters respectively, the latter denoting the light ($\xi_u < 0$, *Beta*) or heavy-tail ($\xi_u > 0$, *Pareto*) behaviour. The important case is

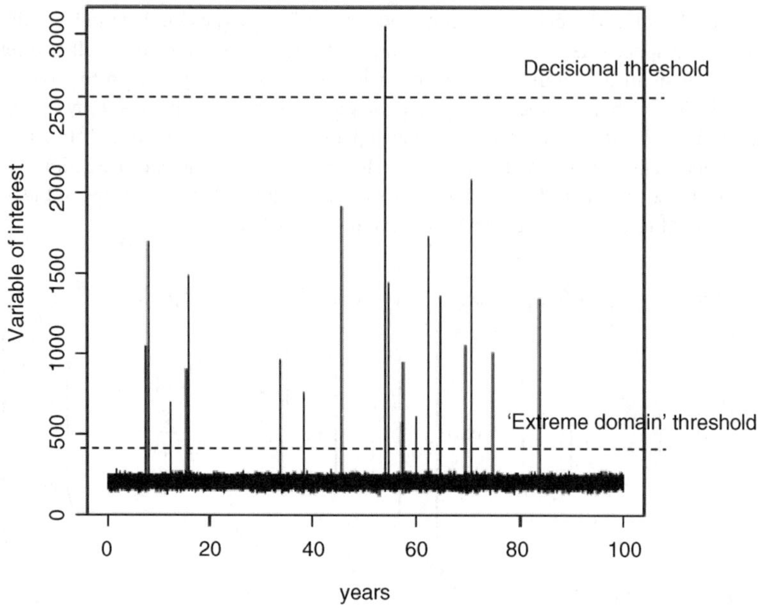

Figure 4.3 *Modelling extreme values.*

where $\xi_u = 0$ corresponds to the *Exponential* distribution. To put it simply, note that those three distribution shapes are the only continuous functions that prove 'POT-stable', that is for which the shape of the distribution of POT is identical to that of the underlying random variable. There is a direct theoretical correspondence between the two approaches that, for instance, leads to formulae linking the parameters for *GPD* or *GEV* (see the end of this section), the shape factor being the same; the sign of which is well known for a number of common base distributions for Z_t. For instance, when Z_t is Gaussian or lognormal, its extreme maxima (respectively its peaks over threshold) follow asymptotically a Gumbel (resp. exponential) distribution (Coles, 2001).

Care has to be taken when handling Fréchet or corresponding Pareto distributions that model '*heavy-tail*' events. Features include, for instance, the fact that the expectation is only defined (i.e. finite) for $0 < \xi < 1$, the variance is only defined for $0 < \xi < {}^1\!/_2$ and the more generally upper moments of order k are only defined provided the shape parameter is small enough so that $0 < \xi < 1/k$. Though most physical, environmental, economic or financial variables of interest would remain bounded by nature, such mathematical models may sketch the empirically-observed behaviours of extreme variability. Think of the cost of catastrophic events, wealth distributions or more generally what Taleb (2007) would refer to as scalable phenomena, for which a single additional observation can disproportionately affect the sum, thus the law of large numbers or the central-limit theorem can be inapplicable. Care is needed for their estimation – still possible through some algorithms though not all of them, see Section 5.2 – and for their later propagation inside risk computation algorithms – again, only possible through some algorithms, such as exceedance frequencies, see Section 7.2.2.

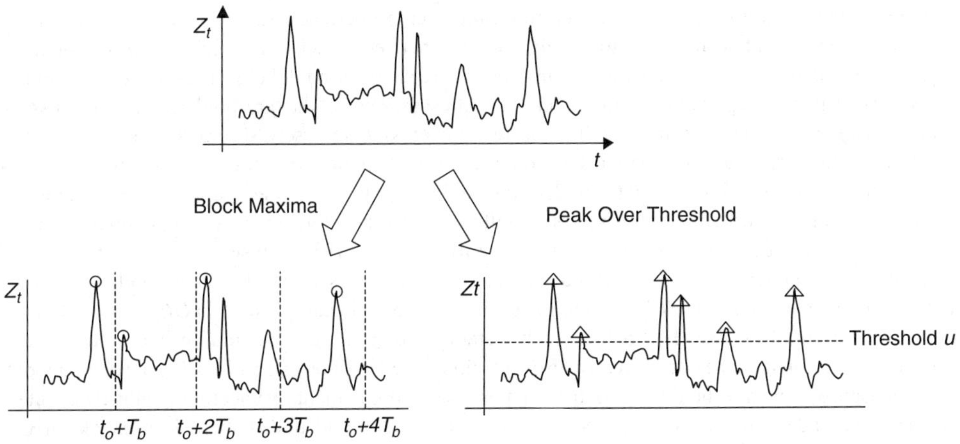

Figure 4.4 *Two basic techniques to recover an i.i.d. sample of extreme values from a time series.*

The straightforward applications of EVT are the situations where there are direct observations on the variable of interest, so that the system and uncertainty model to be estimated are a simple statistical model[4] $Z_t = X_t \sim F_Z$. In order to be able to apply the foundational EVT theorems to real time series, and estimate such a model, there is a crucial need to *transform* the *raw inter-correlated data* into an *i.i.d.* set: this is where the approach involves the above-mentioned simplifications of the uncertainty dynamics. Indeed, a key assumption standing behind the asymptotic results of the two fundamental theorems and the possibility to estimate such models is that the distribution of Z is *i.i.d.* (see the numerical example of a correlated time series in Chapter 5, Section 5.5) which is never the case in practice, as mentioned in Section 4.3.1. Assuming that there is an observed time series, with a small time step (every minute, hour or day) over a large period (a few years or decades)

$$\left(z_j = z_{tj}\right)_{j=1\ldots n} \tag{4.68}$$

the two fundamental techniques are the following (see Figure 4.4):

- either (I) retrieve the maximal value over blocks of fixed time periods (e.g. 1 month, 1 year ...),
- or (II) retrieve the maximal value of peaks over a given threshold.

In each case, the goal is to get rid of auto-correlation assuming that if blocks are long enough in time or if the threshold high enough it is unlikely that the maxima of successive blocks or successive peaks will

[4] More specifically the system model would resemble that of Step One, namely $G(\boldsymbol{x}, \boldsymbol{d}) = z_d - h - z_v$, whereby h is observed statistically, $-z_v$ being penalised deterministically upon pure expertise, and where the action consists of controlling the level of protection z_d, see Chapter 3.

remain correlated. Care is of course needed to ensure such properties, treating for example the cases where an extreme event occurs close to the dividing date between blocks (in which case move the time origin slightly or group into a single maximum); or where a short-term fall below the threshold separates two peaks that are in fact part of the same event (then adjust the level of the threshold accordingly, or keep a single maximum value over the so-called cluster of events). Conversely, the longer the time blocks or the higher the threshold, the smaller the proportion of information kept for later modelling.

Note that such a simplification of the underlying random dynamics in time – hence an apparent loss of information compared with direct time series modelling – is justified not only by its simplicity but also by the fact that the frequent time to time dynamics are irrelevant for risk analysis.

Once such a pre-treatment is achieved, one recovers an *i.i.d.* sample of a static random variable amenable to be modelled by a *GEV* distribution of block maxima in case (I) or by a *GPD* distribution of peaks-over-threshold in case (II). In fact, such a static random variable should be interpreted as the behaviour *conditional on* the occurrence of an event, either fixed in time (every block for (I)) or at random (the exceedance of the threshold for (II)). In the latter case, a complementary uncertainty model is built to represent the underlying random exceedance process $E_t(u)$: provided that there is a check on the independence and equilibrated distribution of time intervals between peaks, this is done generally through a Poisson process estimated straightforwardly through taking the mean frequency of the set of supposedly independent times of threshold exceedance (see Chapter 5, Section 5.2.3). This means that the uncertainty model becomes as described in Table 4.3.

In the case of POT, consider the counting process of the peaks over the threshold and the exceedance probability results as follows:

$$
\begin{aligned}
c_z &= P\left[Max_{[t_f, t_f+\Delta T]} Z_t^u \geq z_s\right] \\
&= 1 - \sum_{k \geq 0} P[N(\Delta T) = k].P(all\ below\ z_s) \\
&= 1 - \sum_{k \geq 0} \exp(-\lambda \Delta T).\frac{(\lambda \Delta T)^k}{k!} F_u(z_s)^k = 1 - \exp(-\lambda \Delta T[1 - F_u(z_s)])
\end{aligned}
\tag{4.69}
$$

Incidentally, this enables one to make explicit the link between the POT and Block Maximum model. Assume that Z_t^u is distributed over a compound Poisson-*GPD* model: then the cdf of the block maximum is distributed over a *GEV* with the following parameters:

$$
\begin{aligned}
F_{\Delta T}(z) = P\left[Max_{[t_f, t_f+\Delta T]} Z_t^u \leq z\right] &= \exp(-\lambda \Delta T[1 - GPD_u(z|\mu_u, \sigma_u, \xi_u)]) \\
&= \exp\left(-\lambda \Delta T\left[1 + \xi_u\left(\frac{z - \mu_u}{\sigma_u}\right)\right]^{-\frac{1}{\xi_u}}\right) \\
&= GEV\left(z \middle| \mu_u + \frac{\sigma_u (\lambda \Delta T)^{\xi_u}}{\xi_u}\left(1 - (\lambda \Delta T)^{\xi_u}\right), \sigma_u (\lambda \Delta T)^{\xi_u}, \xi_u\right)
\end{aligned}
\tag{4.70}
$$

In particular, a Poisson-*GPD* model with a Poisson rate of one per year leads to a *GEV* model with numerically-identical parameters in yearly maximum ($\lambda \Delta T = 1$). More on the associated estimation issues and the relevance of the extreme value theory may be found in Chapter 5.

Table 4.3 The two EVT models.

	Block Maxima	POT
variable of interest	$Z_M = Max_{[tf,tf+\Delta T]}\{Z_t\}$	$Z = Max_{[tf,tf\ +\ \Delta T]}\{Z_t\}$
system model	$Z_M = X$	$Z = Max_{[tf,tf\ +\ \Delta T]}\{1_{Et(u)}X_k\}$
risk measure	$c_z = P(Z_M > z_s) = 1 - F_M(z_s)$	$c_z = P(Max_{[tf,tf\ +\ \Delta T]}\{Z^u{}_t > z_s\})$
	or an associated return period	or an associated return period

Flood example

The dominant component for which the time basis is essential is the hydraulic part, as the system reliability and cost components are relevant only *conditional* upon a large flood event. Within the hydraulic part, the time series of flow is of major interest as it is generally historically monitored (on a daily or even hourly basis for recent years) unlike the associated time series of other characteristics of river flow such as riverbed sedimentation/erosion or friction, which are poorly known beyond limited expertise or rare observation samples. Flow is naturally intensely-correlated on an hourly or daily basis, and is also the subject of strong seasonality that is non-identically distributed throughout the year. Flood studies rely either on yearly maxima (one-year blocks), or a peak-over-threshold. The threshold is chosen high enough to exclude dependence between exceeding values, albeit within a given (say three-day long) flood event that is then clustered into a single maximal flow. Hence, the output of interest becomes the maximal flood extent (either overflow, or associated plant failure or accidental cost) of the next year or any year in the future, or the flood extent for the next large-enough flood (the annual frequency of which can also be also estimated), both being static random variables with *i.i.d.* data available for modelling.

4.4 Choosing an adequate risk measure – decision-theory aspects

This section will provide a closing discussion on the choice of risk measures, taking into account both decision-theory arguments and practical study goals and experience.

4.4.1 The salient goal involved

Let us recall firstly the variety of goals involved in a risk and uncertainty study for which the risk measure is to be chosen:

- *Understand* (*U*) or *Accredit* (*A*) are somewhat upstream goals.
- *Select* (*S*) or *Comply* (*C*) are goals involving full decision-making.

The two former goals may either be part of a larger process targeting eventually one of the two latter goals or may be considered as such within an upstream approach.

When targeting *U* or *A within an upstream approach*, a *quantity of interest* should help in measuring the amount of uncertainty with respect to which it is possible to understand the model and control its accuracy broadly. When nothing else is specified for goal *U*, variance is a typical choice as an efficient quantity of interest with respect to which one may explore the space of uncertain inputs, that is of risk and uncertainty factors. Probability may be viewed more as a space-filling exploratory tool than a true representation of the decision-maker's belief or preference for the space of possible values and the exploration would rather benefit from randomising as many factors as possible, even those for which the choice of distribution is a difficult task (as discussed in Section 4.2). Goal *A* may either rely on:

- (probabilistic) variance-based quantities of interest such as the standard deviation or coefficient of variation, which are practical ways to control the model prediction error;
- (deterministic) maximal ranges, which may be an alternative way to control the model prediction error.

The observations made in Section 4.3 help with understanding the pros and cons of those two possibilities for controlling the model error.

Choosing the risk measure becomes all the more important when considering goals *S* or *C* and they are the prime subject of the debate in the literature. The case is similar when goals *U* or *A* are a pre-requisite for those two, where the quantity of interest to be chosen for ranking or accrediting a model should really be closest to the final risk measure used for the subsequent goal *S* or *C*.

4.4.2 Theoretical debate and interpretations about the risk measure when selecting between risky alternatives (or controlling compliance with a risk target)

Suppose now that the dominant goals are *Select* (*S*) or *Comply* (*C*). This means either selecting between risky alternatives or controlling the risk associated with a given infrastructure or industrial system or investment with respect to '*absolute*' (or say *prior*) risk metrics. Understanding the pros and cons associated with each possible risk measure requires a return to the interpretation given to the probabilities as a tool for representing the uncertainties in the state of the system. The classical set of fundamental axioms of probability theory (see Annex Section 10.2 for a brief recall) allow for different views and interpretations as to their representation of an analyst's or decision-maker's view of uncertainty. Here are the classic choices:

(a) consider probabilities as an '*objective*' reflection of the uncertain (or variable) states of world,
(b) or consider probabilities as a way of representing a rational decision maker's view of the uncertain state of the world, often referred to as '*subjective*',
(c) or even, consider probabilities as a tool to explore (model) the variability of a system model within traceable metrics.

Interpretation (a) makes the understanding of the axioms rather intuitive in basing the probability on the frequency of observed events. Interpretation (b) may be explained briefly from the perspective of this book. A 'rational' decision-maker exhibits a 'rational preference' over the world that follows a given list of properties (axioms). According to the notation of this book, consider that:

- $\boldsymbol{x}$ represents the (uncertain) states of the world;
- $\boldsymbol{d}$ represents the actions that may be taken by the decision-maker;
- $z \mid \boldsymbol{d} = G(\boldsymbol{x},\boldsymbol{d})$ represents the consequences or 'states of the acting subject': this means that z, the variable of interest, has been selected to represent the output of the system which is most relevant to the decision-maker.

Following Savage (1972) or Bedford and Cooke (2001) an *act* corresponding to $\boldsymbol{d}$ would be represented by a function f corresponding to the following:

$$\begin{aligned}&\Omega_x - > \Omega_z \\ &\boldsymbol{x} - > z|\boldsymbol{d} = G(\boldsymbol{x}, \boldsymbol{d})\end{aligned} \tag{4.71}$$

To put it simply, such an approach to rational preference may be understood as follows. A decision-maker having to select actions in the context of uncertain states of the world (implying the uncertain consequences of his actions) exhibits necessarily:

- a (subjective) measure of probability representing relative beliefs in the likelihood of $\boldsymbol{x}$. Probability of events (states of the world) emerges as the relative degree of preference of pure reward-acts, meaning the acts that produce a standard reward if and only if the corresponding event occurs. The posterior probability distribution $f_X^\pi(\boldsymbol{x}|IK, \Xi_n)$could be taken to reflect all background information available to the decision-maker.
- A utility function representing those preferences over the consequences.

Under uncertainty, an act f_A is thus preferred to f_B if and only if the expected utility of its consequences is greater. Mathematically:

$$\begin{aligned}&f_A(\text{or } \boldsymbol{d} = \boldsymbol{d_A}) \text{ is preferable to } f_B(\text{or } \boldsymbol{d} = \boldsymbol{d_B}) \\ &\Leftrightarrow c_z(\boldsymbol{d_A}) = EU|\boldsymbol{d_A} > EU|\boldsymbol{d_B} = c_z(\boldsymbol{d_B})\end{aligned} \tag{4.72}$$

whereby:

$$c_Z(\boldsymbol{d}) = EU|\boldsymbol{d} = E[U(Z|\boldsymbol{d})] = \int_z U(z) f_Z^\pi(z|IK, \Xi_n, \boldsymbol{d})dz \tag{4.73}$$

This may be viewed as an argument in favour of the superiority of the *expected utility* as a risk measure. Note, however:

- Various decision-makers would possibly not only exhibit different utility (according to their risk-aversion, see Chapter 8), but also different probability measures according to their level of information or belief in the likelihood of events. This may not be very practical in the regulation of risk affecting society with widely-differing individuals.
- Expected utility suffering from classical paradoxes as made popular by (Allais, 1953).
- The axioms may not all be natural, particularly in the context of rare or irreversible events (e.g. death of the decision-maker, world catastrophe . . .) for which an ordering of preferences in the sense of Savage's 'rational preference' may not be possible (see Bedford and Cooke, 2001).

4.4.3 The choice of financial risk measures

The choice of the appropriate risk measures has been greatly discussed in the finance or insurance literature. In those contexts, z would represent a net cash flow, a market value, a cumulated cost of claims or a minimal margin to solvency of a portfolio; z being modelled as an aggregate of elementary assets, claims or risk factors represented by the $\boldsymbol{x}$. Decision-making then typically involves controlling the risk that such performance indicators characterising given assets reach undesirably low levels; or alternatively select the best actions to take on the underlying portfolio. In that context, decision making would involve the arithmetic manipulation of the underlying systems formally represented by $z = G(\boldsymbol{x},\boldsymbol{d})$, including, for instance, the merging of two portfolios $z = z_1 + z_2$ or a linear variation of the initial portfolio through associated transactions over the aggregated assets. In those insurance theory or finance contexts, the following list of hypotheses are reckoned classically as necessary to build a 'coherent' risk measure (Artzner *et al.*, 1999), c_z being a functional designed to increase when the level of risk increases and z being formulated so that it increases when the payoff increases:

- monotony
  - if $z(\boldsymbol{x},\boldsymbol{d_A}) \leq z(\boldsymbol{x},\boldsymbol{d_B})$ for every input state $\boldsymbol{x}$ then $c_z(\boldsymbol{d_A}) \geq c_z(\boldsymbol{d_B})$
- sub-additivity
  - $c_z(Z_1) + c_z(Z_2) \geq c_z(Z_1 + Z_2)$
- positive homogeneity
  - $\lambda c_z(Z) = c_z(\lambda Z)$ for any positive scalar λ
- translation invariance
  - $c_z(Z) + a = c_z(Z + a)$ where a is certain (controlled variable)

Consider the value-at-risk introduced in Section 4.2.1, by definition a quantile or more precisely the opposite of the quantile of the payoff $VaR_\alpha = - z^\alpha$ so that it measures the amount that can be *lost* for a given (low) probability α. It is often computed on a v.i. which is a daily log-return, that is differences between log prices. Thus, high *VaR* correspond to high volatility periods, in the sense that volatility is generally understood as the standard deviation of the log-returns on a time series (either daily, monthly, annually . . .). But the *VaR* may more generally be computed on a net cash flow or a portfolio profit/loss cumulated over a longer period of interest, for example a month. Alas, the *sub-additivity* property is violated by the *VaR* which makes it problematic since that property is essential in order to account for the effect of diversification. Merging two assets can only reduce the risk, at worst leave it unchanged if totally correlated (see Section 4.2.1).

As an alternative measure fulfilling all of the four properties mentioned above, an increasingly popular risk measure in insurance or finance is the expected shortfall (ES_α) or tail value at risk ($TVaR_\alpha$):

$$TVaR_{\alpha\%} = E(-Z|-Z > VaR_{\alpha\%}) \tag{4.74}$$

$$ES^\alpha(Z|\boldsymbol{d}) = -\frac{1}{\alpha}\int_0^\alpha F_Z^{-1}(u)du \tag{4.75}$$

Though technically defined differently, both measures coincide when the variable of interest has a continuous density. Their advantage is to focus not only on the (default) probability, but also the consequences of breaching that level. They also fulfil the prescribed features for a risk measure, particularly that of sub-additivity.

4.4.4 The challenges associated with using double-probabilistic or conditional probabilistic risk measures

Chapters 1 and 2 have already shown the use of *double probabilistic* risk measures in risk assessment contexts as an alternative to the single probabilistic risk measures, which are expected utilities in most cases. As already discussed, they keep in evidence the *aleatory-epistemic* distinction inside the risk measure, the level-1 modelling the aleatory (or 'risk') or variability component while the level-2 models the epistemic (or 'uncertainty') component. In the areas of protection against natural risk or integrated probabilistic risk assessment, typical regulatory decision-making criteria under the final goal C may require that the 'yearly frequency of the undesired event of interest is less than *1/T* at an α-level of confidence', the level of confidence being meant to cover the risk of under-evaluating the risk level due to data limitations (small datasets, expert controversies, lack of model validation ...).

Setting aside the particular case where decision making involves an explicitly variability-based output of interest (such as a frequency of failure over a fleet of sub-systems, of disease over a population of individuals or of performance at multiple future time steps), such double probabilistic risk measures used for single systems at a single time horizon have strong disadvantages:

- being quite complex to interpret and communicate,
- failing to provide a complete ranking between risky alternatives.

The practice of risk assessment entails many difficulties in getting the stakeholders to understand properly the meaning of single probabilities, in particular that of a probabilistic risk level (see, for instance, Granger Morgan and Henrion, 1990). While specified explicitly in some risk regulations (e.g. the French nuclear safety rule on external aggressions, cf. Chapter 1), it proves even trickier to handle the 1/1000 yr-70 % figures which typically do not allow for the simple interpretations of 'once on average in a thousand years' and require the consideration of multiple underlying random experiments. Even more difficult is the latter: double probabilistic risk measures lack the core advantage of probabilities for decision making, the ability to rank risky options. Indeed:

- assuming an identical threshold of consequences, it is hard to rank univocally the undesirability of '1/1000 yr-50 %' *vs.* '1/500 yr-70 %' failure likelihoods;
- additionally, it is hard to answer univocally the absolute relevance of selecting any percentage as the appropriate confidence level to cover data limitations: should the targeted level of residual risk (say 10^{-4} per year) be guaranteed with 70 or 95 %, or even 99 % confidence?

While it may be seen as unduly conservative to provide an additional margin for lack of knowledge when dealing with severe risks, rather sophisticated decision theory frameworks become necessary when trying to answer it in a rational way, as evidenced by the long-standing debate in U.S. nuclear regulation. An attractive option is to stick to a *single probabilistic risk measure* which, *in fine*, summarises the analyst's view of the uncertainty in the variables of interest at the prescribed time of interest *after* having clearly *modelled the two* levels at the *estimation stage*. Indeed, while it may be seen as useful at that stage to separate explicitly the aleatory from the epistemic components within a full uncertainty model in order to identify *to what extent uncertainty could be reduced*, it is always possible to eventually summarise adequately both components within a combined risk measure, namely the expected utility averaged over the epistemic component as follows:

$$\begin{aligned} c_z &= E_{\theta_X}[E_X(U(Z)|\boldsymbol{\Theta}_X)|IK, \Xi_n] \\ &= \int_\theta \left[\int_X U(G(\boldsymbol{x}, \boldsymbol{d}) f_X(\boldsymbol{x}|\boldsymbol{\theta}_X) d\boldsymbol{x} \right] \pi(\boldsymbol{\theta}_X|IK, \Xi_n) d\boldsymbol{\theta}_X \\ &= \int_{\underline{X}} UoG(\boldsymbol{x}, \boldsymbol{d}) f_X^\pi(\boldsymbol{x}|IK, \Xi_n).d\boldsymbol{x} \end{aligned} \tag{4.76}$$

Such a risk measure has the desirable properties of the standard expected utility, particularly the full ranking of risky alternatives, while still incorporating the largest account of the level of uncertainty, variability or lack of knowledge remaining about the states of the system.

Besides, one often encounters another type of risk measure that deviates from the standard probabilistic expected utility under goals *C* or *S*, particularly in the field of engineering: the conditional probabilistic risk measures, for example the expected utility *conditional* on a given scenario covering a number of *penalised* inputs inside ***d***, as introduced in Chapter 2:

The observations in Section 4.2.5 have made it clear that such a deterministic penalisation may resolve the tricky controversies associated with the probabilistic representation of complete ignorance about some inputs. However, it similarly reduces the ranking scope of the risk measure as two risky designs or options may only be compared as long as their deterministic conditioning is similar, which clearly excludes the ranking of the impact of actions modifying those uncertain features that are not probabilised. It even reduces the meaning of ranking options regarding the inputs that are probabilised as the change of the penalised value for non-probabilised inputs may lead to inconsistent preferences.

4.4.5 Summary recommendations

A number of potential risk measures are available, whether single or double-probabilistic, or more or less conditioned on deterministic scenarios. Accordingly, various natures of uncertainty may be involved in the state of the system to be predicted, showing to some extent the consequences of building probabilistic or deterministic input uncertainty measures. What can be drawn out of this potentially-confusing situation? First of all, it depends on the dominant goal of the study:

- In upstream stages where 'Understand' or 'Accredit' are dominant, it may prove advantageous to conduct *separate treatments* for different natures of uncertainty, error, variability, lack of knowledge and so on: although it could be seen as making the process more complex, it could also help to value the *reducibility options*.
- In more decisional stages ('Comply' or 'Select'), save the particular case of variability-explicit system performance (or population exposure) control, single probabilistic risk measures conditioned on a core set of deterministic hypotheses (particularly when dependence or choice of *pdf* is too controversial[5]) display key decision-theory advantages, namely their ranking potential and solid formal interpretations.

Regulatory processes may sometimes impose the risk measures and associated settings. Eventually, it is always essential to document clearly the process beyond the final risk measure, including a detailed discussion of the natures of uncertainties, of the significance of the variables of interest for the state of the system (and hence of the sample space), and of sensitivity analyses of all kinds of epistemic estimation or propagation uncertainties: especially when they are eventually to be summarised within a synoptic risk measure. Remember also that while the preliminary planning and decision-making reflections could usefully involve probabilistic risk measures, operational regulation and/or control may require a translation into deterministic guidelines to simplify mass practice and avoid the risk of dispersion in implementation.

Exercises

4.1 There follows a list of exercises referring to the concepts developed in this chapter:
Discuss the link between correlation signs and variance inflation for linear models. Can it be generalised for monotonous models regarding quantiles?

[5] Think also of the case of design codes where there is a risk of discrepancy in probabilistic engineering practice for small structures in particular.

4.2 Discuss the Bertrand paradoxes in the context of time-dependent uncertain inputs: how can the growth of entropy be controlled, for example, through Markovian assumptions (see statistical physics, information theory, etc.); how does this impacts the applicability of maximal entropy to the output variable of interest?
4.3 Discuss the application of the principle of maximal entropy to the dependence model of a bivariate uncertain input (see Chapter 5 for definition). How is it impacted by the characteristics of the system model (e.g. monotony) when maximising entropy into the output variable of interest?
4.4 The standard T_r-level or return period defined in Section 4.3.3 assumes $T_r \geq 1$: how can the definition be extended to $T_r \leq 1$ and for what purpose? Moreover, how do the T_r-levels scale when changing the underlying unit period ΔT (by default, one year)?
4.5 Give examples for which the VaR fails to cover under-additivity and study the associated behaviour of TVaR.

References

Allais, M. (1953) Le comportement de l'homme rationnel devant le risque: critique des postulats et axiomes de l'école américaine. *Economica*, **21**(4), 503–546.

Artzner, P., Delbaen, F., Eber, J.M. and Heath, D. (1999) Coherent measures of risk. *Mathematical Finance*, **9**, 203–228.

Bedford, T. and Cooke, R. (2001) *Probabilistic Risk Analysis – Foundations and Methods*, Cambridge University Press.

Boda, K. and Filar, J.A. (2005) Time consistent dynamic risk measures. *Mathematical methods in Operations Research*, **63**(1), 169–186.

Cacuci, D.G. *et al.* (1980) Sensitivity theory for general systems of nonlinear equations. *Nuclear Science and Engineering*, **75**, 88–110.

Coles, S. (2001) *An Introduction to Statistical Modelling of Extreme Values, Springer Series in Statistics*, Springer Verlag.

Cover, T.M. and Thomas, J.A. (1990, 2006) *Elements of Information Theory*, John Wiley & Sons, Ltd.

de Rocquigny, E. (2006) La maîtrise des incertitudes dans un contexte industriel: 1ère partie – une approche méthodologique globale basée sur des exemples; 2nd partie – revue des méthodes de modélisation statistique, physique et numérique. *Journal de la Société Française de Statistique*, **147**(4), 33–106.

de Rocquigny, E., Devictor, N. and Tarantola, S. (eds) (2008) *Uncertainty in Industrial Practice, A Guide to Quantitative Uncertainty Management*, under publication by John Wiley & Sons, Ltd.

Embrechts, P., McNeil, A. and Straumann, D. (1999) *Correlation and Dependence in Risk Management: Properties and Pitfalls*, Preprint RiskLab/ETH Zürich.

Fisher, R.A. and Tippett, L.H.C. (1928) Limiting forms of the frequency distribution of the largest or smallest member of a sample. *Proceedings of the Cambridge Philosophical Society*, **24**, 180–190.

Gnedenko, B. (1943). Sur la distribution limite du terme maximum d'une série aléatoire (in French). *Ann. Math.* **44**, 423–453.

Granger Morgan, M. and Henrion, M. (1990) *Uncertainty – A Guide to Dealing with Uncertainty in Quantitative Risk and Policy Analysis*, Cambridge University Press.

Helton, J.C. (1994) Treatment of uncertainty in performance assessments for complex systems. *Risk Analysis*, **14**, 483–511.

Helton, J.C., Burmaster, D.E., *et al.* (1996) Treatment of aleatory and epistemic uncertainty, *Special Issue of Reliability Engineering and System Safety*, **54**(2–3).

Helton, J.C. and Oberkampf, W.L. (2004) (eds) Alternative representations of epistemic uncertainty. *Special Issue of Reliability Engineering & System Safety*, **85**(1–3).

Hoffmann, F.O. and Hammonds, J.S. (1994) Propagation of uncertainty in risk assessments: The need to distinguish between uncertainty due to lack of knowledge and uncertainty due to variability. *Risk Analysis*, **14**(5), 707–712.

ISO (1995) Guide to the expression of uncertainty in measurement (G.U.M.), EUROPEAN PRESTANDARD ENV 13005.

Jaulin, L., Kieffer, M., Didrit, O. and Walter, E. (2001) *Applied Interval Analysis, with Examples in Parameter and State Estimation, Robust Control and Robotics*, Springer, London.

Keynes, J.-M. (1921) A treatise on probability.

Knight, F.H. (1921) *Risk, Uncertainty and Profit*, Hart, Schaffner & Marx.

Limbourg, P. and de Rocquigny, E. (2010) Uncertainty analysis using Evidence theory - confronting level-1 and level-2 approaches with data availability and computational constraints. *Reliability Engineering and System Safety*, **95**(5), 550–564.

Nilsen, T. and Aven, T. (2003) Models and model uncertainty in the context of risk analysis. *Reliability Engineering & System Safety*, **79**, 309–317.

Oberkampf, W.L., DeLand, S.M., Rutherford, B.M. *et al.* (2002) Error and uncertainty in modelling and simulation. *Special Issue of Reliability Engineering & System Safety*, **75**(3), 333–357.

Picklands, J. (1975) Statistical inference using extreme value order statistics. *Annals of Statistics*, **3**(1), 119–131.

Reiss, M.D. and Thomas, M. (2001) *Statistical Analysis of Extreme Values – with Applications to Insurance, Finance, Hydrology and Other Fields*, 2nd edn, Birkhäuser.

Saltelli, A., Tarantola, S., Campolongo, F. and Ratto, M. (2004) *Sensitivity Analysis in Practice: A Guide to Assessing Scientific Models*, John Wiley & Sons, Ltd.

Savage, L.H. (1972) *The Foundations of Statistics*, Dover Publication, Inc.

Shannon C.E. (1948) *A mathematical theory of communication*, Bell Systems Technical Journal, **27**, pp. 379–423.

Slob, W. (1994) Uncertainty analysis in multiplicative models. *Risk Analysis*, **14**(4), 571–576.

Taleb N.N. (2007) *The Black Swan: The Impact of the Highly Improbable*, Random House, ISBN 978-1-4000-6351-2.

Zheng, J. and Frey, H.C. (2005) Quantitative analysis of variability and uncertainty in with known measurement error: Methodology and case study. *Risk Analysis*, **25**(3), 663–676.

5

Direct statistical estimation techniques

This chapter addresses the fundamental methods available for the estimation of an uncertainty model when samples and expertise are directly available in order to characterise the uncertain inputs, being either independent (Section 5.2) or dependent (Section 5.3). Simultaneous estimation of both the aleatory and epistemic components is discussed on the basis of classical statistical theorems (such as asymptotic theory) or Bayesian settings (Section 5.4). The importance of physical properties such as plausible bounds or dependence structures is also introduced. In Section 5.5, the use of extreme value distributions is discussed in the light of rare probability computation: the extent to which they may be physically sound is studied in connection with the underlying time process assumptions. The key concept of rarity is discussed. For readers less familiar with probability and statistical modelling, a prior refresher is available in Annex Section 10.1 and formal links to probability theory are discussed in Annex Section 10.2.

5.1 The general issue

The issues developed in the present chapter concern the estimation of the uncertainty model in the case where there are direct observations available. This means that there is no need for the inversion of the observational system model $\Pi_o o M_s(.)$ derived in the general estimation program in Chapter 2, as shown in Figure 5.1 and Equation (5.1).

$$\left.\begin{array}{l}
\boldsymbol{Y}_{mj} = \Pi_o \circ M_s(\boldsymbol{X}_j, \boldsymbol{d}_j, \boldsymbol{U}_j) \\
\boldsymbol{X}_j \sim f_X(\cdot\,|\boldsymbol{\theta}_x),\ \boldsymbol{U}_j \sim f_U(\cdot\,|\boldsymbol{\theta}_u)\ i.i.d \\
\boldsymbol{\Theta}_x, \boldsymbol{\Theta}_u \sim \pi(\cdot\,|\boldsymbol{\zeta}) \\
\text{estimate } \{M_s, \pi(\cdot\,|\boldsymbol{\zeta})\} \\
\text{knowing } \Xi_n = (\boldsymbol{y}_{mj}, \boldsymbol{d}_j)_{j=1\ldots n} \\
\text{and } IK = \left(f_X, f_U, \pi, \boldsymbol{\zeta}^{kn}\right)\ IM = (M_s)_s
\end{array}\right\} \tag{5.1}$$

Modelling Under Risk and Uncertainty: An Introduction to Statistical, Phenomenological and Computational Methods, First Edition. Etienne de Rocquigny.

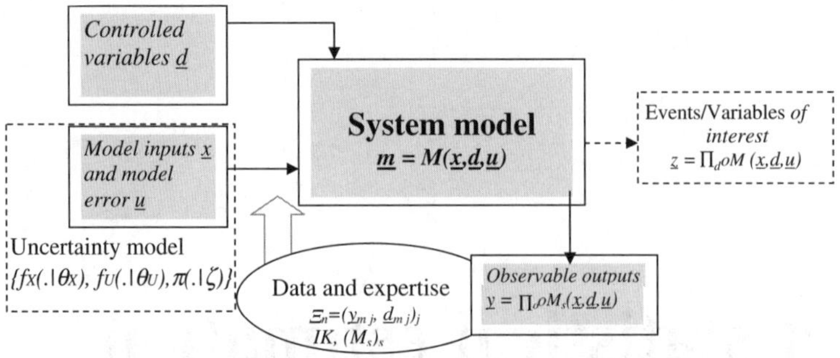

Figure 5.1 *General model estimation program.*

When model inputs are directly observable, as represented in Figure 5.2, that is $\boldsymbol{Y}_{mj} = \boldsymbol{X}_j$ where $\boldsymbol{X}$ may represent any sub-vector of the uncertain input components, variable or discrete event ($\boldsymbol{X}$, $\boldsymbol{E}$) in full notation), the estimation program becomes much simpler. The simplification may even be extended if direct observations of the *outputs of interest* for decision-making are directly available, that is $Y_m = X = Z$. In that most favourable case, $z = G(x,d) = x$ is the (degenerated) system model and the estimation of the risk measure is restricted to a pure statistical issue. Putting aside this ultimate simplification, which impacts only the subsequent propagation phase, the generic statistical program for directly observable inputs is formulated as the following:

$$\left.\begin{array}{l} \boldsymbol{X}_j|\boldsymbol{\theta}_X \sim f_X(\cdot\,|\boldsymbol{\theta}_X)\ i.i.d. \\ \boldsymbol{\Theta}_X \sim \pi(\cdot\,|\zeta) \\ \text{estimate } \pi(\cdot\,|\zeta) \\ \text{knowing } \Xi_n = \left(\boldsymbol{x}_j\right)_{j=1\ldots n} \\ \text{and } IK = \left(f_X, \pi, \zeta^{kn}\right) \end{array}\right\} \tag{5.2}$$

The uncertainty model to be estimated is thus the joint distribution of uncertain inputs and epistemic uncertainty $\{f_X(\cdot\,|\,\boldsymbol{\theta})\cdot\pi(\cdot\,|\,\zeta)\}$. The choice and estimation of the associated distributions returns primarily to the classic statistical problem of fitting distributions to samples. In the context of risk and uncertainty

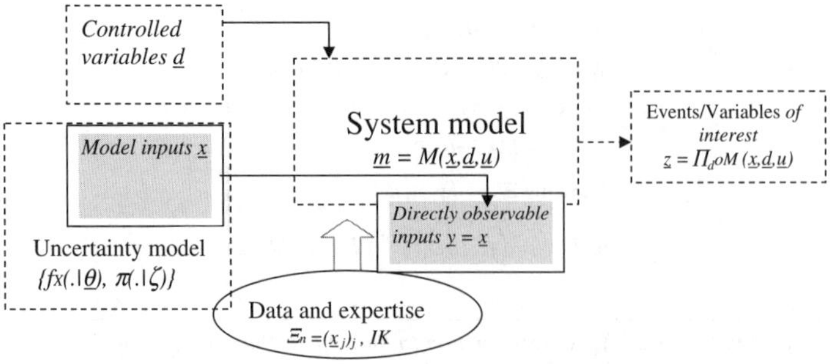

Figure 5.2 *Model estimation program with directly observable inputs.*

analysis, statistical data is however generally insufficient as such, and needs to be completed by expertise. Dependence issues are also quite important, although lacking adequately-sized multi-variate samples. Table 5.1 summarises the type of estimation methods generally encountered in order to tackle this kind of mixed context.

Practices vary not only according to data availability, but also according to the type of input vectors – continuous or discrete events $(\boldsymbol{X}, \boldsymbol{E})$ – encountered across the variety of fields reviewed in Chapter 1. As represented in Table 5.2, the number of input variables may vary from less than five to hundreds; the ratio between continuous input variables and discrete input events, as well as the extent of anticipated dependence also alters greatly. Last but not least, practices differ regarding the representation of a single or double level of uncertainty.

As mentioned in Chapter 2, a *Bayesian approach* offers a convincing setting for the estimation of a comprehensive double probabilistic model in the average situation of small samples complemented by partial expertise. However, as classical estimation appears quite often in practice and as it is also easier to

Table 5.1 Choice and estimation of the laws of input components.

Situation	Choice of laws of X^i	Estimation of $\boldsymbol{\theta}_{XE}$	Commentaries
Standard data sample $(\boldsymbol{X}^i_j)_{j=1\ldots n}$ where $n = 30\text{–}100$	– Choice justified by the physical system and/or standard trade practice	Maximum Likelihood, or Moment Methods	Classic treatments are used in the case of censored data, heterogeneous or unsteady samples *and so on*
	– Comparison of models by standard adjustment tests, or more rarely the selection criteria of Akaike models (AIC), Bayesian models (BIC) and so on		The use of 'modern' non parametric methods, *for example* kernel-based, remains rare.
Mixed data (some data plus expert judgement)	– Choice justified by the physical system and/or standard trade practice	– 'hand picked' choice of parameters suitable for both types of data	This situation is in fact the most industrially current; most often via a 'hand picked' choice
Pure expertise	– Standard choice by default: usually normal, uniform or log normal	– more rarely, a Bayesian framework, with expert judgement setting the a priori $\pi_0\ (\boldsymbol{\theta}_x)$	
	– Use of the Maximal Entropy principle (e.g. uniform if bounded, see §4.2.3)		

Table 5.2 Practices in uncertainty model estimation according to context.

	Size and nature of input vector ($\boldsymbol{X}$, $\boldsymbol{E}$)	Model ($\boldsymbol{\Theta}_X$)	Estimation methods
Natural risk	Very low $p \leq 3$, but detailed choice of pdf	Probabilistic: confidence interval for a few components	MLE on a sample; extreme value theory; model choice testing
	Mostly continuous r.v. with underlying time series		
SRA	Low $p = 3$ to 10, except case of spatial fields	Generally fixed values or deterministic intervals for local sensitivity	Expert judgement
	Mostly continuous r.v. with limited or unknown dependence		Sometimes MLE with standard laws (normal, lognormal, ...), literature review
Systems reliability	Medium to High p ('0 to '00s)	Deterministic or Probabilistic: equivalent lognormal pdfs (PSA)	Expert judgement
	Mostly Boolean or discrete r.v.		Sometimes MLE/Bayesian for standard laws (Poisson, exponential, beta, ...)
	Essential dependence structures (common modes)		
IPRA	Medium to High p ('0 to '00s)	Probabilistic	idem
UASA	Medium $p = 10$–100	Generally fixed values	Expert judgement, entropy principle, literature review

start on a mathematical basis, it will be introduced first in the subsequent sections, with a parametric or non-parametric approach (Section 5.2). Thus, the following key points will be developed:

- choosing the statistical models (Section 5.2) and dependence structures (Section 5.3);
- controlling the epistemic uncertainty $\pi(\boldsymbol{\theta}_X|\zeta)$ through statistical estimation theory, in a classical or Bayesian approach (Section 5.4);
- understanding rare probabilities and the appropriateness of extreme value statistical modelling (Section 5.5).

5.2 Introducing estimation techniques on independent samples

In this section, some basic examples will be given in order to help understand the subsequent issues. Suppose that an observation sample $\Xi_n = \{\boldsymbol{x}_j\}_{j=1\ldots n}$ is made available to estimate the uncertainty model $\{f_X(\boldsymbol{x}|\boldsymbol{\theta}_X)\cdot\pi(\boldsymbol{\theta}_X|IK, \Xi_n)\}$. One will start with the classical estimation procedure for the estimation of a fixed albeit unknown parameter value $\boldsymbol{\theta}_X{}^\circ$, that is we are not modelling the epistemic uncertainty at this stage.

5.2.1 Estimation basics

Statistical estimation of an uncertainty model returns to a key modelling alternative:

- infer a model belonging to a standard class of distribution shapes, such as Gaussian, Lognormal, Gumbel, and so on: this is referred to traditionally as a *parametric* model;
- infer a distribution adhering to data while not belonging to a previously-fixed standard shape: this is referred to traditionally as a *non-parametric* model.

The second option may be more accurately referred to as a 'non-standard parametric' approach, since it still relies on a parameterised distribution fitted to data. However, as will be seen hereafter, the distribution shape resulting from such an estimation approach enjoys much wider degrees of freedom.

As for any modelling choice, both come with a cost that will be illustrated below. Apart from a limited number of situations, it is reasonable to doubt that the distribution of uncertainty surrounding a phenomenological input, possibly the result of complex combinations of sources of variability, follows a given *standard* shape. Hence a parametric model adds uncertainty in choosing the proper distribution shape, or error linked to some unrealistic simplification of a complex shape. Thus, credit is given to non-parametric approaches for sticking to data exclusively. Conversely, a non-parametric approach is unable to produce any extrapolation to distribution tails and also introduces noise due to the smoothening kernels that proves non-negligible in small samples.

The distinction applies firstly to the estimation of the (marginal) distributions of each uncertain input. Assuming the availability of a multi-variate sample, it will apply also to estimating the joint distribution (or dependence model) of an input vector (Section 5.4).

Parametric estimation: maximal likelihood Observations will be made later on a real-valued input noted X. Parametric estimation assumes that a modelling choice has been made so as to infer that the distribution that generated the observed sample $(x_j)_{j=1\ldots n}$ belongs to a *known* family:

$$X \sim f_X(X = x|\boldsymbol{\theta}_X) \tag{5.3}$$

depending on a vector of parameters $\boldsymbol{\theta}_X$, typically mean/variance for Gaussian, or more generally location/scale plus possibly a shape parameter. The purpose of the estimation techniques is to retrieve the 'true' value $\boldsymbol{\theta}_X{}^\circ$ of the input distribution, or more accurately to infer a vector $\hat{\boldsymbol{\theta}}_X$ ensuring that $f_X(\cdot|\,\hat{\boldsymbol{\theta}}_X)$ approximates the unknown behaviour of the uncertain input acceptably.

The *maximal likelihood* estimator (MLE) is the first standard option. It is based intuitively on the following argument: supposing the only pieces of information are the sample and the distribution family, nothing can lead to choosing for $\boldsymbol{\theta}_X$ another value than that for which the probability to observe the observed sample is maximal. Because the observed sample is supposed to follow *i.i.d.* the above-mentioned distribution, such a probability is:

$$P\left[(X_j)_{j=1...n} \in \left([x_j \pm dx]\right)_{j=1...n}\right] = \prod_{j=1...n} f_X[x_j|\boldsymbol{\theta}_X]dx = L\left[(x_j)_{j=1...n}|\boldsymbol{\theta}_X\right] \prod_{j=1...n} dx \tag{5.4}$$

where the likelihood function is defined classically as the product of densities. Hence, the MLE comes as the result of:

$$\hat{\boldsymbol{\theta}}_X = \underset{\theta}{Argmax}\left(L\left[(x_j)_{j=1...n}, \boldsymbol{\theta}_X\right]\right) = \underset{\theta}{Argmax}\left(\prod_{j=1...n} f_X[x_j|\boldsymbol{\theta}_X]\right) \tag{5.5}$$

This may be expressed equivalently as the maximisation of log-likelihood, which turns out to be simpler to handle analytically in many cases:

$$\hat{\boldsymbol{\theta}}_X = \underset{\theta}{Argmax}\left(LL\left[(x_j)_{j=1...n}, \boldsymbol{\theta}_X\right]\right) = \underset{\theta}{Argmax}\left(\sum_{j=1...n} \log f_X[x_j|\boldsymbol{\theta}_X]\right) \tag{5.6}$$

The asymptotic statistical results (see Section 5.4.3 hereinbelow) state the unicity of such an estimator, its convergence in probability towards $\boldsymbol{\theta}_X{}^\circ$ and asymptotic normality when the sample size grows indefinitely, given a number of regularity conditions as well as model being *identifiable*, that is that there exists a unique vector of parameters describing the full distribution:

$$\begin{aligned} &f_X(x|\boldsymbol{\theta}_X) = f_X(x|\boldsymbol{\theta}_X{}^\circ)\ \forall x \\ &\Rightarrow \boldsymbol{\theta}_X = \boldsymbol{\theta}_X{}^\circ \end{aligned} \tag{5.7}$$

In other words, it assumes that the model is not over-parameterised. In the context of MLE, likelihood is assumed to be twice differentiable. Locally at least, identifiability requires that the Hessian matrix be invertible (indeed strictly negative-definite in order to guarantee that it reaches a maximum):

$$\frac{\partial^2}{\partial\boldsymbol{\theta}\partial\boldsymbol{\theta}'} LL\left((\boldsymbol{x}_j)_{j=1...n}, \cdot\right)|_{\boldsymbol{\theta}_X} = \sum\nolimits_{j=1...n} \frac{\partial^2}{\partial\boldsymbol{\theta}\partial\boldsymbol{\theta}'} Log f_X(\boldsymbol{x}_j, .)|_{\boldsymbol{\theta}_X} \tag{5.8}$$

Of course, such estimation also assumes that the distribution which generated the sample is modelled correctly, a significant challenge in practice.

In some cases such as the Gaussian distribution, the MLE enjoys closed-form expressions:

$$L(x_1, \ldots, x_n; \boldsymbol{\theta}) = \frac{1}{(2\pi)^{\frac{n}{2}}\sigma^n} \prod_{j=1}^{n} e^{-\frac{(x_j-\mu)^2}{2\sigma^2}} \tag{5.9}$$

$$\Rightarrow \begin{pmatrix} \hat{\mu} \\ \hat{\sigma}^2 \end{pmatrix} = Argmax L(x_1, \ldots, x_n; \boldsymbol{\theta}) = \begin{pmatrix} \frac{1}{n}\sum_{j=1}^{n} x_j \\ \frac{1}{n}\sum_{j=1}^{n} (x_j - \hat{\mu})^2 \end{pmatrix} \tag{5.10}$$

so that the MLE in the Gaussian case are simply the sample mean and standard deviation. Closed-form estimators may not be the case elsewhere, as for the Gumbel distribution which proves useful in the flood example. Based on the cumulative distribution that was introduced in Chapter 4, the Gumbel density function appears as follows:

$$f_X(x|m,e) = \frac{1}{e}\exp\left\{-\frac{x-m}{e} - \exp\left[-\frac{x-m}{e}\right]\right\} \tag{5.11}$$

The log-likelihood of a sample $\Xi_n = \{\boldsymbol{x}_j\}_{j=1\ldots n}$ appears as follows:

$$LL(\Xi,|m,e) = -n\log e - \sum_{j=1\ldots n}\left\{\frac{x_j-m}{e} + \exp\left[-\frac{x_j-m}{e}\right]\right\} \tag{5.12}$$

Such a distribution is an example of the more general *Generalised Extreme Value* distribution (*GEV*) for richer parameterisation (μ for location, σ for scale and ξ for shape) introduced in Chapter 4. The *cdf* of a *GEV* is as follows:

$$\mathrm{F}_{GEV}(x|m,e,\xi) = \exp\left\{-\left[1+\xi\left(\frac{x-m}{e}\right)\right]^{-\frac{1}{\xi}}\right\} \tag{5.13}$$

of which the Gumbel distribution is a subset corresponding to shape parameter $\xi = 0$. First-order conditions $\frac{\partial}{\partial\theta}LL\left(\left(\boldsymbol{x}_j\right)_{j=1\ldots n},\cdot\right)|_{\boldsymbol{\theta}} = 0$ result in a set of two or three non-linear equations without closed-form solutions. In that case, it proves necessary to implement a numerical optimisation algorithm, generally gradient-based. The convergence of MLE needs to be guaranteed theoretically before entering the numerical resolution. In the case of *GEV* distributions with non-null shape parameter ξ, convergence is theoretically non-obvious as the distribution ranges depend on unknown parameters; however, classical MLE properties still hold for $-{}^1\!/_2 < \xi$. This includes *all Fréchet distributions* ($0 < \xi$), including *heavy-tail* with *infinite k-order moments* ($1/k < \xi$); including only those three parameter-Weibull distributions with modest shape parameter $-{}^1\!/_2 < \xi < 0$, with limited extensions to the cases where $-1 < \xi < -{}^1\!/_2$ (Smith, 1985).

Parametric estimation: moment methods A traditional alternative is the *moment method*. In its simplest form, it assumes that a closed form relation can be derived between the desired parameters and the distribution moments:

$$\begin{aligned}
\boldsymbol{\theta}_X &= \begin{pmatrix}\theta^1\\ \theta^2\\ \theta^3\end{pmatrix} = \begin{pmatrix}m^1(\boldsymbol{\mu})\\ m^2(\boldsymbol{\mu})\\ m^3(\boldsymbol{\mu})\end{pmatrix}\\
\boldsymbol{\mu} &= \begin{pmatrix}\mu^1\\ \ldots\\ \mu^k\end{pmatrix} = \left(E(X)^k\right)_k = \left(\int_x (x)^k f_X(x|\boldsymbol{\theta}_X)dx\right)_k
\end{aligned} \tag{5.14}$$

so that a direct estimation $\hat{\boldsymbol{\mu}}$ of the moments computed from the observed sample (i.e. empiric mean of the x's or of their squares, their third power, etc.) generates an estimator through the closed form functions

$\boldsymbol{m}(\hat{\boldsymbol{\mu}})$. In the Gumbel case, for instance, expectation and variance are related to the distribution parameters as follows (Johnson, Kotz, and Balakrishnan, 1994):

$$\begin{aligned} E(X) &= \mu^1 = m + \gamma \cdot e \\ Var(X) &= \mu^2 - (\mu^1)^2 = \left(\frac{\pi^2}{6}\right) \cdot e^2 \end{aligned} \tag{5.15}$$

where γ stands for the Euler constant ($\gamma \cong 0.577$). Estimation involves their empiric moments, that is mean and standard deviation through the following formulae:

$$\begin{aligned} \hat{M} &= \bar{X} - \gamma \hat{E} = \bar{X} - \gamma \sqrt{\frac{6}{\pi^2} \hat{S}_n^2} \\ \hat{E} &= \sqrt{\frac{6}{\pi^2} \hat{S}_n^2} \end{aligned} \tag{5.16}$$

Naturally, the moment method requires those *moments* to be well-defined, *finite* quantities which may not be the case for some mathematical models, such as the Fréchet (or corresponding Pareto) EVD with a shape parameter such as $1/k < \xi$.

Although classical statistical results state that maximal likelihood is generally more accurate and powerful as an estimation method, the moment method remains quite practical in some cases. More importantly, moment methods can prove more robust when the choice of the distribution shape is uncertain as the estimation of moments does not involve the pdf.

Section 5.4 below will discuss further some of the key features of the two methods: the estimation of essential characteristics on the random vector $\hat{\boldsymbol{\theta}}_X$ or derived quantities, essentially the fact that it is approximately chi-squared or multi-normal with a known variance matrix. In other words, maximal likelihood and moment methods allow also for the estimation of level-2 epistemic uncertainty.

Beyond their simplest scalar form, both methods also generalise to two essential aspects for the purposes of this book:

- multidimensional inputs (see below and in particular the case of dependent variables);
- indirect observation samples (see Chapter 6), whereby the likelihood function incorporates a complex system model.

5.2.2 Goodness-of-fit and model selection techniques

Parametric estimation has so far been undertaken assuming that the model is well-specified, that is that a given distribution shape $f_X(x|\boldsymbol{\theta}_X)$ is representing *acceptably* the variability of the uncertain variable X *given the amount of empiric evidence*. A thorough model-building process should ideally challenge such a fundamental assumption. This is fostered classically by means of two complementary techniques:

- goodness-of-fit testing (e.g. Kolmogorov-Smirnov, Anderson-Darling or Cramer Von Mises tests): this helps in rejecting or not *a given parametric model* on the basis of the residual distance left between the empiric sample distribution and the parametric model;
- model selection techniques (e.g. AIC or BIC, i.e. Akaike or Bayesian Information Criteria): these help in comparing the *alternative parametric models* of a given sample, in particular as to the optimal level of parametrisation.

Goodness-of-fit testing in parametric models *Goodness-of-fit testing* (*GoF*) studies the residual distance left between the empiric sample distribution and a given parametric model. Quantitative misfit is statistically challenged in order to discern the likelihood of it being the result of pure randomness or for providing enough evidence to reject the model. Well within the *Popperian* spirit, goodness-of-fit should never be seen as a means to accepting a 'good' model, but merely to reject a 'bad enough' one.

Mathematically, classical goodness-of-fit tests involve a metric comparing an empiric and a theoretical cumulative distribution function denoted as $F_n(.)$ and $F(.)$ respectively. Given an observed sample $(x_j)_{j=1,...n}$ the empiric *cdf* and *pdf* are the following:

$$\hat{F}_n(x) = \frac{1}{n}\sum_{j=1}^{n} 1_{]-\infty,x[}(x_j) \quad \hat{f}_n(x) = \frac{1}{n}\sum_{j=1}^{n} \delta(x_j) \tag{5.17}$$

whereby $\delta(.)$ stands for the Dirac's distribution. Thus, a goodness-of-fit test involves a so-called *statistic* defined as a functional of the two distributions, or equivalently as a function of the respective empiric and parametric cdfs computed on the ranked observations ($x_{(j)}$ denoting the j-th lowest observed value of the n-sample ranked in an increasing order):

$$S_n = D[F_n, F] = D\left(F_n\left(x_{(j)}\right)_{j=1...n}, F\left(x_{(j)}\right)_{j=1...n}\right) \tag{5.18}$$

Because of the effects of random sampling, even a fully-known distribution generates deviations between an empiric sample and its theoretical curve: such a misfit is all the greater when the sample size is low. Once renormalized by a factor n so as to account for this dependence in sample size, S_n is thus compared to pre-tabulated quantities that represent the α-quantiles s_n^α (e.g. 95 %-quantile) of the distribution of S_n generated by the fluctuations of random sampling. It assumes that the theoretical model is the *true* generator of the observations of finite sample size n. If $S_n > s^\alpha$, the test rejects the hypothesis that the theoretical model is appropriate in order to model the observations given a α-level of confidence. In other words, it is too unlikely (less than *1-α* % chances) that the sole effect of randomness explains as much distance between the empiric and theoretical distribution: the statistical model is unfit. Also useful in practice is the concept of p-value, that is the α-level for which the test becomes critical $S_n = s^\alpha$ for a given set of observations.

The *Kolmogorov-Smirnov test* (KS) involves the maximal distance between both distributions (in other words, a L_∞ norm):

$$K_n = \sqrt{n}D_n = \sqrt{n}\sup_x|F_n(x) - F(x)| = \sqrt{n}\sup_j\left\{\left|F(x_{(j)}) - \frac{j}{n}\right|; \left|F(x_{(j)}) - \frac{j-1}{n}\right|\right\} \tag{5.19}$$

Assuming that the theoretical parameters are known (i.e. neglecting the effect of parameter estimation), the behaviour of the statistic is distribution-free and it is known and tabulated whatever the shape of the theoretical distribution F: for $\alpha = 95\,\%$, $s^\alpha \sim 1.36$ for n large enough. Accounting for the fact that the parameters are estimated leads to a more complex treatment; in the Gaussian case, the critical threshold then falls to $s^\alpha \sim 0.90$.

The *Cramer Von Mises test* involves a quadratic distance between both distributions (in other words, a L_2 norm):

$$W_n^2 = n\omega_n^2 = n\int_{-\infty}^{+\infty}[F_n(x) - F(x)]^2 dF(x) = \frac{1}{12n} + \sum_{j=1}^{n}\left[\frac{2j-1}{2n} - F\left(x_{(j)}\right)\right]^2 \tag{5.20}$$

Like the KS test it is distribution-free when assuming that the theoretical parameters are known, and for $\alpha = 95\,\%$, $s^\alpha \sim 0.46$ for n large enough; accounting for parameter estimation is again more complex, in the Gaussian case, it then falls to $s^\alpha \sim 0.13$.

The *Anderson Darling test* (Anderson and Darling, 1952) involves a weighted quadratic distance between both distributions: the L_2 norm is conspicuously weighted in order to give more importance to the distribution tails (both lower and upper tails):

$$AD_n^2 = n\int_{-\infty}^{+\infty}\frac{[F_n(x)-F(x)]^2}{F(x)(1-F(x))}dF(x) = -n-\frac{1}{n}\sum_{j=1}^{n}(2j-1)\left[\ln\left(F\left(x_{(j)}\right)\right)+\ln\left(1-F\left(x_{(n-j+1)}\right)\right)\right] \quad (5.21)$$

The statistic is not distribution-free in general and the impact of parameter estimation is more complex.

Here are some comparative remarks regarding the tests:

- Kolmogorov-Smirnov concentrates on local distribution shifts while the two others incorporate a 'averaged' vision of the distribution misfits; KS may thus prove more sensitive to outliers, which may or may not be desirable for risk modelling.
- Anderson-Darling is obviously more sensititive to distribution tails, and is thus of interest in many risk applications although most of them would focus essentially on one given tail – that closer to failure or undesired events – instead of both.

The Flood example
Goodness of fit was tested for the 149 samples of flow against three parametric models (see Table 5.3).

More generally, goodness-of-fit tests have many practical limitations:

(i) They detect essentially bad models although never lead legitimately to accepting a 'good one'.
(ii) Tabulated thresholds are theoretically adequate in comparison with 'fixed prior' theoretical distributions with large-enough samples, not to distributions the parameters of which were estimated using the samples nor for very small samples: this adds additional variance and cancels the distribution-free validity of the tests.
(iii) Ranking p-values between models for a given test is not theoretically as simple as testing one model; and it is even more difficult to handle various rankings according to different GoF tests.

In Table 5.3, the results involved the use of estimated parameters as in most real applications. They should thus be interpreted with care. While it is always possible to compute the statistic, its interpretation in relative ranking is not guaranteed; p-values and the associated 5 %-test are firstly computed with respect to the standard critical values (based on known parameters and large-enough samples) given by most statistical software, then completed in some cases with the modified versions of the tests when it is known.

Model selection techniques in parametric models *Model selection techniques* involve a comparison of the penalised log-likelihoods between alternative models of the same family $(M_s)_s$ through a criterion in the following form:

$$kIC(M_s) = -2\left(LL\left(\Xi_n|\underline{\theta}_s, M_s\right) - k(n)\dim[\boldsymbol{\theta}_s]\right) \quad (5.22)$$

Table 5.3 Goodness-of-fit tests on the 149-sample of flows.

		Normal	Gumbel	Lognormal
Kolmogorov	Statistic (relative rank)	0.09 (3rd)	0.04 (1st)	0.09 (2nd)
	p-value and 5 %-test (known parameters)	17 % (not rejected)	94 % (not rejected)	20 % (not rejected)
	Idem (estimated parameters) – when known	<5 % (Reject)	—	—
Cramer Von Mises	Statistic (relative rank)	0.29 (3rd)	0.04 (1st)	0.23 (2nd)
	p-value and 5 %-test (known parameters)	14 % (not rejected)	93 % (not rejected)	22 % (not rejected)
	Idem (estimated parameters) – when known	<5 % (Reject)	—	—
Anderson-Darling	Statistic (relative rank)	2.08 (3rd)	0.25 (1st)	1.44 (2nd)
	p-value and 5 %-test (known parameters)	8 % (not rejected)	97 % (not rejected)	19 % (not rejected)
	Idem (estimated parameters) – when known	—	—	—

The second term of Equation (5.22) multiplies the number of parameters of model M_s (i.e. the size of vector $\boldsymbol{\theta}_s$) by a coefficient $k(n)$ varying with the number n of observations. In other words, improvements in the loglikelihood of observations obtained through a refinement of the statistical model are discounted by a parametrisation cost, essentially the number of parameters involved, in order to find the best modelling compromise. Remember indeed that while an arbitrarily-increasing number of parameters may osculate any empiric density perfectly (hence increasing loglikelihood), the higher the number of parameters to be estimated, the higher their residual estimation variance; thus, the lower the information criterion, the better the fit.

The *Akaike information criterion* (AIC) involves a unit coefficient:

$$AIC(M_s) = -2(LL(\Xi_n|\boldsymbol{\theta}_s, M_s) - \dim[\theta_s]) \tag{5.23}$$

The *Bayesian information criterion* (or Schwarz's Bayesian criterion, denoted as BIC) involves $k(n) = log(n)/2$, thus leading to smaller size of parameterising compromises than AIC whenever $n > 8$:

$$BIC(M_s) = -2\left(LL(\Xi_n|\boldsymbol{\theta}_s, M_s) - \frac{\log(n)}{2}\dim[\boldsymbol{\theta}_s]\right) \tag{5.24}$$

Note that those criteria are defined as a relative statistic that proves meaningful essentially for comparing the increasing number of models of an inclusive family, for example a simpler model is defined through the fixing of one or more parameters of the more complex one.

The Flood model
Flow may be modelled by a *Generalised Extreme Value* distribution (*GEV*) of richer parameterisation Section 5.2.1. On the basis of maximum likelihood estimation for both on the 149-sample of flood flows, AIC and BIC could be computed as follows in Table 5.4.
Adding an extra-parameter improves the loglikelihood insignificantly (an increase of 10^{-3}), which is clearly evidenced by AIC and BIC which both deteriorate (increase). As will be explained in more detail in Section 5.4, this is also obvious from the study of standard errors of estimation which deteriorate for location and scale parameters and show how insignificant a non-zero shape parameter turns out to be (standard error is 30 times the estimated shift from zero). It is better to stick to the two-parameter model, even though the sample size $n = 149$ is already rather large.

Table 5.4 Fitting a two-parameter vs. a three-parameter *GEV* to flood flows (MLE).

	2-parameter Gumbel	3-parameter *GEV*
Loglikelihood	−1177.406	−1177.405
AIC	2358.8	2360.8
BIC	2364.8	2369.8
Loc (± standard error)	1013 (±48)	1013 (±51)
Scale (± standard error)	557 (±36)	557 (±37)
Shape (± standard error)	Fixed at 0	-2.10^{-3} (±0.06)

Keep in mind that both goodness-of-fit and model-selection techniques provide at most partial answers to the fundamental challenge raised by the selection of a credible parametric model. Given the wide (infinite) number of potential alternatives that may never be completely tested, there always remains a fundamental epistemic uncertainty in parametric modelling.

5.2.3 A non-parametric method: Kernel modelling

Non-parametric modelling is an alternative that aims to resolve the epistemic uncertainty associated inevitably with parametric modelling in choosing a prior distribution shape. Amongst these, the popular *kernel modelling* approach may be understood simply as smoothing the raw data. Given an observed sample $(x_j)_{j=1,\ldots n}$ the empiric pdf and cdf (Equation (5.17)) prove somewhat irregular, as illustrated in Figure 5.3.

Smoothing such irregular empiric distributions is achieved by replacing the Dirac's distribution by a kernel distribution (by definition a symmetric, positive, bounded function $\kappa(.)$ with integral 1 and expectation 0), as follows:

$$\hat{f}_n^{\text{ker}}(x) = \frac{1}{nh_n}\sum_{j=1}^{n}\kappa\left(\frac{x-x_j}{h_n}\right) \tag{5.25}$$

The bandwidth h_n supposedly decreases to zero with n in order to gradually yield the place for the purely empiric distribution, though less quickly than n, in order for nh_n to grow indefinitely. Given additional assumptions with regard to the kernel function (e.g. a Parzen-Rosenblatt kernel), convergence in

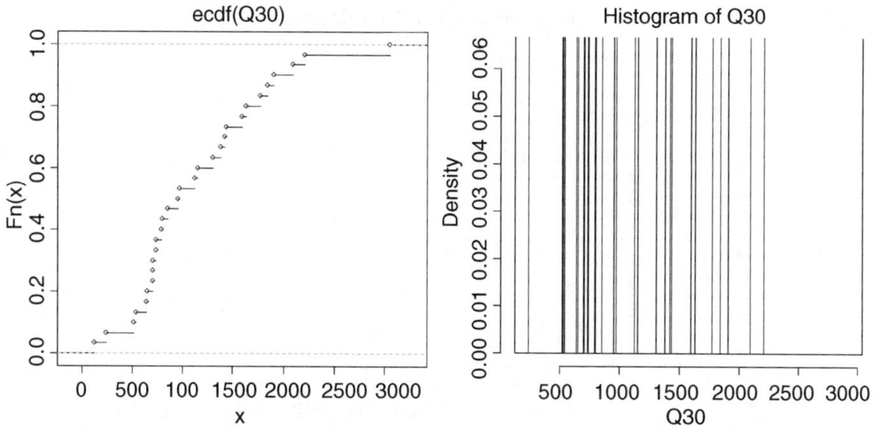

Figure 5.3 *Empiric cdf (left) and pdf (right) of a 30-subsample of riverflows.*

probability to the true distribution of the sample is guaranteed. This is particularly the case with the popular Gaussian kernel, whereby $\kappa(.)$ denotes a standard normal density:

$$\hat{f}_n^{\,\mathrm{ker}}(x) = \frac{1}{nh_n\sqrt{2\pi}} \sum_{j=1}^{n} \exp\left(-\frac{1}{2}\left(\frac{x-x_j}{h_n}\right)^2\right) \tag{5.26}$$

Another interpretation is to see such density as a mixture of narrowly-distributed densities centred on the sample values: think of slight measurement uncertainty u affecting the observed data, and thus the kernel-smoothened density represents the density of the following observable:

$$Y_j = X_j + U_j \quad \text{where } U_j/h_n \sim \kappa(\cdot)\ \text{i.i.d.} \tag{5.27}$$

Theoretical results indicate optimal bandwidth according to specific assumptions, such as the regularity of the unknown density with bounded second-order derivatives, the optimality being pursued according to minimal integrated asymptotic square error. This is the case with the Silverman rule (Silverman, 1986), which recommends the following bandwidth for a Gaussian kernel:

$$h_n^s = \hat{\sigma}_n \cdot n^{-\frac{1}{5}} \tag{5.28}$$

where $\hat{\sigma}_n$ is the empiric standard deviation of the observed sample.
In practice, the key question raised by such kernel estimation concerns the size of bandwidth, which has a large influence on the resulting density. This is illustrated in Figure 5.4 taking the flood example with a sub-sample of flows limited to 30 observations.

An excessively-large bandwidth blurs the observed sample with the information-void kernel density itself, while an excessively-small one adds little to the original sample. This becomes especially critical for the distribution tails. Table 5.5 illustrates the quantiles derived from the various bandwidth choices as compared to those of the empiric sample and of the Gumbel fit, known to be relevant in this case.

Note how badly the kernel model extrapolates beyond the sample size (1/30 being the highest reasonable return period), and to what extent it is sensitive to bandwidth. Clearly, a non-parametric

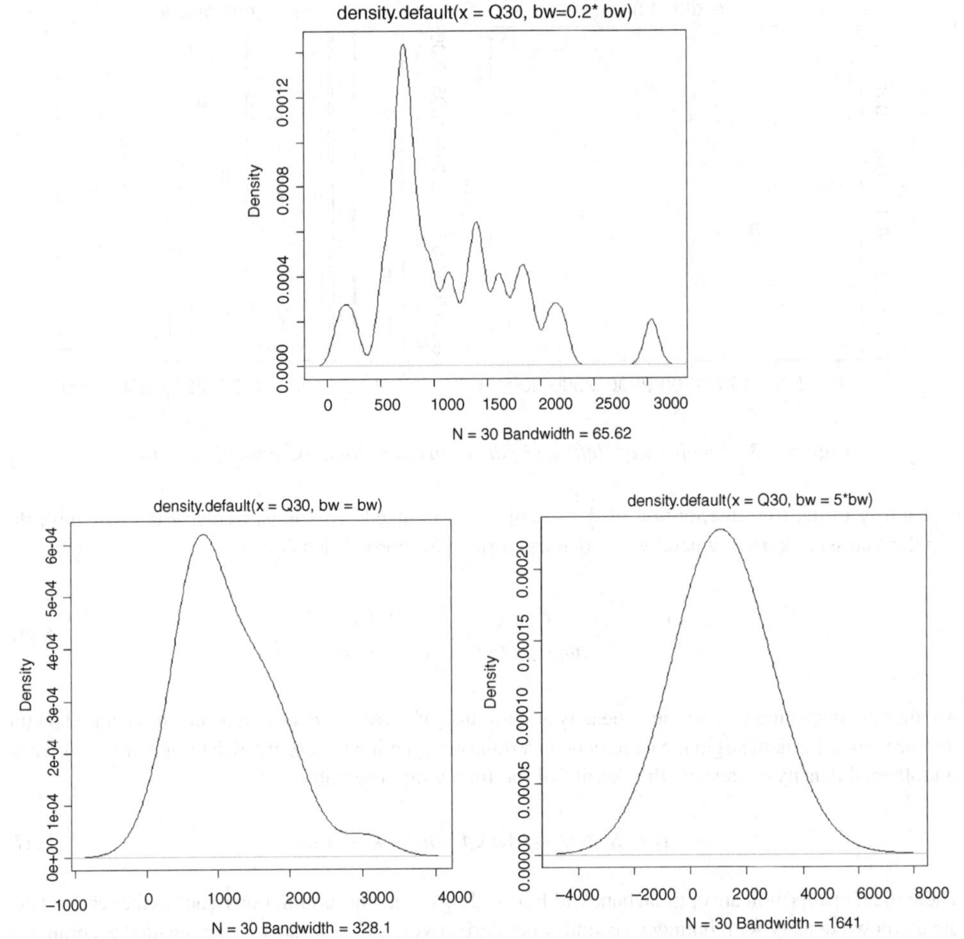

Figure 5.4 *Kernel densities for varying bandwidth –* h_s^n *down – left, 1/5.*h_s^n *up, 5.*h_s^n *down – right (flood example with a 30-sample of flows).*

Table 5.5 Comparing non-parametric quantile prediction (flood example with a 30-sample of flows).

	Kernel - h_s^n	Kernel $-1/5\cdot h_s^n$	Kernel $-5\cdot h_s^n$	Empiric	Gumbel fit
90 % (1/10)	2092	1999	3415	1903	1996
98 % (1/50)	2971	3028	4807	3044	2830
99 % (1/100)	3214	3078	5305	3044	3183
99.9 % (1/1000)	3661	3168	6718	?? > 3044	4348

method is an interpolating device; it is not especially relevant for extrapolating distribution tails, as will be discussed further in the section on extremes.

5.2.4 Estimating physical variables in the flood example

Illustrations will be given in the flood example, helping to understand the impact of the estimation choices on the subsequent risk measure.

Flood flows Consider first the river flow for which the full sample of 149 historical yearly maxima is now employed. The results are given in Table 5.6 and Figure 5.5.

Choosing the best fit is consensual in this example: both the Kolmogorov-Smirnov statistic and a graphical analysis of distribution fit – including a closer look at the fit of high-order quantiles – lead to selecting the Gumbel extreme-value distribution. Hydrological experience confirms that extreme-value distributions generally describe hydrological flows well (see Section 5.5 for further comment).

The *flood example*
Note the significant impact that such fit choices have on risk assessment and design. Step Three of the example introduced in Chapter 3 illustrated the initiator/frequency approach mixing a probabilistic model of the flow variable and deterministic hypotheses into other uncertain inputs. Targeting a 1/100-return protection, normal, Gumbel or lognormal fits lead respectively to designing a dike height of 2.8 m, 3.6 m or 5 m; a 1/1000-return protection requires respectively 3.5 m, 5.2 m or … 8 m! The impact is all the stronger since the only probabilised input is flow and the system model is monotonous (see Chapter 7 for more detail).

Water depths A trickier case involves water depths. Empiric records of water depths $(h_{mj})_{j=1\ldots n}$ (see Figure 5.6) are available in order to estimate an uncertainty model.

Table 5.6 River flow parametric estimation results.

Descriptive statistics	Q $n = 149$ Median = 1256; mean 1335 Min-Max = [122 – 3854]				
	model parameters	Kolmogorov-Smirnov distance	quantiles (90 %)	quantiles (99 %)	quantiles (99,9 %)
Bernoulli model	p – depending on dike level				
Gaussian	$\mu = 1335$ $\sigma = 714$	KS = 0.091	2250	2996	3542
Lognormal	$l\mu = 7.041$ $l\sigma = 0.6045$	KS = 0.087	2478	4660	7394
Gumbel	l = 1013 s = 557	KS = 0.043	2265	3573	4857

Table 5.7 Overflow probability estimated via water depth records ($z_{v\ pn} = 51.0 - z_b = 55.5$ – overflow starts when $H > 4.5$ m)

Descriptive statistics	H_m $n = 123$ Median = 2.4; mean 2.6 Min-Max = [0.5 – 7.4]				
	model parameters	No dike	$h_d = 1\,m$	$h_d = 2\,m$	$h_d = 3\,m$
Bernoulli model	p – depending on dike level	4.9 %	2.4 %	0.81 %	??
Gaussian	$\mu = 2.6$ $\sigma = 1.1$ KS = 0.079	3.4 %	0.3 %	0.01 %	$1.6\ 10^{-6}$
Lognormal	$l\mu = 0.86$ $l\sigma = 0.42$ KS = 0.062	6.3 %	2.3 %	0.82 %	0.31 %
Gumbel	l = 2.1 s = 0.84 KS = 0.057	5.6 %	1.8 %	0.54 %	0.16 %

Thus, it may simply be added to a fixed level for Z_v (for instance a penalised value $z_{v\ pn}$ defined by an expert) in order to generate the Step One inference of the simplest overflow risk measure. The sample size is a bit smaller than that of the flows because of the partial dependability of the gauging devices. The results are given in Table 5.7.

Unsurprisingly, a Gaussian model which does not take into account the empirical asymmetry (as is obvious in the Box plot, Figure 5.6) underestimates the upper quantiles and hence the risk measure so that it leads to an insufficient design. Choosing between the lognormal and Gumbel models is a hard task here as the Kolmogorov-Smirnov distances are close but would rank the comparison differently (Lognormal further than Gumbel) to the observation of the closeness to empiric quantiles (Lognormal closer than Gumbel).

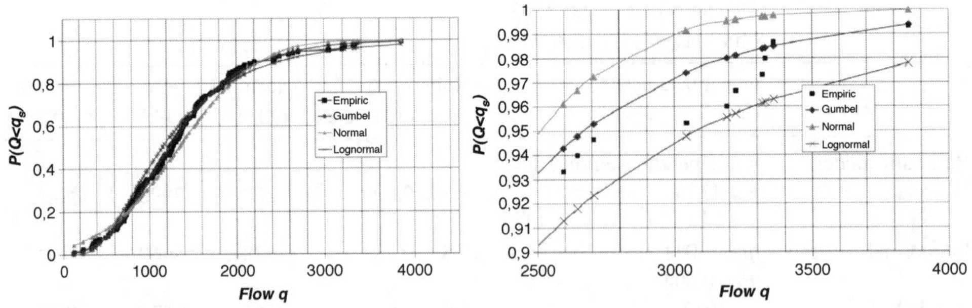

Figure 5.5 *Comparing cdf fits to the empiric 149-sample of flows (up-overall cdf; down-zoom into the higher-end distribution tail.*

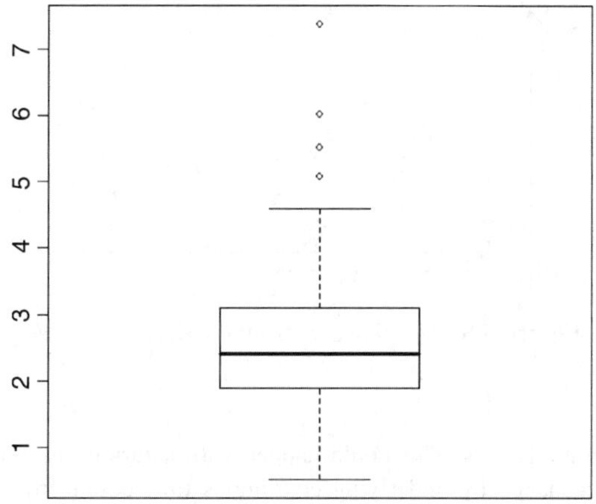

Figure 5.6 *Boxplot of water depth records.*

The latter is retained and *Lognormal* is preferred because additionally it is relatively the more 'conservative' in the sense that the predicted quantiles are the highest although not being such in absolute terms with respect to the empiric distribution (see zoom in Figure 5.7 – right). An explanation might be the following: while input flow follows a pure extreme value distribution, the non-linear transformation and combination to non-extreme variables brought by the hydraulic model takes the output water depth away from a pure extreme value distribution. As a matter of fact, the flow-height transformation is close to the kind of power-product models discussed in Chapter 4, Section 4.2.1 where Lognormal is a natural choice.

5.2.5 Discrete events and time-based statistical models (frequencies, reliability models, time series)

Discrete events: the poisson/exponential lifetime model Consider firstly the case of estimating discrete events introduced in Chapter 4: discrete in the sense that they take place at discrete time dates or that they

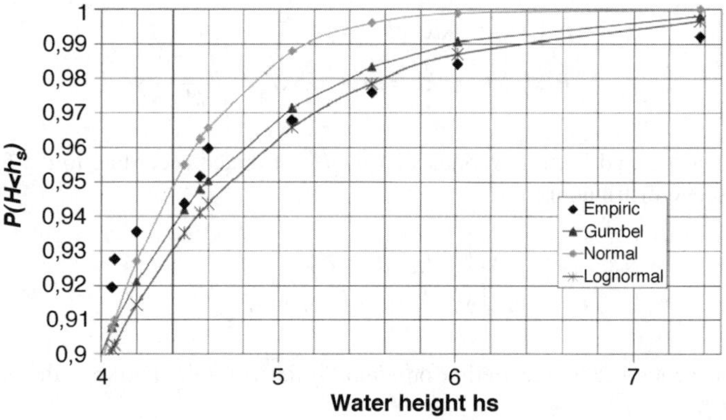

Figure 5.7 *Comparing cdf fits to the empiric 123-sample of water heights, zooming into the higher-end distribution tail.*

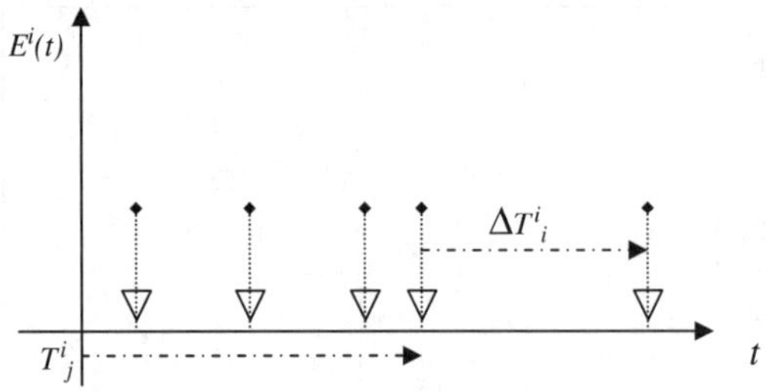

Figure 5.8 *Discrete events and lifetime variables.*

are modelled by discrete modalities. This could happen with naturally-discrete events or with events defined as threshold exceedance by underlying continuous time-based processes, and its associated occurrence times over a given observation period $[t_o, t_o + \Delta T]$:

$$E^i(t) = 1_{ei}(t) \quad (T^i_j)_{j=1,...N} = \{t \in [t_o, t_o + \Delta T] | E^i(t) = 1\} \tag{5.29}$$

Think, for instance, of initiating events in the context of QRA – be it internal (component failures or process temperature exceeding a safety threshold) or external (such as floods, storms exceeding a given threshold, etc.) – or of insurance claims generated by events (firebreak, storm, etc.).

Uncertainty modelling involves the estimation of a distribution (see Figure 5.8) either on:

- the discrete time process ($E^i(t)$) or associated counting process $N^i([t,t + \Delta T])$ of the number of occurrences over a given time period;
- or the series of continous *lifetime* variables ($\Delta T^i_j = T^i_{j+1} - T^i_j$).

The simplest model is a Poisson process, or equivalently an exponential lifetime distribution. As seen in Chapter 4, such a process is stationary so that the distribution of the counting process can be modelled regardless of the starting date by the following single-parametrised discrete distribution:

$$\begin{aligned} &N^i(\Delta T) \sim f_{Ni}(\cdot\,|\boldsymbol{\theta} = \lambda^i) \\ &P[N^i(\Delta T) = k] = \exp(-\lambda^i \Delta T) \cdot \frac{(\lambda^i \Delta T)^k}{k!} \end{aligned} \tag{5.30}$$

Equivalently, the associated lifetime variables are *i.i.d.* distributed according to the following single-parametrised discrete distribution:

$$\begin{aligned} &\Delta T^i_j \sim f_{Ti}(t|\boldsymbol{\theta} = \lambda^i) \quad i.i.d. \\ &P\left[\Delta T^i_j \le t\right] = 1 - \exp(-\lambda^i t) \end{aligned} \tag{5.31}$$

Recall the complementary cdf of the lifetime or *reliability function* (also known as the *survival function*):

$$R^i(t) = 1 - P\left[\Delta T^i_j \le t\right] = \exp(-\lambda^i t) \tag{5.32}$$

In order to estimate such a process, an independent and stationary data sample of time spans $(\Delta T_j^i)_{j=1\ldots n}$ is needed. Loglikelihood appears as follows:

$$LL(\Xi_n|\lambda^i) = n\log\lambda^i - \lambda^i \sum_{j=1\ldots n} \Delta T_j^i \tag{5.33}$$

Thus, the straightforward estimator for the parameter appears as:

$$\hat{\lambda}^i = \frac{n}{\sum_{j=1\ldots n} \Delta T_j^i} \tag{5.34}$$

Such an exponential lifetime parameter is thus estimated as the inverse of the (empiric) mean time between events, also called the mean time to failure (MTF or MTTF) in the context of reliability. It also corresponds in that case to the constant frequency of failures or *failure rate* (expected number of failures per time unit) or frequency of occurrence of the event:

$$\lambda^i = \frac{E[N^i(\Delta T)]}{\Delta T} \tag{5.35}$$

Note that such estimation process can involve data made up of either of the following:

- n successive failure dates for the i-th component of the system, assuming that whenever failure happens, the component is either replaced immediately or repaired 'as good as new'. Timespans between successive failures are used for estimation.
- Failure dates for n (independent and identically-distributed) components running in parallel. Failure dates are then used for estimation.

As will be considered in Section 5.4, it is also essential to control epistemic uncertainty associated with parameter estimators. In the case of exponential lifetime, epistemic uncertainty in $\hat{\lambda}^i$ can be formulated in a closed-form. With the estimator involving the sum of (i.i.d.) exponential lifetimes, a gamma distribution with shape n and scale λ is generated:

$$\sum_{j=1..n} \Delta T_j^i \sim \Gamma(t|\lambda^i, n) = \frac{(\lambda^i t)^n \exp(-\lambda^i t)}{t\,\Gamma(n)} \tag{5.36}$$

Equivalently, the double of the sum of the timespans follows a chi-square distribution with $2n$ degrees of freedom. Thus, a χ^2-based epistemic confidence interval can be built around the estimator:

$$\left[\hat{\lambda}^i \frac{\chi^2_{2n}(\alpha/2)}{2n} ; \hat{\lambda}^i \frac{\chi^2_{2n}(1-\alpha/2)}{2n}\right] \tag{5.37}$$

Such a formula is essential in industrial practice: recall that the typical QRA studies (Chapter 1) take as inputs of the *risk* model the frequencies of failure of components (failure rates) in order to compute the frequency of the undesired event of system failure:

$$f^e = G[(f^i)_i, (p^k)_k, \mathbf{d}] \tag{5.38}$$

The epistemic distribution of each of those inputs is thus involved directly as input uncertainty pdfs for a QRA uncertainty analysis in order to assess a level-2 risk measure such as the 95 % upper quantile of the annual frequency of failure of a given system (say a power plant).

Discrete events: alternative lifetime models The Poisson/exponential model has classical limitations that are of particular concern in the area of reliability when estimating component lifetimes. This is because the model is 'memoryless' or neglects ageing, as measured by the following *failure rate*:

$$h^i(t) = \lim_{dt \to 0} \left[\frac{P(\Delta T_j^i \leq t + dt \middle| \Delta T_j^i > t)}{dt} \right] = \frac{f_{Ti}(t|\lambda^i)}{1 - F_{Ti}(t|\lambda^i)} = -\frac{1}{R^i(t)} \frac{\partial R^i}{\partial t} \tag{5.39}$$

Such a function, also known as the *hazard rate* (or the *force of mortality* noted $\mu_T(t)$ in actuarial science), represents the 'instantaneous' or short-term frequency of failure at time t. The typical heuristic in reliability has it that such a rate would be expected to be higher at the time of initial commissioning (e.g. because of required adaptation before getting to normal operation) or in the longer term (e.g. because of material degradations) in a 'bathtub' curve over time. On the contrary, $h^i(t) = \lambda^i$ constantly in an exponentially-distributed lifetime. The more general Weibull family of distribution can be estimated in order to model such ageing:

$$\Delta T_j^i \sim f_{Ti}(t|\theta = \beta^i, \eta^i) = \frac{\beta^i}{\eta^i} \cdot \left(\frac{t}{\eta^i}\right)^{\beta^i - 1} \cdot e^{-\left(\frac{t}{\eta^i}\right)^{\beta^i}} \tag{5.40}$$

For which the reliability function and the failure rate appear as follows:

$$R^i(t) = e^{-\left(\frac{t}{\eta^i}\right)^{\beta^i}} \quad h^i(t) = \frac{\beta^i}{\eta^i} \cdot \left(\frac{t}{\eta^i}\right)^{\beta^i - 1} \tag{5.41}$$

The scale parameter can be seen as a generalisation of the mean time to failure of the exponential model λ^{-1}. The shape parameter β^i controls the time-dependence ($\beta^i = 1$ returning the exponential model): for $0 < \beta^i < 1$ failure rate decreases with time, modelling 'youth effects'; for $1 < \beta^i$ the failure rate increases with time, modelling 'ageing'. Loglikelihood appears as follows:

$$LL(\Xi_n|\beta^i, \eta^i) = n\log\frac{\beta^i}{\eta^i} + (\beta^i - 1)\sum_{j=1\ldots n} \log\left(\frac{\Delta T_j^i}{\eta^i}\right) - \sum_{j=1\ldots n} \left(\frac{\Delta T_j^i}{\eta^i}\right)^{\beta^i} \tag{5.42}$$

The first-order equations cannot be solved directly so that there is no closed-form for the maximal likelihood estimators, although η^i can be expressed when β^i is known:

$$\hat{\eta}^i = \left[\frac{1}{n}\sum_{j=1\ldots n} \left(\Delta T_j^i\right)^{\beta^i}\right]^{\frac{1}{\beta^i}} \tag{5.43}$$

In other words, η^i represents the mean observed inter-event time, though averaged through a geometrical mean according to the shape parameter.

Estimating continuous time-series More generally, system models may exhibit continuous time-series in their uncertain inputs that need to be modelled and estimated. In Section 4.3.3 of Chapter 4 it was said that time-based risk analysis could typically involve either:

- a simplified approach through a prior transformation of the time series into a discrete process (e.g. through threshold exceedance) as well as time-independent conditional random variables (e.g. modelling the amplitude of the uncertain variable conditional to the event of exceeding the threshold; or of its maximum value over a fixed block);
- or a finer description through an explicitly-modelled time series, such as a Brownian motion or an auto-regressive statistical model.

In the first case, estimation involves the above-mentioned techniques, both with continuous variables and the discrete (e.g. Poisson) process, though they may be coupled in order to find, for instance, the best threshold for modelling the time process (see Section 5.5). The second case involves more sophisticated statistical techniques for time series estimation, generalising the maximal likelihood or moment estimators that have been introduced in the previous sections. You may refer to the specialised literature, for example Brockwell and Davis (1996) for an illustration of windmill production uncertainty (Bensoussan, Bertrand, and Brouste, 2011).

5.2.6 Encoding phenomenological knowledge and physical constraints inside the choice of input distributions

Encoding acceptable distributions in order to represent uncertain inputs or risk factors is generally a challenge. Samples that are large enough to estimate and test a parametric model have been seen to be rare in real-world modelling, all the more so when looking at non-parametric models designed to overcome the errors in choosing a given distribution shape. Expert judgement is often mobilised to complement the rarity of statistical data and Section 5.4 on Bayesian methods will come back to that.

However, phenomenological/physical knowledge may also help in identifying plausible distribution choices. Arguments guiding the choice of *distribution shape* may first arise from a number of phenomenological reasons:

- Symmetry (or asymmetry) may be inferred from underlying phenomena. Metrological deviations, for instance, when limited to small excursions, may be considered plausibly to be equally probable on both signs around the nominal value; conversely, natural events such as flood flows or winds are known empirically to be largely asymmetric (and positively-valued). Consequently, the adequacy of a Gaussian distribution is completely opposite in those two cases.
- A single mode-distribution is plausible when the underlying phenomena are homogeneous or made of a combination of a very large array of heterogeneous phenomena. Conversely, multi-modal distribution is the typical recipe for the mixture of two or a limited number of phenomena, typically two modes of degradation within a physical property (e.g. toughness).
- The review in the literature of analogous types of inputs may guide the choice of distribution. Regional hydrological or atmospheric records help in determining, for instance, the choice of Gumbel or Weibull distribution respectively even if the particular location has not been heavily instrumented.
- An underlying combination of phenomena may help determine some distribution families. Macroscopic uncertainty resulting from the additive combination of many complex but independent

microscopic phenomena results in Gaussian distribution, for example in some metrological devices, cf. ISO (1995). Lognormal is appropriate when the combination is more of a multiplicative type, for example in pollutant transfer systems, whereby each ecological layer transfers to the next an uncertain proportion of the input, cf. Slob (1994). Last but not least, extreme phenomena are often generated by an underlying sequence of maximisation or threshold exceedance (see Section 4.3.5 or Section 5.5).

All such phenomenological knowledge can also be augmented by the rapidly-developing potential of micro-macro simulation: the idea being to model the physical (or physical-chemical) properties of a material at a lower scale and sample them over the sources of micro-variability so as to get a macroscopical distribution. Think of polymers, for which the equilibria between scission and reticulation of the macro-molecules can prove the occurrence of a bimodal distribution of the sizes and hence that of the traction properties.

Physical arguments often not only guarantee non-negative values for inputs but may also offer *plausible bounds* for the domain of variation. Setting precise values for those bounds may be a challenging and risky operation as one deals essentially with *uncertain* phenomena; however, it is possible that over-conservative bounds may generally be safely established, as orders of magnitude are non-infinite in physics. For instance, geophysical arguments may generate plausible maxima of the magnitude of an earthquake given the knowledge of a fault size, though the latter may be imprecisely known. Eventually, more complex *admissible areas in the variability of vectors of physical inputs* might also be inferred. Physical laws may additionally constrain the shared areas of variation of some inputs, excluding some parts of the space that breach physical laws (e.g. conservation laws).

The *flood example*

Plausible physical bounds could be feasible on geological and sedimentological grounds for the variability of upstream z_m and downstream z_v riverbed elevations. In the geologically-short period considered within human risk assessment, the lower bedrocks would not be significantly eroded and hence constitute a robust lower bound. Conversely, sedimentation would surely not exceed the local ground elevation outside of the river though such an upper bound would be over conservative. The consideration of maximum plausible sedimentation levels could possibly further reduce the upper bound, though this would be debatable. Symmetry in the riverbed variability can be argued for on the basis of interannual equilibria between sedimentation and erosion although mean-return could prove more or less rapid on the rising or falling side.

Uncertainty in the Strickler friction coefficient could also be bounded, though possibly in an over-conservative approach: the river should surely *never* flow as smoothly as a concrete channel (say $k_s < 70$). The lower bound is more debatable as to whether the level of obstruction would reliably never exceed that say of a densely-treed forest (say $k_s \sim 6$). Moreover, recall that the Strickler friction coefficient does stand for a simplified model for detailed fluid-bottom interactions involving rugosity, micro-topography, solid suspension and turbulence. Its distribution shape could theoretically be simulated through the modelling of typical river bottom rugosity over which flows are studied with more detailed hydrodynamic codes.

Maximal flows can surely be bounded in lower bounds, for example taking the average normal flow or even a draught flow which obviously constitute a very conservative lower bound for yearly maximal flood flows. Though flows are theoretically finite, the upper bounds would be highly controversial as the hydrological records have seen enormous sudden floods in places, exceeding by far the historically-anticipated values. Asymetry is a known characteristic of the distribution of flood flows.

5.3 Modelling dependence

Section 5.2 discussed the case of independent input components. In the more general case of (possibly) non-independent uncertain inputs, one has to estimate the joint distribution, including $\boldsymbol{\theta}_X$ parameters modelling the dependence structure:

$$X \sim f_X(\cdot|\boldsymbol{\theta}_X) = f_X(x^1, \ldots x^p|\boldsymbol{\theta}_X) \tag{5.44}$$

In general, $f_X(x^1, \ldots x^p| \boldsymbol{\theta}_X)$ cannot be factorised merely into marginal components as is the case – per definition – for independent inputs:

$$f_X(x^1, \ldots x^p|\boldsymbol{\theta}_X) = f_1(x^1|\boldsymbol{\theta}_1)f_2(x^2|\boldsymbol{\theta}_2)\ldots f_p(x^p|\boldsymbol{\theta}_p) \tag{5.45}$$

The methods of increasing complexity may then be contemplated:

- linear (Pearson) correlations between inputs: the correlation (symmetric and diagonal one) matrix then provides the $p(p-1)/2$ additional parameters needed within $\boldsymbol{\theta}_X$;
- rank (Spearman) correlations between inputs: an extension of the previous possibility, again parametrised by a matrix of dependence coefficients;
- copula model: this the most powerful and general approach. An additional function is inferred in order to describe the dependence structure, as well as a set of corresponding dependence parameters.

5.3.1 Linear correlations

Definition and estimation The linear correlation coefficient (or Pearson coefficient) is an elementary probabilistic concept defined pairwise between uncertain inputs as follows:

$$\rho^{ij} = Cor(X^i, X^j) = \frac{Cov(X^i, X^j)}{\sqrt{\text{var } X^i \text{ var } X^j}} = \frac{E[(X^i - EX^i)(X^j - EX^j)]}{\sqrt{\text{var } X^i \text{ var} X^j}} \tag{5.46}$$

Thus, the correlation matrix $R = (\rho^{ij})_{i,j}$ is symmetrical with diagonal 1, is semi-positive-definite and its non-diagonal elements are included in $[-1,1]$ indicating whether correlation is negative or positive, as well as its varying percentage. The correlation matrix can accordingly be estimated by a moment method using the following empiric correlations:

$$\hat{\rho}^{ij} = \frac{\sum_k (x^i_k - \bar{x}^i)(x^j_k - \bar{x}^j)}{\sqrt{\sum_k (x^i_k - \bar{x}^i)^2 \sum_k (x^j_k - \bar{x}^j)^2}} \tag{5.47}$$

While those definitions are general, theory is well developed in the case of Gaussian vectors (i.e. random vectors of Gaussian components *and* for which all linear combinations are Gaussian). In that case:

- The correlation matrix (in combination with the expectations and variances of inputs) fully describes the joint distribution.
- Null non-diagonal elements equate to independence.
- The distribution of the estimator $\hat{\rho}^{ij}$ is known in the case when the true correlation coefficient is null, whence a statistical test can be derived so as to ascertain whether the correlations are significant or not (Kendall and Stuart, 1946).

Table 5.8 Significance of Pearson correlation coefficients according to sample size and test confidence level.

Confidence level	Sample size $N=10$ $(n-2\ df)$	$N=20$	$N=30$	$N=100$
5 %	63 %	44 %	36 %	20 %
1 %	76 %	56 %	46 %	26 %

The Pearson correlation test involves the comparison of the statistic $\hat{\rho}^{ij}$ with a given level of correlation depending on sample size and on test confidence level, as derived from its known distribution conditional upon $\rho^{ij}=0$ (see Table 5.8).

Similar to goodness-of-fit, such a test is to be used in order to *reject* the null-hypothesis of non-significant correlations (or equivalently independence as the pair is supposed Gaussian), not to accept it. Note that small-size samples usually found in risk analysis (a few tens of joint observations at best!) appear challenging in that respect. Correlations as large as 40–50 % cannot be taken as significant and may result merely from the random process between truly independent inputs.

Using the linear correlation coefficients for non-Gaussian inputs is a frequent practice although their properties are less grounded. In that case, the Pearson test and the convergence of moment estimates are generally still guaranteed *asymptotically* in the sense that the sample size becomes limitless by virtue of the central limit theorems and derived asymptotic arguments, given additional hypotheses for a consistent use: see the specialised statistical literature in that respect.

Limitations of linear correlations Linear correlation must be seen as a reflection of the degree of linear dependence between two inputs: in other words, 'narrowness of the cloud' of data along a regressed line (Figure 5.9).

As a classical result of linear regression, the definition can be reformulated as follows:

$$\rho^{ij} = \frac{Cov(X^i, X^j)}{\sqrt{\operatorname{var} X^i \operatorname{var} X^j}} = \sqrt{\frac{\operatorname{var} X^j - Min_{a,b} E[X^j - (aX^i + b)]^2}{\operatorname{var} X^j}} \tag{5.48}$$

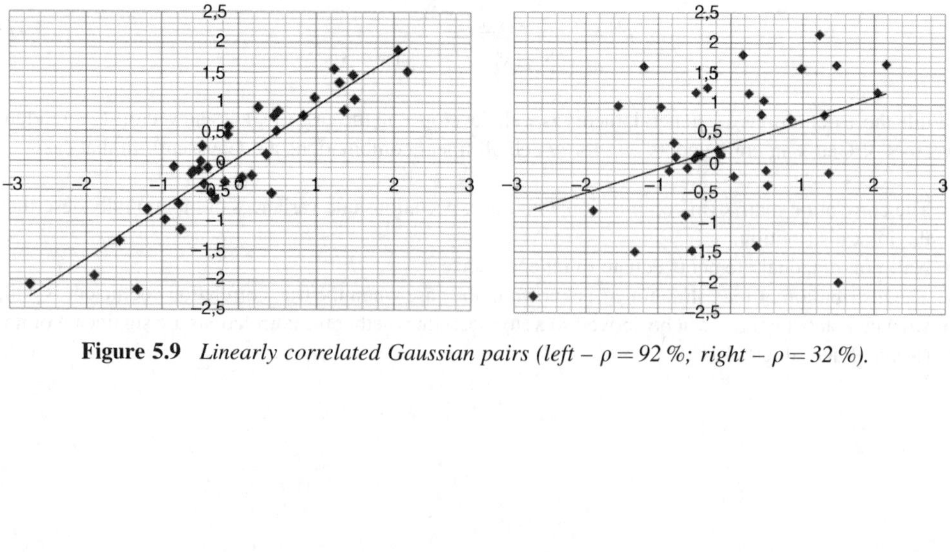

Figure 5.9 *Linearly correlated Gaussian pairs (left – $\rho=92$ %; right – $\rho=32$ %).*

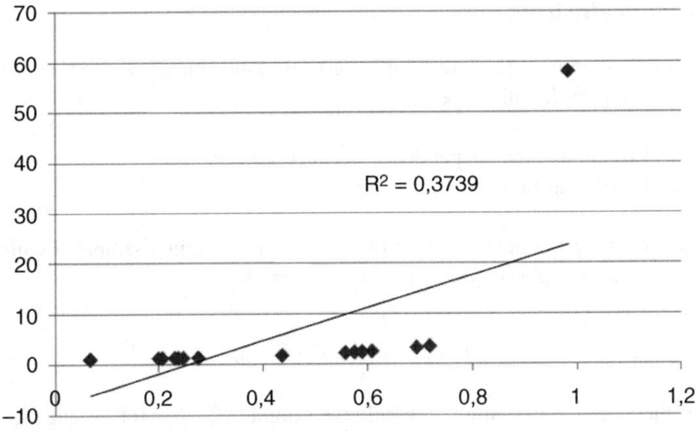

Figure 5.10 *Linear correlation and deterministic dependence between $X \sim U[0,1]$ and $Y = 1/(1-X)$.*

In other words, $\rho^{ij\,2}$ represents the proportion of the variance of X^j that can be at best explained by any linear function of X^i. This means in particular that a perfect deterministic dependence between two inputs, but which is non-linear, will not lead to a 100 % linear correlation. Figure 5.10 provides such an example whereby the linear correlation of a 15-sample of a non-linearly fully dependent pair is limited to 61 % (i.e. almost non-significant at 5 % confidence level).

Correlatively, the numerical value of linear correlation is not independent of monotonous transforms of input components. This proves somewhat counter-intuitive when in discussion with phenomenological experts in order to elicit the degree of dependence of model inputs. For instance, river friction could be judged by an expert to be 'partially' dependent on the level of sedimentation, as some sedimentation mechanisms accompany the development of islands, vegetation growth or channel ramification (higher friction) while others tend to cover with a uniform layer of sediment (lower friction): say 50 %. Yet, friction may be represented by the Strickler coefficient or alternatives that are functionally derived from Strickler, (e.g. through power functions) Sedimentation may also be represented differently whether the average bottom level is taken or the volume of sediment (both also linked through essentially monotone transforms). It is generally irrelevant for an expert in the phenomenological system models to express a distinct percentage quantifying the amount of correlation according to the peculiar representation of those physical phenomena.

As will be clarified in the following section, linear correlations are also distribution-dependent as distribution transformation of inputs is a particular example of monotonous transformations. In practice, when estimating input uncertainties through expertise rather than statistical samples, it also means that the expert should vary the percentage of correlation according to the choice of distribution he suggests for the inputs (e.g. normal or triangular).

Lastly, linear correlation has strong limitations when focusing on distribution tails or quantile predictions. Considering x_i^α and x_j^α as the α-quantiles, tail dependence can be defined through the behaviour of $P(X_j > x_j^\alpha | X_i > x_i^\alpha)$ when α tends to 1. In other words, the conditional probability for one input to be within its $(1-\alpha)$ % upper tail conditional to the fact the other input is also in its $(1-\alpha)$ % upper tail. Linear correlations (except when ρ^{ij} is strictly equal to 1) lead to 0 as the upper limit for tail dependence, that is asymptotic independence. Section 5.3.3 will present richer models accounting for *persistent* tail dependence.

5.3.2 Rank correlations

Interest of rank transformations Rank (or Spearman) correlation coefficients are a useful extension of the linear correlation approach with respect to:

- non-linear while still monotonous dependence between inputs;
- non-Gaussian marginal distributions.

Consider first the following example of a pair of samples. The original sample (Figure 5.11 – left) is transformed into the rank sample (Figure 5.11 – right), that is:

$$(x_j^i, x_j^{i'})_{j=1...n} \rightarrow (r_j^i, r_j^{i'})_{j=1...n} \tag{5.49}$$

where r_k^i denotes the rank of observation x_k^i within the sample$(x_j^i)_{j=1...n}$: for instance $r_k^i = 1$ if x_k^i is the largest of all n observations of the i-th input component.
Rank transformation conspicuously transforms the original monotonous (but non-linear) relation into a close-to-linear relation. This is reflected within the rank correlation coefficient defined as the linear correlation between rank-transformed inputs as follows:

$$\hat{\rho}_S^{ij} = \frac{\sum_k (r_k^i - \bar{r}^i)(r_k^j - \bar{r}^j)}{\sqrt{\sum_k (r_k^i - \bar{r}^i)^2 \sum_k (r_k^j - \bar{r}^j)^2}} \tag{5.50}$$

The rank correlation coefficient rises to $\rho_S^{ij} = 97\%$ in comparison with the linear coefficient $\rho^{ij} = 70\,\%$ evidencing the increased power of inference of values from one input to the other as reflected in the proportion of explained variance.

Definition and properties Indeed, rank correlations may be defined theoretically as follows:

$$\rho_S^{ij} = Cor(F_{X^i}(X^i), F_{X^j}(X^j)) \tag{5.51}$$

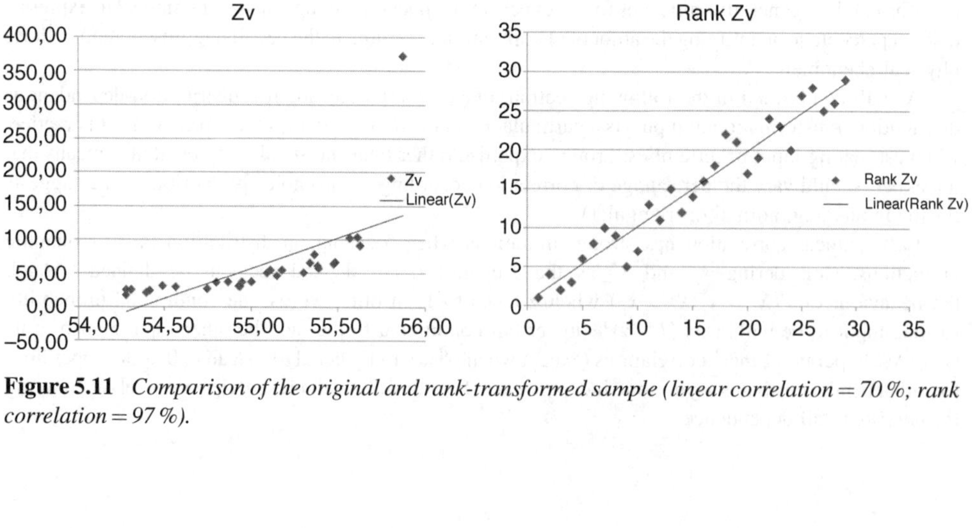

Figure 5.11 *Comparison of the original and rank-transformed sample (linear correlation = 70 %; rank correlation = 97 %).*

where $F_{Xi}(.)$ stands for the marginal cdf of i-th input. A rank correlation matrix is a correlation matrix – symmetric, semi-positive definite with diagonal 1 – applied to modified random variables. Remember that $F_{Xi}(X^i)$ follows a uniform distribution for any input; as explained above on the basis of a sample, the empiric marginal distribution of each input can be defined attributing i/n probability to the exceedance of the *i-th* ranked sample of observations. This results in an empiric sample of the random couple ($F_{Xi}(X^i)$, $F_{Xj}(X^j)$) which is similar to that of $(r_j^i, r_j^{i'})j = 1 \ldots n$ except for the (impactless) factor $1/n$, and thus to the estimation of its linear correlation through the above-mentioned formula.
The essential consequence is that ρ_S^{ij} is stable whatever the monotonous transforms and distribution changes of the inputs (see, e.g. Kendall and Stuart, 1946; Embrechts, McNeil, and Straumann, 1999). Suppose that $T(.)$ is a monotonous (increasing) function and X^1 a random variable whose cdf is $F_{X1}(.)$. Classically, $Y = T(X^1)$ is distributed as follows:

$$Y = T(X^1) \sim F_T(y) = F_{X1} o T^{-1}(y) \tag{5.52}$$

Or equivalently:

$$F_{X1}(X^1) = F_{X1} o T^{-1}(T(X^1)) = F_T(T(X^1)) \tag{5.53}$$

so that the Spearman coefficient between X^1 and any other input X^2 is similar to that between $Y = T(X^1)$ and X^2. This applies in particular to iso-probabilistic changes from a distribution F_X to another F_Y (e.g. Gaussian centred) since $T = F_Y o F_X^{-1}$ is monotonous as the compound of cdfs that are monotonous by definition. This key property resolves some of the limitations of the linear correlation, especially when phenomenology leads one to infer that dependence are essentially monotonous. It enables, for instance, expert elicitation of robust correlations with respect to the type of marginal distributions or to changes of physical inputs.
As mentioned above, estimation of the rank correlation matrix $(\rho_S^{ij})_{i,j}$ is also done through the empiric rank-transformed moments:

$$\hat{\rho}_S^{ij} = \frac{\sum_k (r_k^i - \bar{r}^i)(r_k^j - \bar{r}^j)}{\sqrt{\sum_k (r_k^i - \bar{r}^i)^2 \sum_k (r_k^j - \bar{r}^j)^2}} \tag{5.54}$$

which simplifies into the following, as the ranks are by definition covering the set $\{1, 2 \ldots n\}$:

$$\hat{\rho}_S^{ij} = 1 - \frac{6\sum_k (r_k^i - r_k^j)^2}{n(n^2-1)} \tag{5.55}$$

A test statistic on its significance against zero-rank dependence is also available as the distribution of $\hat{\rho}_S^{ij}$ is known conditional to independence formulated as equiprobability of all rankings: being distribution-free (given reasonable hypotheses), the test is more general than the Pearson correlation test. As exemplified in Table 5.9, the numerical values are close to that of the Pearson test. Again, the

Table 5.9 Significance of Spearman correlation coefficients according to sample size and test confidence level.

Confidence level	Sample size $N = 10$	$N = 20$	$N = 30$	$N = 100$
5 %	65 %	45 %	36 %	20 %
1 %	79 %	57 %	47 %	26 %

significance of moderate correlation levels is challenged by the typically small size of multivariate samples in real-world systems.

> *The flood example*
> A simple illustration is given by empiric correlations between riverbed elevations (Figure 5.12). The 29-joint sample of observations for Z_m and Z_v leads to 0.665 Pearson and 0.656 Spearman correlation coefficients. Although the marginal laws are non-Gaussian, it can be inferred from the size of the sample that the level of correlation is significant, and that the difference between linear or rank correlation is insignificant. Sedimentation/erosion processes shaping the riverbeds along the reaches of the river typically follow a common trend modulated by the local geometry of obstacles, slopes or river bends.

Limitations of rank correlations Rank correlations essentially measure monotonous dependence: limitations appear when non-monotonous dependence is involved. Figure 5.13 illustrates the case of a symmetric U-shaped dependence between two random inputs that the Spearman coefficient fails to capture: think of physical models involving signed (e.g. displacement) and non-signed (e.g. energy) quantities.

More generally, the Spearman coefficient is an example of a much wider family of rank-based statistics that include Kendall's tau, Gini's statistic and so on (see Kendall and Stuart, 1946). They offer partially-different properties although neither is completely satisfactory with respect to non-monotonous complex dependence.

Additional remarks on eliciting correlations By definition, both linear and rank correlation matrices are *semi-positive-definite* matrices. In practice, it is often difficult to work on multi-dimensional samples of uncertain inputs that are large enough, if one is lucky enough to have any. The standard recipe is thus to ask experts to elicit correlation coefficients between inputs in a pairwise questionnaire. It is essential then to

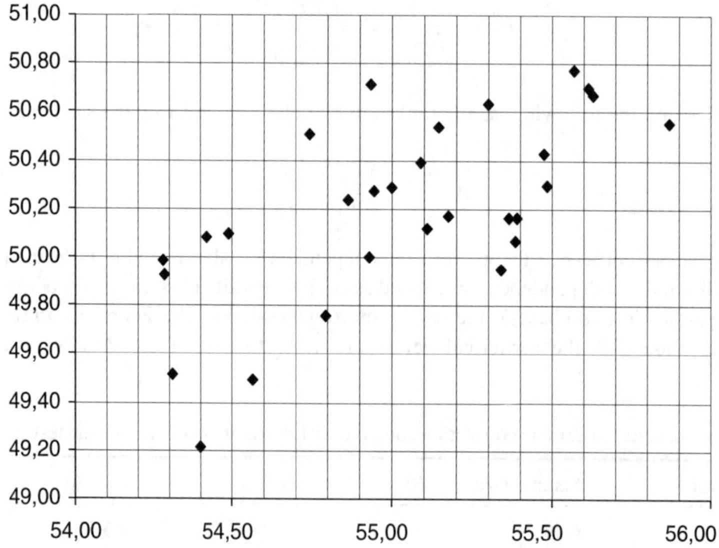

Figure 5.12 *Bivariate 29-sample of Z_m and Z_v.*

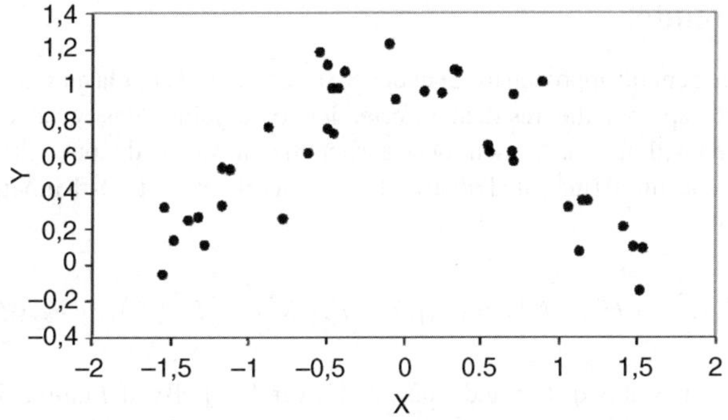

Figure 5.13 *An example of symmetric, non-linear and non-monotonous, dependence neither captured by Pearson* $-\hat{\rho}^{ij} = -0.04$ *nor Spearman* $-\hat{\rho}_S^{ij} = -0.02$ *(for a 40-sample).*

check that the resulting matrices do have such properties. While they are per construction symmetric and real-valued (and thus enjoy spectral decomposition) with diagonal 1 and non-diagonal elements valued within $[-1,1]$, it may happen that the resulting elicited matrix is non-semi definite positive, that is it has at least one strictly negative eigenvalue.

The flood example

The reference correlation matrix (see Annex Section 10.4) results both from the empiric analysis of bivariate $Z_v - Z_m$ sample (rank or linear correlations with similar results ~0.66) and expert elicitation for pairs (K_s, Z_v) and (K_s, Z_m). The values typically encode the following phenomenological facts: (i) friction is partially (negatively) correlated to the riverbed level as sedimentation (i.e. increasing Z_v or Z_m) occurs with obstacles where friction also increases (i.e. decreasing K_s); (ii) river friction shaping the downstream flood level is more likely to be correlated to downstream riverbed Z_v than to the more distant upstream Z_m.

	Q	K_s	Z_v	Z_m
Q	1	0	0	0
K_s	0	1	−0.5	−0.3
Z_v	0	−0.5	1	+0.66
Z_m	0	−0.3	+0.66	1

Such a matrix is semi-positive definite as its eigenvalues prove positive. Conversely, the following one, encoding a very high correlation between K_s and Z_v but neglecting the eventuality of correlation with upstream Z_m, does not fulfil the required semi-positive definiteness.

	Q	K_s	Z_v	Z_m
Q	1	0	0	0
K_s	0	1	−0.9	−0.1
Z_v	0	−0.9	1	+0.66
Z_m	0	−0.1	+0.66	1

5.3.3 Copula model

Copulae model a more general approach to dependence structure. The copula may be introduced in several ways. Essentially, it captures the residual information of a joint density once all marginal input distributions have been withdrawn. As mentioned earlier, the transformed input $F_{Xi}(X^i)$ follows a uniform distribution whatever the initial (marginal) distribution of uncertainty of the *i*-th input. Thus, the following random vector

$$(U^1, U^2, \ldots U^i \ldots U^p) = (F_{X1}(X^1), F_{X2}(X^2), \ldots F_{Xi}(X^i), \ldots F_{Xp}(X^p)) \tag{5.56}$$

is such that all its inputs are distributed uniformly over [*0,1*]. By definition, its joint cumulative distribution denoted as:

$$C(u^1, u^2, \ldots u^i \ldots u^p | \theta_C) = P[U^1 < u^1, U^2 < u^2, \ldots U^i < u^i \ldots U^p < u^p] \tag{5.57}$$

is the *copula* associated with the random vector $\boldsymbol{X}$ or its joint distribution f_X. In fact, it is defined uniquely in the case of continuous multivariate distribution, but can be extended also to non-continuous distributions according to the Sklar theorem (cf. Nelsen, 2006 for a comprehensive introduction). A copula model is the distribution of any random vector the marginals of which are distributed over [*0,1*], the independent case materialising in the simple factorisation $C(u^1, u^2, \ldots u^i \ldots u^p) = u^1 u^2 \ldots u^p$. Dependence may thus be modelled not only by a pairwise coefficient or by a multidimensional matrix, but by the choice of a given function shape as well as of its parameters $\boldsymbol{\theta}_C$ which constitute the copula part of the comprehensive vector of parameters $\boldsymbol{\theta}_X$ of the joint distribution of $\boldsymbol{X}$.

$$\begin{aligned} F_X(\boldsymbol{x}|\boldsymbol{\theta}_X = (\boldsymbol{\theta}_m, \boldsymbol{\theta}_c)) &= P[X^1 < x^1, \ldots X^p < x^p | \boldsymbol{\theta}_X = (\boldsymbol{\theta}_m, \boldsymbol{\theta}_c)] \\ &= C[F_{X^1}(x^1|\boldsymbol{\theta}_{m^1}), \ldots, F_{X^p}(x^p|\boldsymbol{\theta}_{m^p}), \boldsymbol{\theta}_c] \end{aligned}$$

In that respect, linear and rank correlations appear as limited projections in that much larger space of possibilities. Linear correlation underlies, for instance, the implicit choice of a Gaussian copula function:

$$C(u^1, \ldots, u^p | \boldsymbol{\theta}_C) = \Phi_R(\Phi^{-1}(u^1), \ldots, \Phi^{-1}(u^p)) \tag{5.58}$$

where Φ^{-1} stands for the inverse cdf of a standard Gaussian variable and Φ_R stands for the multi-variate cdf of a Gaussian vector of linear correlation matrix R, that is:

$$\Phi_R(\boldsymbol{v}) = F(\boldsymbol{v}|R) = \int_{\boldsymbol{w}<\boldsymbol{v}} [\det(2\pi R)]^{-1/2} \exp\left[-\frac{1}{2}\boldsymbol{w}' R^{-1} \boldsymbol{w}\right] d\boldsymbol{w} \tag{5.59}$$

A large number of alternative copula functions may be used, such as the popular Gumbel copula hereafter expressed in its bivariate formulation:

$$C(u^1, u^2 | \theta_c) = \exp(-[(-\log(u^1))^{1/\theta_c} + (-\log(u^2))^{1/\theta_c}]^{\theta_c}) \tag{5.60}$$

Its unique parameter θ_c takes values within [*0,1*]; $\theta_c = 1$ clearly equates to the independent case while the limit $\theta_c \to 0$ models the case of full dependence. The Gumbel copula, a sub-category of the Archimedean copula family, enjoys valuable tail dependence properties. Unlike the Gaussian copula, it can model persistent dependence even for higher quantiles, which may be a more conservative choice when modelling multi-variate input risk factors. *Tail dependence* can be formalised as the limit behaviour of upper (or lower) quantiles conditional on the other variable reaching a high quantile:

$$\begin{aligned}\lambda_U &= \lim_{v\to 1} P\left[X^1 > F_{X^1}^{-1}(v)\middle|X^2 > F_{X^2}^{-1}(v)\right] \\ &= \lim_{v\to 1} \frac{P\left[X^1 > F_{X^1}^{-1}(v), X^2 > F_{X^2}^{-1}(v)\right]}{P\left[X^2 > F_{X^2}^{-1}(v)\right]} \\ &= \lim_{v\to 1} \frac{P\left[X^1 > F_{X^1}^{-1}(v), X^2 > F_{X^2}^{-1}(v)\right]}{1-v}\end{aligned}$$

The two extreme examples are: (i) the case where inputs are independent (at least in the vicinity of tails), for which it is easy to derive that close to $v=1$, $\lambda_U \sim (1-v)$; (ii) the case where inputs are completely dependent through a deterministic increasing function $x^1 = h(x^2)$ (or at least in the vicinity of tails), for which one also derives the corresponding $\lambda_U \sim 1$. In other words, the tail dependence provides a coefficient between 0 and 1 from none to full deterministic dependence for the upper (or respectively lower) quantiles of each input. Such dependence is essential when dealing with system models that are monotonous as a function of both inputs with a consistent monotonous sign (see Chapter 4).

Estimating the copula function of a given shape may be – theoretically at least – undertaken through maximal likelihood estimation just as any other statistical model as it involves a closed-form expression of the likelihood function combining the copula part and the marginal distributions. In some cases however, simpler moment-based methods of estimation are available. This is the case not only for the Gaussian copula of course, but also for the Gumbel copula parameter which may be related to Kendall's tau.

5.3.4 Multi-dimensional non-parametric modelling

Non-parametric methods (e.g. multi-dimensional kernels) may also be used to estimate the joint distribution (copula function and marginals) without prior specification. The kernel method introduced in Section 5.2.1.1 can be generalised for a joint p-dimensional sample:

$$(\mathbf{x}_j)_{j=1\ldots n} = (x_j^i)_{i=1\ldots p, j=1\ldots n} \tag{5.61}$$

It then involves the following multi-Gaussian kernel density:

$$\hat{f}_X^{\text{ker}}(\mathbf{x}) = \frac{1}{n\cdot(2\pi)^{p/2}\cdot\prod_{i=1}^{p} h_n^i}\sum_{q=1}^{n}\exp\left(-\sum_{i=1}^{p}\left(\frac{x^i - x_j^i}{h_n^i}\right)^2\right) \tag{5.62}$$

with the corresponding Silverman rule:

$$h_n^i = \hat{\sigma}_n^i \cdot n^{-\frac{1}{4+p}} \tag{5.63}$$

Like the independent model, the rich parametrisation involved in such inference processes may, however, lead to extrapolation pitfalls. The reader is invited to consult the considerable research and literature on that subject.

5.3.5 Physical dependence modelling and concluding comments

Dependence is a complex issue. It goes much wider than a set of correlation coefficients. One must remember that independent variables are necessarily non-correlated – either linear or rank correlations – but that the opposite is wrong. The space of plausible dependence structures is of *infinite* dimension as being a *functional* space of copula-type functions, a restricted subset of positively-valued input-wise monotone functions over $[0,1]^p$ abiding by a number of constraints ensuring that it is a multi-dimensional cdf. Such a space of possibilities is all the wider when the input dimension grows: even for the simplified linear correlations, model parameterisation grows as $O(p^2)$ with the input dimension.

Conversely, it has been mentioned that the statistical samples available for uncertainty modelling are generally small (a few tens at most). The situation is much more critical for dependence modelling than it is for the modelling of each (marginal) input distribution as the *joint* observations of realisations of the input vectors are generally a much smaller subset of the input observations available. Flood flow measurements would, for instance, be instrumented distinctively from the observations of the riverbed levels, thus not ensuring the synchronisation of observations. The situation is somewhat different to that of financial portfolios, which contain large amounts of data for highly-dimensional input vectors constituted as portfolio assets. There, the development of the copula has received considerable interest, following a number of well-known allegations of the dangers of limiting dependence modelling to linear correlations within financial or insurance risk analysis (including Embrechts, McNeil, and Straumann, 1999).

In practice, dependence modelling generally involves expert elicitation. Experts would typically be questioned in order to encode correlation coefficients or more elaborate dependence structures. This raises the vast question of controlling the relevance and robustness of such elicitation procedures (Cooke, 1991; Granger Morgan and Henrion, 1990) notably because:

- experts may find difficult to understand the underlying assumptions, and require elaborate training in order to understand the significance of the quantitative answers: all the more so when elaborate (e.g. higher-dimensional dependence, elaborate copula functions, ...) models are involved;
- inconsistency of answers is pervasive: a high-dimensional correlation matrix elicited by pair-wise questions may not result in being semi-positive-definite;
- the problems are mostly ill-posed: a large number of alternative copula functions, if not already consisting of correlation matrices, may be compatible with expert answers;
- as illustrated in Chapter 4, dependence modelling has significant consequences for the risk measure itself: linear correlations applied to linearised system models showed that independence of inputs leads to significantly lower output uncertainty, or a lower value of the risk measure taken as output variance or an upper quantile.

Facing such challenges, phenomenological/physical modelling may sometimes offer an alternative. This means replacing the initial supposedly-dependent input vector $\boldsymbol{X}$ by a sub-system model $g(.)$ based on independent internal inputs $\boldsymbol{W}$:

$$\boldsymbol{X} = g(\boldsymbol{W}) \tag{5.64}$$

In other words, $\boldsymbol{W}$ represent the independent sources that are combined through the phenomena encapsulated in $g(.)$ so as to produce the 'common modes', as referred to in systems reliability.

The flood example

Think, for instance, of the above-mentioned dependence between riverbed upstream Z_m and downstream Z_v level. A sedimentological sub-model would typically distinguish:

- the overall sedimentation phenomena that impact upon the entire river: typically the uncertain occurrence of a series of eroding flows and that of sedimenting flows changing both upstream and downstream riverbeds, modelled inside a single random variable (or time process);
- other secondary phenomena impacting differently upon the upstream and downstream reaches, such as the unplanned arrival of blocking obstacles (vegetation growth, failure of banks or hydraulic works, ...) at various points on the river: those would typically raise riverbed sedimentation upstream while lowering it downstream.

If it is reasonable to assume that each of those sub-phenomena is of independent origin, they can thus be encoded as an independent vector of $\boldsymbol{W}$ explained through $g(.)$ the dependence observed or presumed at the upper scale of $\boldsymbol{X}$. If not, extend the sub-modelling until a chain of independent risk/uncertain factors can be inferred. This of course subsumes additional information: the availability of phenomenological knowledge, as well as the ability of experts to encode distributions for $\boldsymbol{W}$ elementary components. If the observations of $\boldsymbol{X}$ co-exist with such knowledge, the inverse probabilistic methods described in Chapter 6 are required in order to estimate $\boldsymbol{W}$.

5.4 Controlling epistemic uncertainty through classical or Bayesian estimators

So far the estimation techniques have been devoted to the estimation of the conditional (or single-probabilistic) distribution $f_X(\cdot|\boldsymbol{\theta}_X)$ of uncertain inputs $\boldsymbol{X}$. As indicated in Chapter 2, a careful description of the extent of uncertainty should also account for sample limitations. The full aleatory and epistemic uncertainty model involves an additional distribution designed to represent (to some extent) those additional estimation uncertainty sources:

$$\begin{array}{l} \boldsymbol{X}|\boldsymbol{\theta}_X \sim f_X(\cdot|\boldsymbol{\theta}_X) \\ \boldsymbol{\Theta}_X \sim \pi(\cdot|IK,\zeta) \end{array} \qquad (5.65)$$

Such an estimation will be discussed first with a classical approach and then a Bayesian one. The Bayesian approach (see Section 5.4.3) offers a more natural statistical interpretation of the double-level probabilistic structure and of its updating through the introduction of a data sample. However, classical statistics are easier to manipulate in a number of cases, particularly in the Gaussian case where closed-form expressions are available even for very small samples (see Section 5.4.2), or when the sample is large enough for the asymptotic theory to apply (see Section 5.4.3).

5.4.1 Epistemic uncertainty in the classical approach

It is necessary to be more specific with the definition of the associated distributions that will be manipulated. The conditional distribution $f_X(\cdot|\boldsymbol{\theta}_X)$ should first be understood as a model for each observation $\boldsymbol{X}_j$ of the sample $\Xi_n = (\boldsymbol{X}_j)_{j=1...n}$ representing equally *past* observed states of the system, as well as the state of system $\boldsymbol{X}$ at the *future* time of interest. Each of which is supposed to be independent and distributed identically. In a classical approach, the 'true' value $\boldsymbol{\theta}_X{}^\circ$ is the common parameter value driving the variability of any past or future state of the system:

$$\left.\begin{array}{r}\forall j=1,\ldots n \quad \boldsymbol{X}_j \sim f_X(\cdot\,|\boldsymbol{\theta}_X^o) \\ \boldsymbol{X} \sim f_X(\cdot\,|\boldsymbol{\theta}_X^o)\end{array}\right\} \quad i.i.d. \tag{5.66}$$

Albeit unknown to the analyst, $\boldsymbol{\theta}_X{}^\circ$ is understood as taking a *fixed* value in the real world at any time.[1] Hence, it is *not* represented as a random variable. Instead, the random variable models the behaviour of the *estimator* $\hat{\boldsymbol{\theta}}_X$ of $\boldsymbol{\theta}_X{}^\circ$, that is its variability around the true unknown value reflecting the limitations of the sample size. By definition, an estimator $\hat{\boldsymbol{\theta}}_X$ is a known *function of the random observations* $(\boldsymbol{X}_j)_{j=1\ldots n}$:

$$\hat{\boldsymbol{\Theta}}_X = Q(\boldsymbol{X}_1, \ldots \boldsymbol{X}_n) \tag{5.67}$$

so that it stands also as a random variable during the observation process. Its pdf $\pi(.)$ derives from the function $f_X(\cdot|\boldsymbol{\theta}_X{}^\circ)$, thus it is also parameterised by $\boldsymbol{\theta}_X{}^\circ$ albeit in a complex way:

$$\hat{\boldsymbol{\Theta}}_X \sim \pi(\cdot\,|\boldsymbol{\theta}_X^o) \tag{5.68}$$

Think of the Gaussian case where $Q(.)$ is, for instance, the empiric mean, so that its pdf is also Gaussian the parameters of which are deduced from $\boldsymbol{\theta}_X{}^\circ$ (same expectation, variance reduced by factor $1/n$).

The 'true' value $\boldsymbol{\theta}_X{}^\circ$ being unknown, the only practical possibility for making an inference about the state of the system at the *future* time of interest is to compute functions on the basis of the imperfect estimator $\hat{\boldsymbol{\Theta}}_X$. This is particularly the case with the risk measure. Suppose theoretically that there is no epistemic uncertainty so that the risk measure c_Z comes as a direct function of $\boldsymbol{\theta}_X{}^\circ$. Consider the common case of a risk measure defined as an expected utility:

$$c_Z = \int_z U(z) f_Z(z|\boldsymbol{\theta}_X{}^\circ)dz = \int_x UoG(\boldsymbol{x}, \boldsymbol{d}) f_X(\boldsymbol{x}|\boldsymbol{\theta}_X{}^\circ)d\boldsymbol{x} = c_Z(\boldsymbol{\theta}_X{}^\circ, \boldsymbol{d}) \tag{5.69}$$

Such a value is inaccessible as $\boldsymbol{\theta}_X{}^\circ$ is unknown. Thus, the classical approach is to estimate a value for $c_Z(\boldsymbol{\theta}_X{}^\circ, \boldsymbol{d})$ with a given confidence level, that is find $C_Z{}^\beta(\boldsymbol{d})$ so that:

$$P[C_Z^\beta(\boldsymbol{d}) = C(\hat{\boldsymbol{\Theta}}_X, \boldsymbol{d}) \geq c_Z(\boldsymbol{\theta}_Z{}^\circ, \boldsymbol{d})] = \beta \tag{5.70}$$

keeping in mind that such a function $C(.)$ should depend only on $\hat{\boldsymbol{\Theta}}_X$ and $\boldsymbol{d}$ instead of $\boldsymbol{\theta}_X{}^\circ$. The probability should not be read as 'true risk measure has a β % chance to be exceeded by estimated risk measure' but conversely as 'estimated risk measure has a β % chance to exceed the true risk measure', because the random process involves $\hat{\boldsymbol{\Theta}}_X$, not $\theta_X{}^\circ$! Think of the estimated failure probability guaranteed to exceed the 'true' failure probability (i.e. that frequency which would be observed in a very large sample of observations) in β % cases. In other words, there is 95 % confidence that the estimator of the *100-yr*

[1] This model is thus appropriate only over a time period of observations and predictions short enough for the underlying phenomena to remain stationary. Extensions to non-stationary uncertainty models are required for the temperature or sea level variability under climate change.

flood flow is conservative with respect to the (unknown) true risk level. Epistemic uncertainty is modelled here on a unilateral β-confidence level with respect to the *estimation* of a given risk measure $c_Z(\boldsymbol{\theta}_X{}^\circ, \boldsymbol{d})$ on the basis of a random set of observations.

The interpretation of such a double-level risk measure requires care in defining the underlying sample spaces and random experiments. In theory, a verification protocol would require a large number of *repeated* risk assessment *studies*, each of which involves observing a sample $\Xi_n = (\boldsymbol{X}_j)_{j=1...n}$ and later predicting a estimated β-confidence failure probability – plus a reference long-term observation of the same system. Then, on average, for β % of those risk assessment studies, the predicted failure probability would exceed the reference failure probability observed in the long-term (theoretically limitless) observation sample.

5.4.2 Classical approach for Gaussian uncertainty models (small samples)

With Gaussian-based models, epistemic uncertainty $\pi(.)$ and the estimated risk measure with β %-confidence may be computed explicitly as closed-form functions of the observations. Assume a Gaussian distribution for a scalar X:

$$X|\boldsymbol{\theta}_X \sim N(\mu, \sigma^2) \tag{5.71}$$

Table 5.10 shows the distribution of its natural estimators, the empiric mean $\overline{X}$ and the unbiased empiric variances s_{n-1}^2 of a sample (cf. Kendall and Stuart, 1946). The $(n-1)$-df Student (noted t_{n-1}) and chi-square (noted χ_{n-1}^2) distributions have been *defined* historically as the distributions describing respectively the normalised deviation $\sqrt{n}(\overline{X}-\mu)/s_{n-1}$ of an empiric mean from the theoretical expectation and the normalised ratio of the unbiased empiric variance to theoretical variance $(n-1)\frac{s_{n-1}^2}{\sigma^2}$. Table 5.10 also yields the closed-form expressions for the estimated risk measure with β %-confidence.

Note that an unbiased estimation of the standard deviation requires a corrective factor B_n to be applied to the square root of s_{n-1} of s_{n-1}^2 because $E(\sqrt{X}) \neq \sqrt{E(X)}$. Note also how epistemic uncertainty decreases with the increase of sample size. The epistemic uncertainty describing, for instance, the error made in estimating an expectation on the basis of the empiric mean decreases as:

$$\overline{X}-\mu \sim \frac{s_{n-1}}{\sqrt{n}} t_{n-1}(\cdot). \tag{5.72}$$

where t_{n-1} has a variance of:

$$Var(T_{n-1}) = \frac{n-1}{n-3} \tag{5.73}$$

Assuming that the empiric variance s_{n-1} maintains a stable order of magnitude in spite of inevitable statistical fluctuations, this means that *epistemic variance decreases* at a speed close to *1/n*. Note that a *stable empiric variance* means, in other words, *consistency in the injection of additional observations*. Lack of knowledge modelled by epistemic uncertainty decreases with the injection of information providing that it does not depart too much from the previous inference.

This result is straightforward in the case of Gaussian distribution and remains true even at low sample sizes; it is generalised considerably *asymptotically* by virtue of the well-known Central-Limit Theorem to the empiric mean of *any random variable*, given a finite variance and a large-enough sample.

Table 5.10 Epistemic uncertainty of Gaussian estimators.

θ	Unbiased estimator $\hat{\theta}_X$	Estimated β %-quantile for the risk measure C_z^β	Asymptotic approximation
$E(X) = \mu$	$\overline{X} = \frac{1}{n}\sum X_i$	$\overline{X} + \frac{s_{n-1}t_{n-1}(\beta)}{\sqrt{n}}$ where $t_{n-1}(\beta)$ is the quantile of *n-1d.f.*-Student law	$\overline{X} + \frac{s_n\Phi^{-1}(\beta)}{\sqrt{n}}$ where $\Phi^{-1}(\beta)$ is the inverse Gaussian cdf
$Var(X) = \sigma^2$	$s_{n-1}^2 = \frac{\sum_i (X_i - \overline{X})^2}{n-1}$	$s_{n-1}^2 \frac{n-1}{\chi_{n-1}^2(1-\beta)}$	$s_n^2\left[1 + \sqrt{\frac{2}{n}}\Phi^{-1}(\beta)\right]$
Standard deviation (σ)	$s_{n-1}B_n$ where $B_n = \sqrt{\frac{n-1}{2}}\frac{\Gamma\left(\frac{n-1}{2}\right)}{\Gamma\left(\frac{n}{2}\right)}$ or approximately if $n > 10$: $B_n \approx \frac{4n-4}{4n-5}$	$s_{n-1}\sqrt{\frac{n-1}{\chi_{n-1}^2(1-\beta)}}$	$s_n\left[1 + \sqrt{\frac{1}{2n}}\Phi^{-1}(\beta)\right]$

The Gaussian assumption makes explicit the additional *penalty* that has to be added to the central estimator of an expectation in order to cover the epistemic uncertainty at a β-confidence level: $\frac{s_{n-1}t_{n-1}(\beta)}{\sqrt{n}}$. It decreases at a speed close to $1/\sqrt{n}$ as the β-quantile of a $(n-1)$-df Student distributed r.v. does not vary much except for a very small n, thus converging quickly to the Gaussian β-quantile. Taking $\beta = 0.95$, $t_{n-1}(\beta)$ decreases from 2.1 (for $n = 5$) to 1.8 (for $n = 10$) and down to 1.65 (for a limitless n).

Beyond such simple Gaussian models and possibly a few other examples, it proves more difficult to control epistemic uncertainty through closed-form distributions. Statistical asymptotic theory is then a generic recipe for approximating such epistemic distribution π as a multivariate Gaussian co-variance matrix. The asymptotic Gaussian distribution for the epistemic uncertainty in the Gaussian case has been included in the third column of Table 5.10. Nevertheless, large sample sizes are necessary so as to justify such an approximation, as will be discussed in more depth in the next section.

5.4.3 Asymptotic covariance for large samples

Asymptotic variance of estimators Whatever the distribution shapes (Gaussian or not), when the estimation is based on the maximal likelihood or moment methods, classical statistical results state that the behaviour of the random variable:

$$\hat{\mathbf{\Theta}}_X = Q(\mathbf{X}_1, \ldots \mathbf{X}_n) \tag{5.74}$$

is *asymptotically* Gaussian, that is approximated well by a Gaussian multi-variate distribution as n grows to inifinity and that its covariance matrix may be approximated through the information matrix. Consider the case of maximal likelihood, or more conveniently its equivalent normalised log-likelihood formulation (introduced in Section 5.2 for the 1D-case):

$$\frac{1}{n} LL\left(\left(\mathbf{X}_j\right)_{j=1\ldots n}, \boldsymbol{\theta}_X \right) = \frac{1}{n} \sum_{j=1\ldots n} Logf_X\left[\mathbf{X}_j|\boldsymbol{\theta}_X\right] \tag{5.75}$$

whereby the MLE is the following function:

$$\hat{\mathbf{\Theta}}_n = \underset{\theta}{Arg\max}\left(LL\left[\left(\mathbf{X}_j\right)_{j=1\ldots n}, \boldsymbol{\theta}_X \right] \right) \tag{5.76}$$

The normalised log-likelihood of the observed sample appears as an empiric mean of *i.i.d.* random variables and thus may converge to its expected value (expected according to the unknown true distribution noted $f_X{}^\circ$) under some conditions:

$$\frac{1}{n} LL\left(\left(\mathbf{X}_j\right)_{j=1..n}, \boldsymbol{\theta}_X \right) = \frac{1}{n} \sum_{j=1..n} Logf_X\left[\mathbf{X}_j|\boldsymbol{\theta}_X\right] \underset{n\to\infty}{\longrightarrow} E_{f_X{}^\circ}(Logf_X[\mathbf{X}|\boldsymbol{\theta}_X]) \tag{5.77}$$

As a special application of the *Law of Large Numbers* and *Central Limit Theorem* respectively, the following asymptotic properties can be guaranteed (proof is classical, cf., for instance, Cramér, 1946):

$$\hat{\mathbf{\Theta}}_n \xrightarrow[n\to\infty]{P} \boldsymbol{\theta}_X{}^{\circ} (\text{convergence in probability}) \tag{5.78}$$

$$\sqrt{n}\left(\hat{\mathbf{\Theta}}_n - \boldsymbol{\theta}_X{}^{\circ}\right) \xrightarrow[n\to\infty]{L} N\left(0, I_F^{-1}(\boldsymbol{\theta}_X{}^{\circ})\right) \quad (\text{convergence in law}) \tag{5.79}$$

whereby $I_F^{-1}(\boldsymbol{\theta}_X{}^{\circ})$denotes the inverse of the true Fisher information matrix (for one sample) computed on the true value $\boldsymbol{\theta}_X{}^{\circ}$:

$$I_F(\boldsymbol{\theta}_X{}^{\circ}) = E\left[-\frac{\partial^2}{\partial\boldsymbol{\theta}\partial\boldsymbol{\theta}'} Logf_X(\mathbf{X}, .)\bigg|_{\boldsymbol{\theta}_X{}^{\circ}} \right] \tag{5.80}$$

provided a number of conditions, including notably the following[2]:

- *Regularity*: C^2 continuous differentiability of the log-likelihood function, that is of the density (as a function of $\boldsymbol{\theta}_X$), as well as the finiteness of the expectation of the log-likelihood and of its gradient and Hessian matrix (2nd order derivative).
- *Identifiability*: identifiability of the model, as well as inversibility of the Hessian matrix.
- *i.i.d.* observations[3]: the $\mathbf{X}_j$ need to be distributed identically and independently from one observation j to the other j' (though vector $\mathbf{X}$ itself may contain dependent input components, as in Section 5.3).

Positive-definiteness of the (symmetric) Fisher information matrix is required for the optimisation program to have a local maximum and guarantee invertibility at the same time. The identifiability condition ensures the uniqueness of the maximum because of Kullback's inequality.

Thus, the variance of the MLE is estimated approximately as a result of the second property, through the estimated Fisher information matrix computed on the estimated parameter as follows:

$$\begin{aligned} I_F(\boldsymbol{\theta}_X{}^{\circ}) &= E\left[-\frac{\partial^2}{\partial\boldsymbol{\theta}\partial\boldsymbol{\theta}'} Logf_X(\mathbf{X}, .)\bigg|_{\boldsymbol{\theta}_X{}^{\circ}} \right] \\ &\approx \frac{1}{n}\sum\nolimits_{j=1\ldots n} -\frac{\partial^2}{\partial\boldsymbol{\theta}\partial\boldsymbol{\theta}'} Logf_X(\mathbf{x}_j, .)\bigg|_{\boldsymbol{\theta}_X{}^{\circ}} = \hat{I}(\boldsymbol{\theta}_X{}^{\circ}) \\ &\approx \frac{1}{n}\sum\nolimits_{j=1\ldots n} -\frac{\partial^2}{\partial\boldsymbol{\theta}\partial\boldsymbol{\theta}'} Logf_X(\mathbf{x}_j, .)\bigg|_{\hat{\boldsymbol{\theta}}_X} = \hat{I}\left(\hat{\boldsymbol{\theta}}_X\right) = \frac{1}{n}\hat{I}_n\left(\hat{\boldsymbol{\theta}}_X\right) \end{aligned} \tag{5.81}$$

[2] As well as a number of other technical conditions; complications arise, for instance, in the case when the area of definition of the density depends on some of its parameters, as is the case for the 3-parameter Weibull distribution.

[3] Extensions resolving the third condition partly can be developed for independent though non-identically distributed observations, as in inverse algorithms required for Chapter 6 (see Section 6.5.1) or for non-independent identically distributed observations, as for correlated time series (see Section 5.2.3).

yielding the following estimator of the asymptotic epistemic variance:

$$Var_{asympt}(\hat{\Theta}_X) = \frac{1}{n}\hat{I}\left(\hat{\theta}_X\right)^{-1} = \hat{I}_n\left(\hat{\theta}_X\right)^{-1} = \left[\frac{\partial^2}{\partial\theta\partial\theta'}(-LL)|_{\hat{\theta}_X}\right]^{-1} \tag{5.82}$$

Numerical analysts can understand it intuitively as the reverse of the 2nd-order derivative of the cost function to be minimised $-LL$, see Annex Section 10.5. Note that such an expression, though asymptotically true, generally *underestimates* the epistemic variance for a finite sample. Indeed, a classical result states that the inverse of the true Fisher information matrix $I_F(\theta_X{}^\circ)^{-1}$ is the Cramér-Rao *lower bound* of the variance $Var(\hat{\Theta}_X)$ for any unbiased estimator $\hat{\theta}_X$. From that perspective, MLE is said to be asymptotically efficient as its epistemic uncertainty variance converges to the lowest possible bound.

In general, epistemic uncertainty as measured traditionally by the standard deviation decreasing at a speed of $1/\sqrt{n}$ with the increase of sample size because the Fisher information matrix for the n-sample ($\hat{I}_n$) grows in order of magnitude linearly with n (i.e. the average Fisher information matrix per sample ($\hat{I}$) maintains the same order of magnitude). Think, for instance, of a Gaussian (unidimensional) model:

$$\theta = \begin{pmatrix} \mu \\ \sigma^2 \end{pmatrix} \quad \hat{\theta}_X = \begin{pmatrix} \bar{x} \\ s_n^2 \end{pmatrix}$$

$$Logf(x_1, \ldots, x_n; \theta) = -\frac{n}{2}\log\left(2\pi\sigma^2\right) - \frac{1}{2\sigma^2}\sum_j \left(x_j - \mu\right)^2 \tag{5.83}$$

$$\hat{I}\left(\hat{\theta}_X\right) = \frac{1}{n}\sum\nolimits_{j=1\ldots n} -\frac{\partial^2}{\partial\theta\partial\theta'} Logf_X\left(\mathbf{x}_j, .\right)|_{\hat{\theta}_X} = \begin{pmatrix} \frac{1}{s_n^2} & 0 \\ 0 & \frac{1}{2s_n^4} \end{pmatrix} \tag{5.84}$$

Hence:

$$Var_{asympt}(\hat{\Theta}_X) = \frac{1}{n}\begin{pmatrix} s_n^2 & 0 \\ 0 & 2s_n^4 \end{pmatrix} \tag{5.85}$$

The MLE of the Gaussian expectation is identical to the unbiased estimator, though its asymptotic variance differs slightly from the exact (Student-based) variance (Table 5.10); conversely, the MLE of the Gaussian variance s_n^2 is biased, its variance also differing from the exact (χ^2-based) variance of the unbiased estimator s_{n-1}^2.

In general, the expression for the asymptotic variance matrix of the estimates involves the Hessian or second-order derivative of the likelihood function computed at the maximal point. In many cases, there is no closed-form for such a maximum so that it needs to be estimated numerically through finite differences, a standard by-product of ML estimation (called 'standard error of estimates' in statistical toolboxes) because maximal likelihood generally involves gradient-based optimisation. Moment methods can also be an alternative in these cases so as to yield a convenient closed-form expression.

Application to epistemic uncertainty – multi-normal confidence intervals In other words, the results above mean that epistemic distribution $\pi(.)$ of estimator $\hat{\mathbf{\Theta}}_X$ is approximately multivariate Gaussian with a null expectation (i.e. unbiased with respect to $\boldsymbol{\theta}_X{}^{\circ}$) and a covariance matrix given by Equation (5.82). Thus, a simple technique for inferring the epistemic uncertainty in the risk measure involves an additional asymptotic assumption. Consider the following function (the conditional risk measure to a given parameter vector):

$$c_Z(\boldsymbol{\theta}_X, \boldsymbol{d}) = \int_x UoG(\boldsymbol{x}, \boldsymbol{d}) f_X(\boldsymbol{x}|\boldsymbol{\theta}_X) d\boldsymbol{x} \tag{5.86}$$

It can be linearised for small variations of $\boldsymbol{\theta}_X$, so that the random variable defined by computing such function with the estimator of $\boldsymbol{\theta}_X$ as follows:

$$h(\hat{\mathbf{\Theta}}_n, \boldsymbol{d}) = c_Z(\hat{\mathbf{\Theta}}_n, \boldsymbol{d}) \tag{5.87}$$

is again asymptotically Gaussian with an epistemic distribution centred on $c_Z(\boldsymbol{\theta}_X{}^{\circ}, \boldsymbol{d})$ and of known covariance matrix:

$$Var_{asympt} h(\hat{\mathbf{\Theta}}_n, \boldsymbol{d}) \approx \nabla_{\hat{\mathbf{\Theta}}} c_Z Var_{asympt}(\hat{\mathbf{\Theta}}_X) \nabla_{\hat{\mathbf{\Theta}}} c_Z^t \tag{5.88}$$

from which a β-confidence risk measure can be estimated approximately.

$$c_Z^{\beta}(\boldsymbol{d}) = h(\hat{\mathbf{\Theta}}_n, \boldsymbol{d}) + \Phi^{-1}(\beta)\sqrt{Var_{asympt} h(\hat{\mathbf{\Theta}}_n, \boldsymbol{d})} \tag{5.89}$$

Note again that the additional margin brought by such a β-confidence risk measure on top of the expected risk measure $h(\hat{\mathbf{\Theta}}_n, \boldsymbol{d})$ would also typically decrease in $1/\sqrt{n}$.

Application to epistemic uncertainty – profile likelihood confidence intervals The previous approximate confidence intervals should always be considered conservatively as the lower bounds of the epistemic uncertainty in various respects. An alternative approach based on *profile likelihood*, though still asymptotic, may produce more accurate confidence intervals (Coles, 2001). Profile log-likelihood can be defined as a partially-maximised log-likelihood function as follows:

$$PL(\Xi_n, \theta_X^i) = \max_{(\theta_X^h)_{h \neq i}} (LL[\Xi_n, \boldsymbol{\theta}_X]) \tag{5.90}$$

In other words, maximise log-likelihood along all components but the i-th of the uncertainty parameter vector. Then, it can be proved under suitable regularity conditions that its deviation to the fully-maximised likelihood follows a χ_1^2 – distribution:

$$D_P(\theta_X^i) = 2(LL[\Xi_n, \boldsymbol{\theta}_X] - PL[\Xi_n, \theta_X^i]) \xrightarrow[n \to \infty]{L} \chi_1^2 \tag{5.91}$$

This result generates confidence intervals. Then reparameterise the uncertainty vector $\boldsymbol{\theta}_X$ replacing, for instance, the first component by a risk measure such as a α-quantile (or T-return level):

$$\boldsymbol{\theta}_X = (\theta_X^1, \theta_X^2, \ldots \theta_X^p) \tag{5.92}$$

$$\boldsymbol{\theta}'_X = (c_z, \theta_X^2, \ldots \theta_X^p) \tag{5.93}$$

Remembering that a quantile is a known regular function of the unknown parameters (e.g. a linear combination with positive coefficients), the maximal likelihood of the newly-parameterised model generally coincides with that of the original. Thus, the β-confidence risk measure appears as follows:

$$C_Z^\beta(\boldsymbol{d}) = \max\{c_Z | D_P(c_Z) \le \chi_1^2(1-\beta)\} \tag{5.94}$$

Keep in mind that neither these nor the previous ones account for the wider *distributional uncertainty*, that is the lack of knowledge of the plausible shape of distribution, though more will be said about this below.

The example of the Gumbel model Take the example of a Gumbel distribution, defined conveniently through its double-exponential cdf involving two parameters:

$$F_X(x|\boldsymbol{\theta}_X = (m, e)) = \exp\left(-\exp\left(-\frac{x-m}{e}\right)\right) \tag{5.95}$$

Suppose that a method of estimation of the asymptotictype has generated estimators for $\boldsymbol{\theta} = (m,e)$ as well as its Gaussian covariance matrix. Then, for instance, a β-confidence prediction for a risk measure defined as a α-quantile can be computed. Indeed, the α-quantile of a Gumbel-distributed r.v. appears as the known following linear combination of its parameters:

$$c_Z(\boldsymbol{\theta}_X{}^\circ, \boldsymbol{d}) = x_\alpha = m - \ln(-\ln(\alpha)) \cdot e \tag{5.96}$$

Thus, the random variable defined by computing the α-quantile on the basis of the estimators of $\boldsymbol{\theta} = (m,e)$:

$$c_Z(\hat{\boldsymbol{\Theta}}) = \hat{M} - \ln(-\ln(\alpha))\hat{E} \tag{5.97}$$

is also asymptotically Gaussian distributed, with an unbiased mean and the following variance:

$$Var\left[c_Z(\hat{\boldsymbol{\Theta}})\right] = \text{var}\,\hat{M} + \ln(-\ln(\alpha))^2 \text{var}(\hat{E}) - 2\ln(-\ln(\alpha))\text{cov}(\hat{M}, \hat{E}) \tag{5.98}$$

so that a (unilateral) β-confidence level risk-measure can be computed as follows:

$$c_{z\beta} = \widehat{m} - \ln(-\ln(\alpha)) \cdot \widehat{e} + \Phi^{-1}(\beta)\sqrt{Var\left[c_Z(\hat{\boldsymbol{\Theta}})\right]} \tag{5.99}$$

As mentioned above, Maximal Likelihood lacks a closed-form solution: finding the ML estimators involves numerical optimisation. Estimating the information matrix is thus achieved by a finite difference estimate of the Hessian matrix. Conversely, moment methods issue closed-form estimates of the parameters based on the empiric mean and variances (see above). Thus, the α-quantile of a Gumbel-distributed r.v. appears as a linear combination of the empiric moments:

$$c_Z(\hat{\boldsymbol{\Theta}}_X) = \bar{X} - [\gamma + \ln(-\ln(\alpha))]\sqrt{\frac{6\hat{S}_n^2}{\pi^2}} \tag{5.100}$$

Such a variable is also asymptotically Gaussian distributed, with an unbiased mean and the following variance:

$$Var\hat{C}_Z = Var\bar{X} + \frac{6}{\pi^2}(\gamma + \ln(-\ln(\alpha)))^2 Var\sqrt{\hat{S}_n^2} - \sqrt{\frac{6}{\pi^2}}(\gamma + \ln(-\ln(\alpha)))Cov(\bar{X}, \sqrt{\hat{S}_n^2}) \tag{5.101}$$

When X is distributed according to a distribution as asymmetric as Gumbel, sampling shows that the variance of the empiric variance estimator S_n^2 is different from – significantly larger than – the asymptotic Gaussian variance $(2 \cdot S_n^2/n)$ used above in Table 5.10. Moreover, the empiric mean $\bar{X}$ and variance S_n^2 prove to be highly correlated unlike the Gaussian statement of asymptotic independance. The asymptotic results are thus trickier to implement for the epistemic uncertainty, and Bootstrap will be required (see next section).

> The *flood example*
> The 149-sample of flow observations was shown to best fit to a Gumbel distribution. Both moment and maximal likelihood methods may be used to estimate cdf parameters and thus predict quantiles on this basis. Epistemic uncertainty can be easily estimated in the maximal likelihood case with asymptotic results. Note the width of the resulting *95 %* confidence intervals, showing about *10 %* imprecision on both sides of the *1/100* and *1/1000* central quantile estimates (see Table 5.11).
> Note also the influence of the sample size. In the case of a smaller/less instrumented river, typically providing no more than 50 years of observations, the standard deviations would be 70 % higher ($\sim\sqrt{3}$ as the asymptotics typically converge at a $\sqrt{n}$ speed), thus resulting in close to 20 % additional margin for design if natural risk regulations base the risk measure on the upper bound of such interval, as discussed in Chapter 1.

Note that the previous developments in estimating a β-confidence prediction in a risk measure accounting for epistemic uncertainty have been undertaken in *asymptotic approximations*. This means

Table 5.11 Flood flow quantiles and epistemic uncertainty

	Moment method	Maximum likelihood
Central estimate for m and associated stdd deviation	1014	1013 ± 48 (5 %)
Central estimate for e and associated stdd deviation	557	557 ± 36 (6 %)
Correlation of estimates	*(asymptotics less relevant)*	0.31
0.99 quantile and associated stdd deviation	3575	3573 ± 184
95 % confidence interval for the quantile	*(asymptotics less relevant)*	[3212 – 3934]
0.999 quantile and associated stdd deviation	4859	4858 ± 264
95 % confidence interval for the quantile	*(asymptotics less relevant)*	[4340 – 5375]

that the sample size n has to be large enough for epistemic uncertainty to be small enough to render acceptable a Gaussian approximation. It is a hard task in practice to assess whether such an approximation is legitimate or to what extent it underestimates the extent of lack of knowledge. Yet, note that the asymptotic theory involves linearising a number of functions involved in the process: not only the likelihood function, but also the system model $G(.)$ when turning to predicting a risk measure. This provides the first evaluation of the appropriateness of the associated assumptions. Chapter 7 will show that this stands in close connection with the Taylor approximate propagation method.

5.4.4 Bootstrap and resampling techniques

Bootstrapping and more generally resampling techniques (Efron and Tibshirani, 1993) provide intuitive and quite powerful means to estimate the confidence or epistemic uncertainty in risk measures. Assume that each observation X_j of the sample $\Xi_n = (X_j)_{j=1\ldots n}$ represents equally *past* observed states of the system, as well as the state of the system X at the *future* time of interest. Each of these helps in estimating the risk measure. Considering then the fluctuation that would arise when using another sample of observations to estimate the risk measure, that is the epistemic uncertainty, it is fair to assume that such a sample for the same phenomenon will resemble closely the observed one. Assuming independent and identically distributed samples within the original data, a virtual n-sample composed of n values resampled independently from the original sample (with replacement) is a reasonable simulation of what could happen with natural data fluctuation. It can then be used to replicate a risk study and generate new distribution estimates and associated risk measures.

Such a procedure, known as *non-parametric Bootstrap*, is a natural generator of empiric distributions for the distribution parameters and associated risk measures that enable the estimation of epistemic uncertainty; it has also desirable convergence properties. While obviously bearing the same limitations as those of the original sample, such as potential biases in the representativity of natural variability, lack of data and so on, it brings useful distribution-free hypotheses for epistemic uncertainty (unlike Gaussian asymptotics). *Parametric Bootstrap* involves a resampling within a simulated Gumbel distribution the parameters of which are estimated initially with the original sample; then a Gaussian fit and the subsequent quantile prediction are performed for each replicate.

The *flood example*

Resampling of the 149-sample of flow observations was undertaken together with a fit to a Gumbel distribution and hence generate a simulation of the epistemic distribution of its parameters (see Table 5.12). Non-parametric or parametric bootstrap provide estimates of standard errors that are comparable to those obtained through asymptotic results. Moment methods give slightly larger epistemic uncertainty than ML. Bootstrap-generated epistemic distributions of estimates turn out to be close to Gaussian although slightly asymmetric; equal results are obtained for the epistemic distribution of quantiles, a linear combination of model parameters in the Gumbel model. Note, however, that bootstrap involves random sampling and thus gives slightly random results: pushing higher than 10 000 resamples might change the estimates slightly. All of these alternative techniques end up in the ccdf (complementary cdf) representation of the aleatory and epistemic uncertainty combination as suggested by Helton (1993), cf. Figure 5.14 in the case of parametric bootstrap, in natural (left) or log scale (right).

Note that an alternative provided by the R *evd* package involves parametric boostrap of replicated 149-samples, but then computes the quantiles of the order statistics (i.e. the fluctuation in the 1st, 2nd, ... ranked value of each 149-sample), which prove more conservative in some cases.

Table 5.12 Flood flow quantiles and epistemic uncertainty.

	Asymptotic maximum likelihood	Non-param bootstrap of MM	Non-param bootstrap of ML	Param bootstrap of ML
Central estimate for m and associated stdd deviation	1013 ± 48 (5 %)	1016 ± 49 (5 %)	1014 ± 49 (5 %)	1014 ± 48 (5 %)
Central estimate for e and associated stdd deviation	557 ± 36 (6 %)	553 ± 41 (7 %)	555 ± 34 (6 %)	554 ± 35 (6 %)
Correlation of estimates	0.31	0.18	0.36	0.30
0.99 quantile and associated stdd deviation	3573 ± 184	3562 ± 203 (5.7 %)	3567 ± 182 (5.1 %)	3561 ± 183 (5.1 %)
95 % confidence interval for the 0.99 quantile	[3212 – 3934]	[3167 – 3960]	[3221 – 3930]	[3205 – 3927]
0.999 quantile and associated stdd deviation	4857 ± 264	4839 ± 295 (6.1 %)	4847 ± 260 (5.4 %)	4839 ± 263 (5.4 %)
95 % confidence interval for the 0.999 quantile	[4340–5375]	[4263–5417]	[4352–5367]	[4329–5365]

5.4.5 Bayesian-physical settings (small samples with expert judgement)

Bayesian estimation basics A Bayesian setting offers a traceable process designed to mix the encoding of engineering expertise and the observations inside an updated epistemic layer; such a setting proves

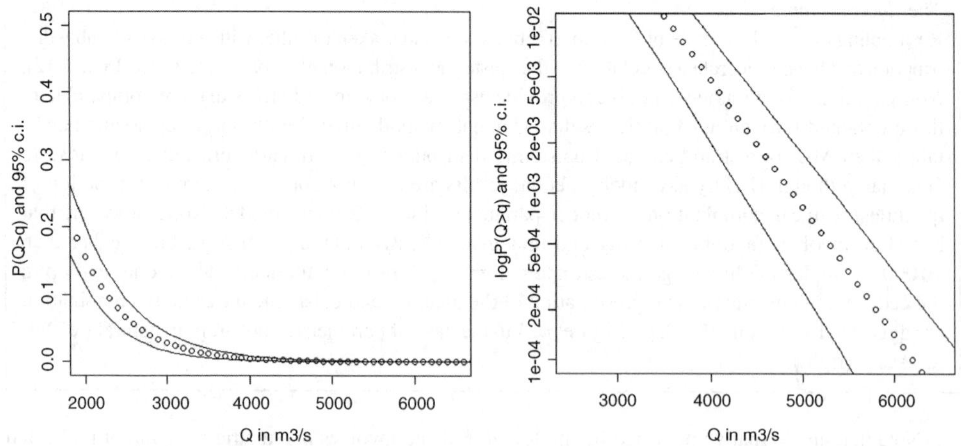

Figure 5.14 *Flood flow quantiles and epistemic uncertainty.*

mathematically consistent even when dealing with very low-size samples. It includes by definition the conditional (or double-level) probabilistic structure:

$$\begin{aligned} \boldsymbol{X}|\boldsymbol{\theta}_X &\sim f_X(\cdot|\boldsymbol{\theta}_X) \\ \boldsymbol{\Theta}_X &\sim \pi_o(\cdot) = \pi(\cdot|IK,\zeta) \end{aligned} \tag{5.102}$$

whereby experts represent inside the distribution shape f_X as well as the *prior* distribution π_o their knowledge of the plausible values for $\boldsymbol{\Theta}_X$. To be more specific, the uncertainty model should also reflect the modelling of the statistical observations of Ξ_n – that is $(\boldsymbol{X}_j)_{j=1\ldots n}$ representing *past* states of the system – as compared to that of the state of the system $\boldsymbol{X}$ at the *future* time of interest.

$$\left.\begin{aligned} &\boldsymbol{\Theta}_X \sim \pi(\cdot|\zeta) \\ &\forall j = 1\ldots n \quad \boldsymbol{X}_j|\boldsymbol{\theta}_X \sim f_X(\cdot|\boldsymbol{\theta}_X) \\ &\boldsymbol{X}|\boldsymbol{\theta}_X \sim f_X(\cdot|\boldsymbol{\theta}_X) \\ &(\boldsymbol{X}_1,\ldots\boldsymbol{X}_j,\ldots\boldsymbol{X}_n,\boldsymbol{X})|\boldsymbol{\theta}_X \; cond.\, i.i.d. \end{aligned}\right\} \tag{5.103}$$

The basic assumption is that all such states of the system are *conditionally i.i.d.*, meaning that:

- all past states of the system, as reflected in the vector of inputs $\boldsymbol{x}$, are one another identically distributed and also their distribution is also identical to that of the future state of the system,
- all past states as well as the future state are conditionally independent, that is independent in the sub-space conditional upon the random vector $\boldsymbol{\Theta}_X$ being fixed to a given value.

Both properties have a straightforward meaning for the modeller: in other words, this means that the observations are deemed relevant to inferring likely values for the future states and that they are independent pieces of information *in the real world* while they are linked to each other within the *analysts' view on reality.* The second property is a Bayesian expression of the formal frontier established between the lack of knowledge of the analyst – reflected in the uncertainty in $\boldsymbol{\Theta}_X$ – and the true variability of the states of the system – governed by a *fixed* (unknown) value of $\boldsymbol{\theta}_X{}^\circ$ driving the natural variability of the system. One single value $\boldsymbol{\theta}_X{}^\circ$ – constant in time – drives both the past and future variability of the system.

Thus, the estimation process incorporates the statistical sample of Ξ_n to produce the *posterior* distribution π_1 for $\boldsymbol{\Theta}_X$ through the following integration, based on Bayes' law:

$$\pi_1(\boldsymbol{\theta}_X) = \pi(\boldsymbol{\theta}_X|IK,\Xi_n) = \frac{\prod_j f_X(\underline{x}_j|\boldsymbol{\theta}_X)\cdot\pi_0(\boldsymbol{\theta}_X)}{\int_{\boldsymbol{\theta}_X}\prod_j f_X(\boldsymbol{x}_j|\boldsymbol{\theta}_X)\pi_0(\boldsymbol{\theta}_X)d\boldsymbol{\theta}_X} \tag{5.104}$$

ending up with the *posterior* uncertainty model for the future state of the system $\boldsymbol{X}$ as well as the vector of epistemically-uncertain parameters $\boldsymbol{\Theta}_X$, incorporating both Ξ_n and *IK*:

$$\begin{aligned} \boldsymbol{X}|\boldsymbol{\theta}_X &\sim f_X(\cdot|\boldsymbol{\theta}_X) \\ \boldsymbol{\Theta}_X|\Xi_n &\sim \pi_1(\cdot) = \pi(\cdot|IK,\Xi_n) \end{aligned} \tag{5.105}$$

An additional important property arises with the Bayesian estimation procedure. Consider now the risk measure *conditional* upon a given value of the epistemically-uncertain vector of parameters; it is a known – although computationally complex – function:

$$c_Z(f_Z|\boldsymbol{\theta}) = c_Z(\boldsymbol{\theta}, \boldsymbol{d}) \tag{5.106}$$

By definition, the Bayesian estimator of such a risk measure is the *posterior expectation* of the function, that is:

$$C_Z^*(IK, \Xi_n) = E_{\pi_1}(c_Z(\boldsymbol{\Theta}, \boldsymbol{d})) = \int_{\boldsymbol{\theta}_X} c_Z(\boldsymbol{\theta}_X, \boldsymbol{d})\pi_1(\boldsymbol{\theta}_X)d\boldsymbol{\theta}_X = \int_{\boldsymbol{\theta}_X} c_Z(\boldsymbol{\theta}_X, \boldsymbol{d})\pi(\boldsymbol{\theta}_X|IK, \Xi_n)d\boldsymbol{\theta}_X \tag{5.107}$$

In other words, it is the risk measure 'averaged over the epistemic uncertainty given all information available' (see Chapter 2). Such an estimator appears to be that choice of estimator – amongst the family of estimators, that is of functions $C(\Xi_n, G(.))$ of the observed sample and of the system model when considering uncertain outputs – which minimises the *Bayesian risk* (to be understood as a risk of mis-estimation) defined as follows:

$$\begin{aligned} R(C, \pi_0) &= E_{\pi_0}\left(E\left([C(\Xi_n, G(\cdot)) - c_Z(\boldsymbol{\Theta}, \boldsymbol{d})]^2\right)\right) \\ &= \int_{\boldsymbol{\theta}_X}\left(\int_{\Xi_n}[C(\Xi_n, G(.)) - c_Z(\boldsymbol{\theta}_X, \boldsymbol{d})]^2 f(\Xi_n|\boldsymbol{\theta}_X)d\Xi_n\right)\pi_o(\boldsymbol{\theta}_X)d\boldsymbol{\theta}_X \end{aligned} \tag{5.108}$$

In the formula, Ξ_n stands for the i.i.d. vector sample $(\boldsymbol{X}_j)_{j=1\ldots n}$ the distribution of which is the product of the conditional distributions $f_X(\boldsymbol{x}|\boldsymbol{\theta}_X)$. In other words, the Bayesian risk is the mean squared error in estimating the conditional risk measure, then averaged over the epistemic distribution. From that perspective, assuming that all available background information has been well encoded inside the prior model, the Bayesian estimator of the risk measure is that which entails the minimal risk of error in a quadratic MSE approach.

Bayesian modelling raises practical challenges that will be introduced briefly below:

- eliciting background expertise, particularly in the selection of the appropriate prior distribution faithfully representing the extent of lack of knowledge;
- computing the posterior distribution, involving multiple integration (Equation (5.104)).

So-called non-informative priors are an important concept for the former challenge. The idea is to reflect consistently the extent of lack of knowledge of the prior distribution in order to disrupt as little as possible the value of data to be incorporated later. Similar to the use of the entropy maximisation principle (Chapter 4) the concept is paradoxical to some extent as the mere specification of a random variable structure represents as such a piece of information. However, it is quite useful in practice and offers a link to classical maximal likelihood estimation, as will be shown in subsequent examples.

The latter challenge has long motivated the use of conjugate distributions, referring to those distribution shapes yielding closed-form solutions to the posterior integration formulae. Modern approaches involve advanced simulation techniques that are connected to those used in computing the risk measure (Chapter 7). Beyond the basic techniques provided below, the reader is referred to the specialised literature for more detail on Bayesian methods (Robert, 2001).

Gaussian bayesian settings As was the case in a classical approach regarding the estimation of epistemic uncertainty, Gaussian settings have the advantage of yielding closed-form expressions in many cases. They play an important role in uncertainty analysis, metrological applications and so on though not necessarily being the most adequate choice in risk analysis where extreme events generally depart from the symmetrical central parts of variability distributions, A one-dimensional Gaussian example will serve to illustrate the essential features of a Bayesian setting, particularly the impact of incorporating more statistical information into both aleatory and epistemic uncertainty components.

A Gaussian assumption is made on the distribution shape f_X, that is on the aleatory uncertainty model of X, parameterised by $\boldsymbol{\theta}_X = (m, \Sigma^2)$ the mean and variance respectively:

$$\begin{array}{c} X|\boldsymbol{\theta}_X \sim N(m, \Sigma^2) \\ \boldsymbol{\Theta}_X|IK \sim \pi_o(\cdot\,|IK) \end{array} \tag{5.109}$$

Assuming that (m, Σ^2) follows, a standard (multi-variate) Gaussian distribution is not desirable as it allows the uncertain variance to have negative values and fails to issue closed-form expressions in the posterior distributions. A normal-gamma setting offers a natural conjugate distribution when aleatory uncertainty is sought to be (conditionally) Gaussian (see, e.g. Droesbeke, Fine, and Saporta, 2002), with the following features:

- Conditional upon a given variance, the expectation parameter $m|\Sigma^2$ follows a Gaussian law.
- The variance parameter Σ^2 follows an inverse-gamma law defined as follows:

$$\begin{aligned} m|\Sigma^2 &\sim N\left(\mu, \frac{\Sigma^2}{B_o}\right) = N\left(\mu, \tau^2 \cdot \frac{\Sigma^2}{\sigma^2}\right) \\ \Sigma^2 &\sim \Gamma^{-1}\left(\sigma^2 \frac{\nu_o}{2}, \frac{\nu_o}{2}\right) \end{aligned} \tag{5.110}$$

When incorporating an observed sample, the two important statistics are the empiric mean $\overline{X}$ and standard deviation s^2. Thus, the Student law (denoted as $T(.)$) characterises the combined unconditional uncertainty model, both on prior and posterior predictive distribution f_X^π:

$$X|IK \sim T(\mu, \sigma^2 + \tau^2, \nu_o) \tag{5.111}$$

$$X|IK, \Xi_n \sim T(m_n, \sigma_n^2 + \tau_n^2, \nu_o + n) \tag{5.112}$$

where

$$\begin{aligned} m_n &= \frac{\sigma^2}{\sigma^2 + n\cdot\tau^2}\mu + \frac{n\cdot\tau^2}{\sigma^2 + n\cdot\tau^2}\overline{X} \\ \tau_n^2 &= \frac{\sigma_n^2\tau^2}{\sigma^2 + n\cdot\tau^2} \\ \sigma_n^2 &= \frac{[(n-1)s^2 + \nu_o\sigma^2 + D_n]}{\nu_o + n} \\ D_n &= B_o\mu^2 + n\overline{X}^2 - (B_o + n)m_n^2 \end{aligned} \tag{5.113}$$

the prior and posterior combined aleatory and epistemic variance being as follows:

$$Var(X|IK) = (\sigma^2 + \tau^2) \cdot v_o/(v_o - 2) \tag{5.114}$$

$$Var(X|IK, \Xi_n) = (\sigma_n^2 + \tau_n^2)(v_o + n)/(v_o + n - 2) \tag{5.115}$$

Expertise elicitation involves four parameters instead, as shown in Table 5.13.

Note that:

- The posterior expectation m_n of X (i.e. the Bayesian estimator of the expectation) corrects the (expert-judged) central value μ by the observed empiric mean, with a increased (resp. decreased) weight upon the observed mean if prior epistemic uncertainty τ^2 is larger (resp. smaller) than prior intrinsic uncertainty σ^2 and if the number of observations n is large (resp. small):

$$m_n = \frac{\sigma^2}{\sigma^2 + n.\tau^2}\mu + \frac{n.\tau^2}{\sigma^2 + n.\tau^2}\bar{X} \tag{5.116}$$

  Uncertainty distribution shapes (prior and posterior marginal laws of the observable X) follow Student laws in the second model, but this resembles a Gaussian shape as the number of observations increase the number of d.o.f. (all the more so when v_o is fixed to high values, that is intrinsic uncertainty is reckoned to be assessed precisely by an expert).
- Unconditional Variance aggregates epistemic (τ^2 *or* τ_n^2) and aleatory (σ^2, s^2 *or* σ_n^2) components; however, at a low number of observations, variance is inflated by a non-negligible multiplicative common epistemic uncertainty factor $(v_o + n)/(v_o + n - 2)$.
- Posterior aleatory uncertainty σ_n^2 appears essentially as a weighting between the prior best-estimate for intrinsic uncertainty σ^2 and the empiric variance s^2, weighted by the expert-judged precision v_o of σ^2;

Table 5.13 Requirements for expert elicitation in both models.

Values to be elicited	Interpretation
μ	Best-estimate for the physical variable central value (prior expectation for both models)
σ^2	Best-estimate for the intrinsic uncertainty *(N.B. in the second model, prior expectation of the intrinsic uncertainty is proportional to σ^2, but shifted by v_o if v_o is small)*
τ^2 (or Bo $= \sigma^2/\tau^2$ in second model)	Estimate for the epistemic uncertainty first component (unique component in first model, uncertainty of the 'expectation bias' for the second) Small τ^2: very precise expertise (on μ only in second model) Large τ^2: low-informative expertise (id.)
v_o (only for second model; $v_o > 2$ for non-degeneracy)	Estimate of the epistemic uncertainty second component (uncertainty of the intrinsic uncertainty for the second only) Small v_o: low-informative expertise on intrinsic uncertainty Large v_o: very precise expertise

however it also includes a 'bias' contribution D_n which is not negligible if prior best-estimate μ departs significantly from true expectation:

$$\sigma_n^2 = \frac{[(n-1)s^2 + \nu_o\sigma^2]}{\nu_o + n} + \frac{D_n}{\nu_o + n} \tag{5.117}$$

Asymptotically, both models tend to the classical maximal likelihood estimates, the expert judgement component becoming negligible.

Flood example
Simple numerical examples are provided below for the Strickler coefficient. A sample of observations ($n = 3$, 10 or 30) is first generated by random simulation according to a $N(27,7^2)$ normal distribution. Thus, the learning process is tested as well as the following scenarios of expert judgement: (case 1) prior expert judgement is supposed to be of the right order of magnitude and only slightly biased, epistemic uncertainty being fixed at moderately large values ($\nu_o = 3$ is poorly informative); (case 2) identical to (case 1) but with a largely over-estimated prior central value; (case 3) identical to (case 1) but with a largely-overestimated prior intrinsic uncertainty.
Posterior complete Standard Deviation (SD) is computed from the posterior full variance of K_s. As a matter of comparison, the classical upper bound of a one-sided χ^2 confidence interval on the s^2 estimator of σ^2 is given in Tables 5.14 and 5.15.
It can be observed in (case 1) and (case 2) that including expert judgments improves generally the purely empiric variance at a low number of observations. An erroneous best-estimate for the expectation is rapidly overset by the empiric mean and therefore the estimates do not deteriorate by too much. A full Bayesian model leads to a larger variance than the simple one (which is more conservative in safety studies), especially under an erroneous best-estimate for expectation or intrinsic variance. Excessive prior variances, which would tend to represent conservative approaches in safety studies with low information, have a moderate impact on posterior central values, but do lead to large posterior variance. The observations do reduce the overestimation but at a relatively slow pace (slower than the classical upper bound χ^2 confidence interval in case 3). However, this classical c.i. is highly fluctuating at a low number of observations, and the result could differ on another sample. This can be interpreted as maintaining a significant epistemic component on top of the true intrinsic uncertainty, which is acceptably conservative.

Exponential bayesian setting Returning to the exponentially-distributed lifetime variables of a Poisson process distributed *i.i.d.* according to the following single-parametrised discrete distribution:

$$\Delta T_j \sim f_T(t|\boldsymbol{\theta} = \lambda) \quad i.i.d. \tag{5.118}$$

A Bayesian conjugate (hence closed-form) model can be built given a Gamma prior distribution for the epistemically-uncertain failure rate (here $\boldsymbol{\theta} = \lambda$):

$$\Lambda \sim \pi_o(\boldsymbol{\theta}|\varsigma) = \Gamma(\lambda|\varsigma_1, \varsigma_2) = \frac{(\varsigma_1\lambda)^{\varsigma_2}\exp(-\varsigma_1\lambda)}{\lambda\Gamma(\varsigma_2)} \tag{5.119}$$

Table 5.14 Hypotheses of the three examples.

	Real values	(1) Prior Expert judgement	(2) Prior Expert judgement	(3) Prior Expert judgement
Mostly observed interval $\pm 2\sigma$	*13–41 (95 %)*	*18–42*	*48–72*	*10–70*
Full uncertain interval $\pm 2\sqrt{\sigma^2+\tau^2}$		*10–50*	*40–80*	*0–80*
Expectation	*27*	*30*	*60*	*40*
Intrinsic SD (σ)	*7*	*6*	*6*	*15*
Full SD $\sqrt{\sigma^2+\tau^2}$		*10*	*10*	*20*
Epistemic (τ)	*0*	*8*	*8*	*13*
Epistemic (v_o)	—	*3*	*3*	*3*

Thus, the posterior distribution yields:

$$\begin{aligned} \Lambda|\Xi_n &\sim \pi_1(\theta|IK,\Xi_n) = \Gamma(\lambda|\varsigma'_1,\varsigma'_2) \\ \varsigma'_1 &= \varsigma_1 + n\overline{\Delta T} = \varsigma_1 + \frac{n}{\hat{\lambda}} \\ \varsigma'_2 &= \varsigma_2 + n \end{aligned} \tag{5.120}$$

Thus, the Bayesian estimator for the failure rate is:

$$\begin{aligned} \Lambda^*(IK,\Xi_n) &= \frac{\varsigma_2+n}{\varsigma_1+n\overline{\Delta T}} = \alpha\frac{\varsigma_2}{n} + (1-\alpha)\hat{\lambda} \\ \alpha &= \left(1+n\frac{\overline{\Delta T}}{\varsigma_1}\right)^{-1} \end{aligned} \tag{5.121}$$

It can simply be interpreted again as a weighed average between the classical estimator $\hat{\lambda}$ and a correction due to the prior, the correction tending to zero as the observed n-sample grows in size. Such an estimator is

Table 5.15 Case results.

		Case 1				Case 2				Case 3			
Nb of observations		0	3	10	30	0	3	10	30	0	3	10	30
$\overline{K_s}$		—	*25,3*	*30,2*	*26,2*	—	*25,3*	*30,2*	*26,2*	—	*25,3*	*30,2*	*26,2*
s		—	*4,5*	*5*	*6,6*	—	*4,5*	*5,0*	*6,6*	—	*4,5*	*5,0*	*6,6*
s + 95 % χ^2			*19,9*	*8,2*	*8,4*		*19,9*	*8,2*	*8,4*		*19,9*	*8,2*	*8,4*
m_n		*30*	*26,1*	*30,2*	*26,3*	*60*	*30,8*	*31,8*	*26,9*	*40*	*29,7*	*31,3*	*26,8*
Full bayesian	τ_n	*8*	*2,7*	*1,6*	*1,2*	*8*	*5,8*	*2,4*	*1,4*	*13*	*5,9*	*2,6*	*1,4*
	Posterior complete SD	*17,3*	*7,1*	*5,8*	*6,8*	*17,3*	*15,2*	*9,0*	*8,1*	*34,6*	*16,7*	*10,0*	*8,5*

quite useful in reliability studies involving highly-reliable components where typical samples are small but some expertise can be derived on analogous behaviours and thus encoded in Bayesian priors.

General bayesian settings and the MCMC algorithm In non-Gaussian cases, estimating the posterior distributions requires numerical integration or simulation of the following posterior epistemic distribution:

$$\pi_1(\boldsymbol{\theta}_X) = \pi(\boldsymbol{\theta}_X|IK, \Xi_n) = \frac{\prod_j f_X(\boldsymbol{x}_j|\boldsymbol{\theta}_X) \cdot \pi_0(\boldsymbol{\theta}_X)}{\int_{\boldsymbol{\theta}_X} \prod_j f_X(\boldsymbol{x}_j|\boldsymbol{\theta}_X)\pi_0(\boldsymbol{\theta}_X)d\boldsymbol{\theta}_X} \tag{5.122}$$

enabling the later use of the joint input distribution:

$$(\boldsymbol{X}, \boldsymbol{\Theta}_X) \sim f_X(x|\boldsymbol{\theta}_X)\pi(\boldsymbol{\theta}_X|IK, \Xi_n) = f_X(\boldsymbol{x}|\boldsymbol{\theta}_X)\pi_I(\boldsymbol{\theta}_X) \tag{5.123}$$

or else the posterior predictive distribution:

$$\begin{aligned} \boldsymbol{X} \sim f_X^{\pi}(\boldsymbol{x}) &= E_{\boldsymbol{\theta}_X}\left[f_X(\boldsymbol{x}|\underline{\Theta}_X)|IK, \Xi_n\right] = \int_{\boldsymbol{\theta}_X} f_X(\boldsymbol{x}|\boldsymbol{\theta}_X)\pi(\boldsymbol{\theta}_X|IK, \Xi_n)d\boldsymbol{\theta}_X \\ &= \frac{\int_{\boldsymbol{\theta}_X} f_X(\boldsymbol{x}|\boldsymbol{\theta}_X) \prod_j f_X(\boldsymbol{x}_j|\boldsymbol{\theta}_X) \cdot \pi_0(\boldsymbol{\theta}_X)d\boldsymbol{\theta}_X}{\int_{\boldsymbol{\theta}_X} \prod_j f_X(\boldsymbol{x}_j|\boldsymbol{\theta}_X)\pi_0(\boldsymbol{\theta}_X)d\boldsymbol{\theta}_X} \end{aligned} \tag{5.124}$$

An extensive literature has discussed the use of simulation techniques such as Monte-Carlo Markov Chains (MCMC) or Gibbs sampling (e.g. Robert and Casella, 1999). MCMC are used to generate a random sample $(\boldsymbol{y}_k)_{k=1..N}$ following a potentially complex distribution for which there may be only partial knowledge in the sense that the density f_Y is only known through quotients of likelihoods,

$$L(\boldsymbol{y}_1, y_2) = f_Y(y_1)/f_Y(y_2) = q(y_1)/q(y_2) \tag{5.125}$$

while the normalising factor C in $f_Y(\boldsymbol{y}) = Cq(\boldsymbol{y})$ may not be known or be too expensive to compute. This is typically the case with the posterior epistemic distribution $\pi(\boldsymbol{\theta}_X|IK, \Xi_n)$, defined above, where the normalising denominator:

$$\int_{\boldsymbol{\theta}_X} \prod_j f_X(\boldsymbol{x}_j|\boldsymbol{\theta}_X)\pi_0(\boldsymbol{\theta}_X)d\boldsymbol{\theta}_X \tag{5.126}$$

is an *a priori* unknown multiple integral. On the contrary, the likelihood ratio only involves a numerator that is generally closed-form (or quickly computable):

$$L(\boldsymbol{\theta}_1, \boldsymbol{\theta}_2) = \prod_j \frac{f_X(\boldsymbol{x}_j|\boldsymbol{\theta}_1)}{f_X(\boldsymbol{x}_j|\boldsymbol{\theta}_2)} \cdot \frac{\pi_0(\boldsymbol{\theta}_1)}{\pi_0(\boldsymbol{\theta}_2)} \tag{5.127}$$

MCMC do so on the basis of a chained simulation that randomly noises the previous result and accepts/ rejects it with a probability that is connected to the desired density. Though many variants have been developed, a typical simple implementation (the Metropolis algorithm, Metropolis, 1953) involves the following step:

> Choose a seed y_o
> Iterate the following loop: (at $k + 1$-th iteration, given y_k)
> - Add a random noise $\boldsymbol{u}_k$ to $\boldsymbol{y}_k$ (typically a centred Gaussian or uniform r.v., i.i.d. and independent from all other variable) thus defining the candidate $\boldsymbol{y}'_{k+1} = \boldsymbol{y}_k + \boldsymbol{u}_k$ for the next sample.
> - Compute the likelihood ratio $r = L(\boldsymbol{y}'_{k+1}, \boldsymbol{y}_k)$ of the candidate $k + 1$-th to the k-th.
> - Accept $\boldsymbol{y}_{k+1} = \boldsymbol{y}'_{k+1}$ if $r > 1$, or else with a probability r. If not, keep $\boldsymbol{y}_{k+1} = \boldsymbol{y}_k$.
> - Go to next iteration.

The third step of the algorithm can be implemented through the ancillary sampling of a random acceptance indicator, a random variable R_{k+1} uniformly distributed over $[0,1]$, leading to acceptance if $R_{k+1} < r$. A key tuning factor is the distribution and variance of the random perturbation: they are critical to the convergence of the algorithm. If the perturbation is large compared to the target distribution width, then the likely domain is largely explored but acceptance is rare so that the resulting sample tends to build up slowly. Conversely, perturbation that is too small allows for a larger acceptance ratio but leaves large autocorrelation in the sample and limited exploration of the likely domain, so that convergence is also unsatisfactory.

The algorithm generates a chain of $(\boldsymbol{y}_k)_{k=1\ldots N}$ samples that is not *independent* but *Markovian* in the sense that $\boldsymbol{y}_{k+1}$ depends only on $\boldsymbol{y}_k$ while not in the previous samples. Convergence to desirably stable properties of the generated sample – particularly its stationarity – is a key issue that ha generated a great deal of discussion, as well as many heuristics (such as targeting a mean acceptance ratio of $r \sim 0.25$). It generally requires care, possibly discarding the first samples or testing different seeds at the same time and often the number or iterations may need to reach 10^4 to 10^7 before desirable convergence is achieved.

5.5 Understanding rare probabilities and extreme value statistical modelling

This section discusses the particular issues raised by the handling of rare risk measures from a statistical point of view. It will first come back to the concept of extreme value theory introduced in Chapter 4, Section 4.3.5 to underline both its advantages and limitations in the handling of limited datasets. Further, there is discussion of the sensitivity of extremely low probabilities to the choices made in the uncertainty model and the associated questions on the level of information that is truly shown.

5.5.1 The issue of extrapolating beyond data – advantages and limitations of the extreme value theory

Chapter 4 showed that in many risk studies, the risk measure involves *rare* probabilities or quantiles. Start with the case where the variable or event of interest may be directly observed: rarity should then be assessed from a *statistical* point of view, meaning that the desired quantile reaches the limits of the available sources of information or even beyond. The section on non-parametric estimation made it clear that it becomes tricky to estimate a quantile around $\alpha = 1/n$, and impossible beyond that on the sole basis of an n-sample, as shown in Figure 5.15 on flood flows.

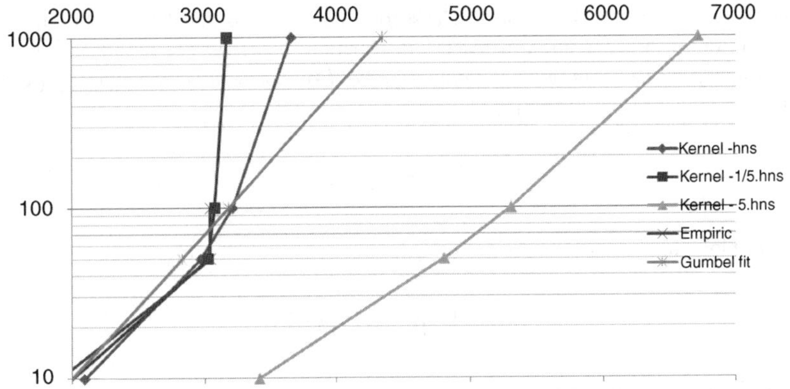

Figure 5.15 *Extrapolating beyond the empiric histogram – kernel vs. parametric model (the flood flows example: 10–1000 yearly flood return periods inferred on the basis of 30 observations in m3/s).*

Beyond such an order, a parametric model becomes necessary so as to extrapolate higher-order quantiles. Here arises the difficult issue of choosing the appropriate distribution shape as the results become very sensitive by extrapolation to such a choice, as illustrated by the water heights in Section 5.2.2. There may be up to three metres of difference for the same 1/1000yr-return period according to whether a Gaussian or lognormal model is chosen. The goodness-of-fit tests introduced above relied also on the observations below the desired quantile so that it may be questionable to base a choice of model for the distribution tail on those tests.

Extrapolating extremes of a reproducible stochastic process This is where the extreme value theory concepts introduced in Chapter 4 become very attractive, and they are a popular tool amongst risk analysts, especially when dealing with natural aggressions such as flood, storms or earthquakes. Guidance for the three distribution shapes plausible for the tails – in fact only one when parameterized with the ξ shape parameter – proves to be very attractive as it allows for the use of a part of the dataset (the maximal values or values above a threshold) in order to extrapolate. It therefore reduces greatly the size of datasets required by the comparison of a larger set of potential distributions, or *a fortiori* by distribution-free kernel techniques.

However, a number of key assumptions need to be examined in practice. The first assumption is that there is an *underlying reproducible stochastic process*, such as:

- A time series: the most frequent case encountered for meteorological, climatological or environmental variables of interest (wind, temperature, water flows, flood or sea levels, earthquake . . .). The variable of interest exhibits random trajectories X_t in time.
- A spatial process: such as happens when considering flaws, bits of corrosion, strengths of material and so on. The variable of interest is distributed over random fields X_r in space.

In such a context, the sampling of random (maxima or minima) extrema constitutes a tail distribution. It is much harder to imagine a sound application of the EVT to an isolated (hardly-reproducible) event such as the 9/11 terrorist attack or the destruction of Earth, or even to theoretically-reproducible events that have never been observed more than, say, a few times.

Independence and auto-correlation issues The second assumption is that there is enough independent sampling going on in the generation of the tail distribution. *Independence* is indeed crucial for two reasons. It is firstly *necessary* in order *to apply maximal likelihood estimation*; starting from an auto-correlated sample:

$$(x_j = x_{tj})_{j=1\ldots n} \tag{5.128}$$

the process of block maximisation (of size T_b):

$$x_k = Max_{[k \cdot Tb,(k+1)Tb]}\{x_t\} \tag{5.129}$$

or of threshold filtering (above u):

$$\begin{cases} (t_l)_l = \{t_j | x_j \geq u\} \\ x_l = x_{t_l} \end{cases} \tag{5.130}$$

is assumed to generate i.i.d. samples so that maximal likelihood may be applied. Adaptive choices of block size or threshold levels prove necessary in practice in order to ensure the reasonable independence of the filtered observations, a process which may be tested through a few elementary tests (see Reiss and Thomas, 2001). As shown in Section 5.4.3.1, the convergence of the estimation relies on the law of large numbers, formulated classically on *i.i.d.* observations; extensions to correlated samples are available, for instance assuming that auto-correlation decays quickly enough so that $\sum_{r\geq 0}|Cov(X_t, X_{t+r})| < +\infty$ – referred to as *short-memory* time series; note that it slows down convergence with an inflated (epistemic) estimation variance.

The second reason for requiring some independent underlying sampling is linked to the asymptotic behaviour assumed by the EVT. Recall, for instance, that a basic formulation of the theorem on the limit laws of the maxima (Fisher and Tippett, 1928) requires that Z be the maximum of a limitless set of *i.i.d.* random variables:

$$\begin{aligned} Z &= Max_{j=1\ldots n} Z_j \\ Z_j &\sim F_z(z|\theta) \quad i.i.d. \end{aligned} \tag{5.131}$$

in order for it to converge to the *GEV* family:

$$\left| F_n(z) - GEV\left(\frac{z-\mu_n}{\sigma_n}\right) \right| \underset{n\to\infty}{\longrightarrow} 0 \tag{5.132}$$

Suppose that a set of daily observations over 10 years is made available. If there is reasonable day-to-day independence and identity of distribution, such a sample amounts to a 3660-sample of i.i.d. random variables, assumed to be large enough to allow for asymptotic behaviour. Conversely, assuming very schematically that each block of 100 successive days is close to being completely correlated within, but independent from each other, it is easy to understand that there are no more than about 37 independent values to deal with; so that it would presumably be hard to argue that the distribution of maxima has reached the limit shape. A more realistic assumption might be that there is close to a full correlation from one day to another but that one has to wait for 100 days ahead to pull inter-correlation down to a non-significant level: the approximate equivalent would be somewhere between the two figures.

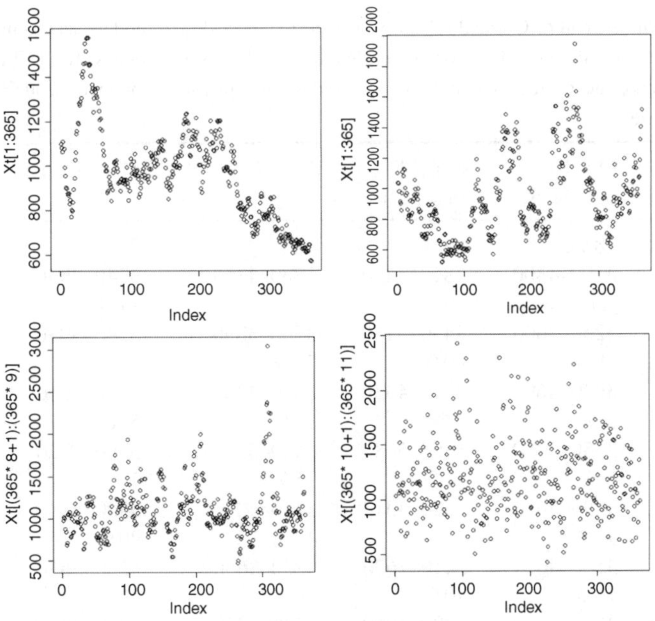

Figure 5.16 *Auto-correlated time series: daily sampling over 1 year (left up – $\tau = 100dy$, right up – $\tau = 30dy$, left down – $\tau = 10dy$, right down – $\tau = 1dy$).*

Figure 5.16 and Table 5.16 illustrate the phenomena in sampled data. The model is taken as autocorrelated lognormal daily deviations, the logs of which follow a simple auto-regressive Gaussian model with variable correlation length (from $\tau = 1$ to 100 days where $\phi = exp(-1/\tau)$):

$$\begin{aligned} \log X_{t+1} &= \phi \log X_t + (1-\phi)\mu + U_t \\ U_t &\sim N(0,\sigma^2) \quad i.i.d. \end{aligned} \tag{5.133}$$

Elementary derivations lead to the following moments of the stationarily time-correlated distribution of $\log X_t$:

$$\begin{aligned} E[\log X_t] &= \mu \\ Cov[\log X_t, \log X_{t+r}] &= \sigma^2 \frac{\phi^r}{1-\phi^2} = \frac{\sigma^2}{1-\exp\left(-\frac{2}{\tau}\right)} \exp\left(-\frac{r}{\tau}\right) \end{aligned} \tag{5.134}$$

Auto-correlation (numerically close for X_t or $\log X_t$) thus represents ~*60 %*, *37 %* and *13 %* at $\tau/2$, τ and 2τ intervals. While not being real hydrological data (e.g. because it does not represent the key hydrological *seasonality* (see Collet, Epiard, and Coudray, 2009), the asymmetry of the lognormal distribution, the return to the mean and decreasing correlation behaviour could possibly calculate the dynamics of flow after rainfall events; correlation lengths of $\tau = 1$ (resp. $\tau = 10$) could illustrate a medium (resp. large) river. Table 5.16 compares goodness-of-fit according to the lognormal model (which is appropriate at a daily scale) and Gumbel (which should be appropriate at block maxima); it should be seen as a mere sampling

Table 5.16 Goodness-of-fit of Gumbel (G-up) vs Lognormal (LN-down) models according to the p-value of Kolmogorov-Smirnov test, depending correlation length and block size (in days) for 150 yrs of daily data. Prediction of quantiles according to models is compared to true quantiles (obtained through empiric quantiles with very long series).

		M = 10 (n = 5475)	M = 100 (n = 547)	M = 365 (150)	M = 547 (100)	Quantile 1/100-1/1000
$\tau = 0$	G	38 %	84 %	98 %	75 %	3700–4300
		3790–4400	3820–4430	3760–4350	3740–4330	
	LN	<1 %	19 %	47 %	17 %	
		3250–3600	3450–3780	3560–3890	3590–3930	
$\tau = 1$	G	18 %	40 %	70 %	58 %	3650–4250
		3920–4600	3730–4330	3640–4200	3570–4080	
	LN	5 %	9 %	16 %	5 %	
		3350–3750	3360–3680	3450–3770	3500–3820	
$\tau = 10$	G	2 %	42 %	95 %	77 %	3500–4000
		3890–4630	3620–4310	3450–4070	3500–4130	
	LN	95 %	44 %	44 %	90 %	
		3500–4070	3210–3620	3230–3600	3280–3640	
$\tau = 100$	G	16 %	70 %	94 %	90 %	3070–3800
		3530–4220	3290–4060	3240–4050	3220–4040	
	LN	50 %	85 %	76 %	80 %	
		3280–3890	3060–3680	3040–3630	3060–3660	

example (with one fixed sample for each correlation length) with random results. With 150 years of daily data, both models are generally acceptable, while the block size clearly needs to be large enough with respect to the correlation length to get an acceptable fit and good quantile prediction of the Gumbel model (the lognormal being closer to data for highly correlated maxima), although blocks that are too large then miss the asymptotics because of the decreasing size of the resulting sample of maxima.

However, convergence is not yet clear: longer samples representing 1000 years of data are much more conclusive (for the case $\tau = 1$ at least, less for the larger correlation lengths) with a clear attraction to Gumbel. Such data length is very rare in practice, save possibly the Nile river in hydrology although the latter is a well-known example of a *long-memory* time series breaching the standard hypotheses of auto-correlation decay.

Remember eventually that even with an uncorrelated *i.i.d.* time series of Gaussian variables, the asymptotic rate of convergence is only in $(loglogn)^2/logn$ (Leadbetter *et al.*, 1983).

Homogeneity and stationarity issues A third assumption is that the underlying time series is *identically distributed*, as is obvious within the formulation of the two foundational theorems (see Annex Section 10.1 and Chapter 4, Section 4.3.5). In other words, the tail distribution of extreme observations needs to reflect random sampling within the same phenomenology in order to be able to infer something about its likely shape. If the data reports a panel of highly different extreme records all generated by very different physical phenomena, it is understandably hard to trust an argument based on statistical regularity. However, if the extreme events that have been observed belong to a small set of different situations, each of which enjoys a significant number of repeated extreme observations, some extensions may be contemplated. While the basic application shown in Chapter 4 evidenced the use of time series that originated supposedly from a fully stationary and homogeneous phenomenon, more elaborate applications of EVT could typically build *a mixture* of identically distributed *sub-distributions* as follows:

- Treat the *seasonality*, or seasonal *non-stationarity*: subdivide the time series into fixed time slices corresponding typically to seasons, within which the times processes are then assumed to be identically distributed. Think of the four calendar seasons for wind or temperatures extremes.
- Treat the heterogeneity of underlying events: subdivide the time series into subpopulations of identically distributed events, with the similar assumption of identical distribution within each sub-group. Think of the distinct types of low pressure perturbations generating extreme rainfall events such as the Atlantic low-pressures or the Mediterranean in Southern France.
- Treat the potential *long-term non-stationarity* of data, due typically to long-term climate or environmental change: this latter effect is potentially the more challenging as the data are rarely long enough to clearly establish non-stationarity effets in its extreme values. Conversely, if non-stationarity is not implausible, a terrifying dilemma consists in either dropping parts of the past data sample for fear of misrepresenting the present and future system or keeping it for fear of downsizing the data sample by too much and thus inflating the estimation of uncertainty. For extreme temperatures with a plausible type of long-term climate change, a model could try to incorporate a slow trend in the parameters of the EVD as has been done with nuclear risk assessment (Parey *et al.*, 2007). Sequential or point detection techniques also have potential when more *abrupt changes* of phenomenology are not implausible (Basseville and Nikiforov, 1993).

Returning to the generic modelling framework introduced in this book, these extensions may be viewed as the building of a more elaborate system model whereby the basic case $Z_t = X_t$ is replaced by:

$$\begin{aligned} Z_t &= \sum_{i\ seasons} 1_{t\in\Delta T_i} X_t^i \\ \text{or } Z_t &= \sum_{j\ sub-populations} 1_{E_j} X_t^j \end{aligned} \tag{5.135}$$

An illustrative example is given in Figure 5.17 where two extreme events are visibly outside the range of a first *GEV* (Gumbel) fit. The two alternatives come either with a 3-parameter *GEV* fit, or more with a mixture of two Gumbel distributions representing typically two sub-populations of common floods and

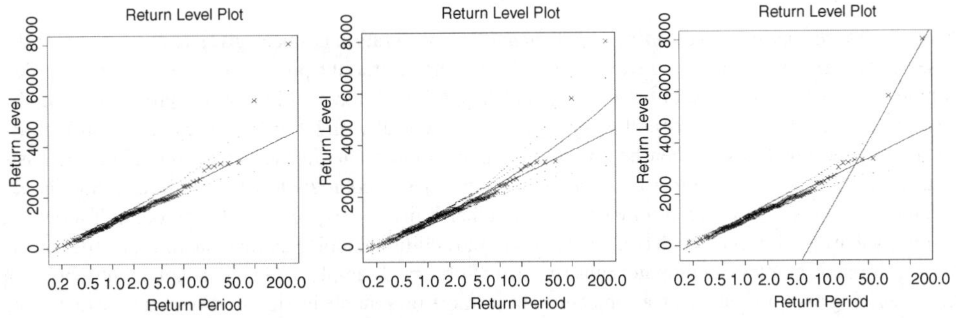

Figure 5.17 *Alternative fits for extreme values; (left) Gumbel fit; (centre) 3-parameter GEV fit; (right) fitting a mixture of two Gumbel distributions. The data sample is modified version of the flood example, including upper outliers sampled through a distinct distribution.*

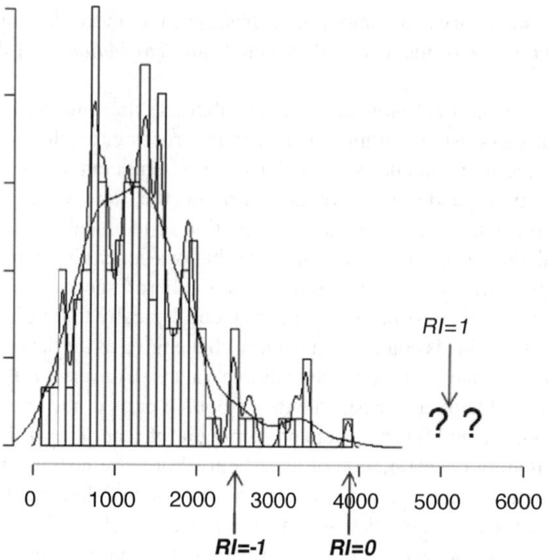

Figure 5.18 *Extrapolating a rare quantile.*

rare but severe ones. The hydrological evidence supported to some extent such phenomena whereby severe rainfall events may saturate the watershed and yield stronger runoff above a certain threshold (cf. the Gradex/Schadex approaches, Paquet, Gailhard, and Garçon, 2006).

Regional approaches are an attractive field for development of the extreme value theory, for example in the hydrological community through the recovery of larger data sets from nearby rainfall or riverflow gauges. Associated variables of interest are assumed to be homogeneous enough for simple scaling relationships to allow for a mixture of samples. More elaborate models combining spatially-distributed variables of interest raise the challenging issue of *spatial dependence of extremes*. This recently developing area of research (e.g. Northrop and Jonathan, 2011) is in fact relevant not only for a stabler estimation of local characteristics through enriched samples. It is also essential for the estimation of risk models involving distributed industrial systems, such as the vulnerability of a power transport grid to continental-wide storm events.

Extensions to the output of a complex system model Consideration has been given to the case where the observations are very close in nature to the variable of interest, although possibly through a mixture model. A much more challenging question will arise in Chapter 7 with the handling of low probabilities in the non-observable output of a complex system model, typically involving physics-based differential equations. The process will then be based on running simulations in the system model on the basis of uncertainty models *estimated* on the *input* side, in order to *compute* low probabilities in the *output* variable of interest. However, it may be asked whether it may be expected that such output follows an extreme value distribution: which could then provide an additional tool to extrapolate rare quantiles on the basis of limited samples of computer runs (instead of observed samples). Given the comments above on the underlying assumptions, such a conjecture seems more reasonable in the case where at least one of the uncertain inputs as well as the output variable of interest are time (or spatial) processes:

$$\boldsymbol{Z}_t = G(\boldsymbol{X}_t, \boldsymbol{d}) \tag{5.136}$$

A simple case is that of a linear system model $G(.)$ noised by a Gaussian uncertain input process $\mathbf{X}_t$. It is well known that $\mathbf{Z}_t$ will again be a Gaussian process in this case, so that the maxima of Z over a large-enough time interval are expected to be Gumbel-distributed (or exponentially-distributed when considering its conditional distribution above a threshold). To the knowledge of the author, this question has received too little discussion in the literature in the broad general cases where $G(.)$ is non-linear and/or inputs are non-Gaussian.

5.5.2 The significance of extremely low probabilities

Given the previous comments on the importance of sound estimation within risk analysis, is it reasonable to handle risk measures involving extra-low probabilities? To what extent should this be undertaken, given a fixed level of information? Preliminary comments will be made in this section on this challenging question.

It is useful to introduce the following simple *rarity index* at this point:

$$RI = Log_{10}\frac{1}{c_s} - Log_{10}n \tag{5.137}$$

In other words, RI counts the number of orders of magnitude by which a risk measure c_s, a threshold exceedance $c_s = 10^{-b}$, is covered by the size n of the data sample. Straightforward statistics show how tricky it becomes to estimate a probability below $\alpha = \rho/n$ where ρ is a factor 3 to 10, in other words above $RI = -1$ to -0.5, and impossible through empiric estimators above $RI = 0$ (see Figure 5.18). With the help of extreme value theory, it is common, however, to *extrapolate* a bit further. A standard flood risk study could, for instance, infer *100-yr* return period on the basis of only a few decades of flow data, that is a few tenths of yearly-block maxima samples. The largest flood studies involved in the design of protections for nuclear plants or large cities would infer *1000-yr* return period with up to a century of data; or a little more with the so-called historical information, meaning censored data on the highest floods of the previous centuries (Miquel, 1984). This amounts to estimating as low as $\alpha = 1/(\rho \cdot n)$ where ρ is a factor 3 to 10 beyond the data available, or $RI = 0.5$ to 1. At that level of extrapolation, epistemic uncertainty encapsulated in parametric confidence intervals is already rather high; it is much worse even when accounting for the uncertainty in the shape of distribution, which may not even be of the *GEV/POT* type due to slow convergence (as illustrated in Table 5.16). Table 5.17 illustrates the typical statistical features according to *RI*.

When moving on to risk analysis in a large industrial facility, much lower probabilities may be encountered. Recall the figures introduced in Chapter 1 which amounted to 10^{-5} for a plant-year or 10^{-9} per flight-hour in aerospace. As a gross average, they may still be comparable to the previous orders of

Table 5.17 The rarity index.

Rarity index	Domain	Statistical features
$RI \leq -2$	Frequent	Quasi-asymptotics
$-2 \leq RI \leq -1$	Moderately rare	Domain of robust methods
$-1 \leq RI \leq 0$	Rare	
$0 \leq RI \leq 1$	Very rare	Domain of degraded to poor control
$1 \leq RI$	Extreme	

magnitude as, for instance, there are about 10 000 nuclear plant-years of data. However, these gross averages mask a large heterogeneity between the systems, which reduces the ratio of relevant data for assessing a given facility. They also occlude the fact that reliability and risk are being controlled in sub-systems for which even more stringent figures may be required. Targeted failure probabilities for sub-systems may be as low as 10^{-8} per year in the nuclear field, for instance; some structural reliability papers mention results down to 10^{-10} or less for the computed failure probability (Harris and Lim, 1982).

At that level, design mostly relies on modelling and computing probability through a system model. Direct estimation is impossible as the data are too scarce to reach such levels, or may be simply non-existent when investigating newly built or young systems. The level of credit that may be given to the stable *computation* of such extremely low levels will be left to Chapter 7. Keeping in mind the increasing sensitivity of the output risk measure to the *estimation* of the input uncertainty model when probability decreases, one may seriously question the significance of such figures with respect to the availability of information.

A simple model may help in understanding to some extent the significance of such values. Consider a system model as simple as the product of independent input events. In other words, the output undesired event of interest occurs if and only if all input events occur, that is a parallel system of redundant components in reliability.

$$z = \prod_{j=1 \ldots p} e^j \tag{5.138}$$

so that the risk measure simply results as the product of p elementary failure probabilities:

$$c_z = \Pi_{j=1 \ldots p} P(e^j) \tag{5.139}$$

Assuming that some data is available on each of the input events with an average size of n, the rarity index should now be computed at the level of elementary events at which the estimation is undertaken:

$$RI(e^j) = Log_{10} \frac{1}{P(e^j)} - Log_{10} n \approx \frac{1}{p} Log_{10} \frac{1}{c_s} - Log_{10} n \tag{5.140}$$

the above-mentioned ratio now turns out to be $\alpha = 1/n^p$ meaning, for instance, that if ten samples are available on each of the five main input events, a system failure probability could be assessed down to about 10^{-5}. Which is obviously very far from a ratio of $\alpha = 1/n$ where $n = 50$ would be taken to gauge very grossly the total available information on the risk factors affecting the system. The key difference lies firstly in the fact that all events are supposedly known to be independent: this is a courageous assumption, particularly when dealing with rarely observed events that were possibly *never observed* in common operation. Additionally, it requires that the system is simple enough for it to fail only upon the common occurrence of one event per factor, excluding any other more subtle combination of input factors. The latter could happen in more elaborate system models $G(\mathbf{x},\mathbf{d})$ where input uncertainties are described by continuous variables and/or non-multiplicative combined effects. Besides, one could imagine drawing in pure expert judgement replacing missing statistical data in order to assess the elementary probabilities for some of these events: this means that the ratio would be pushed further depending upon the trustworthiness of those experts.

While this example is obviously a very elementary simplification of real complex systems and uncertainty models, it illustrates the amount of confidence in phenomenological knowledge encoded inside the models required to extrapolate low probability risk measures beyond pure statistical data. Eventually, model-based decision-making is all about relative likelihoods: if not low enough to be fully-confident about predictions, rarity indices associated with competing options should at least be comparable.

Exercises

There follows a list of exercises referring to the concepts developed in this chapter:

5.1 Study numerically the shape of the distribution of the ML estimators of a Gumbel distribution. How fast does it converge to its asymptotic normal limit? How much error is made in assuming it to be normal on a 5–95 % confidence interval with $n = 149$? How important is it in comparison to other level-2 uncertainty sources?

5.2 Investigate the impact of mixing a population of sub-systems (each with a constant failure rate λ_i which varies from one sub-system to another) into the distribution of the overall population. In comparison with an overall average exponential lifetime, what would it achieve to sub-model various types of uncertainty in the failure dates: manufacturing discrepancies, degradation processes fuelled by differing operational conditions, ... ? How does it compare to a Bayesian model representing epistemic uncertainty in the failure rate of an (aleatory) exponential model?

5.3 Discuss the choice of alternative copula models for the Z_m/Z_v couple that better fits to the empiric data. Compare it to a simple model of the underlying physical process reflecting the phenomenological hints about sedimentology, adjusted to data (use the techniques in Chapter 6 to calibrate such a model). How does the estimation of uncertainty behave by comparison?

5.4 Simulate simple time processes for the hourly or daily flow and study numerically the convergence of annual maxima or peak-over-threshold distributions towards theoretical extreme value distribution. What feature accelerates the convergence (shape of autocorrelation function, length of autocorrelation, boundedness of flows or their variations,...) ? How much error is made in modelling those maxima or POT through EVD distributions?

5.5 Formulate the aggregation of epistemic uncertainty when estimating the parameters of a Poisson model and those of the Peak-Over-Threshold distribution of exceedances over a given time series of observations (see Chapter 4, Section 4.3.5). Apply it to the flood example, on the basis of a sample of the log auto-regressive model (Section 5.5.1).

5.6 Estimate (through bootstrap or asymptotic theory) epistemic uncertainty in the correlation indices in the flood example. How does it compare, in orders of magnitude, to correlation tests?

5.7 Assume that a given uncertain input is modelled through an inappropriate distribution shape and estimated through maximal likelihood or alternatively moment methods. How does the parametric epistemic uncertainty cover the deviations in the (unknown) appropriate distribution? Simulate it through a normal or lognormal fit of the series of flow observations. Are there any applicable results from the statistical estimation theory regarding the pseudo-true values of ill-specified statistical models?

5.8 Study the deviations between a Bayesian estimator and a classical estimator and their associated estimates of epistemic uncertainty in the Gaussian, exponential and Gumbel cases, as a function of the number of observations and of the width of the prior distributions. How do the deviations of central estimates compare in order of magnitude with the epistemic uncertainty associated with each?

5.9 Study the impact of acquiring more friction coefficient data in a risk measure taken as a quantile of the output water level (see Chapters 3 and 7): both through a classical approach and through a Bayesian approach in the case where expertise is also available. Simulate through the sample cases provided in Section 5.4.5.2. How conservative are the initial assumptions regarding the *pdf* of friction (low or high variance suggested by the expert, risk of prior mis-estimation of the expectation...)?

References

Anderson, T.W. and Darling, D.A. (1952) Asymptotic theory of certain 'goodness of fit' criteria based on stochastic processes. *Annals of Mathematical Statistics*, **23**, 193–212.

Basseville, M. and Nikiforov, I.V. (1993) *Detection of Abrupt Changes: Theory and Application,* Prentice Hall Information and System Sciences Series, Prentice Hall, Inc., Englewood Cliffs, NJ, pp. xxviii + 528.

Bensoussan, A., Bertrand, P.R. and Brouste, A. (2011) Forecasting the Energy produced by a Windmill on a yearly basis. *Stochastic Environmental Risk* xxx, Under review.

Brockwell, P.J. and Davis, R.A. (1996) *Introduction to Time Series and Forecasting*, Springer, New York.

Coles, S.G. (2001) *An Introduction to Statistical Modelling of Extreme Values*, Springer.

Collet, J., Epiard, X. and Coudray, P. (2009) Simulating hydraulic inflows using PCA and ARMAX. *European Physical Journal Special Topics*, **174**, 125–134.

Cooke, R.M. (1991) *Experts in Uncertainty; Opinion and Subjective Probability in Science*, Oxford University Press.

Cramér, H. (1946) *Mathematical Methods of Statistics*, Princeton Univ. Press.

Droesbeke, J.-J., Fine, J. and Saporta, G. (2002) *Méthodes bayésiennes en statistique*, Ed Technip.

Efron, B. and Tibshirani, R. (1993) *An Introduction to Bootstrap*, Chapman & Hall.

Embrechts, P., McNeil, A. and Straumann, D. (1999) Correlation and Dependence in Risk Management: Properties and Pitfalls. Preprint RiskLab/ETH Zürich.

Fisher, R.A. and Tippett, L.H.C. (1928) Limiting forms of the frequency distribution of the largest or smallest member of a sample. *Proc. Camb. Phil. Soc.*, **24**, 180–190.

Granger Morgan M. and Henrion M. (1990), Uncertainty – A Guide to Dealing with Uncertainty in Quantitative Risk and Policy Analysis, Cambridge University Press.

Harris, D.O. and Lim, E.Y. (1982) Applications of a fracture mechanics model of structural reliability to the effects of seismic events on reactor piping. *Progress in Nuclear Energy*, **10**(1), 125–159.

Helton, J.C. (1993) Uncertainty and sensitivity analysis techniques for use in performance assessment for radioactive waste disposal. *Reliability Engineering and System Safety*, **42**, 327–367.

ISO (1995), Guide to the expression of uncertainty in measurement (G.U.M.), EUROPEAN PRESTANDARD ENV 13005.

Johnson, N.L., Kotz, S. and Balakrishnan, N. (1994) *Continuous Univariate Distribution, Wiley Series in Probability and Mathematical Statistics*, Wiley.

Kendall, MG. and Stuart, A. (1946) *The Advanced Theory of Statistics*, 1st edn, vol. **2**, London, Griffin.

Leadbetter M.R., Lindgen G. and Rootzen H. (1983) Extremes and Related Properties of Random Sequences and Series. *New York: Springer.*

Metropolis, N., Rosenbluth, A.W., Rosenbluth, M.N. *et al.* (1953) Equations of state calculations by fast computing machine. *Journal of Chemical Physics*, 1087–1091.

Miquel J. (1984), *Guide pratique d'estimation des probabilités de crue*, Collec. DER, Eyrolles.

Nelsen, R.B. (2006) *An Introduction to Copulas*, 2nd edn, Springer.

Northrop, P.J. and Jonathan, P. (2011) Threshold modelling of spatially dependent non-stationary extremes with application to hurricane-induced wave heights. *Environmetrics*. doi:

Paquet, E., Gailhard, J. and Garçon, R. (2006) Evolution de la méthode du Gradex: approche par type de temps et modélisation hydrologique. *Colloque Société Hydrologique de France 'Valeurs rares et extrêmes de débit' – Lyon – Mars* 2006.

Parey, S., Malek, F., Laurent, C. and Dacunha-Castelle, D. (2007) Trends and climatic evolution: Statistical approach for very high temperatures in France. *Climatic Change* (2007) **81**, 331–352.
Reiss, R.D. and Thomas, M. (2001) *Statistical Analysis of Extreme Values*, Ed. Birkhäuser.
Robert, C.P. (2001) *The Bayesian Choice*, Springer.
Robert, C. and Casella, G. (1999) *Monte-Carlo Statistical Methods*, Springer-Verlag.
Slob, W. (1994) Uncertainty analysis in multiplicative models. *Risk Analysis*, **14**(4), 571–576.
Smith R. (1985) Maximum likelihood estimation in a class of non regular cases, *Biometrika*, **72**, 67–90.

6

Combined model estimation through inverse techniques

The general modelling framework (Chapter 2) has shown the need for estimating the uncertainty model as a prior step before the computation of risk measures and sensitivity and importance indices (Chapter 7) or optimisation in support of decision-making (Chapter 8). Following upon the previous chapter where direct estimation was discussed, this chapter develops the estimation theory in the case where available modelling information may be *indirect*, in the sense that the system model needs some form of inversion in order to retrieve the useful variables. In other words, the inference of uncertainty *pdf* concerns the hidden input variables of large physical models for which only the outputs are observable. It will be seen that this also embraces the essential issues of the calibration and validation of system models as against data and expertise, enabling thereby some form of control of model uncertainty. Inverse probabilistic problems include data assimilation or regression settings for calibration and full probabilistic inversion so as to identify the intrinsic variability of inputs, a spectrum of quite different motivation-based techniques for which clear distinctions will be made hereafter and challenging research algorithms reviewed. One of the big issues in practice is to limit to a reasonable level the number of (usually large CPU-consuming) physical model runs inside the inverse algorithms.

6.1 Introducing inverse techniques

6.1.1 Handling calibration data

Recalling the general estimation program introduced in Chapter 2 and sketched in Figure 6.1:

Data and expertise, denoted by Ξ_n and IK respectively, should best represent the extent of knowledge and lack of knowledge within the combined system and uncertainty model for the subsequent inference of the risk measures. In the simplest situation, covered in Chapter 5, data and expertise enabled the direct estimation of an uncertainty model on the $\boldsymbol{x}$ inputs. In many real-world situations, data and expertise are available on *observable* variables (the $\boldsymbol{y}$) that *differ from the inputs* (the $\boldsymbol{x}$) required by the phenomenological system models at hand.

Modelling Under Risk and Uncertainty: An Introduction to Statistical, Phenomenological and Computational Methods, First Edition. Etienne de Rocquigny.

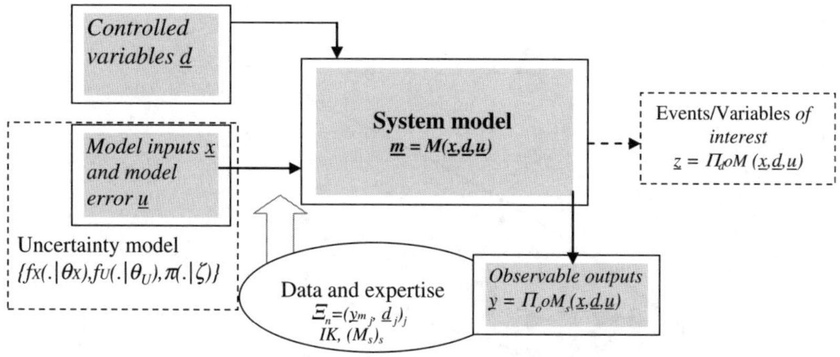

Figure 6.1 *Generic context for risk/uncertainty modelling – role of data and expertise.*

For the sake of simplicity, take a scalar example. Consider a simple physical system (e.g. river section, mechanical device) for which both an experimental setting and a corresponding numerical model relate a single observable y (e.g. water level, strain) to a scalar uncertain property x (e.g. friction coefficient, Young modulus) and known environmental conditions $\mathbf{d}$ (such as reference flow, temperature, pressure ... or even time of the experiment, if sedimentary evolution or material ageing is considered, for instance).

$$y = H(x, \mathbf{d}) \tag{6.1}$$

$$y_m = y + u \tag{6.2}$$

where y, y_m and x are scalars, but $\mathbf{d}$ can be a vector and the observable system model H is not necessarily linear; u represents the model-measurement residual, encompassing potential measurement error and/or model imprecision.

Suppose we have acquired a data sample Ξ_n of n measurements for given environmental conditions:

$$\Xi_n = \left(y_{mj},\ \mathbf{d}_j\right)_{j=1\ldots n} \tag{6.3}$$

A common goal would be to calibrate the model in order with such data to then predict z $(= y)$ for other environmental conditions $\mathbf{d}$ than those measured:

$$z = G(x, \mathbf{d}) = y = H(x, \mathbf{d}) \tag{6.4}$$

This simple example assumes that the prediction system model G and the identification system model H coincide, which may not be the case in general.

The Flood example

Figure 6.2 illustrates such a case in the flood example with a bivariate sample of flows (q_j)/water heights (y_{mj}) upon which was superimposed the system model computed for the three given values of the friction coefficient $x = k_s$ to be calibrated (other uncertain inputs being taken at their best-estimate values):

$$y_m = h + u = \left(\frac{q}{k_s * \sqrt{\frac{z_m - z_v}{l}} * b} \right)^{3/5} + u \tag{6.5}$$

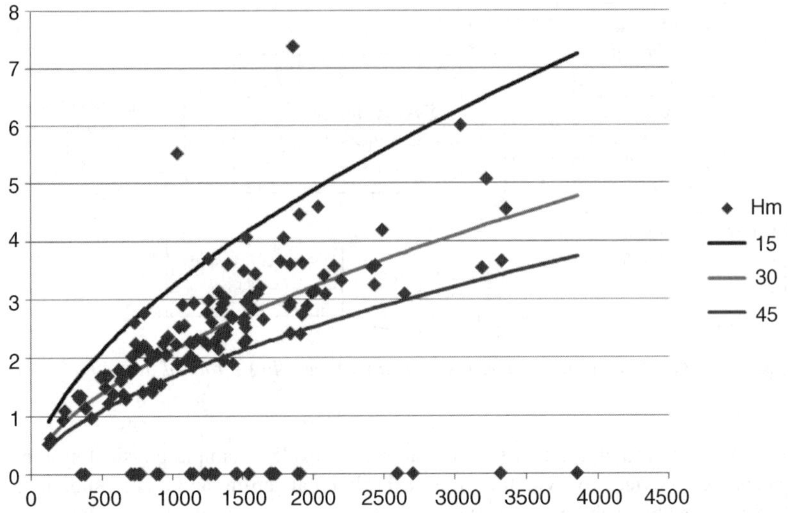

Figure 6.2 *Flood example – calibration of Ks against flow-water height measurements (note the importance of missing data).*

$$x = k_s \quad \boldsymbol{d}_j = (q_j, z_{mbe}, z_{vbe}) \tag{6.6}$$

The missing data refer to cases where the metrological device failed: thus the useful sample is reduced from *149* flows to a size of *123* bivariate observations. However, it is obviously tempting to build a model relating the bivariate observations as the deviations seem to leave a regular average trend.

Inverse algorithms provide different ways and algorithms designed to best value such observational data within the system model. Does the data suggest a different model than the one contemplated initially, with a more or less complex input parameterisation (vector $\boldsymbol{x}$)? Should a single best value be chosen for the unknown model input $\boldsymbol{x}$, or rather variable values (a random variable $\boldsymbol{X}$) so as to better account for the empiric variability? How are the residuals u being conceptually understood, either as limitations to the acceptability or accuracy of the system model, or as metrological noise corrupting the calibration of a correct model? How should the residuals be modelled: with a more or less stable variance across the sample, Gaussian or non-Gaussian distributed, more or less correlated? Should epistemic uncertainty be formulated to account for the limited size of calibration data? Can a linear behaviour of the model be assumed acceptably?

6.1.2 Motivations for inverse modelling and associated literature

More generally, various inverse algorithms respond to differing motivations. Typical uses of such data and expertise as described in previous section may be:

(i) To *calibrate* or *update* the system model, by identifying appropriate fixed values for its input parameters (some components of $\boldsymbol{x}$) on the basis of observational data. This involves a fully-specified and well-established system model. Updating may take place within a dynamic context, as in data assimilation or forecast modelling, meaning that new data are integrated from time to time yielding an update of the future prediction.

(ii) To *qualify* or *validate* the system model, meaning assessing the model error (or model uncertainty) in comparison with reference data, either experimental or generated by a reference model, and possibly controlling it up to a minimal (resp. maximal) threshold of model accuracy (resp. error or uncertainty). The system model is generally fully-specified and calibration may be included in the process, although it is generally desirable to distinguish clearly between data for calibration and data for validation.

(iii) To *build* or *select* an appropriate system model amongst competing structures (denoted as $(M_s)_s$) in order to best fit to data. This refers to upstream modelling stages where system models are not yet completely-specified and potentially draws in a whole range of estimation and model selection techniques.

(iv) To *identify* unobservable input uncertainty through the inversion of a system model. This refers to a more refined approach than the calibration of best fitting *constant* model inputs, whereby model inputs may be represented as intrinsically varying throughout the observed sample. Ξ_n is used along with model inversion to infer the uncertainty model $\{f_X(.|\theta_X), f_U(.|\theta_U), \pi(.|\zeta)\}$. A full formulation may also include model selection as developed in (iii).

These different motivations should be considered as complementary or progressive steps within an overall model development process. The terminology may also vary a great deal in the literature, where one term, such as parameter identification or data assimilation, may refer specifically to (i) or sometimes more generally to (iii), or even cover all three steps. Motivation (i) is somewhat basic engineering practice; it is not always as rigorously specified in probabilistic terms as that which is done in data assimilation, an area that has seen a growing application since the late 1980s, particularly in meteorology (Talagrand and Courtier, 1987). Motivation (ii) is standard practice in some sectors such as nuclear engineering under the popular name of *Verification & Validation* (*V&V*). Motivation (iii) is well known to statisticians as well as being long-standing practice in environmental and biological modelling. As a formalised mathematical technique, it is less disseminated in fields of engineering dominated by deterministic numerical modelling. Motivation (iv) is much more recent as it is tied specifically to explicit uncertainty modelling. Key difficulties arise with the highly-limited sampling information directly available on uncertain input variables in real-world industrial cases. Real samples in industrial safety or environmental studies, if any, often fall below the critical sizes required for stable statistical estimation. Thus, it is acknowledged to have quite a large industrial potential in the context of the rapid growth of industrial monitoring and data acquisition systems. In a nuclear reactor, large flows of data are generated for multiple physical monitoring (on pressure, temperature, fluence, etc.) thereby documenting the intrinsic physical variability of the systems. Proper model calibration and inverse algorithms could theoretically supply information on the unobservable sources of uncertainties that are at the root of the process, for example thermal exchange coefficients, basic nuclear data and so on. Here are a few examples:

- Flood monitoring generates data on maximal water elevations or velocities rather than on uncertain friction coefficient or riverbed topology.
- Monitoring of nuclear reactors generates data on external neutron fluxes, or temperatures/pressure at inlet/outlet rather than fuel rod temperatures or internal thermal exchange coefficients.
- Or the monitoring and control of vibrations of steam turbine blades rather than various stiffness coefficients, which is the case that motivated the application illustrated below.

These motivations imply different algorithms and statistical models. While being closely connected and only sometimes contrasting with respect to terminology, data assimilation, parameter identification, model calibration techniques, or inverse uncertainty identification algorithms incorporate distinctive features. This in regard to the way uncertainty sources are acknowledged conceptually and modelled mathematically on unknown model parameters or on model uncertainty.

While inverse probabilistic techniques are already old (cf. Beck and Arnold, 1977; Tarantola, 1987), it may not have been until quite recently (de Crécy, 1997; Kurowicka and Cooke, 2002) that full probabilistic inversion was considered, in the sense that the distribution of *intrinsic* (or irreducible, aleatory) input uncertainty is sought as in motivation (iv). On the contrary, classical data assimilation (Talagrand and Courtier, 1987) or parameter identification techniques (Beck, 1987; Walter and Pronzato, 1997) typically involve the estimation of input parameters (or initial conditions in meteorology) that are unknown but physically fixed, as is generally the case for motivations (i), (ii) or (iii). Model residuals are given a formal probabilistic description in algorithms crafted for motivation (ii) and possibly even more for (iii). Bayesian-inspired formulations and the increasingly-popular Gaussian process-based techniques (Beven and Binley, 1992; Kennedy and O'Hagan, 2002 for a start) develop the model uncertainty description carefully, including the key inter-correlation structure of model-measurement residuals.

6.1.3 Key distinctions between the algorithms: The representation of time and uncertainty

Key distinctions help in choosing amongst the plethora of inverse probabilistic algorithms available. The common basis for most techniques involves the following ingredients:

- A sample $\Xi_{tn} = (\mathbf{y}_m(t_j), \mathbf{d}(t_j))_{j=1\ldots n}$ available at the time of analysis, made up of observations at n past dates $(t_j)_{j=1\ldots n}$ of degrees of freedom of the system operating under known actions or experimental conditions.
- The ambition to 'tune' models of the system under uncertainty, on the basis of all such information available, and later infer measures $c_z(t_f)$ about the uncertain variable of interest at a future time t_f of interest (or period of interest $[t_f, t_f + \Delta T]$) for known actions or conditions $\mathbf{d}(t_f)$.
- The use of the two above-mentioned observational and predictive system models, used respectively for the past dates and for the time of interest:

$$\mathbf{y}_m(t_j) = H(\mathbf{x}(t_j), \mathbf{d}(t_j), \mathbf{u}_j) \quad (6.7)$$

$$z(t_f) = G(\mathbf{x}, \mathbf{d}(t_f), \mathbf{u}) \quad (6.8)$$

A study involving indirect estimation can be viewed virtually as an *uncertainty study* targeting a *time of interest* lying in the *past*. Observations at past dates are being used for the best-possible estimation of some features of the state of the system at that time, features that remain uncertain because they cannot be observed directly. From that perspective, the process is called the *analysis* in the field of data assimilation, and inevitably leaves estimation uncertainty in the so-called 'true' states of the system. For some algorithms, it will be shown that such uncertainty is modelled as epistemic in the sense that a theoretically large enough set of past observations could reduce it to a quasi-perfect description of the state; Chapter 4 has commented upon the limitations of that view with respect to the controversy in considering a single 'true' value for any feature. Other algorithms deliberately represent those past features as intrinsically variable from one past observation to the other, and thus completely irreducible. Nevertheless, whatever the status given to that part, such a *past-orientated* uncertainty study is generally a preliminary to another uncertainty study focusing on a variable of interest for a future period of interest, the inference of which gains from the insights and limitations acquired from the analysis of the past states. However, important distinctions concern the way time and uncertainty are represented in the model inputs for that purpose: compare first what may be called the *risk* vs. *forecast* modelling approaches.

Risk or forecast modelling *Forecast models* typically target the prediction of a variable of interest over a period of interest in the *near future*. The variable of interest could be a discrete event (e.g. an undesired event of biological proliferation or wind storm in the next 6h to 48h. . .) or a continuous performance variable (e.g. flow of the next day, oil price, hourly electricity demand for the next week, or daily demand for the next year . . .). The knowledge of the past and present states of inputs is usually taken into account through an additional *forecast layer* developed in the system model. Such a layer values explicitly the *correlations* between the recent states of the system and the future state at the time of interest through a noised functional dependence (for known actions or scenarios $\boldsymbol{d}_f$):

$$X^f(t_f) = \boldsymbol{M}[(x(t_j))_{j=1\ldots n,}\ \boldsymbol{d}_f,\ U_f] \tag{6.9}$$

While such a model may incorporate complex differential stochastic equations, it could be viewed in the simplest case as a sequential noised prediction based on the present state:

$$X^f(t_{j+1}) = M[x(t_j),\ \boldsymbol{d}(t_{j+1})] + V_{j+1} \tag{6.10}$$

In that case, the inverse techniques in the observational models help to estimate the uncertain input vector $\boldsymbol{x}(t_j)$ that combines generally components representing time-varying states of the system as well as time-independent model parameters or system properties. The predictive model then converts the predicted state (e.g. temperature, electricity demand . . .) into the final variable of interest for decision-making (e.g. gross power production, net profit . . .) if not identical to the predicted state. The forecast model quality is summarised classically within a matrix of confusion, recording the percentages of true/false predictions as compared to observations summarised into discrete classes. However, such a performance matrix is generally used in order to assess the quality of prediction (and sometimes calibrate the forecast model) with respect to maximising the frequency of good predictions, that is the prediction of frequent events. More recent research has developed *probabilistic forecasting*, *density forecasting* or quantile prediction, meaning that the prediction of the near future includes not only the prediction of the expected value of the v.i. but also that of some of its quantiles representing the uncertainty in the prediction, and the associated validation issues of such quantile prediction (Tay and Wallis, 2000).

Risk models predict a risk level in a *remote future*, generally beyond the short-term correlations involved in forecasting. They are developed in order to predict a *risk measure*, for example the probability of a maximal event or of variables of interest rarely (if at all) observed in the past (dike overflows). They are less relevant for the prediction of an expected value of observable input risk factors (e.g. flood flows) for which they generally bring little more than the present state, save a deterministic trend in some cases. Indeed, in contrast to forecast modelling, short-term correlations between the present state and risky events are assumed to be so negligible that: (i) there is no forecast model; and (ii) the core assumption of the risk models is that the parameterisation involves uncertain inputs that are *independent* amongst the past and between the past and future states of the system:

$$\begin{aligned} &\boldsymbol{Y}_{mj} = H(\boldsymbol{X}_j, \boldsymbol{d_j}, \boldsymbol{U}_j) \\ &\boldsymbol{X}_j|\boldsymbol{\theta}_x \sim f_X(.|\boldsymbol{\theta}_x) \quad \boldsymbol{X}_{t_f}|\boldsymbol{\theta}_x \sim f_X(.|\boldsymbol{\theta}_x) \quad i.i.d.|\boldsymbol{\theta}_x \\ &\boldsymbol{Z}_{t_f} = G(\boldsymbol{X}_{t_f}, \boldsymbol{d}_{t_f}, \boldsymbol{U}) \end{aligned} \tag{6.11}$$

Unknown in the sense of being only indirectly observed, $(\boldsymbol{X}_j)_{j=1\ldots n}$ represent *past* states of the system at time t_j that are, *given* $\boldsymbol{\theta}_X$, *conditionally* independent amongst them and from the system $\boldsymbol{X}_{tf}$ at the *future* time of interest. Think of the state of the riverbed that affected the observed maximal water levels during the major floods of the last century, as compared to the state in which the riverbed would be in during the next major flood, possibly twenty years after of the time of analysis. Refer to sections Sections 4.3.5 and 5.5 for the pre-treatments of raw data (with short-term temporal correlations) possibly required in order to

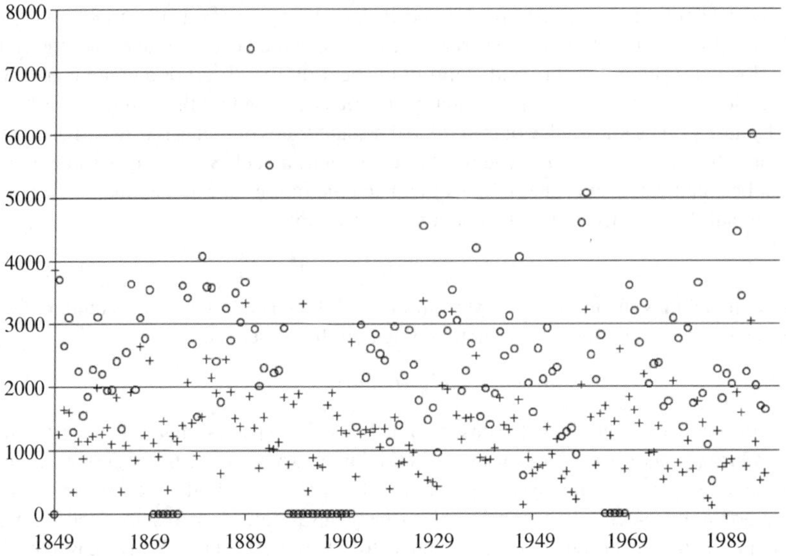

Figure 6.3 *Flood example – Flows d_j ('+' in m3/s) and water levels y_{mj} ('o' in mm) as a function of time (the year).*

legitimate this classical assumption in risk analysis. Based on *i.i.d.* uncertain inputs, some risk models are even used to predict without change the risk measure at a *random* time inside the period of interest; say the next big flood over the coming decade.

The Flood example

Figure 6.3 shows the absence of any trend that could be forecast over the time basis. Both flows and water levels do not seem to display straightforward auto-correlations, but rather an *i.i.d.* behaviour: this is a clear rationale for choosing a risk model rather than a forecast model.

Though the time-dependence is generally absent from the description of the risk model inputs, note that the *dynamic nature* of the prediction is not limited to forecast modelling, but can be accounted for differently in some risk models. The predictive model $G(.)$ can first incorporate deterministic dynamics, such as long-term trends reflected in some components of $\boldsymbol{d}$ that affect the system's response to input uncertainty (e.g. known ageing trend for materials, climate warming, ...). It could include also the description of the complex short-term dynamics of the consequences of the accident conditional upon the random initiator $\boldsymbol{X}_{tf}$, for example z_{tf} being then the maximal reactor temperature following the accident. The input vector can itself be replaced by a time series over, say, a past observed or future projected reference timespan ΔT (e.g. a year of operation). Moreover, $G(.)$ may incorporate stochastic risk models required to assess the risk measures as mentioned above for time-varying system reliability (e.g. Markov chains to compute the reliability over ΔT). However, the identification of the uncertain input parameterisation still assumes independence between past observed and future projected dynamics (e.g. no link between the past and future state of failure of the sub-system components).

For the purpose of understanding rather than strictly classifying them, Table 6.1 provides some of the respective features of the risk and forecast modelling approaches.

Table 6.1 Risk modelling vs. forecast modelling.

	Risk modelling	Forecast modelling
Future date for prediction	Remote (possibly random)	Close (and fixed)
Prediction output	Probabilistic quantity of interest at a *remote* future date	(mostly in standard forecasting) central prediction at a *close* future date (more recently with density forecasting) probabilistic q.o.i at a *close* future date,
Dependency/correlation between present and future state	Generally none	Generally significant and explicitly modelled within the forecast trend
Forecast model	Generally none (or a deterministic trend included in the predictive model)	Explicit forecast model with prediction uncertainty
Treatment of the observational sample	Global	Either global (4DVar) or sequential (KF)
Input variables	Either fixed x or variable (stationary) X	A time series x_t of fixed but unknown states of the system
Typical identification algorithms	3DVar, EM, ECME, …	4DVar, KF, EKF…

Forecast modelling is not the purpose this book: save for a few remarks about dynamic data assimilation in Section 6.4.6, most of the following comments in this chapter will address the identification issues related to the risk modelling approach. The key differences lie in the fact that *forecast model identification* requires the representation of auto-correlation or time series statistics, while *risk model identification* mostly involves standard random variables (while the risk model prediction could still involve the propagation of time series through the system model). In theory, one could contemplate the sufficient development of the forecast models in order for them to be instrumental also in predicting the (high-threshold) risk measures, for example in updating the likelihood of short-term extreme flood risk or earthquake on the basis of present environmental states. However, this assumes firstly that short-term correlations do bring significant information on rare events; secondly, that such inference can be proven reliably in spite of the statistical challenges brought by, for instance, the estimation of auto-correlated non-Gaussian tail distributions with small samples.

The Flood example (*continued*)

Figure 6.4 illustrates the difference between the two approaches in the pseudo-hydrological process introduced in Chapter 5 Section 5.5.1: a (log-AR1) auto-regressive lognormal process with an exponential time correlation (τ in days), being simulated over a month ahead with predictions of the risk model and forecast model at future times of interest every three days (the forecast being renewed at mid-period). For the sake of simplicity, strong assumptions have been made in *suppressing epistemic* uncertainty: (i) the forecast model available to the analyst is tuned to the true process of the system, that is $logX_t + 1 = \phi logX_t + (1 - \phi)\mu + U_t$ with accurate parameters ϕ,μ,σ^2; and (ii) the risk model available is also tuned to the true statistical parameters of the (lognormal) variability of the process, that is $logX_t \sim N(\mu,\sigma^2/(1 - \phi^2)$. However, it already displays some of the key features of the two approaches: the fact that when correlations are negligible at the time scale of interest (right),

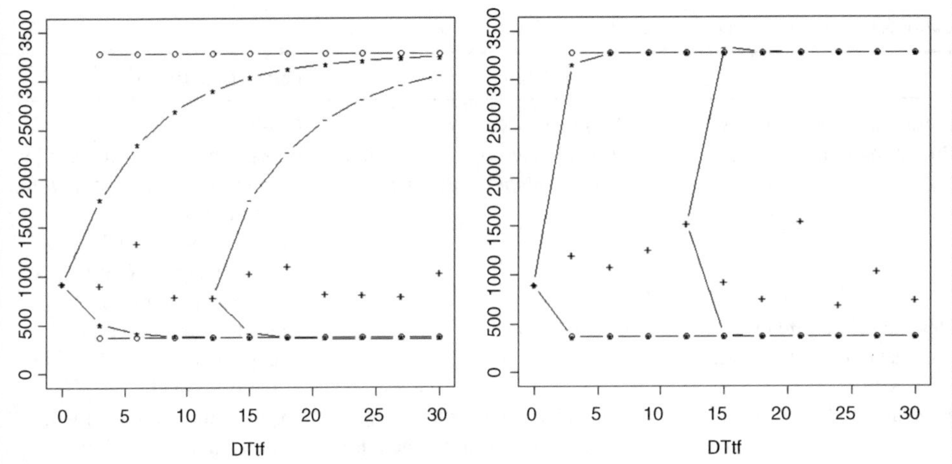

Figure 6.4 *Comparing forecast and risk modelling on the pseudo-hydrological log-AR1 process, with 1/20-yr return confidence bands over a month ahead, with* $^+$ *= realisation;* $^\circ$ *= risk model;* * *= forecast at* $t_o = 0$*;* $^-$ *= forecast at* $t_o = 146$*; (left) – long correlation,* $\tau = 10$ *days; (right) – short correlations,* $\tau = 1$ *day.*

the forecast model does not say more than the risk model, the state of the system at the time of analysis does not provide any information for the inference of the state at future times of interest which follow the long-term variability; when correlations are comparatively high (left), the forecast gives a much more focused confidence interval valuing the initial information and thus reducing the conservatism of (long-term) the risk model. This does not always mean that the confidence bands (here taken at a 1/20-yr return level each side, that is a daily confidence bands of 1–1/3650) associated with the forecast are narrower depending on the state at the time of analysis (in the case of short correlations forecast with a mid-term flow at time of analysis larger than the average).

Note the limitations of this illustration with respect to real models. Besides having neglected estimation uncertainty and the availability of the true statistical parameters to the analyst, the particular example also featured a complete consistency between the forecast and the risk model in the sense that the iterated forecast over increasing time steps converged statistically to the risk model. Real models would generally be tuned to their dominant goal, the short-term forecast or the long-term average risk, such consistency not being always easily at hand. Even more importantly, the risk measure associated with the risk model would differ from that of a forecast model, typically targeting a return-level for the maximum over a yearly period instead of return-levels for the value on a given day, as illustrated in the example (see, e.g. Krzysztofowicz, 1999 for forecasting in the hydrological context).

Dynamic updating process Note that *both kinds of models* exist within an essentially *dynamic updating* perspective over time (Figure 6.5), as there is generally a pre-existing description of uncertainties that gets updated by the injection of new information, a process which can happen more than once. Remember indeed that the risk measure arises as a functional of the input uncertainty model or a function of the information available (cf. Chapter 2, Equation (2.49)):

$$c_z = \mathcal{F}[f_Z^{\pi}(.)] = \mathcal{G}[f_X^{\pi}(.|IK, \Xi_n), G(., \boldsymbol{d})] = C_z(\boldsymbol{d}, IK, \Xi_n) \tag{6.12}$$

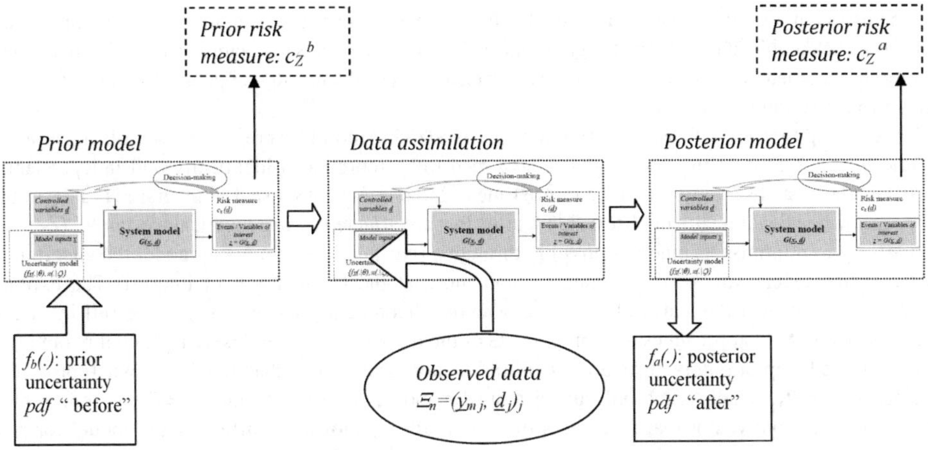

Figure 6.5 *A dynamic updating process.*

Estimation through the inverse techniques presented in this chapter would first generate the predictive input distribution $f_X^{\pi}(.|IK,\ \Xi_n)$ on the basis of the information available at the time of analysis t_n and thus the risk measure for the future time of interest t_f. With a new sample of observations $(\boldsymbol{y}_{mj},\boldsymbol{d}_j)_{j=n+1,...m}$, a later analysis would update the risk measure for t_f on the basis of Ξ_m.

This shows that these approaches may often be gathered under the cover of the terminology of *data assimilation* which uses the subscripts *b* and *a* before/after the incorporation of new data, although the term may sometimes refer to a restricted class of inverse methods, as will be explained below. The prior model may be only implicit in many cases though it becomes explicit in the important Bayesian settings, as will be seen.

The input representation of uncertainty The representation of uncertain inputs $\boldsymbol{x}$ in models leads to an additional distinction between *calibration* or identification and *full inversion*.

To put it simply, *calibration* means choosing a *single* $\boldsymbol{x}$ input value that fits best to observed data. Measurement and model error are more or less combined in the form of a residual error; given a large-enough data sample, the uncertainty in $\boldsymbol{x}$ is assumed to be negligible in the sense that a single value can explain the whole data sample or, in other words, that *all past observations* are related to a *single* and *constant*, though unknown, feature or *state of the system*. Formally, the uncertain input model is a Dirac distribution around the unknown parameter $\boldsymbol{x} = \boldsymbol{\theta}_X$ thus simplifying Equation (6.11) into:

$$\begin{aligned} \boldsymbol{Y}_{mj} &= H(\boldsymbol{X}_j, \boldsymbol{d}_j, \boldsymbol{U}_j) = H(\boldsymbol{x}, \boldsymbol{d}_j, \boldsymbol{U}_j) \\ \boldsymbol{X}_j|\boldsymbol{\theta}_x &\sim f_X(.|\boldsymbol{\theta}_x) = \delta_{\boldsymbol{x}=\boldsymbol{\theta}_x} \end{aligned} \tag{6.13}$$

Full inversion is a more elaborate modelling choice recognising more explicitly that intrinsic variability affected the input parameters $\boldsymbol{x}$ as past observations related to the physically-distinct states of the system represented by $\boldsymbol{X}(t_j)$. The full model of Equation (6.11) is kept: $(\boldsymbol{X}_j)_{j=1...n}$ model *past* states of the system at time t_j that are assumed to be, *conditionally to* $\boldsymbol{\theta}_X$, i.i.d. amongst them and it is further assumed that the state of the system $\boldsymbol{X}$ at the *future* time of interest is also i.i.d, conditional upon $\boldsymbol{\theta}_X$.

Subsequent observations will illustrate that the distinction consists essentially of a modelling choice and could be combined for different degrees of freedom or components of vector $\boldsymbol{x}$: calibration being best for model parameters while inversion can be a powerful (though data-costly) approach for the inference of the intrinsic variability of states.

Returning to the risk/forecast distinction, *risk models* essentially represent irreducible uncertainty in the *inputs* $\boldsymbol{X}$ forcing the system that is assumed to be reproduced identically and independently between the past observations and the future time of interest. In that sense, full inversion is a desirable approach when dealing with unobservable factors that are affected by irreducible uncertainty. *Forecast models* try to minimise as much as possible through a (partially) deterministic model the uncertainty between the recent values of the states $\boldsymbol{X}$ and those at the future times of interest. The explicit correlations between the states of the system render theoretically more complex the full inversion approaches. In both approaches the components of the model inputs $\boldsymbol{x}$ representing model parameters are identified appropriately through a calibration approach. Note that the level-two (epistemic) uncertainty is $\boldsymbol{\theta}_X$, which is the only uncertainty for model parameters since $\boldsymbol{x} = \boldsymbol{\theta}_X$, is represented by a single random vector that represents the equal lack of knowledge affecting the model for past observations or future inference, and is thus completely dependent.

Other practical features distinguishing the relevance of the competing algorithms will be the following:

(i) The representation of any background or prior information that may be available: Bayesian formulations, for instance, represent explicitly probabilistic priors.
(ii) The type of simplification that can be made in the system and uncertainty model, typically linear and/or Gaussian features that will prove to turn many algorithms into closed-form computations.

6.2 One-dimensional introduction of the gradual inverse algorithms

To clarify the distinctive features and mathematical properties of inverse algorithms, this section will introduce them within simplified one-dimensional formulations.

6.2.1 Direct least square calibration with two alternative interpretations

Returning to the example provided at the beginning of the chapter:

$$y = H(x, \boldsymbol{d}) \tag{6.14}$$

$$y_m = y + u \tag{6.15}$$

where y, y_m and x are scalars, but $\boldsymbol{d}$ can be a vector and the observable system model H is not necessarily linear; u represents the model-measurement residual, encompassing potential measurement error and/or model imprecision. Suppose we have acquired a data sample Ξ_n of n measurements for given environmental conditions:

$$\Xi_n = \left(y_{mj}, \boldsymbol{d}_j\right)_{j=1\ldots n} \tag{6.16}$$

The easiest way to proceed is with a straightforward least-square calibration of the uncertain x, which means that calibration is assumed to equate to the selection of a proper point value for x:

$$\left.\begin{aligned} \hat{x} &= \underset{x}{Arg\,Min}\left[C(x,\Xi_n)\right] \\ C(x,\Xi_n) &= \frac{1}{2}\sum_j \left(y_{mj} - H(x,\boldsymbol{d_j})\right)^2 \end{aligned}\right\} \tag{6.17}$$

In fact, if it is reasonable to assume that the uncertainty (or variability) involved in measurement-model difference u is random, independent and identically distributed amongst the sample,[1] this first inverse problem receives a very straightforward statistical interpretation:

$$\left.\begin{aligned} Y_{mj} &= H(x,\boldsymbol{d_j}) + U_j \\ U_j &\sim N(\mu,\sigma^2) \end{aligned}\right\} \tag{6.18}$$

for which the loss (or cost) function $C(x,\Xi_n)$ appears to be closely connected to the log-likelihood of the residual of the homoscedastic non-linear regression model. In fact, the standard regression notations would rather be $\boldsymbol{X_j}$ instead of $\boldsymbol{d_j}$ and θ instead of x, ending up with a model denoted as $Y_{mj}|\boldsymbol{X_j} \sim N(H(\theta,\boldsymbol{X_j}) + \mu,\sigma^2)$. Neglecting the bias μ, log-likelihood of the sample Ξ_n goes:

$$\begin{aligned} LL\left[(y_{mj},\boldsymbol{d_j})_j|x,\sigma^2\right] &= -\frac{n}{2}\log(2\pi\sigma^2) - \frac{1}{2\sigma^2}\sum_j \left(y_{mj} - H(x,\boldsymbol{d_j})\right)^2 \\ &= -\frac{n}{2}\log(2\pi\sigma^2) - \frac{1}{\sigma^2}C(x,\Xi_n) \end{aligned} \tag{6.19}$$

Maximising the likelihood is equivalent to least square minimisation (Equation (6.17)) for the x component, adding if necessary the bias component μ inside the sum of square deviations; meanwhile, the σ^2 component can be determined separately through the first-order condition.

As it is classical in regression or experimental design, note that Y_j is an independent but *non-identically distributed* sample, and the log-likelihood (or least-squares) maximisation requires special conditions in order to converge (see Jennrich, 1969; Hoadley, 1971), concerning typically the regularity of $H(.)$.

The linearised Gaussian case Closed-form solutions are available when supposing that H is linear (or close to linear) with respect to x, and that residuals are homoscedastic and Gaussian. Then Equation (6.18) becomes more explicit:

$$\left.\begin{aligned} &U_j \sim N(0,\sigma^2) \\ &Y_{mj} = H(x_m,\boldsymbol{d_j}) + U_j \approx H'_j.(x_m - x_o) + H_j^o + U_j \end{aligned}\right\} \tag{6.20}$$

which can equally be expressed as the following Gaussian model for the observed sample:

$$Y_{mj} \sim N(H'_j.(x_m - x_o) + H_j^o,\sigma^2) \tag{6.21}$$

[1] Such an assumption is classical though not easy to guarantee in practice. Assume at least that all systematic errors have been treated, and that there is some 'experimental re-cast' between any observations. Random and independent but not identically-distributed measurement errors can be treated in a generalised heteroscedastic setting, see the end of the section.

where $H_j^o = H(x_o, \boldsymbol{d_j})$ and derivatives $H_j' = \partial H/\partial x|_{x=x_o,d=\boldsymbol{d_j}}$ are computed at a *known* linearisation point x_o. Log-likelihood becomes:

$$LL[(y_{mj}, \boldsymbol{d_j})_j \,|\, x_m, \sigma^2] = -\frac{n}{2}\log(2\pi\sigma^2) - \frac{1}{2}\sum_j \frac{1}{\sigma^2}\left(y_{mj} - H_j'(x_m - x_o) - H_j^o)\right)^2 \tag{6.22}$$

Simple algebra leads then to the following closed-form maximum likelihood estimators (refer for a demonstration to the general solution of a quadratic program given in Annex Section 10.5.2):

$$\left.\begin{aligned}\hat{x}_m &= x_o + \frac{\sum_j H_j'[y_{mj} - H_j^o]}{\sum_j (H_j')^2} = x_o + \frac{\overline{H_j'[y_{mj} - H_j^o]}}{\overline{(H_j')^2}} \\ \hat{\sigma}^2 &= s_u^2 = \frac{1}{n}\sum_j \left(y_{mj} - H_j'(\hat{x}_m - x_o) - H_j^o\right)^2 = \overline{\left(y_m - H'_j(\hat{x}_m - x_o) - H_j^o\right)^2}\end{aligned}\right\} \tag{6.23}$$

When supposing additionally that experimental conditions $\boldsymbol{d}_j$ do not vary by much, so that H_j' are constant (or approximately constant) and equal to H' the first estimator simplifies into:

$$\hat{x}_m = x_o + \frac{1}{H'}\left[\overline{y_m - H^o}\right] \tag{6.24}$$

These formulae get a straightforward physical interpretation:

(a) The optimal parameter value for the model input is the weighted average of the observed outputs, simply adjusted for position (H_j^o) and scale (H_j') changes due to the model transformation.
(b) The residuals, representing measurement-model error (or model uncertainty), equate to the standard error of the model best-estimate in predicting the observed sample.

The MLE estimator for the residual variance is biased when n is small, a situation in which the following unbiased formulation is more adequate:

$$\hat{\sigma}_u^2 = \frac{n}{n-2}s_u^2 = \frac{1}{n-2}\sum_j \left(y_{mj} - H_j'(\hat{x}_m - x_o) - H_j^o\right)^2 \tag{6.25}$$

This is a special application of the classical result of multiple regression with $p + 1 = 2$ parameters to be estimated (x_m and σ^2) on the basis of n observations. Representing less than 10% of error whenever $n > 12$, the correction $n/(n-2)$ will often be neglected. Note that the existence of a *non-zero* residual in Equation (6.23) is necessary for the model to be identifiable in the sense that it will be more explicit in multi-dimensional extensions.

Linearisation and model bias Note that the linearised program above (Equation (6.20)) assumed implicitly that the linearised model was unbiased. This may not be the case; in fact, even the non-linear model may prove to be biased with respect to the observed sample. In that case, an additional bias parameter μ has to be estimated as follows:

$$\left.\begin{aligned}&U_j \sim N(\mu, \sigma^2) \\ &Y_{mj} = H(x_m, \boldsymbol{d_j}) + U_j \approx H_j'.(x_m - x_o) + H_j^o + U_j\end{aligned}\right\} \tag{6.26}$$

Log-likelihood now involves three parameters that need to be estimated:

$$LL[(y_{mj}, \boldsymbol{d_j})_j | x_m, \mu, \sigma^2] = -\frac{n}{2}\log(2\pi\sigma^2) - \frac{1}{2}\sum_j \frac{1}{\sigma^2}\left(y_{mj} - H_j'(x_m - x_o) - H_j^o - \mu\right)^2 \tag{6.27}$$

Simple algebra leads then to the following closed-form maximum likelihood estimators:

$$\left.\begin{aligned} \hat{x}_m &= x_o + \frac{\overline{\left(H_j' - \overline{H_j'}\right)\left(y_{mj} - H_j^o - \overline{y_{mj} - H_j^o}\right)}}{\overline{\left(H_j' - \overline{H_j'}\right)^2}} = x_o + \frac{\operatorname{cov}\left(H_j', y_{mj} - H_j^o\right)}{\operatorname{var} H_j'} \\ \hat{\mu} &= \overline{\left(y_m - H_j'(\hat{x}_m - x_o) - H_j^o\right)} \\ \hat{\sigma}^2 &= \overline{\left(y_m - H_j'(\hat{x}_m - x_o) - H_j^o - \mu\right)^2} \end{aligned}\right\} \tag{6.28}$$

that correspond to the standard estimators of a linear regression whereby:

$$\left.\begin{aligned} &X_j = (1, H_j') \quad \boldsymbol{X} = (X_j)_j \\ &\boldsymbol{\beta} = (\mu, x_m - x_o)' \\ &\boldsymbol{U} = (U_j)_j \\ &\boldsymbol{Y} = (Y_{mj} - H_j^o)_j \end{aligned}\right\} \Rightarrow \boldsymbol{Y} = \boldsymbol{X}.\boldsymbol{\beta} + \boldsymbol{U} \tag{6.29}$$

Bias and residual variance are in fact simply the empiric bias and variance obtained after subtracting the calibrated model predictions from the observed sample. The Equation (6.28) MLE estimators are defined provided of course that the model is *identifiable* in the sense defined in Chapter 5. Here, this means in particular that experimental conditions $\boldsymbol{d_j}$ should vary enough for H_j' not to be constant so that the denominator of $\hat{x}_m$ is defined: otherwise, if all gradients are equal to H', there is overparametrisation between $\hat{x}_m$ and the bias μ and the above-mentioned unbiased model whereby $\hat{x}_m = x_o + \frac{1}{H'}\left[\overline{y_m - H^o}\right]$ offers the solution.

Non-linear generalisation with two interpretations: simple parameter identification, or complete model calibration The non-linear generalisation does not lead to closed-form estimators but the interpretation will stay the same, as will be explained below. Parameter identification adjusts essentially the parameter value so as to best predict on average the cloud of data sample/model predictions, as is meant by the quadratic cost function. Note in fact that two different interpretations follow according to the status given to σ^2.

In a *first interpretation* σ^2 is considered to be well known, representing, for instance, a measurement error considered much more significant than the deviation due to an assumingly *solid* model. Simple least square $\hat{x} = Arg\,Min_x\left[C(x, \Xi_n)\right]$ now lacks a closed form solution; it is fully equivalent to likelihood maximisation and appears as a simple ***parameter identification*** for x. Attached to the identification model $y_{mj} = H(\hat{x}, \boldsymbol{d}) + u_j$ only, σ^2 will normally *not* be *included* in the subsequent prediction model $z = G(x, \boldsymbol{d}) = H(\hat{x}, \boldsymbol{d})$.

In a *second interpretation* σ^2 has to be calibrated so as to represent the unknown measurement-model error. After least square minimisation (Equation (6.17)) yielding $\hat{x}$, σ^2 is estimated by putting to zero the derivative of (Equation (6.19)) with respect to σ^2 which leads classically to the mean square of residuals for the MLE $\hat{\sigma}$:

$$\partial LL[(y_{mj}, \boldsymbol{d_j})_j \mid (x, \mu, \hat{\sigma}^2)]/\partial\sigma^2 = 0 \quad \Leftrightarrow -\frac{n}{2\hat{\sigma}^2} + \frac{1}{\hat{\sigma}^4} C(x, \Xi_n) = 0$$
$$\Leftrightarrow \hat{\sigma} = \sqrt{\frac{2}{n} C(x, \Xi_n)} = \sqrt{\frac{1}{n} \sum_j (y_{mj} - H(x, \boldsymbol{d_j}) - \mu)^2} \tag{6.30}$$

In what can be viewed as a more ***complete model calibration***, σ^2 is not only part of the identification model, but would also remain normally as part of the prediction model after calibration.

$$y_{mj} = H(\hat{x}, \boldsymbol{d}) + u_j$$
$$Z = G(x, \boldsymbol{d}) = H(\hat{x}, \boldsymbol{d}) + U, \quad U \sim N(\mu, \hat{\sigma}^2) \tag{6.31}$$

In both cases, estimation of the bias is undertaken as the residual empiric bias.

$$\partial LL[(y_{mj}, \boldsymbol{d_j})_j | (x, \mu, \hat{\sigma}^2)]/\partial\mu = 0 \Leftrightarrow \mu = \frac{1}{n} \sum_j (y_{mj} - H(x, \boldsymbol{d_j})) \tag{6.32}$$

Looking to variance decomposition, the raw variance of the observed sample is accounted for as follows:

$$\begin{aligned}\mathrm{var}(y_{mj})_j &= \overline{(y_{mj} - \overline{y_{mj}})^2} = \overline{\left(H_j(\hat{x}_m) - \overline{H_j(\hat{x}_m)}\right)^2} + \overline{\left(H_j(\hat{x}_m) - y_{mj}\right)^2} + \overline{(y_{mj} - H_j(\hat{x}_m))^2} \\ &= m^2 + \hat{\mu}^2 + \hat{\sigma}^2\end{aligned}$$

The first term represents the variance of the model prediction m^2, that is the square variation of the output of interest because of (known) varying experimental conditions as reflected in the (deterministic) model: the well-known r^2 coefficient computes the proportion of such explained variance to the overall sample variance.

$$r^2 = \frac{\overline{\left(H_j(\hat{x}_m) - \overline{H_j(\hat{x}_m)}\right)^2}}{\overline{\left(y_{mj} - \overline{y_{mj}}\right)^2}} = 1 - \frac{\hat{\mu}^2 + \hat{\sigma}^2}{\mathrm{var}(y_{mj})_j}$$

The second term accounts for the potential bias. The third term represents the residual error: under the first interpretation, it is later neglected within the prediction model as representing measurement error; under the second interpretation, it estimates the residual noise to be added to the mean model in order to account for its uncertain predictions.

In both cases, under convenient assumptions that include identifiability (see Jennrich, 1969; Hoadley, 1971) the estimators are asymptotically Gaussian and their variances may be (asymptotically) estimated to measure their accuracy similar to the discussion in Chapter 5 on epistemic uncertainty:

$$\mathrm{var}\,\hat{\boldsymbol{\theta}} = \mathrm{var}\begin{bmatrix}\hat{x} \\ \hat{\mu} \\ \hat{\sigma}^2\end{bmatrix} \cong \frac{1}{n}\hat{I}_F^{-1} \cong \hat{V}_n = \left[-\frac{\partial^2 LL}{\partial\theta\partial\theta'}\right]^{-1} \tag{6.33}$$

Closed-form expressions can be derived for this variance matrix, even in the non-linear case. Noting $H_j = H(\hat{x}, \boldsymbol{d_j})$ $H_j'' = \partial^2 H/\partial x^2|_{x=\hat{x}, d=\boldsymbol{d_j}}$ and considering the case without bias, first-order conditions result in somewhat simple expressions when computed at the estimator values $(\hat{x}, \hat{\sigma}^2)$, assuming that the residual is non-zero (refer to Annex Section 10.5.2 for a demonstration):

$$\left.\frac{\partial^2 LL}{\partial x^2}\right|_{\hat{x},\hat{\sigma}} = -\frac{1}{\hat{\sigma}^2}\sum_j \left[H_j''(\hat{x})\left(H(\hat{x},\mathbf{d}_j)-y_{mj}\right) + H_j'^2(\hat{x})\right] = -\frac{n}{\hat{\sigma}^2}\overline{H_j''\left(H_j-y_{mj}\right)+H_j'^2}$$

$$\left.\frac{\partial^2 LL}{\partial(\mu)^2}\right|_{\hat{x},\hat{\sigma}} = -\frac{n}{2\hat{\sigma}^2};\quad \left.\frac{\partial^2 LL}{\partial x\partial(\mu)}\right|_{\hat{x},\hat{\sigma}} = -\frac{n}{2\hat{\sigma}^2}\overline{H_j'} \tag{6.34}$$

$$\left.\frac{\partial^2 LL}{\partial(\sigma^2)^2}\right|_{\hat{x},\hat{\sigma}} = -\frac{n}{2\hat{\sigma}^4};\quad \left.\frac{\partial^2 LL}{\partial x\partial(\sigma^2)}\right|_{\hat{x},\hat{\sigma}} = 0;\quad \left.\frac{\partial^2 LL}{\partial \mu\partial(\sigma^2)}\right|_{\hat{x},\hat{\sigma}} = 0$$

In particular, for the parameter identification case, where σ^2 is considered to be well known and the bias neglected:

$$\operatorname{var}\hat{x} \cong \left[-\left.\frac{\partial^2 LL}{\partial x^2}\right|_{\hat{x},\hat{\sigma}}\right]^{-1} = \sigma^2\left[\frac{\partial^2 C}{\partial x^2}\right]^{-1} = \frac{\sigma^2}{n}\overline{H_j''\left(H_j-y_{mj}\right)+H_j'^2}^{-1} \tag{6.35}$$

Such variances for the estimators quantify the estimation inaccuracy:

(a) Estimation inaccuracy increases with the observational noise σ^2, be it purely measurement noise, or model-measurement residuals.
(b) Estimation inaccuracy decreases with the sample size at the classical speed *1/n*.
(c) In the linear case where H_j'' is zero, the variance of x is inversely proportional to the quadratic mean of the derivatives $\frac{1}{n}\sum_j H_j'^2(x)$. Interpretable as the mean square sensitivity of the model prediction to x, the parameter to be calibrated. Intuitively enough, accuracy increases with model sensitivity.
(d) The estimator of the unknown parameter and that of the measurement-model error are (asymptotically) uncorrelated, their covariance being null.

It is essential to remark here that the underlying statistical model assumes that $\mathbf{x}$ *has just one physically accurate value* $\mathbf{x_o}$, albeit unknown before calibration, that would be estimable with limitless precision given a very large number of measurements. In fact, given reasonable assumptions on the regularity of H, $\hat{x}$ converges almost surely to x_o. Its variance should be seen as the consequence of the *lack of knowledge only* (or indeed measurements), and not the physical variability of the material property x for which there is supposedly just one correct value. It represents an epistemic uncertainty attached to the estimation incompleteness. Indeed, providing some physically-reasonable assumptions on H (twofold differentiable with bounded non-zero values for its first derivatives) it can be proved to be tending to zero as the number of observations n increase.

Consider the variance of $\hat{x}$:

$$\hat{V}_n^x = \operatorname{var}\hat{x} \cong \frac{\hat{\sigma}^2}{n}\cdot\frac{1}{\frac{1}{n}\sum_j\left[H_j''(\hat{x})\left(H(\hat{x},\mathbf{d}_j)-y_{mj}\right)+H_j'^2(\hat{x})\right]} \tag{6.36}$$

The denominator is a mean of local derivatives. For it to tend to zero it is enough that, for any sample size n:

(a) the mean residual ($\hat{\sigma}^2$) remains bounded superiorly;
(b) the terms summed in the denominator remain bounded inferiorly.

Note that the variance of $\hat{\sigma}^2$ tends to zero by simply assuming the first of those two conditions. Note also that in the first interpretation (simple parameter identification), this condition drops because σ^2 is considered to be well known, and thus fixed whatever the sample size.

These conditions appear to be physically sound. The first condition simply says that the mean residual does not increase too much, that is that the model is close enough to observations on average, that is that (infinitely) *gross prediction errors remain rare throughout the sample*. The second condition is easier to grasp in the linear case, that is H is linear with respect to x, but not necessarily to $\boldsymbol{d}_j$. In that case H_j'' is zero and the denominator is simply the quadratic mean of the derivatives $\frac{1}{n}\sum_j H_j'^2(x)$, that is model sensitivities, as already explained. Assuming that *most observations do bring significant information* translates into the fact that *for most j values the sensitivity is superior to a minimal bound*. Thus, the denominator, the mean of terms all positive and mostly superior to a given bound also remains inferiorly bounded (say by half the bound).

Note that a non-linear case can be turned into a linear one through an adaptive linearisation technique. Linearise on some starting value for x, solve the subsequent linear problem, re-linearise around the linear solution and so on: the technique will be described in greater detail in Section 6.5.3. In summary:

(a) In simple parameter identification, the representation of uncertainty is limited to the purely epistemic calibration of the imprecision of x, tending necessarily to zero when more data is made available.

(b) In the second interpretation, complete model calibration, besides the epistemic estimation uncertainty of x, $\hat{\sigma}^2$ carries some model uncertainty within the predictive model (Equation (6.8)). There is no reason in general to anticipate that such model uncertainty will decrease with more data. However, unlike that which will be introduced below, this model uncertainty is not specifically apportioned to any intrinsic variability of the physical property.

Generalisations to heteroscedastic cases Note also that the development introduced above is generalised classically to the case of the *heteroscedasticity* of the residual measurement-model error. This refers to the situation where it is necessary to account for a varying accuracy of the model with respect to the observations or when the experiments involve various captors or imprecision depending upon the environmental conditions $\boldsymbol{d}$. Think of temperature measurements influenced by ambient temperature, or flow measurement inaccuracy increasing with the water level. A heteroscedastic formulation means then that variance for each observation σ_j^2 can vary instead of being always σ^2, so that the initial probabilistic model (Equation (6.18)) is slightly changed (assuming it to be unbiased for the sake of simplicity):

$$\left.\begin{aligned} Y_{mj} &= H(x, \boldsymbol{d}_j) + U_j \\ U_j &\sim N\left(0, \sigma_j^2 = (\gamma_j \sigma)^2\right) \end{aligned}\right\} \tag{6.37}$$

whence log-likelihood maximisation (Equation (6.19)) becomes:

$$\begin{aligned} LL\left[(y_{mj}, \boldsymbol{d}_j)_j \mid \left(x, (\sigma_j^2)\right)\right] &= -\frac{1}{2}\sum_j \log(2\pi\sigma_j^2) - \frac{1}{2}\sum_j \frac{1}{\sigma_j^2}\left(y_{mj} - H(x, \boldsymbol{d}_j)\right)^2 \\ &= -\frac{1}{2}\sum_j \log(2\pi\sigma_j^2) - C_h(x, \Xi_n) \end{aligned} \tag{6.38}$$

Regarding x, the maximisation results in a new least square cost function C_h appearing as a weighed average, normalising the square residuals according to measurement inaccuracies.

$$C_h(x, \Xi_n) = \frac{1}{2\sigma^2}\sum_j \frac{1}{\gamma_j^2}\left(y_{mj} - H(x, \boldsymbol{d}_j)\right)^2 \tag{6.39}$$

This new estimation program has comparable properties of convergence, given reasonable properties for the given measurement variances, such as boundedness of the whole suite $(\sigma_j^{\ 2})_j$. As with the homoscedastic model under its first interpretation, an asymptotic variance can be computed as in Equation (6.33) and tends to zero when the sample size becomes infinite, thus there is only epistemic uncertainty.

The *linearised case* is then:

$$LL[(Y_{mj}, \boldsymbol{d}_j)_j \mid (x_m, (\sigma_j^2)_j)] = -\frac{1}{2}\sum_j \log(2\pi\sigma_j^2) - \frac{1}{2}\sum_j \frac{1}{\sigma_j^2}\left(y_{mj} - H_j'(x_m - x_o) - H_j^o)\right)^2 \tag{6.40}$$

which still results in closed-form MLE, including also the variable residual variances within the weighted average in a physically-intuitive way: the higher the variance (inaccuracy), the lesser the weight attached to the j-th observation:

$$\left.\begin{aligned} \hat{x}_m &= x_o + \frac{\sum_j \frac{H_j'}{\gamma_j^2}\left[y_{mj} - H_j^o\right]}{\sum_j \left(\frac{H_j'}{\gamma_j}\right)^2} \\ \hat{\sigma}^2 &= s_u^2 = \frac{1}{n}\sum_j \frac{1}{\gamma_j^2}\left[y_{mj} - H_j'\left(\hat{x}_m - x_o\right) - H_j^o\right]^2 \end{aligned}\right\} \tag{6.41}$$

In the *non-linear heteroscedastic case*, the MLE of σ^2 is still the mean square residuals weighted by relative accuracies of observations; the MLE of x no longer has a closed-form. However, the first-order condition still yields a straightforward physical interpretation:

$$\partial LL\left[(y_{mj}, \boldsymbol{d}_j)_j \middle| x, (\sigma_j^2)_j\right] / \partial x = 0 \Leftrightarrow \sum_j \frac{H_j'(x)}{\gamma_j^2}\left(y_{mj} - H(x, \boldsymbol{d}_j)\right) = 0 \tag{6.42}$$

The optimal value for x is that which leads to a null average residual, weighted by the model sensitivities (H_j') and accuracies $(1/\gamma_j^2)$ of the observations.

6.2.2 Bayesian updating, identification and calibration

The *Bayesian updating* formulations of Equation (6.17) with prior and posterior variance influenced by expertise on top of the data sample are current in the data assimilation or parameter identification literature, under the keywords of Bayesian identification or updating (see Beck on structural dynamics and seismic analysis).

In such a formulation, it is assumed that some prior expert knowledge is available. One thus infers that the likely values for the unknown parameter x follows a given (prior) distribution $\pi_o(x)$.

While slightly modifying Equation (6.19) and that of the estimator variance with the addition of a term representing the impact of prior expertise, this does not alter that fundamental difference regarding the type of uncertainty covered.

Linearised gaussian case with Known Variance To simplify, suppose such a prior distribution is Gaussian:

$$X \sim \pi_o = N(x_b, b^2) \tag{6.43}$$

where b^2 stands for the prior expert uncertainty (variance) about the true value of x, while x_b represents the expert's prior best-estimate. Consider the Gaussian linearised (heteroscedastic) observational model:

$$\left.\begin{array}{l} Y_{mj}|x = H(x, \mathbf{d_j}) + U_j \\ U_j \sim N(0, \sigma_j^2) \end{array}\right\} \tag{6.44}$$

Then, the Bayesian (posterior) estimate for x, including both expertise and data sample, appears to be the following:

$$\hat{x}_m = x_o + \frac{\frac{1}{b^2}[x_b - x_o] + \sum_j \frac{H_j'}{\sigma_j^2}[y_{mj} - H_j^o]}{\frac{1}{b^2} + \sum_j \left(\frac{H_j'}{\sigma_j}\right)^2} \tag{6.45}$$

with the posterior uncertainty (variance) being:

$$a^2 = \left[\frac{1}{b^2} + \sum_j \left(\frac{H_j'}{\sigma_j}\right)^2\right]^{-1} \tag{6.46}$$

This is easily demonstrated through the consideration of the posterior distribution of x (i.e. the combined likelihood of the observed sample and the prior expertise):

$$\pi_1(x) = f(x|\Xi_n) = \frac{\prod_j f_y(y_{mj}|x).\pi_0(x)}{\int_x \prod_j f_y(y_{mj}|x).\pi_0(x)dx} \tag{6.47}$$

The log of such distribution in the Gaussian case yields:

$$Log\, f[(x|\Xi] = -\frac{1}{2}\sum_j \frac{1}{\sigma_j^2}\left(y_{mj} - H_j'(x - x_o) - H_j^o\right)^2 - \frac{1}{2b^2}(x - x_b)^2 - Cst(x) \tag{6.48}$$

where the remaining terms do not depend upon x. Maximising such a function provides the *mode* of the posterior distribution which also equals the *posterior expectation* since the model is Gaussian. Thus, the Bayesian estimate appears to be the result of the optimisation of a compound cost-function aggregating both a quadratic *distance to data* and a quadratic *distance to the prior best-estimate*.

Supposing additionally that the experimental conditions $\mathbf{d}_j$ do not vary much so that H_j' equal to H' and supposing homscedastic measurement error σ^2, the first estimator simplifies into:

$$\hat{x}_m = x_o + \frac{\frac{1}{b^2}[x_b - x_o] + n\frac{H'}{\sigma^2}\left[\overline{y_m - H^o}\right]}{\frac{1}{b^2} + n\frac{H'^2}{\sigma^2}} = x_o + \frac{\sigma^2[x_b - x_o] + nH'^2 b^2 \frac{1}{H'}\left[\overline{y_m - H^o}\right]}{\sigma^2 + nH'^2 b^2} \tag{6.49}$$

$$a^2 = \frac{\sigma^2 b^2}{\sigma^2 + nH'^2 b^2} \tag{6.50}$$

These formulae have straightforward physical interpretations:

(a) The optimal value $\hat{x}_m$ for the model input is the weighted average between observed data and prior expertise.
(b) The larger the observed sample (n), the model sensitivity (H'), the prior inaccuracy (b^2), or the measurement accuracy ($1/\sigma^2$) the less influential is the prior best-estimate (x_b). And conversely …

(c) The posterior variance, equal to prior variance when the data sample is non-existent, tends to zero when data sample grows.
(d) The classical estimator is identical to the Bayesian one supposing that prior variance is infinite (or very large).

Generalisations: non-linear, unknown variance, non Gaussian The consideration of full Bayesian calibration, that is including also the updating of an unknown variance u with a corresponding prior, leads to more complex formulae, but still enjoys closed forms in the linear Gaussian case through the use of Gaussian-gamma conjugate distributions. In the non-linear or non-Gaussian case, the posterior distribution above needs to be computed using more sophisticated algorithms (see Section 5.4.4, e.g. MCMC) as closed-form solutions are no longer available.

6.2.3 An alternative identification model with intrinsic uncertainty

Suppose now that, for some physical reason, it is legitimate to infer that within the n-sample Ξ_n of measurements the material property to be identified does in fact change slightly. This may be because those measurements do involve a *population* of mechanical samples, with some inevitable manufacturing variability between them, albeit with a similar specification program. This means that on top of a prior lack of knowledge of the value of x, there is also physically some *intrinsic* variability. It would probably be desirable to include all of that in the prediction model $z = G(x, \mathbf{d})$ for other environmental conditions, in order not to miss a potentially important physical behaviour (or alternatively to prove that it is not important).

Assuming a probabilistic setting is acceptable in order to describe not only the measurement error but also the intrinsic physical variability. One may consider the following new model:

$$\left.\begin{aligned} Y_{mj} &= H(X_j, \boldsymbol{d}_j) + U_j \\ X_j &\sim f_X(x|\boldsymbol{\theta}_x) \quad iid \\ U_j &\sim N(\mu, \sigma_j^2) \end{aligned}\right\} \tag{6.51}$$

Note that x is no longer a *parameter* in the sense of having an unknown but fixed true value; it is a *random variable* with different x_j realisations for each measurement, following i.i.d. an unknown distribution (or a known distribution with unknown parameters θ_x). Residuals are assumed to be normal, independent, with known but potentially heteroscedastic variances. This new model proves to have very different – and in practice interesting – properties than the previous ones.

The linearised Gaussian case To begin to understand this, suppose that X follows a normal distribution around the unknown expectation x_m with an unknown variance v^2, and that H is linear with respect to x, and that the residuals are homoscedastic. Suppose also that the model is unbiased. Then Equation (6.51) becomes more explicit:

$$\left.\begin{aligned} &X_j \sim N(x_m, v^2) \\ &U_j \sim N(0, \sigma^2) \\ &Y_{mj} = H(X_j, \boldsymbol{d}_j) + U_j = H_j'.(X_j - x_o) + H_j^o + U_j \end{aligned}\right\}$$
$$\Rightarrow Y_{mj} \sim N(H_j'.(X_j - x_o) + H_j^o, H_j'^2 v^2 + \sigma^2) \tag{6.52}$$

Log-likelihood becomes:

$$LL[(y_{mj}, \boldsymbol{d_j})_j | x_m, v^2] = -\frac{n}{2}\log(2\pi) - \frac{1}{2}\sum_j \log\left(\sigma^2 + H'^2_j v^2\right) - \frac{1}{2}\sum_j \frac{1}{\sigma^2 + H'^2_j v^2}\left(y_{mj} - H'_j(x_m - x_o) - H^o_j\right)^2 \tag{6.53}$$

Suppose now that experimental conditions $\boldsymbol{d}_j$ do not vary by much so that H'_j are approximately constantly equal to H'. Simple algebra leads then to the following closed-form maximum likelihood estimators:

$$\left.\begin{aligned} \hat{x}_m &= x_o + \frac{1}{H'}\left[\overline{y_m - H^o}\right] \\ \hat{v}^2 &= \frac{1}{H'^2}\left[\overline{\left(y_m - H'(\hat{x}_m - x_o) - H^o_j\right)^2} - \hat{\sigma}^2\right] \end{aligned}\right\} \tag{6.54}$$

It is easy to see that x_m equates to the x of linear parametric identification (Equation (6.20)). Its estimation variance appears as follows:

$$\operatorname{var}\hat{x}_m \cong \frac{1}{n}\overline{\left(y_m - H'(\hat{x}_m - x_o) - H^o_j\right)^2} = \frac{1}{n}\left[v^2 + \frac{\sigma^2}{H'^2}\right] \tag{6.55}$$

The main difference with calibration comes with a new variance component: the intrinsic variance v^2 of X proves to be very different to the variance given by Equation (6.33) as it is now irreducible, that is non-decreasing to zero with n. Decomposition of the raw variance of the observed sample now accounts for this additional component:

$$\begin{aligned} \operatorname{var}(y_{mj})_j &= \overline{\left(H'(\hat{x}_m - x_o) + H^o_j - \overline{y_{mj}}\right)^2} + \overline{\left(y_m - H'(\hat{x}_m - x_o) - H^o_j\right)^2} \\ &= \overline{\left(H'(\hat{x}_m - x_o) + H^o_j - \overline{y_{mj}}\right)^2} + H'^2 v^2 + \sigma^2 \end{aligned} \tag{6.56}$$

The variance explained by this new model incorporates not only the deterministic part $(H'(\hat{x}_m - x_o) + H^o_j - \overline{y_{mj}})^2$ (i.e. the variance explained by the *mean* linearised model), but also a proportion of output variance $H'^2 v^2$ which originates in the intrinsic variance of X. An extended coefficient r_e^2 can now compute the proportion of both:

$$r_e^2 = \frac{\overline{\left(H'(\hat{x}_m - x_o) + H^o_j - \overline{y_{mj}}\right)^2} + H'^2 v^2}{\operatorname{var}(y_{mj})_j} \tag{6.57}$$

In such a case, the combined identification of v^2 and of σ^2 (or rather of $H'^2 v^2$ and σ^2 so as to remain physically-homogeneous) is a non-identifiable problem as there is nothing within the mere sample of observations to fix their proportions in the residual variance of the measured output after removing the variance of the mean model. Note conversely that one may want to calibrate the mean input x_m of a model with known intrinsic variance v^2: the estimator for x_m would remain the same, but then the variance decomposition would yield an estimator for the residual noise $\hat{\sigma}^2$.

When the H'_j vary significantly the more general formula for x_m and for v^2 loses any closed-form, and is therefore different from the traditional calibration program. Numerically, a major difference appears between this maximisation program and the others. When H_j' vary, Equation (6.53) does not involve a quadratic sum with known weights to be minimised: the denominator of the weights includes an unknown to be estimated (v^2). This will become even more difficult in higher dimensions (see Section 6.4). While becoming implicit, the residual variance after subtraction of the variance of the mean model continues to be explained both by the intrinsic and measurement variances:

$$\operatorname{var}(y_{mj})_j - \overline{\left(H_j'(\hat{x}_m - x_o) + H_j^o - \overline{y_{mj}}\right)^2} = \overline{\left(y_m - H_j'(\hat{x}_m - x_o) - H_j^o\right)^2} \tag{6.58}$$

Generalisations The *non-linear* case becomes here much more complex because the expectation and variance resulting from the transformation by H_j of that of X has no closed-form in general, as shown by the log-likelihood:

$$LL\left[(Y_{mj}, \mathbf{d}_j)_j / (x_m, v^2)\right] =$$
$$-\frac{n}{2}\log(2\pi) - \frac{1}{2}\sum_j \log(\sigma^2 + \operatorname{var} H_j(X)) - \frac{1}{2}\sum_j \frac{1}{\sigma^2 + \operatorname{var} H_j(X)} \left(y_{mj} - EH_j(X) - \mu\right)^2 \tag{6.59}$$

Bayesian formulations may be developed as in the case of calibration (Section 6.2.2) although this becomes more complex.

6.2.4 Comparison of the algorithms

Table 6.2 summarises a number of features distinguishing the algorithms reviewed above in the 1D-case.

A rational modelling process would typically chain gradually the increasingly sophisticated algorithms:

- Pure parameter identification concentrates on estimating the empiric data so as to identify a *mean* or representative value for the model parameter. As non-linear identification lacks closed-form, linear formulations help to understand how such a value is found. Roughly speaking, it consists of dividing the mean output by the mean gradient and the accuracy of such identification increases at $\sqrt{n}$ speed with the data sample.
- Full calibration yields the same result for the parameter value, but adds on a formal estimate for model-measurement error. Linearisation again yields closed-form results. Of course, a thorough check should then be made of the shape of the residual to make sure such a statistical model of model error makes sense, including a breakdown of the overall empiric sample into a learning sample and a remaining validation sample. Bias and a heteroscedastic formulation of residual variance may be estimated if necessary.
- Intrinsic variability identification consists of apportioning at least partially the residual model-measurement variance into input variance. Thus, the input parameter gets not only an estimated expected value, but also a whole (Gaussian) distribution. The estimate for the central value is generally different to that of the two previous algorithms though it remains identical if the gradient is constant and may result in practice as close (see the flood example). A choice has to be made as to the amount of non-explained variance σ^2 which becomes non-identifiable if an intrinsic variance is identified.

Sections 6.3 and 6.4 will develop n-dimensional generalisations of the previous results: salient differences will be proven to remain, although identifiability issues become quite complex.

Table 6.2 Compared features of the inverse algorithms in 1D.

	Pure parameter identification (or data assimilation)	Model calibration	Intrinsic variability identification (full probabilistic inversion)
Cost function	$\frac{1}{2}\sum_j \left(y_{mj}-H(x,d_j)\right)^2$	$\frac{n}{2}\log(\sigma^2)+$ $\frac{1}{2\sigma^2}\sum_j \left(y_{mj}-H(x,d_j)\right)^2$	$\frac{n}{2}\log(\sigma^2+\mathrm{var}H(X,d_j))+$ $\sum_j \frac{1}{2[\sigma^2+\mathrm{var}H(X,d_j)]}\left(y_{mj}-E\left[H(X,d_j)\right]\right)^2$
Outputs of the identification	$\hat{x}_m$	$\hat{x}_m;\ \hat{\sigma}^2$	$\hat{x}_m \cong EX;\ \hat{v}^2 \cong \mathrm{var}\,X$ σ^2 has to be fixed to ensure identifiability
Variance and natures of uncertainty	Estimation var. (reducible/data)	Residual var. $\hat{\sigma}^2$ (reducible/model) Estimation var. (reducible/data)	Intrinsic var. $\hat{v}^2$ (irreducible) Residual var. σ^2 (reducible/model) Estimation var. (reducible/data)
r^2 (variance explained)	$\frac{\overline{\left(H_j(\hat{x}_m)-\overline{H_j(\hat{x}_m)}\right)^2}}{\mathrm{var}(y_{mj})}$	Idem	Adds the output image of $\hat{v}^2$
Estimators in the *linearised* case with *constant gradient* H'_j	$\hat{x}_m = x_o + \frac{1}{H'}\left[\overline{y_m - H^o}\right]$	$\hat{x}_m = x_o + \frac{1}{H'}\left[\overline{y_m - H^o}\right]$ $\hat{\sigma}^2 = s_u^2 = \overline{\left(y_m - H'(\hat{x}_m - x_o) - H_j^o\right)^2}$	$\hat{x}_m = x_o + \frac{1}{H'}\left[\overline{y_m - H^o}\right]$ $\hat{v}^2 = \frac{1}{H'^2}\left[\overline{\left(y_m - H'(\hat{x}_m - x_o) - H_j^o\right)^2} - \hat{\sigma}^2\right]$
Idem, with varying gradients H'_j	$\hat{x}_m = x_o + \frac{\overline{H'_j\left[y_{mj}-H_j^o\right]}}{\overline{(H'_j)^2}}$	$\hat{x}_m = x_o + \frac{\overline{H'_j\left[y_{mj}-H_j^o\right]}}{\overline{(H'_j)^2}}$ $\hat{\sigma}^2 = s_u^2 = \overline{\left(y_m - H'_j(\hat{x}_m - x_o) - H_j^o\right)^2}$	No closed form
Variance of the estimators (same case)	$\mathrm{var}\,\hat{x}_m \cong \frac{\sigma^2}{n}\overline{H'^2_j}^{-1}$	$\mathrm{var}\,\hat{x}_m \cong \frac{\sigma^2}{n}\overline{H'^2_j}^{-1}$ $\mathrm{var}\,\hat{\sigma}^2 \approx \frac{2\hat{\sigma}^4}{n}$	No closed form

(NB: to simplify, all cases were assumed to be homoscedastic and unbiased, and constant terms dropped from the cost functions).

6.2.5 Illustrations in the flood example

The various algorithms were tested on the basis of a bivariate 123-sample Ξ_{123} representing flow-water height measurements (see Figure 6.2), against which the flood model expressed in water height is first calibrated in K_s only, Z_m and Z_v being taken at their best-estimate values: the results are given in Table 6.3.

The results in the linearised case whereby the gradient is computed for each $\boldsymbol{d}_j$ are very close to full non-linear optimation, with 10 times less calls to the system model provided that the linearisation is close to the ($x_o = 30$). Even more economically, the results are close in the linearised case whereby the

Table 6.3 Results of the inverse algorithms on the bivariate flow-water height sample.

	Pure parameter identification (or data assimilation)	Model calibration	Intrinsic variability identification (full probabilistic inversion)
Outputs of the linearised algorithms	Linearised in $x_o = 15$: $\hat{x}_m = 22.4$ (constant gradient) and 22.7 (varying gradients). Linearised in xo = 30: $\hat{x}_m = 26.68$ (constant gradient) and 26.63 (varying gradients)	Identical to identification σ^2 is higher than for the linear case with constant gradient (0.75^2 for $x_o = 15$), but almost identical with varying gradients or for a fair linearisation point ($x_o = 30$)	Linearised in $x_o = 30$: $\hat{x}_m = 26.6$ and $\hat{v}^2 = 11.5^2$ for $\sigma^2 = (0.66/2)^2$. $\hat{x}_m = 26.6$ and $\hat{v}^2 = 12.6^2$ for $\sigma^2 = (0.66/4)^2$ $c_V(x_m) = 5\%$, $c_V(v^2) = 14\text{–}20\%$, *with less than 1% of correlation between estimators.*
Number of calls for linearised algorithms	$124 = 123 + 1$ for constant gradients (computed once for the mean flow) and 123 different d_j $246 = 2^* 123$ for varying gradients	124 (constant gradients) 246 (varying gradients)	61 iterations, but all based on the initial $246 = 2^* 123$ gradients and central estimates
Outputs of the non linear algorithms	$\hat{x}_m = 27.0$	$\hat{x}_m = 27$ $\sigma^2 = 0.66^2$	$\hat{x}_m \sim 30$ and $\hat{v}^2 \sim 8.5^2$ for $\sigma^2 = (0.66/2)^2$
Variances	$\mathrm{Var}x_m \sim 1$, $c_V = 4\%$	$\mathrm{Var}x_m \sim 1$, $c_V = 4\%$ $\mathrm{Var}\sigma^2 \sim 3.10^{-3}$, $c_V = 13\%$	More difficult to estimate, say in order of magnitude, $c_V(x_m) >$ 15%, $c_V(v^2) > 30\%$
Number of calls[a] to $H(.)$	$\sim 3{,}5.10^3 = 25\text{–}30$ iterations* 123 different d_j	$\sim 3{,}5.10^3$	$\sim 2.10^6 = 30$ iterations* 500 random samples* 123 different d_j

[a]The optimisation algorithm used was the default one suggested by the R statistical software (Nelder and Mead, 1965), simplex-based: it only calls the function and does not use the gradient for descent. Add to those figures around ~5 additional calls to compute the Hessian by finite differences for getting epistemic uncertainty. According to the cases there is a (limited) dependence on starting points in the number of iterations.

gradient is only computed for the mean $\boldsymbol{d}_j$, although with a slightly higher residual variance. These observations, of course, attach to the particular flood example.

In the example, the biased formulations for calibration were proven not to change the result by much though they increased the variance of the estimators. Indeed, either non-linear or linearised (with a fair starting point) calibration assuming no bias resulted, for instance, in comparable output biases of only $\sim 3.\ 10^{-3}$ (the figure is to be compared with the output standard deviation of *0.66*). Biased formulation reduced only slightly the output bias with a modification of the calibration x_m from *26.6* to *27.3* although its variance increased greatly.

Intrinsic variability identification apportions an increasing share of observed variability into input variance as the share of measurement noise σ^2 is reduced with respect to the calibration residual of 0.66^2. While the value of x_m remains stable and close to the calibration results, multiple solutions are thus possible for v^2. It is up to the analyst to choose the model, hopefully with external information about the order of magnitude of measurement uncertainty. Assuming that such metrological uncertainty represents $\sigma^2 \sim (0.66/2)^2$, the *linearised solution* appears as $\hat{x}_m = 26.6$ and $\hat{v}^2 = 11.5^2$. Since the variance of Z_m and Z_v is known elsewhere, it is tempting to explain a part of the input variance through the contribution of such variance to that of Y_{mj}. This can be done through a slight modification of the linearised likelihood:

$$LL\left[(y_{mj}, \boldsymbol{d}_j)_j | x_m, v^2\right] = -\frac{n}{2}\log(2\pi) - \frac{1}{2}\sum_j \log\left(\sigma^2 + H'^2_j v^2\right) - \frac{1}{2}\sum_j \frac{1}{\sigma^2 + H'^2_j v^2}\left(y_{mj} - H'_j(x_m - x_o) - H^o_j\right)^2 \quad (6.60)$$

Introducing, besides the unknown $H'^2_j v^2$, a known contribution to the variance of Y_{mj}:

$$w_j^2 = \operatorname{var}\left(H(x^o, Z_v, Z_m, \boldsymbol{d}_j)\right) \sim (dH_j/dz_v)^2 \operatorname{var} Z_v + (dH_j/dz_m)^2 \operatorname{var} Z_m$$

so that the modified likelihood accounting for an approximate contribution of the variance to the out variance appears as follows:

$$\begin{aligned} LL\left[(y_{mj}, \boldsymbol{d}_j)_j | x_m, v^2\right] &= -\frac{n}{2}\log(2\pi) - \frac{1}{2}\sum_j \log(\sigma^2 + H'^2_j v^2 + w_j^2) \\ &\quad - \frac{1}{2}\sum_j \frac{1}{\sigma^2 + H'^2_j v^2 + w_j^2}\left(y_{mj} - H'_j(x_m - x_o) - H^o_j\right)^2 \end{aligned} \quad (6.61)$$

Unfortunately, the computations show that such an effect contributes merely $w_j^2 \sim 0.01\text{-}0.02$ with respect to an average total output variance of $0.66^2 = 0.44$, because of the relatively low importance of variables Z_m and Z_v as evidenced by their sensitivity indices (see Section 6.3.7). Assuming again $\sigma^2 \sim (0.66/2)^2$, it therefore only reduces slightly, for instance, $\hat{v}^2 = 11.5^2$ down to 11.3^2, a change that is mostly negligible with respect to estimation uncertainty (the coefficient of variation for $\hat{v}^2$ can be assessed through a Hessian of the log-likelihood at 15–20%).

The *non-linear algorithm* involves costlier random sampling in order to estimate the log-likelihood at each iteration of optimisation (see Annex § 10.4.5 for the details of a process that required $1.8\ 10^6$ calls to the system model). Again, with $\sigma^2 = (0.66/2)^2$, the non-linear algorithm leads to a minimum of log-likelihood for a lower intrinsic variance for k_s, closer to 8^2–9^2, and a slightly higher value for the expectation of k_s, around 29–31, although more sampling would be required in order to stabilise and estimate the associated epistemic uncertainty.

Looking now at later prediction: the calibration ends up in generating the following double-level model predicting the water level conditional upon a given d (i.e. a given flow), when considering the unbiased case:

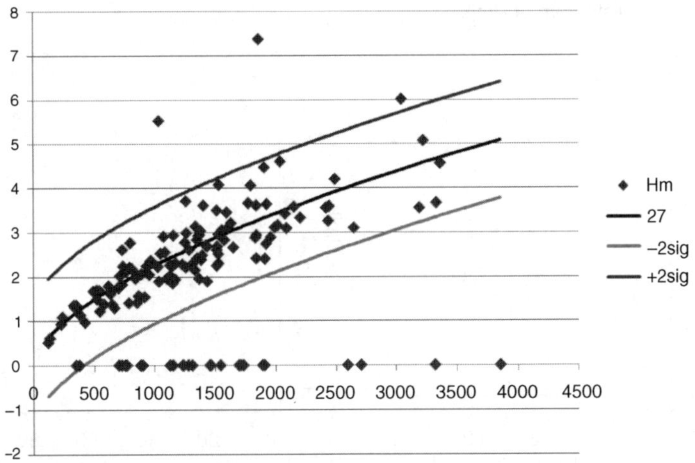

Figure 6.6 *Flood example – mean prediction and confidence bounds after linearised calibration.*

$$Z \mid \boldsymbol{d}, (\boldsymbol{\Theta}_x = x_m, \boldsymbol{\Theta}_u = \sigma^2) \sim N(H(x_o, \boldsymbol{d}) + \partial H/\partial x|_{x=x_o,d}(x_m - x_o), \sigma^2)$$
$$\boldsymbol{\Theta}_u \sim N\left(\hat{\sigma}^2, \frac{2\hat{\sigma}^2}{n}\right) \quad \boldsymbol{\Theta}_x \sim N(\hat{x}_m, \text{var}\,\hat{x}_m) \tag{6.62}$$

This may be represented in Figure 6.6 through confidence bounds around the mean prediction (averaged first-level normal quantiles over the second-level epistemic uncertainty): being Gaussian, ± 2 standard deviations give the approximate 95% confidence bounds.

The limitations of such a homoscedastic model are obvious regarding the variability for a given flow predicted to be uniform whatever the flow. Empiric evidence shows to the contrary that the variability of observations, that is the residual between the calibrated model and the observations (see Figure 6.6), tend to increase with flow.

The resulting histogram (Figure 6.7-left) is clearly asymmetric and a Shapiro-Wilks test unsurprisingly rejects the normality of the residuals without any hesitation (*p-value* $= 10^{-12}$). This is still the case even if one removes from the sample the two observations standing more than 3σ apart from the rest to be outliers (which is debatable): then *p-value* $= 0.004$. Such behaviour of the residual violates the basic assumptions of the calibration approach, both in the homoscedasticity (see Figure 6.7-right) and normality of variance: it should be corrected by a heteroscedastic version of the calibration algorithm, modelling the variance as a function of $\boldsymbol{d} = q$.

Alternatively, intrinsic variability identification ends up with the following modified model:

$$Z \mid \boldsymbol{d}, \boldsymbol{\Theta}_x = (x_m, v^2), \boldsymbol{\Theta}_u = \sigma^2$$
$$\sim N(H(x_o, \boldsymbol{d}) + \partial H/\partial x|_{x=x_o,d}(x_m - x_o), \sigma^2 + (\partial H/\partial x|_{x=x_o,d})^2 v^2)$$
$$\boldsymbol{\Theta}_u \sim N\left(\hat{\sigma}^2, \frac{2\hat{\sigma}^2}{n}\right) \quad \boldsymbol{\Theta}_x \sim N(\hat{x}_m, \text{var}\,\hat{x}_m) \tag{6.63}$$

Again, confidence bounds may be computed, now taking into account the fact that X is uncertain with variance v^2 and averaging over the second-level epistemic uncertainty.

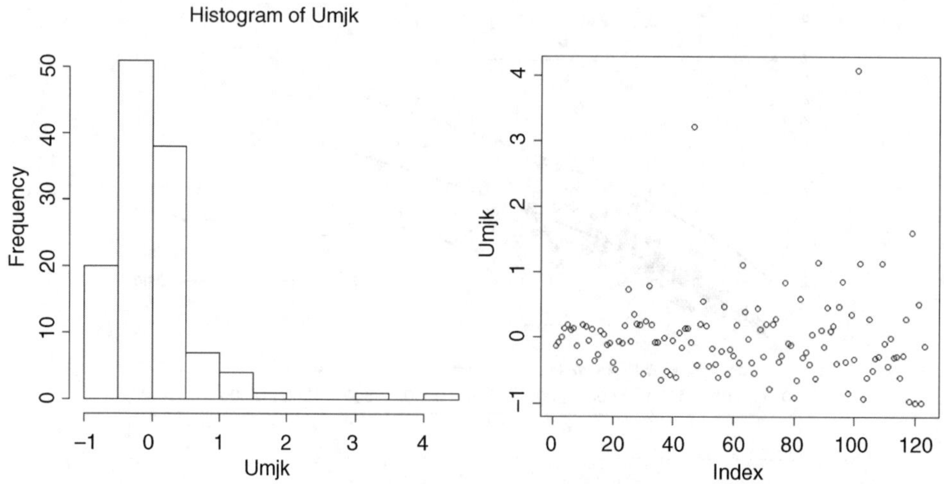

Figure 6.7 *Flood example – distribution of calibration residuals.*

Because uncertainty was apportioned to the input component, the distribution of uncertainty conditional upon a given flow increases with the flow in a much more satisfactory way, *cf.* Figure 6.8. Note, however, that the 95% confidence bounds may be slightly overestimated as there should be on average $5\%^{*}123 = 6$ samples outside such bounds, instead of the two observed ones.

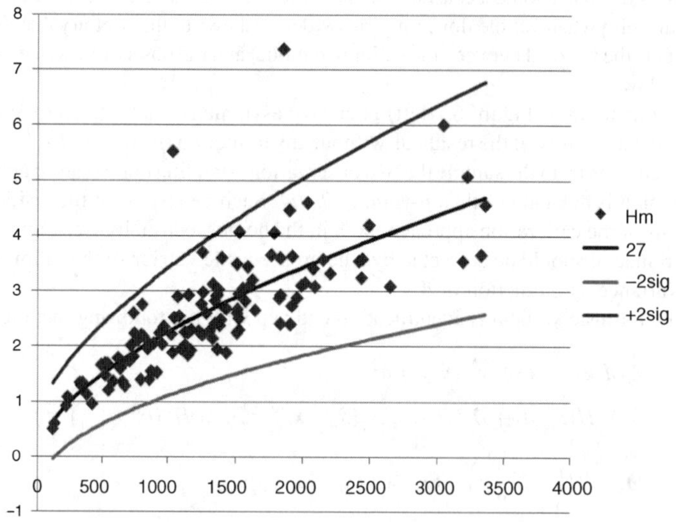

Figure 6.8 *Flood example – Confidence bounds and mean prediction after linearised intrinsic variability identification.*

6.3 The general structure of inverse algorithms: Residuals, identifiability, estimators, sensitivity and epistemic uncertainty

This section will generalise in the *multi*-dimensional case the settings introduced above, recalling firstly a number of generic features of the inverse problem. A multiple-dimensional generalisation is twofold:

- Multidimensionality of $\boldsymbol{x}$ (of dimension $p > 1$): suppose that more than one physical property, for instance, needs to be estimated within the model, vector $\boldsymbol{x}$ would then encompass all those.
- Multidimensionality of $\boldsymbol{y}$ (of dimension $q > 1$): suppose that more than one physical observable is measured at each measurement, vector $\boldsymbol{y}$ would then encompass all those.

6.3.1 The general estimation problem

As will be described in this chapter, most inverse algorithms responding to the above-mentioned motivations are included formally within the general model estimation program introduced in Chapter 2:

$$\left.\begin{array}{l} \boldsymbol{Y}_{mj} = \Pi_o \circ M_s(\boldsymbol{X}_j, \boldsymbol{d}_j, \boldsymbol{U}_j) \\ \boldsymbol{X}_j|\boldsymbol{\theta}_X \sim f_X(.|\boldsymbol{\theta}_X) \quad \boldsymbol{U}_j|\boldsymbol{\theta}_{uj} \sim f_U(.|\boldsymbol{\theta}_{uj}) \\ \boldsymbol{\Theta}_x, \left(\boldsymbol{\Theta}_{uj}\right)_j \sim \pi(.|\zeta) \\ \text{estimate} \quad \{M_s, \pi(.|\zeta)\} \\ \text{knowing} \quad \Xi_n = \left(\boldsymbol{y}_{mj}, \boldsymbol{d}_j\right)_{j=1...n} \\ \text{and} \quad IK = (f_x, f_U, \pi) \quad IM = (M_s)_s \end{array}\right\} \tag{6.64}$$

The following Table 6.4 provides a synoptic view of some of the most common variants according to the definitions of the components of the observational model and uncertainty distributions f_X , f_U (with potentially heterodastic residuals θ_{uj}) and π that will be studied in more detail in the remainder of this chapter. Each category can be used for most of the above-mentioned motivations, although the first and last categories would classically be involved for motivations (i) and (iv) respectively.

6.3.2 Relationship between observational data and predictive outputs for decision-making

Recalling the discussion in Chapter 2, some detail will be provided hereafter as to the potential scope of model inputs or outputs and their potential for linking estimation to prediction confidence. Two categories were distinguished within the potentially large number of raw outputs of the system model:

- the model output *variables of interest* that are important for the assessment of the risk measure: the vector $\boldsymbol{z} = (z^l)_{l=1...r}$
- the model *observable outputs*, upon which some data is available as they refer to physical variables or degrees of freedom that can be instrumented easily or observed albeit potentially noised by measurement error: the vector $\boldsymbol{y} = (y^k)_{k=1...q}$.

Table 6.4 Variants of the generic estimation program.

Type	Form of the observational model	Parametrisation of the uncertainty model	
		1st level «variability»	2nd level «epistemic»
Data assimilation/ parameter identification	$\Pi_o \circ M_s(.) = H(\boldsymbol{x}, d_j) + U_j$	$\boldsymbol{x} \sim \delta_{x=\theta_x}$ to be updated $\boldsymbol{U}_j \sim f_U(.\|R_j)$ with R_j known	$\boldsymbol{X} = \boldsymbol{\Theta}_X \sim \pi(.\|IK, \Xi)$ often taken Gaussian and noted as $N(x_a, A)$
Phenomenological modelling (e.g. biostats)/model calibration	$\Pi_o \circ M_s(.) = H_s(\boldsymbol{\theta}_x, d_j) + U_j$	$\boldsymbol{\theta}_x$ and $\boldsymbol{U}_j \sim f_U(.\|\theta_{uj})$ to be calibrated; s competing models to be selected	$(\boldsymbol{\Theta}_X, (\boldsymbol{\Theta}_{uj})_j) \sim \pi(.\|IK, \Xi)$
Full variability identification/ probabilistic inversion	$\Pi_o \circ M_s(.) = H(\boldsymbol{X}_j, d_j) + U_j$	$\boldsymbol{X}_j \sim f_X(.\|\boldsymbol{\theta}_x)$ $\boldsymbol{U}_j \sim f_U(.\|\boldsymbol{\theta}_u)$ Components of $(\boldsymbol{\theta}_x, \boldsymbol{\theta}_u)$ partially known, the rest need to be calibrated	$(\boldsymbol{\Theta}_X, \boldsymbol{\Theta}_U) \sim \pi(.\|IK, \Xi)$ $\boldsymbol{\Theta}_X$ generally of low dimension to keep identifiable

Both are outputs from a common system model:

$$\boldsymbol{x},\ \boldsymbol{d} \Rightarrow \boldsymbol{m} = M(\boldsymbol{x},\ \boldsymbol{d},\ \boldsymbol{u}) \tag{6.65}$$

formulated either in a predictive (or decision-making) model or in an observational (or estimation) model, respectively (Figure 6.9):

$$\boldsymbol{x},\ \boldsymbol{d} \Rightarrow \boldsymbol{z} = \Pi_d\, o\, M(\boldsymbol{x},\ \boldsymbol{d},\ \boldsymbol{u}) = G(\boldsymbol{x},\ \boldsymbol{d},\ \boldsymbol{u}) \tag{6.66}$$

$$\boldsymbol{x},\ \boldsymbol{d} \Rightarrow \boldsymbol{y} = \Pi_o\, o\, M(\boldsymbol{x},\ \boldsymbol{d},\ \boldsymbol{u}) = H(\boldsymbol{x},\ \boldsymbol{d},\ \boldsymbol{u}) \tag{6.67}$$

where Π_d and Π_o refer to deterministic operators that can be somewhat more complex than simple projections. In many risk studies, post-treatments are compounded with the raw model outputs $\boldsymbol{m}$ in ways that differ between the $\boldsymbol{z}$ and the $\boldsymbol{y}$. For Π_d, the variable of interest is defined as the maximal value of an

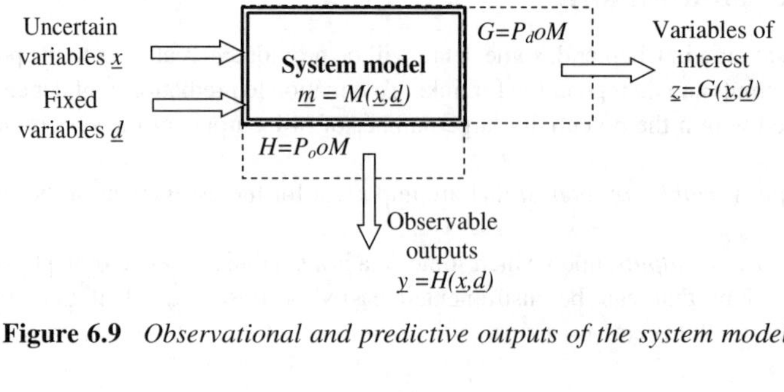

Figure 6.9 *Observational and predictive outputs of the system model.*

output spatial field $\boldsymbol{m}$, standardised to a reference scale: think of a fracture mechanics criterion being the minimal relative margin stress intensity-toughness. For Π_o, the observable outputs differ from the $\boldsymbol{m}$ for various reasons:

- Involuntary: instead of observing the point value of a physical property, there is inevitable volume averaging imposed by a metrological device.
- Deliberate: the raw observation of many df $\boldsymbol{m}$ is possible, but it issues highly-inter-correlated raw measurands in order to produce 'clean' and independent observables, similar to that which was undertaken with time series under the extreme value theory (Chapters 4 and 5).

To what extent does the difference between observables $\boldsymbol{y}$ and prediction variables of interest z matter? From the point of view of the eventual decision-making process *the closer the observable and predictive degrees of freedom, the better.* Indeed, what matters is to provide the best-informed predictions about the likely ranges of z, as measured by the risk measure: *any modelling intermediation* or incompleteness *may* bring *undesirable* noise (i.e. model uncertainty). By order of desirability, one may then find the following cases.

(a) The ideal case when $\boldsymbol{y}$ and z do coincide (cf. Droguett & Mosleh, 2008).
Yet such a case is rarely complete or else there is no need to use a system model to infer the likely values of z: refer to Chapter 5 for a direct uncertainty model that can be built with the sample of observations $(y_{m\,j} = z_j)_{j=1\ldots n}$.

(b) A functional dependency between z and $\boldsymbol{y}$

$$y = G_2(\boldsymbol{x}, \boldsymbol{d}, \boldsymbol{u}) \tag{6.68}$$

$$z = G_1(y, \boldsymbol{d}) \tag{6.69}$$

This still allows for the full incorporation of the residual model uncertainty of $\boldsymbol{y}$ into the prediction of z although it is not then possible to control and model the uncertainty attached to an inference of z from $\boldsymbol{y}$ on the basis of model $z = G_1(y, \boldsymbol{d})$.

(c) A functional dependency between z and $\boldsymbol{y}$ plus other independent components images of $\boldsymbol{x}$.

$$z = G_1(y,\ G_2(\boldsymbol{x}, \boldsymbol{d})) \tag{6.70}$$

Think of fracture mechanics regarding toughness and stress intensity (cf. Section 1.2). Failure modelling involves firstly a complex finite-element thermo-mechanical model $\boldsymbol{m} = M(\boldsymbol{x}_1, \boldsymbol{d})$ predicting stress and temperature fields as a function of numerous variables (properties of materials, flaw characteristics, the thermodynamics of accidental transients, the radiation received over time and the resulting fragilisation, etc.) as well as design or operational conditions $\boldsymbol{d}$ (such as temperature and pressure limits, recovery times, etc.). Thus, the failure margin z, representing the *variable of interest*, is computed by subtracting a stress intensity factor (denoted as $K_I(\boldsymbol{m},\boldsymbol{d})$) from a toughness function (denoted as $K_{Ic}(\boldsymbol{m},\boldsymbol{x}_2,\boldsymbol{d})$) defining altogether the failure function $G(\boldsymbol{x},\boldsymbol{d}) = [K_I(\boldsymbol{m},\boldsymbol{x},\boldsymbol{d}) - K_{Ic}(\boldsymbol{m},\boldsymbol{x},\boldsymbol{d})$. In such a case K_{Ic} is observable as well as $\boldsymbol{x}_1$ while $K_I(\boldsymbol{m},\boldsymbol{d})$ is not completely so. Again, the model inaccuracy brought by $G_2(\boldsymbol{x},\boldsymbol{d})$ cannot be incorporated.

(d) The worst case consisting of the absence of direct dependency between z and $\boldsymbol{y}$, which means that each reflects separate degrees of freedom of $\boldsymbol{m}$ as follows:

$$z = \Pi_d(\boldsymbol{m}) = G(\boldsymbol{x},\ \boldsymbol{d},\ \boldsymbol{u}) \tag{6.71}$$

$$y = \Pi_o(\boldsymbol{m}) = H(\boldsymbol{x},\ \boldsymbol{d},\ \boldsymbol{u}) \tag{6.72}$$

The vector z involves degrees of freedom of $\boldsymbol{m}$ that are not observable at all. Model inaccuracies shown in measurement-model residuals affecting the $\boldsymbol{y}$ (i.e. $\boldsymbol{u}_j = \boldsymbol{y}_{mj} - \boldsymbol{y}_j$), that blur the observation of the corresponding degrees of freedom, cannot be accounted for in predicting z. There is a risk of misrepresenting the extent of uncertainty.

6.3.3 Common features to the distributions and estimation problems associated to the general structure

Notation As the remainder of this chapter concentrates on the observational model $\Pi_o \, o \, M$ required for estimation, the notation will henceforth be simplified as follows:

$$\boldsymbol{Y}_{mj} = \Pi_o \circ M_s(\boldsymbol{X}_j, \boldsymbol{d}_j, \boldsymbol{U}_j) = H_s(\boldsymbol{X}_j, \boldsymbol{d}_j, \boldsymbol{U}_j) \tag{6.73}$$

Keep in mind the meaning of the notation above:

- $\boldsymbol{Y}_{mj}$ and $\boldsymbol{d}_j$ are *always* known as part of the sample.
- $\boldsymbol{X}_j$ and $\boldsymbol{U}_j$ *always* need to be estimated, *except* when their sub-components are constantly *null* (as a deliberate consequence of the formal definitions of observational models, see Table 6.1). According to the cases, $\boldsymbol{X}_j$ may figure as different realisations of i.i.d. random variables throughout the sample; it may also denote a single common random variable $\boldsymbol{X}_j = \boldsymbol{X}$ or even a single fixed deterministic parameter $\boldsymbol{X}_j = \boldsymbol{x}$. All of this may also vary according to the component considered within the input vectors.
- All of these vectors may combine discrete and/or real-valued components.

The subscript s will be skipped when there is a single model structure to be estimated or calibrated. Note that it might also be skipped formally when differing model structures are compounded within a single (super-) model $H(.)$ through the integration of the subscript s as an additional discrete (super-) parameter. In a model selection process, s is not known *a priori* nor is it an observation residual (unlike $\boldsymbol{U}_j$), but it stands as a super-input of the supermodel branching between competing model structures to be selected, *viz.* estimated. Thus, it may be attached as a $p+1$-th component of an enlarged vector $\tilde{\underline{X}}_j = (\underline{X}_j, s)$ as follows:

$$\begin{aligned} \boldsymbol{Y}_{mj} &= H(\tilde{\boldsymbol{X}}_j, \boldsymbol{d}_j, \boldsymbol{U}_j) \\ &= H((\boldsymbol{X}_j, s), \boldsymbol{d}_j, \boldsymbol{U}_j) \\ &= H_s(\boldsymbol{X}_j, \boldsymbol{d}_j, \boldsymbol{U}_j) \end{aligned} \tag{6.74}$$

Alternatively, *Ensemble modelling* approaches (cf. Section 6.4.5) prefer to keep competing models together and combine them in the prediction of $\boldsymbol{Z}$ within an averaged prediction weighed uniformly or possibly through likelihood-derived coefficients.

Probabilistic structure of the observational model It is worth understanding in detail the structure of the probabilistic model to be estimated on the basis of the sample $\Xi_n = (\boldsymbol{y}_{mj}, \boldsymbol{d}_j)_{j=1\ldots n}$:

$$\begin{aligned} &\boldsymbol{Y}_{mj} = H(\boldsymbol{X}_j, \boldsymbol{d}_j, \boldsymbol{U}_j) \\ &\boldsymbol{X}_j | \boldsymbol{\theta}_X \sim f_X(.|\boldsymbol{\theta}_X) \quad \boldsymbol{U}_j | \boldsymbol{\theta}_{uj} \sim f_U(.|\boldsymbol{\theta}_{uj}) \end{aligned} \tag{6.75}$$

Consider the common case where $\boldsymbol{X}_j$ and $\boldsymbol{U}_j$ are assumed to be distributed independently from each other as well as *i.i.d.* over all of the sample $j = 1 \ldots n$ with an additive residual error model:

$$Y_{mj} = H(\boldsymbol{X}_j, \boldsymbol{d}_j, \boldsymbol{U}_j) = H(\boldsymbol{X}_j, \boldsymbol{d}_j) + \boldsymbol{U}_j \tag{6.76}$$

This equation together with the probabilistic definition of $\boldsymbol{X}_j$ and $\boldsymbol{U}_j$ may be re-expressed as follows:

$$P(Y_{mj} \leq y_{mj} | \boldsymbol{x}_j, \boldsymbol{d}_j, \boldsymbol{\theta}_u) = P(U_j \leq y_{mj} - H(\boldsymbol{x}_j, \boldsymbol{d}_j) | \boldsymbol{x}_j, \boldsymbol{d}_j, \boldsymbol{\theta}_u) = F_U(y_{mj} - H(\boldsymbol{x}_j, \boldsymbol{d}_j) | \boldsymbol{x}_j, \boldsymbol{d}_j, \boldsymbol{\theta}_u) \tag{6.77}$$

hence:

$$Y_{mj}, |\boldsymbol{x}_j, \boldsymbol{d}_j, \boldsymbol{\theta}_X, \boldsymbol{\theta}_U \sim f_U\,(y_{mj} - H(\boldsymbol{x}_j, \boldsymbol{d}_j) | \boldsymbol{\theta}_U) \tag{6.78}$$

Assuming theoretically that $\boldsymbol{X}_j$ were observable together with $\boldsymbol{Y}_{mj}$ and $\boldsymbol{d}_j$, the distribution of $\boldsymbol{Y}_{mj}$ conditional upon the realisation $\boldsymbol{x}_j$ would be simply that of the measurement-model residuals. Thus, a conditional density could be estimated similar to a standard regression model. However, in the most general case defined above, $\boldsymbol{X}_j$ cannot be observed so that one requires the distribution of $\boldsymbol{Y}_{mj}$ *unconditional* upon $\boldsymbol{X}_j$. By definition, it comes theoretically as the conditional density *expected over the density of* $\boldsymbol{X}_j$:

$$\begin{aligned} f_{Y_j}(Y_{mj} = y_{mj} | \boldsymbol{d}_j, \boldsymbol{\theta}_{uj}, \boldsymbol{\theta}_X) &= E_{f_X | \boldsymbol{d}_j, \boldsymbol{\theta}_{uj}, \boldsymbol{\theta}_X}\left[f_U(y_{mj} - H(\boldsymbol{X}_j, \boldsymbol{d}_j) | \boldsymbol{\theta}_u)\right] \\ &= \int_{\boldsymbol{x}} f_U(y_{mj} - H(\boldsymbol{x}, \boldsymbol{d}_j) | \boldsymbol{\theta}_u) f_X(\boldsymbol{x} | \boldsymbol{\theta}_X) d\boldsymbol{x} \end{aligned} \tag{6.79}$$

Such a density will be the basis of the estimation process, for example through the maximisation of its likelihood:

$$L\left[(y_{mj})_{j=1\ldots n} \middle| (\boldsymbol{d}_j, \boldsymbol{\theta}_u)_{j=1\ldots n}, \boldsymbol{\theta}_X\right] = \prod_j \int_{\boldsymbol{x}} f_U(y_{mj} - H(\boldsymbol{x}, \boldsymbol{d}_j) | \boldsymbol{\theta}_u) f_X(\boldsymbol{x} | \boldsymbol{\theta}_X) d\boldsymbol{x} \tag{6.80}$$

whereby f_U is known as well as $H(.)$ but the latter is a complex and CPU-costly function to simulate. f_X has a known shape, but obviously its parameters $\boldsymbol{\theta}_X$ are unknown, as being the purpose of the maximisation process. In general, Maximal Likelihood Estimation (MLE) consists of maximising Equation (6.80) through the adjustment of at least some components of the vectors $\boldsymbol{\theta}_X$ and possibly of some components of $(\boldsymbol{\theta}_{Uj})_j$. An alternative is to use Moment Methods, that is adjusting $\boldsymbol{\theta}_X$ until there is an acceptable fit between the theoretical moments of Y_m conditional upon such a value of $\boldsymbol{\theta}_X$ and the empiric moments of the sample measurements Y_{mj} (see below § 6.5.4).

As a major difference with Chapter 5 where a direct uncertainty model was estimated, such a likelihood involves critically a large integral and will therefore generally need simulation techniques in order to compute. It involves the image distribution $f_{Y_j}(y_j | \boldsymbol{d}_j, \boldsymbol{\theta}_{uj}, \boldsymbol{\theta}_X)$ of each j-th model output $y_j = H(\boldsymbol{x}, \boldsymbol{d}_j)$ conditional upon the knowledge of the parameters $\boldsymbol{\theta}_X$, which is in general completely unknown as being the output of a complex physical model. Thus, direct maximisation is simply impossible. This means that a maximum likelihood technique will couple optimisation and simulation, resulting in an elaborate computational process in contrast with Chapter 5.

Practical separation of the component estimation problems Bear in mind that the vector notation gathers formally a whole range of possibly heterogeneous estimation problems inside a single likelihood integral. Some components of $\boldsymbol{Y}_{mj}$, may refer to direct observations of the inputs:

$$\forall j = 1 \ldots n,\ Y^{i}_{mj} = X^{i}_{j} \tag{6.81}$$

others may refer to observations of an observation model with a fixed-input:

$$\forall j = 1, \ldots n,\ Y^{i'}_{mj} = H^{i'}(x^{i'}, \boldsymbol{d}^{i'}_{j}) + U^{i'}_{j} \tag{6.82}$$

while others may refer to indirect observations of the randomly variable inputs of an observation model:

$$\forall j = 1, \ldots n,\ Y^{i''}_{mj} = H^{i''}(\boldsymbol{X}^{i''}_{j}, \boldsymbol{d}^{i''}_{j}) + U^{i''}_{j} \tag{6.83}$$

As long as each is observed independently from one another – mathematically the quadruples X^{i}_{j}, $U^{i'}_{j}$, $X^{i''}_{j}$ and $U^{i''}_{j}$ and thus also the triples Y^{i}_{mj}, $Y^{i'}_{mj}$, $Y^{i''}_{mj}$ are independent from one another – the single likelihood breaks down into a product of sub-likelihood integrals as do the corresponding joint densities $f_{Xi\ Ui'\ Ui''\ Xi''} = f_{Xi} \ldots f_{Ui'}\, f_{Ui''}.f_{Xi''}$. With each of those sub-likelihoods involving separate parameters $\boldsymbol{\theta}_{Xi}$, $\boldsymbol{\theta}_{Ui''}$, $\boldsymbol{\theta}_{Ui''}$, $\boldsymbol{\theta}_{Xi''}$ for their maximisation, the global maximisation results simply in a set of distinct smaller maximisation problems.

Level-2 uncertainty in the observational model So far, distributions have been described *conditionally upon* the known *parameters* θ_X and $(\theta_{uj})_j$ of the uncertainty model $\{f_X(.|\ \boldsymbol{\theta}_X), (f_U(.|\ \boldsymbol{\theta}_U))_j\}$. Similar to that which has been derived in Chapter 5, the description of the uncertainty model is further developed to account for the epistemic uncertainty $\pi(.)$, in connection with the description of the sample space for the later prediction phase. The parameter $\boldsymbol{\theta}_X$, being the common value driving the variability of any past or future state of the system, it is modelled similarly with either a classical or Bayesian approach. In the classical approach, thc asymptotics are the only means of approaching the epistemic distribution $\pi(.)$ of estimator $\hat{\boldsymbol{\Theta}}_X$, approximately Gaussian multi-variate with a null expectation (i.e. unbiased with respect to $\boldsymbol{\theta}_X$) and a covariance matrix given by the Fisher information matrix that will require numerical simulation. The Bayesian approach involves the computation of the posterior distribution $\pi_I(.)$, in an extension of the Bayesian setting introduced in Chapter 5 for direct observations:

$$\begin{aligned}
\pi_1(\boldsymbol{\theta}_X) &= \pi(\boldsymbol{\theta}_X \quad | IK, \Xi_n) \\
&= \frac{\prod_j f_{Y_j}(y_{m\,j}|\boldsymbol{d}_j, \boldsymbol{\theta}_{uj}, \boldsymbol{\theta}_X).\ \pi_0(\boldsymbol{\theta}_X)}{\int_{\boldsymbol{\theta}_X} \prod_j f_{Y_j}(y_{m\,j}|\boldsymbol{d}_j, \boldsymbol{\theta}_{uj}, \boldsymbol{\theta}_X)\pi_0(\boldsymbol{\theta}_X)d\boldsymbol{\theta}_X} \\
&= \frac{\prod_j \left[\int_{\boldsymbol{x}} f_U(y_{m\,j} - H(\boldsymbol{x}, \boldsymbol{d}_j)|\boldsymbol{\theta}_{uj}) f_X(\boldsymbol{x}|\boldsymbol{\theta}_X)d\boldsymbol{x}\right].\pi_0(\boldsymbol{\theta}_X)}{\int_{\boldsymbol{\theta}_X} \prod_j \left[\int_{\boldsymbol{x}} f_U(y_{m\,j} - H(\boldsymbol{x}, \boldsymbol{d}_j)|\boldsymbol{\theta}_{uj}) f_X(\boldsymbol{x}|\boldsymbol{\theta}_X)d\boldsymbol{x}\right].\pi_0(\boldsymbol{\theta}_X)d\boldsymbol{\theta}_X}
\end{aligned} \tag{6.84}$$

Computing the posterior distribution results in a challenging computational task; the MCMC methods introduced in Chapter 5, Section 5.4.5 enable the computation to be limited to the ratios of the numerator of Equation (6.84), although this is still expensive in the general case.

6.3.4 Handling residuals and the issue of model uncertainty

The concept of residuals and preliminary checks Most inverse algorithms require the explicit modelling of residuals. Residuals are defined experimentally as deviations between the observations and the model outputs for known inputs ($\boldsymbol{x}_j$, $\boldsymbol{d}_j$), either in an additive way:

$$\begin{aligned} \boldsymbol{u}_j &= \boldsymbol{y}_{mj} - \boldsymbol{y}_j = H(\boldsymbol{x}_j, \boldsymbol{d}_j) \\ \Leftrightarrow \boldsymbol{y}_{mj} &= H(\boldsymbol{x}_j, \boldsymbol{d}_j) + \boldsymbol{u}_j \end{aligned} \tag{6.85}$$

or in a multiplicative way if more appropriate:

$$\begin{aligned} \boldsymbol{u}_j &= (\boldsymbol{y}_{mj} - \boldsymbol{y}_j)/\boldsymbol{y}_j \\ \Leftrightarrow \boldsymbol{y}_{mj} &= H(\boldsymbol{x}_j, \boldsymbol{d}_j).(1 + \boldsymbol{u}_j) \end{aligned} \tag{6.86}$$

or alternatively in any functional combination that best standardises model-data misfits:

$$\begin{aligned} \boldsymbol{u}_j &= D(\boldsymbol{y}_{mj}, \boldsymbol{y}_j) \\ \Leftrightarrow \boldsymbol{y}_{mj} &= U[H(\boldsymbol{x}_j, \boldsymbol{d}_j), \boldsymbol{u}_j] \end{aligned} \tag{6.87}$$

Care should be taken with the handling of residuals as they critically limit the acceptability of the algorithms. Residuals aggregate a variety of uncertainties that often cannot be discerned from each other: model inadequacy, that is conceptual limitations or possibly numerical solving errors, metrological noise and artefacts, ambiguous or differing definitions of the measured observable and the modelled observable (e.g. with respect to space or time location, volume or interval averaging) and so on. While some algorithms may theoretically allow one to further refine a model of residuals as a compound $\boldsymbol{U} = \boldsymbol{U}_{mes} + \boldsymbol{U}_{mod,}$ when adding to measurement error $\boldsymbol{U}_{mes}$ a calibration residue $\boldsymbol{U}_{mod}$, their estimation would generally require assuming a known pdf for one of the two components in order for the model to remain statistically identifiable. Such an assumption is mostly unverifiable, however, save in precisely-specified sensor and experimental calibration procedures.

A number of checks are required to handle residuals legitimately in a probabilistic calibration approach. Well-known to practitioners of statistical regression, these checks depend on the type of probabilistic model retained for the residuals. *Graphical control* is a mandatory starting point, possibly completed by formal *statistical tests* designed to check:

- balanced and erratic-like deviations around the mean (in other words, with null expectations and non-correlated deviations);
- acceptably-described variance, either stable across the residuals (homoscedastic case) or otherwise properly-modelled through a dedicated variance function (heteroscedastic case);
- acceptably-described shape of deviations, generally through a Gaussian goodness-of-fit test.

It is essential to keep in mind that the *checks of the residuals* is a key recipe for selecting, validating and calibrating properly an observational model, which *does not mean* in general *simply minimising* them, for example, through a minimal quadratic sum. Consider the two competing models M_1 and M_2 (which could be a more or less refined parameterisation of the same phenomenological model):

$$Y_{mj} = M_1(\boldsymbol{X}_j, \boldsymbol{d}_j) + \boldsymbol{U}_j \quad \boldsymbol{X}_j, \boldsymbol{U}_j \sim f_{1,X}(.|\boldsymbol{\theta}_{1,X}).f_{1,U}(.|\boldsymbol{\theta}_{1,U}) \tag{6.88}$$

$$Y_{mj} = M_2(\boldsymbol{X}_j, \boldsymbol{d}_j) + \boldsymbol{U}_j \quad \boldsymbol{X}_j, \boldsymbol{U}_j \sim f_{2,X}(.|\boldsymbol{\theta}_{2,X}).f_{2,U}(.|\boldsymbol{\theta}_{2,U}) \tag{6.89}$$

Model M_1 can be deemed preferable to M_2 (or uncertainty parameterisation $f_{1,X}$ to $f_{2,X}$) despite leaving a greater residual error $\sigma^2_1 > \sigma^2_2$ if the behaviour of the residuals provides a better check of the assumptions.

The relative superiority of one of those could also be evidenced by the use of the model selection techniques that watch out for a model's over-parameterisation and excessive fit to data (AIC and BIC information criteria introduced in Chapter 5).

As is standard practice in the larger field of data analysis or machine learning, note that a sound calibration or model selection process should always *distinguish* a *learning sub-sample* from a *validation sub-sample* in checking the behaviour of the residuals. Intuitively, the learning sub-sample leads generally to better-controlled residuals since the parameter estimation and model selection process have adapted to data the essential check of the model validity requiring an independant sample from that which served for parameter estimation or at least for model selection. Given a limited data sample, partition the sample into two fixed parts, or even undertake at random through cross-validation techniques.

A challenge arises with the fact that residuals are *per definition* observable only within the observed ranges of behaviour of the system in the past so that the control of model-measurement deviations makes more sense for interpolation than for extrapolation. On the contrary, risk modelling would often involve the consideration of extrapolated behaviours beyond the observed features.

Residual modelling and validation Modelling the residuals $\boldsymbol{u}$ could at its simplest take the form of a Gaussian random vector *i.i.d.* throughout the sample as follows:

$$\left.\begin{aligned} \boldsymbol{Y}_{mj} &= H(\boldsymbol{X}_j, \boldsymbol{d}_j) + \boldsymbol{U}_j \\ \boldsymbol{X}_j|\boldsymbol{\theta}_x &\sim f_X(.|\boldsymbol{\theta}_x) \\ \boldsymbol{U}_j|\boldsymbol{\theta}_u &\sim N(\mu, R) \end{aligned}\right\} \tag{6.90}$$

Rather strong assumptions for the residuals would need to be checked in the sample and calibrated model. Residuals need to be erratic-like around the mean without correlations for close values of $\boldsymbol{d}_j$; mean deviations should not vary significantly whatever the $\boldsymbol{d}_j$, and the distribution should be symmetrical and Gaussian-like (Figure 6.10 – top left). If any of those checks fail, the prediction of the risk measure using the calibrated model and residuals would misrepresent the amount of uncertainty. For instance, if residuals remain always positive for a certain region of $\boldsymbol{d}_j$ (Figure 6.10 – bottom left), model predictions for those type of actions $\boldsymbol{d}_j$ will underestimate the variable of interest in a way that is not represented by adding a symmetric confidence band around Y.

More refined techniques can account for: (i) a heteroscedastic behaviour, meaning that the spread of measurement-model deviations increases or decreases for large or small values for $\boldsymbol{d}_j$ (Figure 6.10 – top right); and/or (ii) non-Gaussian distribution, meaning that the spread of deviations for a given $\boldsymbol{d}_j$ does not follow a bell curve, for example being assymetrical. The former case consists of substituting for the simple model $U_j \sim N(\mu,R)$ a model $U_j(\boldsymbol{d}_j) \sim N(\mu,R(\boldsymbol{d}_j))$ where the variance follows a known functional dependence to $\boldsymbol{d}_j$. In the latter case, the residual model becomes $U_j \sim f_U(.,\theta_U)$, f_U being a non-Gaussian distribution of known shape (e.g. uniform or lognormal), whence the model parameter θ_U can also become a function of $\boldsymbol{d}_j$ in a non-Gaussian heteroscedastic combination.

These refinements still assume the independence of residuals U_j from one observation (Y_{mj}, $\boldsymbol{d}_j$) to another ($Y_{mj'}$, $\boldsymbol{d}_{j'}$). Much more complex is the case where the inter-correlations of the residuals can no longer be neglected, meaning that the random variables $\boldsymbol{U}(\boldsymbol{d}_j)$ and $\boldsymbol{U}(\boldsymbol{d}_{j'})$ become correlated when $\boldsymbol{d}_j$ and $\boldsymbol{d}_{j'}$ are close enough (Figure 6.10 – bottom right). This happens when the observational process involves a dense-enough design of experiment or neighbouring degrees of freedom, for instance when $\boldsymbol{d}_j$ represent a frequency band in noise calibration or the spatial location in soil properties. A powerful tool is then the representation of the residuals as a stochastic process over the space indexed by $\boldsymbol{d}$ such as the popular Kriging or Gaussian process techniques developed in the field of geostatistics or machine learning. The

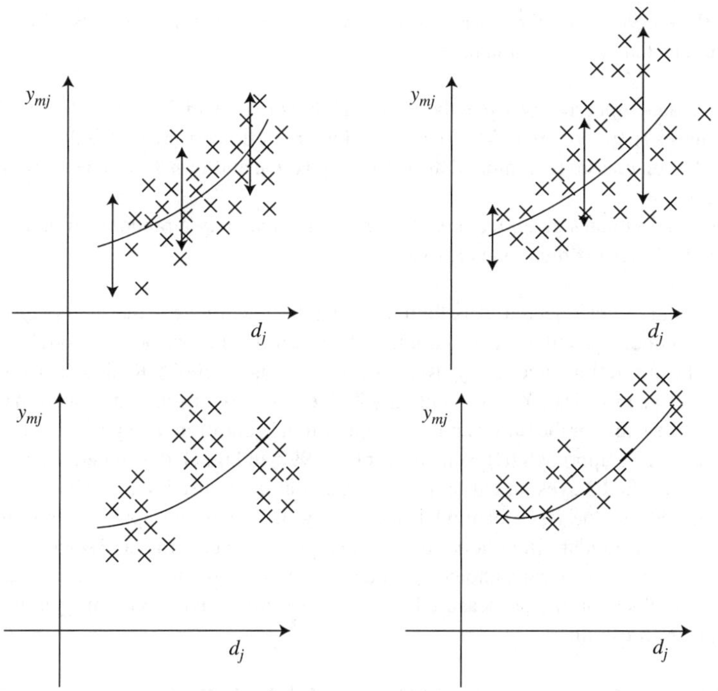

Figure 6.10 *Handling residuals.*

vector θ_U becomes then not only the bias and variance of residuals as function of $\boldsymbol{d}$ but also its ($\boldsymbol{d}$ - spatial) covariance structure:

$$[U(\boldsymbol{d}_j),\ U(\boldsymbol{d}_{j'})]|\theta_U \sim N((\mu(\boldsymbol{d}_j),\ \mu(\boldsymbol{d}_{j'})),\ C(\boldsymbol{d}_j,\ \boldsymbol{d}_{j'})) \tag{6.91}$$

whereby $C(d_j,\boldsymbol{d}_{j'}) = \Sigma\ c(\boldsymbol{d}_j,\boldsymbol{d}_{j'})$ combines the point variance (if U is a vector with multiple observed dof, it becomes a matrix denoted as Σ) with a correlation function typically decreasing with the distance between $(d_j,\boldsymbol{d}_{j'})$ with respect to a given norm and correlation length:

$$c(\boldsymbol{d}_j,\ \boldsymbol{d}_{j'}) = \exp(-||\boldsymbol{d}_j - \boldsymbol{d}_{j'}||_D) \tag{6.92}$$

Look, for instance, at Chapter 5, Section 5.5.1 for a L_I-norm defining an exponential auto-correlated process of flood flows. Though essential in recent inverse modelling or Bayesian updating techniques, Kriging modelling and associated techniques are beyond the scope of this book and the reader should refer to the specialised literature (Kennedy and O'Hagan, 2002).

The issue of model uncertainty Eventually, although the concept of modelling residuals, that is unexplained variability, is paradoxical to some extent and controversial in the literature (Aven, 2003; Droguett and Mosleh, 2008), its probabilistic control through such a calibration approach accounts to some extent for the model prediction inaccuracies. A *Popperian* view on *scientific* model building as advocated by Beck (1987) would require us to specify the observational conditions by which the system model may be rejected. This could happen:

- when the residual uncertainty exceeds a certain threshold, which may be controlled either in continuous terms (e.g. more than 5 % likelihood that prediction error exceeds 3 % or 5 °C etc.) or in discrete prediction classes (e.g. more than α % false positive and/or β % false negative predictions of a given event),
- when the residual violates given features defining mathematically a random variable and associated modelling hypotheses (as mentioned above).

Provided the model is not rejected, it is thought valuable to estimate – that is model in a statistical approach – the model inaccuracy and include it later within the predictive model in order not to underestimate the prediction uncertainty (Kennedy and O'Hagan, 2002). Residuals become a true part of the later predictive models of $\boldsymbol{Y}$ (and eventually $\boldsymbol{Z}$ given the links discussed in Section 6.1.2) in order to assess the risk measure for the future actions or operational conditions $\boldsymbol{d}$ beyond the observed sample.

The popular GLUE approach ((Beven and Binley, 1992 and much developed since), which can be viewed as the simplified Bayesian calibration approach already discussed in Chapter 2, is intended interestingly to address the issue of model uncertainty. It rejects the approaches specifying a full residual model with a standard likelihood function and prefers to generate and select a set of so-called 'behavioural solutions', meaning a choice of model structures and parameter sets that are acceptably consistent with the observations, as measured for instance by a cutoff of model-measurement deviations exceeding a given threshold....

6.3.5 Additional comments on the model-building process

Besides the knowledge of the shape of f_X, note that expertise may sometimes allow the prior inference of some of the components of $\boldsymbol{\theta}_{X.}$ or $(\boldsymbol{\theta}_{Uj})_j$. Consider, for instance, the unbiased and uncorrelated Gaussian homoscedastic case for residuals:

$$\boldsymbol{U}_j \sim f_U(\boldsymbol{U}_j = \boldsymbol{u}|\boldsymbol{\theta}_u) = N(\mu, \Sigma) \tag{6.93}$$

Vector $\boldsymbol{\theta}_U$ would contain a known component of the vector of parameters of null biases $\mu = \boldsymbol{0}$, while the variance matrix would be assumed to be diagonal, the parameters to be estimated being the diagonal terms, that is residual variance per type of observable degree of freedom.

This may also be the case in a heteroscedastic formulation of error whereby relative weights ${\gamma^2}_j$ are considered to be well known as a function of $\boldsymbol{d}_j$.

$$\boldsymbol{U}_j|\boldsymbol{\theta}_{uj} \sim f_U(.|\boldsymbol{\theta}_{uj}) = N(\mu, \gamma_j^2\Sigma) \tag{6.94}$$

Indeed, it may be both practical to simplify the estimation or even necessary in order to secure the *identifiability* of the model when $\boldsymbol{\theta}_X$ is of a large dimension: thus, the maximisation would not involve those parameters kept fixed at their best-estimates.

The formulation above may be extended to non-additive measurement errors involving a modified likelihood.

Thus, the key questions to be asked in such model-building are the following:

- Checking the identifiability of the model: is the parameterising adequate to the data available?
- Controlling convergence of the estimation procedure, both regarding statistical consistency and computational possibilities.

- Checking the model assumptions, in particular regarding residuals: bias, homoscedasticity, independence and randomness, goodness-of-fit of the parametric distribution assumption (generally Gaussian in those complex models).
- Using model selection techniques in order to compare various input parameterisations with increasing power to fit observed data but rising estimation uncertainty – the AIC/BIC techniques introduced in Chapter 5, Section 5.2.2 can be reused directly.

Recall also that before moving into more sophisticated inverse algorithms, it may be necessary to challenge the original system model with other plausible ones with a model selection approach (see Leroy *et al.*, 2010):

- Transform the observable variables $\mathbf{y}$ through functional shifts of some of its components – typically a log which may help normalising the residuals.
- Transform the observable variables $\mathbf{y}$ through a grouping of its components that may be too dependent – think of the joint measurement of the series of fine frequency bands over the entire spectrum for a given observation in signal processing, whereby an averaging into a smaller number of large frequency bands may reduce intercorrelation of the residuals.
- Transform the vector $(\mathbf{x},\mathbf{d})$ of calibrated inputs according to any of the above-mentioned shifts.

A number of general remarks will be made hereafter, particularly in the Gaussian linear cases. More sophisticated variants of the general problem, such as data assimilation or intrinsic variability identification will be described later (Sections 6.4 and 6.5).

6.3.6 Identifiability

The general issue Identifiability, as defined in Chapter 5 for a statistical model, means that the density of the observed measurements is characterised uniquely by one value for $\boldsymbol{\theta}_X$:

$$f_{Ym}(y_m|\boldsymbol{\theta}_X) = f_{Ym}(y_m|\mathring{\boldsymbol{\theta}}_X) \quad \forall y_m \quad \Rightarrow \boldsymbol{\theta}_X = \mathring{\boldsymbol{\theta}}_X \tag{6.95}$$

in other words that the parametrisation brings a true and univoque change in the distribution of observations: there is no *over-parameterisation* of the model. While it may be difficult to prove this theoretically in the context of a complex model *H*, it is reasonable in general to undertake the following heuristic checks:

- Compare the dimension of $\boldsymbol{\theta}_X$ which should be low enough with respect to the degrees of freedom of the sample.
- Simulate the likelihood for various values of $\boldsymbol{\theta}_X$ (which is generally part of the estimation process anyway).
- Perform the linear checks, as will be defined hereafter, generally involving the full-rankedness of matrices associated with the gradients of the observational model *H*.

The linear and gaussian cases Much more may be said with the following additional Linear and Gaussian assumptions:

- *X* follows a normal distribution around the unknown expectation x_m with an unknown covariance matrix *V*.
- *H* behaves linearly (or close to linearly) with respect to $\mathbf{x}$: note that it may not be the case with respect to $\mathbf{d}_j$, and that the partial derivative with respect to $\mathbf{x}$ may still vary considerably according to the experiment $\mathbf{d}_j$.

Thus, problem becomes the following:

$$\left.\begin{array}{l} Y_{mj} = H(X_j, \boldsymbol{d}_j) + U_j = \mathbf{H}_j\left(X_j - x_o\right) + H_j^o + U_j \\ \boldsymbol{X}_j \sim N(x_m, V) \\ U_j \sim N(\mu, R_j) \\ identify \quad (x_m, \mu, V) \quad knowing \quad \Xi_n = \left(y_{mj}, \boldsymbol{d}_j\right)_{j=1\ldots n} and \quad partially \quad \left(R_j\right)_{j=1\ldots n} \end{array}\right\} \tag{6.96}$$

where $H_j^o = H(\boldsymbol{x}_o, \boldsymbol{d}_j)$and $\boldsymbol{H}_j = \nabla_{\boldsymbol{x}_o} H|_{d=\boldsymbol{d}_j}$ denote the linearisation of $H_j(\boldsymbol{x}) = H(\boldsymbol{x}, \boldsymbol{d}_j)$ with respect to $\boldsymbol{x}$ at a known linearisation point $\boldsymbol{x}_o$. That point is usually fixed at the best *prior* estimate for the unknown expectation *E[X]*. Thus, the observable outputs are also Gaussian, with the following moments:

$$Y_{mj} \sim N(\mathbf{H}_j.(x_m - x_o) + H_j^o + \mu, \mathbf{H}_j V \mathbf{H}'_j + R_j) \tag{6.97}$$

Log-likelihood (*LL*) becomes:

$$\begin{array}{l} LL(\Xi_n|(x_m, \mu, V)) = -\frac{1}{2}\sum_j \log\left(\det\left(2\pi\left[R_j + \mathbf{H}_j V \mathbf{H}'_j\right]\right)\right) \\ -\frac{1}{2}\sum_j (y_{mj} - \mathbf{H}_j(x_m - x_o) - H_j^o - \mu)'(R_j + \mathbf{H}_j V \mathbf{H}'_j)^{-1}(y_{mj} - \mathbf{H}_j(x_m - x_o) - H_j^o - \mu) \end{array} \tag{6.98}$$

According to the cases, the parameters that may require identification are the following: $\boldsymbol{x}$ as well as biases μ possibly (a vector for the biases may differ according to the observed components); the intrinsic input variance *V*; or even the measurement-model residual variances R_j or some of their components.

A very general ***necessary*** requirement for identifiability is that the *Hessian matrix of the log-likelihood is invertible*, which will be shown to imply the *full-rankedness* of matrices involving the *gradients* $\mathbf{H}_j = \nabla_{\boldsymbol{x}_o} H|_{d=\boldsymbol{d}_j}$ of the system model. In some cases, this constitutes also a *sufficient* condition for identifiability.

Identifiability requirements in (linear gaussian) parameter identification Consider firstly the *parameter identification* case, that is where the parametrisation consists of a *fixed* $\boldsymbol{x}$ as well as its possible output biases μ, while the input intrinsic variance *V* is neglected and the output observational variance *R* is supposedly known.

An intuitive condition is that the number of observations (*nq*) should at least be higher than the number of parameters to be identified ($p + q$), including the *p* components of $\boldsymbol{x}$ and *q* components of the bias. This condition proves to be necessary but not sufficient. The model needs to be responding 'differently enough' to the various parameters within the various experimental conditions for the identification to be possible.

Concatenate the *nq* observations as follows:

$$Y = \begin{pmatrix} y_{m1} - H_1^o \\ .. \\ y_{mn} - H_n^o \end{pmatrix} \quad X = \begin{pmatrix} I^q & \nabla H_1 \\ .. & .. \\ I^q & \nabla H_n \end{pmatrix} \quad \beta = \begin{pmatrix} \mu \\ \boldsymbol{x}_m - \boldsymbol{x}_o \end{pmatrix} \quad U = \begin{pmatrix} U_1 \\ .. \\ U_n \end{pmatrix} \quad R = \begin{pmatrix} R_1 & 0 & 0 \\ .. & .. & .. \\ 0 & 0 & R_n \end{pmatrix} \tag{6.99}$$

Y and *U* are *nq*-vectors; β is a $(q + p)$-vector; *X* and *R* are $nq^*(q + p)$- and nq^*nq-matrices respectively and I^q stands for the q^*q identity matrix. The aggregate observational model may be reformulated as a *Gaussian multi-linear regression problem* (with correlated errors of known variance R).

$$Y = X.\beta + U \quad where \quad U \sim N(0, R) \tag{6.100}$$

the likelihood of which is re-written as follows:

$$LL[Y|X,\beta] = -\frac{nq}{2}\log(2\pi) - \frac{1}{2}\log(\det R) - \frac{1}{2}(Y - X\beta)'R^{-1}(Y - X\beta) \tag{6.101}$$

the maximisation of which yieds:

$$\beta = Arg\max LL[Y|X,\beta] = Arg\min\left[\frac{1}{2}(Y - X\beta)'R^{-1}(Y - X\beta)\right] = Arg\min \|Y - X\beta\|_{R^{-1}} \tag{6.102}$$

Intuitively the matrix X needs to be *injective* for the solution β (i.e. the q components of bias and p components of $\boldsymbol{x}_m$) to be unique. X being a linear application, there would otherwise exist a non-null kernel vector β_n such that $X(\beta + \beta_n) = X.\beta$ for any β. Thus, if β is a solution, the two sets of parameters $\beta + \beta_n$ and β would be indiscernible from the point of view of the observations, standing as equal solutions to the program. Identifiability therefore requires X to be injective, that is full-rank. Which, in particular, requires $q + p < nq$ as a necessary but not sufficient condition.

In fact, the injectiveness of X proves also to be a *sufficient* condition for identifiability, or solution existence and unicity. This is a very general result of linear algebra when looking at identification as a least-square projection of Y over the vector space of X according to the norm R^{-1}. Note that R^{-1} is a positive-definite being the inverse of the block-diagonal concatenation of positive-definite error variance matrices R_j. Indeed, an injective X (see Annex Section 10.5.1) ensures that $X'R^{-1}X$ is invertible in that case and the general result of least-square projection yields the following unique solution (see Annex Section 10.5.2):

$$\beta = (X'.R^{-1}.X)^{-1}.X'.R^{-1}.Y \tag{6.103}$$

In fact, the condition of full-rankedness of X is equivalent to:

- the fact that the 1st order condition of the likelihood maximisation leads to an invertible linear program;
- the fact that the Hessian of the cost function (or opposite Hessian of the loglikelihood) is a definite-positive matrix;
- the fact that the epistemic variance is non null.

Indeed, differentiation results detailed in Annex Section 10.5.2 show that the Hessian matrix of Eq (6.83) log-likelihood appears as follows:

$$\Delta_\beta LL = -X'.R^{-1}.X$$

When undertaking *parameter identification neglecting the biases*, note that the condition becomes the full-rankedness of the nq^*p matrix of model gradients:

$$X = \begin{pmatrix} \nabla H_1 \\ \nabla H_n \end{pmatrix} \tag{6.104}$$

In such a case, only p parameters need to be estimated. Intuitively, it means that variations of different components of the p-vector of inputs (i.e. in rows) result in (linearly) independent variations of the nq-observed values (in columns).

Identifiability requirements in (linear gaussian) calibration Consider now the full *calibration* case whereby residual variance has to be estimated, the above-mentioned condition proves *necessary but not sufficient*. Simple versions of calibration generally assume a known shape for the R_j matrices. Suppose, for instance, the average measurement-model error is unknown prior to calibration, but that the relative inaccuracy of the model between types of physical variables observed (components of vector $\mathbf{y}_j$) for each observation sets (j) is known:

$$R_j = s^2\,\Gamma_j \tag{6.105}$$

where all Γ_j are given and only σ^2 needs to be estimated additionally. Think of relative orders of magnitude of prediction error between water level and flow speed or even correlations between those errors (resulting in non-diagonal Γ_j), while those may be heterogeneous between observation sets j and j' because some of them correspond to extreme experimental conditions. Such a setting is a q-dimensional generalisation of the heteroscedastic calibration introduced in Section 6.2.

An extended *necessary* condition for identifiability of this setting that involves $q+p+1$ parameters to be estimated is the full-rankedness of the following larger $nq^*(1+q+p)$ matrix;

$$(Y,X) = \begin{pmatrix} \cdots & \cdots & \cdots \\ y_{mj}-H_j^o & I^q & \nabla H_j \\ \cdots & \cdots & \cdots \end{pmatrix}_{j=1\ldots n} \tag{6.106}$$

In other words, add a new column gathering the nq observations to the former X matrix. The full-rankedness of (Y, X) requires classically that of X, so that the previous necessary condition is included in this new one. Yet, if X is full rank while (Y, X) is not, there exists a linear combination of coefficients β such that:

$$Y = X.\beta \tag{6.107}$$

leading to a zero residual inside the least square program $Arg\min \|Y-X\beta\|_{\Gamma^{-1}}$ (denoting as Γ the block matrix gathering the Γ_j matrices in diagonal as in the previous section for R) and hence a null variance. Conversely, the requirement is also *sufficient* in that case as the full-rankedness of (Y, X) guarantees both the existence of the $\hat{\beta} = (X'.R^{-1}.X)^{-1}.X'.R^{-1}.Y$ estimator and the non-null variance $\hat{\sigma}^2 = \|Y-X\hat{\beta}\|^2_{\Gamma^{-1}}$ that fulfils the first-order condition.

However, sufficient identifiability conditions still need to be studied in the more general cases of calibration which include the estimation of more inside the residual matrices $R_j = \sigma^2\,\Gamma_j$ than just their mean variance σ^2.

Identifiability requirements in (linear gaussian) full inversion Consider now the full inversion case, where x_m, μ and V need to be estimated with the following log-likelihood (assuming the R_j are known):

$$\begin{aligned} LL(\Xi_n|(x_m,\mu,V)) &= -\frac{1}{2}\sum_j \log\left(\det\left(2\pi\left[R_j + \mathbf{H}_j V \mathbf{H}'_j\right]\right)\right) \\ &-\frac{1}{2}\sum_j (y_{mj}-\mathbf{H}_j(x_m-x_o)-H_j^o-\mu)'(R_j+\mathbf{H}_j V\mathbf{H}'_j)^{-1}(y_{mj}-\mathbf{H}_j(x_m-x_o)-H_j^o-\mu) \end{aligned} \tag{6.108}$$

The previously-mentioned identifiability conditions still hold as a *necessary* condition Indeed, the differentiation results detailed in *Annex* Section 10.5.2 show that the Hessian matrix of this new log-likelihood with respect to $\beta = \binom{\boldsymbol{\mu}}{\boldsymbol{x}_m - \boldsymbol{x}_o}$ appears as follows:

$$\Delta_\beta LL = -X'.Q^{-1}.X \tag{6.109}$$

Where the new variance of the observable output $Q_j = R_j + H_j VH_j'$ is gathered in block notation as above in Q. Thus, a necessary condition is that the following Jacobian matrix has a full rank:

$$X = \begin{pmatrix} I^q & \nabla H_1 \\ .. & .. \\ I^q & \nabla H_n \end{pmatrix} \tag{6.110}$$

As already mentioned, this implies $p + q \leq nq$, that is that there are at least as many scalar observations as there are uncertain inputs + biases to be estimated. Note, however, that this is *not sufficient* in practice to ensure that enough data is available for estimation. Take the $p = q = 1$ example neglecting the bias. At least $nq = 2$ observations are necessary in order to be able to estimate x_m and the associated one-dimensional variance matrix $V = \sigma^2$. With $nq = 1$ observation, only the mean can be estimated, and a singular null variance. More generally, a supplementary $2p \leq nq$ condition (neglecting the biases) should presumably apply in the Gaussian case for the estimation to be feasible, although this needs to be further investigated. Of course, avoiding excessive epistemic uncertainty would probably require much more, say $10p \leq nq$ at least (see Section 6.3.7).

Eventually, the estimation of the model-measurement variances – or at least of an average scalar as in $R_j = \sigma^2 \Gamma_j$ where all Γ_j are given and only σ^2 needs to be estimated – poses an stronger issue of identifiability.

The flood example

Returning to the flood case with the observed couples of water heights/flows, the uncertain inputs to be calibrated can now be considered in a vector formulation including not only the friction coefficient but also the waterbed elevations:

$$y_{mj} = H_j(\boldsymbol{x}) + u_j = \left(\frac{q_j}{k_s * \sqrt{\frac{z_m - z_v}{l}} * b} \right)^{3/5} + u_j \tag{6.111}$$

$$\boldsymbol{x} = (k_s, z_m, z_v) \quad \boldsymbol{d}_j = q_j \tag{6.112}$$

In such an example $q = 1$, $p = 3$ and n is of the order of 100. Because of the power structure of the model, the Jacobian matrix (including bias) appears as follows:

$$X = \begin{pmatrix} I^q & \nabla H_1 \\ .. & .. \\ I^q & \nabla H_n \end{pmatrix} = \begin{pmatrix} 1 & -\tfrac{3}{10} H_1(\boldsymbol{x}^\circ). \left(\frac{2}{k_s^\circ} \quad \frac{1}{z_m^\circ - z_v^\circ} \quad \frac{-1}{z_m^\circ - z_v^\circ} \right) \\ .. & .. \\ 1 & -\tfrac{3}{10} H_n(\boldsymbol{x}^\circ). \left(\frac{2}{k_s^\circ} \quad \frac{1}{z_m^\circ - z_v^\circ} \quad \frac{-1}{z_m^\circ - z_v^\circ} \right) \end{pmatrix} \tag{6.113}$$

Given that observations are made for various flows (q_j are not constant), *rank*$(X) = 2$ so that only one parameter within $\boldsymbol{x}$ can be identified (plus the bias) in a parameter identification approach: waterbed elevation cannot be identified on top of friction. Full calibration, including a residual variance, requires the vector $\left(y_{mj}-H_j(x^\circ)\right)_{j=1\ldots n}$ – gathering the erratic deviations of the water heights to the flow-height curve taken at linearization point, cf. Figure 6.6 – to be linearly-independant from vector $\left(H_j(x^\circ)\right)_{j=1\ldots n}$– representing the flow height curve at linearisation point. It should also be so with respect to the constant vector of ones (for biases). It is conspicuously so in Figure 6.7-right. Alternatively, full inversion of a single input (e.g. friction) proves to be also identifiable, but not at the same time as an output residual variance (see Section 6.2.5).

This would change if the observations had been made on the flood elevations instead of water depths:

$$y^*_{mj} = H^*_j(\boldsymbol{x}) + u_j = z_v + \left(\frac{q_j}{k_s^* \sqrt{\frac{z_m - z_v}{l}} * b}\right)^{3/5} + u_j \tag{6.114}$$

Partial derivatives with respect to z_v would then no longer be collinear to the two other gradients: either (k_s, z_v) or (z_m, z_v) could then be identified, as well as a constant bias and a residual variance.

$$\nabla H^*_j = \left({}^{-3}\!/\!_{10} . H^*{}_j(x^\circ)\frac{2}{k_s^\circ} \quad {}^{-3}\!/\!_{10} . H^*{}_j(x^\circ)\frac{1}{z_m^\circ - z_v^\circ} \quad 1 + {}^{-3}\!/\!_{10} . H^*{}_j(x^\circ)\frac{1}{z_m^\circ - z_v^\circ} \right) \tag{6.115}$$

Extended identifiability requirements in non-linear or non-gaussian models Invertibility of the loglikelihood Hessian should remain a necessary condition, as being in fact a basic requirement for the maximal likelihood estimation to converge. Yet, the difficulty is now that in a non-linear model, such a matrix changes according to the differentiation point so that it becomes more difficult to guarantee it beyond the local area, particularly since such an appropriate location is by definition to be estimated. In the non-linear Gaussian case, a necessary requirement can still be formulated in closed-form using the log-likelihood differentiation with respect to x_m: it means simply replacing the above-mentioned constant gradients H_j by the gradients functions of the search point $\boldsymbol{x}$.

Handling the biases Beyond Section 6.3.6, the consideration of model biases can be included formally inside an extended system model through the following:

$$\tilde{H} = (Id^q + H) \quad \tilde{x} = \begin{pmatrix} \mu \\ \boldsymbol{x}_m \end{pmatrix} \quad \tilde{V} = \begin{pmatrix} V & 0 \\ 0 & 0 \end{pmatrix} \tag{6.116}$$

So that the biased formulation with a p-dimensional input:

$$\left.\begin{aligned} Y_{mj} &= H(X_j, \boldsymbol{d}_j) + U_j \\ \boldsymbol{X}_j &\sim N(x_m, V) \\ U_j &\sim N(\mu, R_j) \end{aligned}\right\} \tag{6.117}$$

Is replaced by the following unbiased formulation with an extended input dimension $p+q$:

$$\left.\begin{aligned} Y_{mj} &= \tilde{H}(\tilde{X}_j, \boldsymbol{d}_j) + U_j \\ \tilde{\boldsymbol{X}}_j &\sim N(\tilde{x}_m, \tilde{V}) \\ U_j &\sim N(0, R_j) \end{aligned}\right\} \tag{6.118}$$

The formulation is thus formally unchanged: note in particular that $R_j + \tilde{\mathbf{H}}_j \tilde{V} \tilde{\mathbf{H}}'_j = R_j + \mathbf{H}_j V \mathbf{H}'_j$ for the linearised case. As the formulation is close, the tilde superscript will be generally omitted hereinafter: the biased cases will require implicitly that the vector of input parameter has been extended although with null variance.

In the homoscedastic case where the R_j are constant, the notation can be simplified even further through the incorporation of the residuals inside the uncertain inputs as follows:

$$\tilde{H} = (Id^q + H) \quad \tilde{x} = \begin{pmatrix} \mu \\ \boldsymbol{x}_m \end{pmatrix} \quad \tilde{\tilde{V}} = \begin{pmatrix} V & 0 \\ 0 & R \end{pmatrix} \quad Y_{mj} = \tilde{H}(\tilde{X}_i, \boldsymbol{d}_j) \qquad \tilde{X}_i \approx N(\tilde{x}_m, \tilde{\tilde{V}}) \tag{6.119}$$

From that perspective, the residuals can be incorporated inside the global picture of input uncertainty of the extended model. As mentioned in Chapter 2, this last notation will be used only for the prediction model $Z = G(\boldsymbol{X}, \boldsymbol{d})$ whereby $\boldsymbol{X}$ incorporate all of the uncertainty description inside the later propagation steps; it proves generally meaningful to keep a keen eye on the residuals in the calibration/estimation phase.

6.3.7 Importance factors and estimation accuracy

Intuitively, estimation accuracy – as measured by the variance of the estimators – should be linked to the sensitivity of the observational models to the input coefficients to be estimated, that is to the relative impact of the input coefficients in explaining the observed variability of the observation sample. This can be explained in the linear Gaussian case.

As a general result of maximal likelihood estimation (see Chapter 5), the variance of the estimators can be estimated through the inverse of the Fisher information matrix of the n-sample, that is the inverse of the opposite Hessian of the log-likelihood:

$$\operatorname{var} \hat{\boldsymbol{\theta}} \cong \hat{I}_n(\hat{\boldsymbol{\theta}})^{-1} = \left[-\frac{\partial^2 LL}{\partial \boldsymbol{\theta} \partial \boldsymbol{\theta}'} (\Xi_n, \hat{\boldsymbol{\theta}}) \right]^{-1} \tag{6.120}$$

Consider the likelihood of the general linear Gaussian case introduced in Section 6.3.2.2, including any of the variants suitable for parameter identification, calibration or intrinsic variability. Reformatting the notation with the expectation and variance of the observables Y_j:

$$\begin{aligned} \boldsymbol{\theta} &= (\boldsymbol{x}_m, \boldsymbol{\mu}, V, R) \\ EY_{mj}(\boldsymbol{\theta}) &= H_j(\boldsymbol{x}_m) + \boldsymbol{\mu} \\ VarY_{mj}(\boldsymbol{\theta}) &= R_j + H_j V H_j' \end{aligned} \tag{6.121}$$

Log-likelihood is rewritten as follows:

$$LL\left[(y_{mj})_{j=1..n} | \boldsymbol{\theta} \right] = -\frac{1}{2} \sum_j \log(\det(2\pi \operatorname{var} Y_{mj}(\boldsymbol{\theta}))) - \frac{1}{2} \sum_j \left(y_{mj} - EY_{mj}(\boldsymbol{\theta}) \right)' \left(\operatorname{var} Y_{mj}(\boldsymbol{\theta}) \right)^{-1} \left(y_{mj} - EY_{mj}(\boldsymbol{\theta}) \right) \tag{6.122}$$

Consider now the input parameters $\boldsymbol{x}_m$, the unknown value for fixed inputs $\boldsymbol{x}$ or the expectation of the input $\boldsymbol{X}$ in the case of intrinsic variability. The information matrix coefficients appear closely linked to partial derivatives:

$$-\frac{1}{n}\frac{\partial^2 LL}{\partial x^{i2}} = \frac{1}{n}\sum_j \left(\frac{\partial H_j}{\partial x^i}\right)'(\mathrm{var}\,Y_{mj})^{-1}\left(\frac{\partial H_j}{\partial x^i}\right) = \overline{\left(\frac{\partial H_j}{\partial x^i}\right)'(\mathrm{var}\,Y_{mj})^{-1}\left(\frac{\partial H_j}{\partial x^i}\right)} \tag{6.123}$$

The formula may be interpreted as the ratio of squared partial derivatives to overall observed variance, averaged over the $j = 1_{\dots} n$ trials. As a matter of fact, this may be further interpreted in terms of linear *sensitivity indices,* as:

$$-\frac{\partial^2 LL}{\partial x^{i2}} = n\frac{\overline{S^i}}{\mathrm{var}\ X^i} \tag{6.124}$$

Where:

$$S_j^i \cong \left(\frac{\partial H_j}{\partial x^i}\right)'(\mathrm{var}\,Y_j)^{-1}\left(\frac{\partial H_j}{\partial x^i}\right)\mathrm{var}\ X^i \tag{6.125}$$

reading in dimension $q = 1$:

$$S_j^i \cong \frac{\left(\frac{\partial H_j}{\partial x^i}\right)^2 \mathrm{var}\,X^i}{\mathrm{var}\,Y_j} \tag{6.126}$$

This denotes the classical Taylor (or linear) sensitivity index of the function $Y_j = H_j(\boldsymbol{X})$, measuring an approximate percentage of importance of the variable X^i in explaining Y_j (see Chapter 7, Section 7.5). In order to end up with the variance of the estimator of the i-th input parameter $x^i{}_m$, one needs to capture the i,i-coefficient of the inverse matrix which is not simply the inverse of the above-derived i,i-coefficient if the Hessian matrix is not diagonal:

$$\mathrm{var}\,\hat{x}_m^i \cong \left[-\frac{\partial^2 LL}{\partial\boldsymbol{\theta}\partial\boldsymbol{\theta}'}(\Xi_n, \hat{\boldsymbol{\theta}})\right]^{-1}|_{i,i} \neq \left[-\frac{\partial^2 LL}{\partial x^{i2}}\right]^{-1} \tag{6.127}$$

Non-diagonal terms in the matrix are quite complex in the general case. However, assuming that the derivatives are much larger for the i-th input than for the others, the i,i-coefficient of the Hessian matrix is dominant enough for the direct inversion of the i,i-coefficient to become a good approximate:

$$\mathrm{var}\,\hat{x}_m^i \cong \left[-\frac{\partial^2 LL}{\partial x^{i2}}\right]^{-1} \tag{6.128}$$

Then:

$$\frac{\mathrm{var}\,\hat{x}_m^i}{\mathrm{var}\,X^i} \cong \frac{1}{n}\overline{S_j^i}^{-1} \tag{6.129}$$

A clear and fruitful interpretation follows: *when the sensitivity index is large,* the 'relative imprecision' of the estimator of the mean (as normalised by the input uncertainty variance) is small: *estimation accuracy is fair.* However if, within $var\, Y_{mj}$, measurement noise R_j is high compared to X^i-induced variance, then the sensitivity index is lower and estimation variance increases. Intuitively, a factor *1/n* additionally drives the speed of epistemic decrease of the estimation variance.

Remember nevertheless that $var\, X^i$ and $var\, Y_{mj}$ are not known *a priori* in an estimation process: only the derivatives can be computed, supposing the linearisation point can be inferred without too much error. Thus, the sensitivity index may not be accessible. Expertise may, however, infer orders of magnitude of $var\, X^i$.

Further research is needed to clarify when these approximations apply and to investigate the estimation variance for the other parameters to be estimated, such as the estimators of input variances in the intrinsic variability.

6.4 Specificities for parameter identification, calibration or data assimilation algorithms

As shown in the previous sections, parameter identification or calibration algorithms identify the 'central value' of the input variables, $\boldsymbol{X}$ being replaced by a fixed (although unknown) parameter $\boldsymbol{x}_m$, representing, for instance, system properties in a mechanical model or initial conditions in a meteorological model. A review of the particular features will be given for those cases, expanding to the general domain of data assimilation and parameter identification, and the main existing algorithms.

6.4.1 The BLUE algorithm for linear Gaussian parameter identification

Consider the simplest formulation where the system model is *linear* (or linearised), the uncertainty model is Gaussian and where *data only* is assimilated to estimate the value $\boldsymbol{x}_m$ in absence of any prior expertise available. The program to be estimated is the following:

$$\left.\begin{array}{l} Y_{mj} = \mathbf{H}_j(\boldsymbol{x}_m - \boldsymbol{x}_o) + H_j^o + U_j \\ U_j \sim N(0, R_j) \\ identify \quad \boldsymbol{x}_m \quad knowing \quad \Xi_n = \left(\boldsymbol{y}_{m\,j}, \boldsymbol{d}_j, R_j\right)_{j=1\ldots n} \end{array}\right\} \tag{6.130}$$

The BLUE estimator The log-likelihood is the following quadratic function of $\boldsymbol{x}_m$:

$$LL(\Xi_n|x_m) = -\frac{1}{2}\sum_j \log\left(\det\left[2\pi R_j\right]\right) - \frac{1}{2}\sum_j (y_{m\,j} - \mathbf{H}_j(x_m - x_o) - H_j^o)' R_j^{-1} (y_{m\,j} - \mathbf{H}_j(x_m - x_o) - H_j^o) \tag{6.131}$$

the maximisation of which enjoys the following closed-form solution (refer for a demonstration onto the general solution of a quadratic program given in Annex Section 10.5.2):

$$\hat{x}_m = x_o + \left(\sum\nolimits_{j=1\ldots n} H_j' R_j^{-1} H_j\right)^{-1} \left(\sum\nolimits_{j=1\ldots n} H_j' R_j^{-1} \left(y_{mj} - H_j^o\right)\right) \tag{6.132}$$

It may be seen as an algebraic n-dimensional generalisation of the estimator that was observed in Section 6.2.1. Beyond being the maximal likelihood estimator (MLE), this solution is demonstrated classically to be also the Best Linear Unbiased Estimator (BLUE), meaning that:

- it depends linearly upon the observations y_{mj};
- its expectation equals the true unknown value of the parameter x to be identified;
- its estimation variance proves to be minimal amongst the class of estimators that enjoy those two properties.

The variance of the estimator of $\boldsymbol{x}_m$ appears as follows:

$$\mathrm{var}\,\hat{x}_m = \left(\sum\nolimits_{j=1\ldots n} H_j' R_j^{-1} H_j\right)^{-1} \tag{6.133}$$

Straightforward matrix inversion yields such variance although it is often preferred to approximate it through an optimisation (called *variational*) process, in order to avoid the inversion of potentially-high dimensional matrices. The following quadratic program:

$$Min_{x_m} \frac{1}{2}\sum_j (y_{mj} - \boldsymbol{H}_j(x_m - x_o) - H_j^o)' R_j^{-1} (y_{mj} - \boldsymbol{H}_j(x_m - x_o) - H_j^o) \tag{6.134}$$

is solved in meteorology using *variational* approaches under the name of *3D-VAR*. 3D refers to the fact that $\boldsymbol{x}_m$ would generally embody the initial conditions of the atmosphere in three dimensions of space albeit in a *instantaneous* formulation, in opposition to the 4D-Var which is a *dynamic* formulation over the temporal trajectory of observations at different dates: see Section 6.4.5 for more on dynamic data assimilation.

Importance factors and estimation accuracy As a consequence of the general remarks given above on sensitivities and estimation variance, *estimation accuracy* of the components of the BLUE are closely related to the *derivatives of the observational model*. Consider the above-mentioned estimation variance:

$$\mathrm{var}\,\hat{x}_m = \left(\sum\nolimits_{j=1\ldots n} \boldsymbol{H}_j' R_j^{-1} \boldsymbol{H}_j\right)^{-1} \tag{6.135}$$

Consider the case of scalar observations ($q = 1$) that help to visualise the property clearly. The notation may be simplified as follows:

$$h_j^i = \left.\frac{\partial H}{\partial x^i}\right|_{dj} = \frac{\partial H_j}{\partial x^i} \text{ and } R_j = r_j^2 \tag{6.136}$$

Hence:

$$\mathrm{var}\,\hat{x}_m = \left(\sum_j \frac{h_j^r h_j^s}{r_j^2}\right)^{-1} = L_H^{-1} \tag{6.137}$$

In the case of BLUE, the output variance is restricted to the known R_j so that the q-dimensional general formula established in Section 6.3.7 becomes the following:

$$\frac{\partial^2 LL}{\partial x^{i2}} = -\sum_j \left(\frac{\partial H_j}{\partial x^i}\right)' (\mathrm{var}\,Y_{mj})^{-1} \left(\frac{\partial H_j}{\partial x^i}\right) = -\sum_j \left(\frac{\partial H_j}{\partial x^i}\right)' R_j^{-1} \left(\frac{\partial H_j}{\partial x^i}\right) \tag{6.138}$$

Thus, suppose that the derivative of the observational model with respect to the p-th input x^p is higher than for other components, that is coefficients $\left(h_j^p\right)_{j=1\ldots n}$ are much larger than others, say by a scaling factor m:

$$\forall r, s \neq p \quad \left|\sum_j \frac{h_j^r h_j^p}{r_j^2}\right| \Big/ \left|\sum_j \frac{h_j^r h_j^s}{r_j^2}\right| \geq m \tag{6.139}$$

Then, matrix L_H belongs to the following family:

$$\begin{pmatrix} s_{11} & s_{1j} & ms_{1p} \\ s_{j1} & s_{jj} & ms_{jp} \\ ms_{p1} & ms_{pj} & m^2 s_{pp} \end{pmatrix} \tag{6.140}$$

the inverse of which is in fact:

$$\begin{pmatrix} s_{11} & s_{1j} & ms_{1p} \\ s_{j1} & s_{jj} & ms_{jp} \\ ms_{p1} & ms_{pj} & m^2 s_{pp} \end{pmatrix}^{-1} = \begin{pmatrix} t_{11} & t_{1j} & 1/mt_{1p} \\ t_{j1} & t_{jj} & 1/mt_{jp} \\ 1/mt_{p1} & 1/mt_{pj} & 1/m^2 t_{pp} \end{pmatrix} \tag{6.141}$$

Where:

$$\left(s_{ij}\right)^{-1} = \left(t_{ij}\right) \tag{6.142}$$

so that the variance of the estimator is downscaled by a factor m^2:

$$\operatorname{var} \hat{x}_m^p = \left(-\frac{\partial^2 LL}{\partial \mu^r \partial \mu^s}\right)^{-1}\Bigg|_{p,p} = \frac{1}{m^2} t_{pp} \tag{6.143}$$

Quite intuitively, variance decreases – *that is* estimation accuracy increases – if the observational model becomes more sensitive to a given component x^p. Conversely, observations Y_j made under experimental circumstances $\boldsymbol{d}_j$ for which large changes in x^i values entail little changes in the model output $y = H(\boldsymbol{x},\boldsymbol{d}_j)$ will not help much in the identification of such a parameter.

The estimation variance: purely epistemic uncertainty As mentioned in the previous paragraph, the estimation variance computed by the BLUE algorithm is essentially an indicator of the accuracy of the identification process. This variance cannot, however, be interpreted as the intrinsic variability of a *changing* physical value $\boldsymbol{X}$: it represents only the statistical fluctuation of the estimate of a *fixed* physical value $\boldsymbol{x}_m$, that is purely *epistemic* or reducible uncertainty.

As demonstrated in Section 6.2.1 in the non-linear 1-dimensional case, it may be proved that this diminishes to zero when the sample size increases, in properly-defined non-degenerated cases. Intuitively, the arguments put forward in the 1D case still hold as the estimator variance:

$$\operatorname{var} \hat{x}_m = \left(\sum\nolimits_{j=1\ldots n} \boldsymbol{H}'_j R_j^{-1} \boldsymbol{H}_j\right)^{-1} \tag{6.144}$$

appears as the inverse of the sum of an ever-increasing number of inversible matrices. In non-degenerate cases, all observations have a minimal accuracy. This means that R_j, the covariance of measurement errors, has positive diagonal terms superiorly bounded by a maximal error variance, or conversely that R_j^{-1} has positive diagonal terms inferiorly bounded by a given positive value. As explained in the previous section, H_j embodies the model sensitivities: once again, non-degenerate cases mean that we are trying to estimate parameters to which a minimal sensitivity is guaranteed, so that the scalar product $H_j'H_j$ brings positive terms inferiorly bounded by a minimal square sensitivity. Hence, the ever-increasing sum of inferiorly-bounded positive terms brings the norm down to zero.

6.4.2 An extension with unknown variance: Multidimensional model calibration

The previous data assimilation settings have assumed that the model-measurement error variance matrices R_j are known, which is the situation when sensor imprecision is assumed to be dominant and the model is assumed to be good enough. Similar to the one-dimensional case, extensions may lead to estimating those variances also, supposing they are unknown because the model needs to be fully calibrated, including its residual error. This leads into the field of (multidimensional) model calibration.

In order for the problem to be identifiable, there needs to be some hypotheses, typically assuming a form for the matrices as was suggested in Section 6.3.2 with the following model:

$$R_j = \sigma^2 \Gamma_j$$

where all Γ_j are assumed to be known, only the averaged variance σ^2 needing estimation. Similar to the one-dimensional cases studied in Section 6.2.1, this does not change the estimator and associated variance for $\boldsymbol{x}_m$ but simply adds on a very intuitive estimator for the variance σ^2. The new model is the following:

$$\left.\begin{array}{l} Y_{mj} = \mathbf{H}_j(\boldsymbol{x}_m - \boldsymbol{x}_o) + H_j^o + U_j \\ U_j \sim N(0, \sigma^2\Gamma_j) \\ \textit{identify} \quad (\boldsymbol{x}_m, \sigma^2) \quad \textit{knowing} \quad \Xi_n = \left(\boldsymbol{y}_{mj}, \boldsymbol{d}_j, \Gamma_j\right)_{j=1\ldots n} \end{array}\right\} \tag{6.145}$$

Log-likelihood becomes the following (remember that q represents the dimension of $\boldsymbol{y}_{mj}$, that is the number of types of physical variables observed n times):

$$\begin{aligned} LL(\Xi_n|x_m, \sigma^2) &= -\frac{1}{2}\sum_j \log\left(\det\left[2\pi R_j\right]\right) - \frac{1}{2}\sum_j (y_{mj} - \mathbf{H}_j(x_m - x_o) - H_j^o)' R_j^{-1} (y_{mj} - \mathbf{H}_j(x_m - x_o) - H_j^o) \\ &= -\frac{1}{2} qn\log\sigma^2 - \frac{1}{2}\sum_j \log\left(\det\left[2\pi\Gamma_j\right]\right) - \frac{1}{2\sigma^2}\sum_j (y_{mj} - \mathbf{H}_j(x_m - x_o) - H_j^o)'\Gamma_j^{-1}(y_{mj} - \mathbf{H}_j(x_m - x_o) - H_j^o) \end{aligned} \tag{6.146}$$

Maximisation is done separately for $\boldsymbol{x}_m$ and for σ^2 and yields the following closed-form solution:

$$\begin{aligned} \hat{x}_m &= x_o + \left(\sum\nolimits_{j=1\ldots n} \boldsymbol{H}_j'\Gamma_j^{-1}\boldsymbol{H}_j\right)^{-1} \left(\sum\nolimits_{j=1\ldots n} \boldsymbol{H}_j'\Gamma_j^{-1}\left(y_{mj} - H_j^o\right)\right) \\ \hat{\sigma}^2 &= \frac{1}{n}\sum_j (y_{mj} - \boldsymbol{H}_j(x_m - x_o) - H_j^o)'\Gamma_j^{-1}(y_{mj} - \boldsymbol{H}_j(x_m - x_o) - H_j^o) \end{aligned} \tag{6.147}$$

Associated variances are estimated asymptotically as follows, the estimators being again asymptotically uncorrelated (refer for a demonstration to the general solution of a quadratic program given in Annex Section 10.4.6):

$$\begin{aligned} \operatorname{var}\hat{x}_m &\cong \hat{\sigma}^2\left(\sum_{j=1\ldots n}\boldsymbol{H}_j'\Gamma_j^{-1}\mathbf{H}_j-1\right. \\ \operatorname{var}\hat{\sigma}^2 &\cong \frac{2\hat{\sigma}^4}{qn} \end{aligned} \tag{6.148}$$

This means that the estimator for $\boldsymbol{x}_m$ is similar to that of simple parameter identification under known variance and that the estimator is simply the mean residual (weighted according to known relative model inaccuracies) after model calibration. It is a direct multi-dimensional generalisation of the heteroscedastic model calibration introduced in Section 6.2.1.

Similar to the BLUE algorithm, such an estimation variance delineates purely epistemic uncertainty, again decreasing to zero as the sample size increases. The only uncertainty that may remain after incorporation of big data samples is encapsulated in the model-measurement error term U_j; or simply model prediction error, if measurement error is relatively negligible. It is not apportioned to input intrinsic variability.

Similar to the 1D-case, note that the MLE estimator for the residual variance is biased when nq is small, a situation in which the following unbiased formulation is more adequate:

$$\hat{\sigma}_u^2 = \frac{nq}{nq-p-1}\hat{\sigma}^2 \tag{6.149}$$

It is a classical result of multiple regression with $p + 1$ parameters to be estimated ($\boldsymbol{x}_m$ and σ^2) on the basis of nq observations (see Annex Section 10.4.6) though such a correction will often be neglected.

6.4.3 Generalisations to non-linear calibration

In some cases, it is no longer legitimate to approximate the models' H_j by their linearised approximations, through local derivatives: the non-linearity becomes significant at the scale of the prior lack of knowledge. Thus, Equation (6.145) becomes the following:

$$\left.\begin{aligned} &Y_{mj} = H_j(\boldsymbol{x}_m) + U_j \\ &U_j \sim N(0, R_j) = N(0, \sigma^2\Gamma_j) \\ &\textit{identify} \quad (\boldsymbol{x}_m, \sigma^2) \quad \textit{knowing} \quad \left(\boldsymbol{y}_{m\,j}, \boldsymbol{d}_j, \Gamma_j\right)_{j=1\ldots n} \end{aligned}\right\} \tag{6.150}$$

Log-likelihood becomes then a non-linear function of $\boldsymbol{x}_m$ and σ^2:

$$LL(\Xi_n|x_m,\sigma^2) = -\frac{1}{2}qn\log\sigma^2 - \frac{1}{2}\sum_j \log\left(\det\left[2\pi\Gamma_j\right]\right) - \frac{1}{2\sigma^2}\sum_j (y_{m\,j}-H_j(x_m))'\Gamma_j^{-1}(y_{m\,j}-H_j(x_m)) \tag{6.151}$$

the maximisation of which no longer enjoy a closed-form solution. Optimisation algorithms become necessary. A classical approach involves adaptive linearisation:

- infer a prior value $x^{(k)}$;
- compute the derivatives of H at that point;
- launch the linearised algorithm, which enjoys a closed-form solution;
- take the linear estimate as the new prior value $x^{(k+1)}$ for the new iteration.

Such an iterative process involves $N_{it}{}^{*}(n+np)$ computations, a more or less competitive approach as compared to the direct non-linear optimisation algorithm in the log-likelihood.

An equivalent rewriting more familiar to statisticians would be the following:

$$\left.\begin{array}{l} \boldsymbol{Y}_{mj}\big|\boldsymbol{d}_j,\boldsymbol{\theta} = H(\boldsymbol{d}_j,\boldsymbol{\theta}) + \boldsymbol{U}_j \\ \boldsymbol{U}_j \sim f_U(\boldsymbol{u}\big|\sigma^2\Gamma_j) = N(0,\sigma^2\Gamma_j) \\ \textit{identify} \quad (\boldsymbol{\theta},\sigma^2) \quad \textit{knowing} \quad \left(\boldsymbol{y}_{mj},\boldsymbol{d}_j,\Gamma_j\right)_{j=1\ldots n} \end{array}\right\} \tag{6.152}$$

the unknown but fixed parameter $\boldsymbol{x}$ being treated as the parameter $\boldsymbol{\theta}$ of a regression model. Non-linear calibration can equally be viewed as a non-linear hetero-scedastic regression. Note that the estimation variance may still be assessed by the information matrix involving the gradients and Hessians of the observational model.

6.4.4 Bayesian multidimensional model updating

Bayesian extensions are simple in the linear Gaussian case with known variance. Supposing that prior knowledge is incorporated in the following distribution:

$$X \sim N(\boldsymbol{x}_b, B) \tag{6.153}$$

where $\boldsymbol{x}_b$ stands for the vector of best-estimates for the unknown parameters $\boldsymbol{x}$, and B for the covariance representing the prior uncertainty about those values. Note that non-diagonal terms inside B embody the potential correlations between the errors of appreciation by the expert of parameter values. The estimator changes slightly to incorporate the prior expertise:

$$\hat{x}_a = x_o + \left(B^{-1} + \sum\nolimits_{j=1\ldots n} \boldsymbol{H}_j' R_j^{-1} \boldsymbol{H}_j\right)^{-1} \left(B^{-1}(x_b - x_o) + \sum\nolimits_{j=1\ldots n} \boldsymbol{H}_j' R_j^{-1}\left(y_{mj} - H_j^o\right)\right) \tag{6.154}$$

the variance of which becomes the following:

$$\text{var } \hat{x}_a = \left(B^{-1} + \sum\nolimits_{j=1\ldots n} \boldsymbol{H}_j' R_j^{-1} \boldsymbol{H}_j\right)^{-1} \tag{6.155}$$

These formulae are algebraically-rearranged extensions of the '*BLUE with prior*' estimator that is very popular in data assimilation (cf. next section). Suppose that the linearisation point is being taken at the prior best-estimate $\boldsymbol{x}_o = \boldsymbol{x}_b$, and non-varying experimental conditions, then the formulae become the following:

$$\hat{x}_a = x_b + K.(\boldsymbol{y} - H(x_b)) \quad A = \text{var } \hat{x}_a = \left(Id_p - K.\mathbf{H}\right).B \tag{6.156}$$

where K stands for the *optimal gain* that weighs the correction brought by the observation sample with minimal variance.

$$K = \left(B^{-1} + \mathbf{H}'.\left(\frac{R}{n}\right)^{-1}.\mathbf{H}\right)^{-1}.\mathbf{H}'.\left(\frac{R}{n}\right)^{-1} = B.\mathbf{H}'.\left(\mathbf{H}.B.\mathbf{H}' + \frac{R}{n}\right)^{-1} \tag{6.157}$$

Note two important properties of this Bayesian estimator: it can firstly be interpreted as a classical estimator where a *direct* observation $\boldsymbol{x}_b$ noised with B (i.e. with an observational model $H_o = Id_p$ and $R_o = B$) of the uncertain input $\boldsymbol{x}$ would be added to the indirect observations through $(\boldsymbol{y}_{mj})_{j=1\ldots n}$; secondly, it can be viewed as the solution of a variational problem with the following modified cost function:

$$Min_{x_m}\begin{bmatrix} \frac{1}{2}\sum_j (y_{mj} - \mathbf{H}_j(x_m - x_b) - H_j^o)' R_j^{-1}(y_{mj} - \mathbf{H}_j(x_m - x_b) - H_j^o) \\ + \frac{1}{2}(x_m - x_b)' B^{-1}(x_m - x_b) \end{bmatrix} \tag{6.158}$$

Extensions with unknown variance, non-linear or non-Gaussian models may also be developed.

6.4.5 Dynamic data assimilation

Note that data assimilation techniques are used less in the risk model formulation developed so far than in *forecast modelling-like* cases where an additional forecast equation is made available between the time of the analysis and the time of interest, usually in the short-term ahead. In such a formulation, $\boldsymbol{x}$ refers to the state $\boldsymbol{x}(t)$ of the system at time t, or initial conditions at time $t_o = 0$ for which there are two sub-models, the noised observational model at time t considered up to now:

$$Y_{mj} = H[x(t_j), \boldsymbol{d}(t_j)] + U_j = H_j[x_j] + U_j \tag{6.159}$$

as well as a noised forecasting model:

$$X_f(t_{j+1}) = M_{j->j+1}[x(t_j), \boldsymbol{d}(t_{j+1})] + V_{j+1} = M_j[x_j] + V_{j+1} \tag{6.160}$$

predicting an approximate forecast $X_f(t_{j+1})$ (f as forecast) for the state of the system $x(t_{j+1})$ at time $j + 1$, with a forecasting error V_{j+1} conditional upon the knowledge of the true state $x(t_j)$ at time j. Variances of the observational errors U_j and forecasting error V_j (noted R_j and Q_j respectively) are generally assumed to be known, though prior calibration processes could be undertaken similarly to the above-mentioned algorithms.

Such a forecasting setting is not the central topic of this book, and will only be reviewed briefly: the reader may refer to Talagrand and Courtier (1987) and Talagrand (1997). In short, the classical alternatives for estimating the trajectory are the following (Figure 6.11):

- Sequential algorithms (*Kalman filters – KF*): step by step, t_j-state is predicted on the basis of the forecast of the previous best estimate at time $t_j - 1$ and then corrected with the t_j -observations through a cost function similar to the Bayesian 3DVar, or 'BLUE with prior' algorithm mentioned earlier.

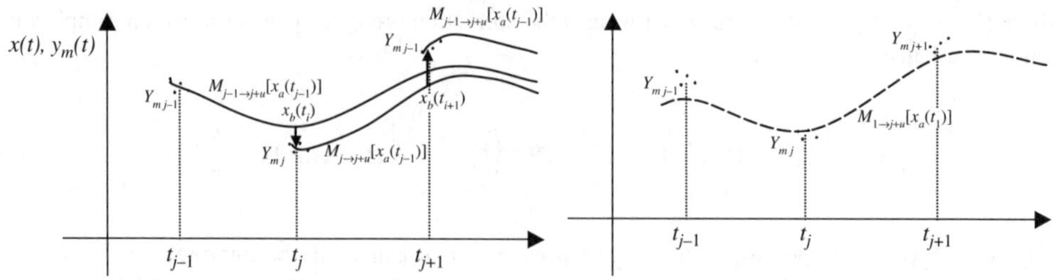

Figure 6.11 *Unknown true states and estimated states (assuming direct observations H = Id): Sequential Kalman filter (left) and 4D-Var algorithms (right).*

- Variational formulations (*4DVar and derived*): the whole trajectory is assimilated at once within a time-compounded cost function in order to perform a deeper re-analysis.

Sequential algorithms update the prediction whenever a new piece of observation becomes available, while the 'future observations' are logically not taken into account for the calibration which fits logically the *forecasting* studies. On the contrary, 4DVar includes the observation along a whole trajectory to calibrate at once the model, typically helping the so-called '*re-analysis*' studies of, for example, the weather dynamics and models.

Here is a simple comparison of the cost functions that are minimised by each competing algorithm: for the sake of simplicity, linearization is assumed to be done at the origin. The basic *KF* incorporates within a 'BLUE with prior' the group of observations corresponding to time t_j as well as a prior issued by the forecast of the t_{j-1} analysis in order to update the state of the system at the same time and then forecast again on that basis as follows:

$$\begin{aligned} x_b(t_j) &= x_f(t_j) = M_{j-1\to j}\left[x_a(t_{j-1})\right] \\ x_a(t_j) &= Arg\min\begin{bmatrix} J(x) = \frac{1}{2}(y_m(t_j)-\mathbf{H}_j x)'R_j^{-1}(y_m(t_j)-\mathbf{H}_j x) \\ +\frac{1}{2}(x-x_b(t_j))'B^{-1}(x-x_b(t_j)) \end{bmatrix} \\ x_f(t_{j+1}) &= M_{j\to j+1}\left[x_a(t_j)\right] \end{aligned} \tag{6.161}$$

The matrices of covariance of the estimators can be deduced from the formulae given in the earlier section, with the following classic notation for *KF*:

$$\begin{aligned} P_j^f &= \text{var } \hat{x}_b(t_j) \\ P_j^a &= \text{var } \hat{x}_a(t_j) = \left(Id_p - K_j.\mathbf{H}_j\right).P_j^f \\ K_j &= P_j^f.\mathbf{H}'_j.\left(\mathbf{H}_j.P_j^f.\mathbf{H}'_j + R_j\right)^{-1} \\ P_{j+1}^f &= \text{var } \hat{x}_f(t_{j+1}) = M_{j\to j+1}P_j^a M_{j\to j+1}' + Q_{j+1} \end{aligned} \tag{6.162}$$

Note that the so-called 'true' state of the system at time t_j is modelled through a fixed vector $\boldsymbol{x}_j$, though it is unknown and directly unobservable. Its estimate is thus affected by residual uncertainty which is purely *epistemic*: given a very large group of quasi-simultaneous observations of the system around time t_j the variance of X_j^a could decrease to zero. However, this is never the case in the historical applications of the scheme in meteorology where the dimension of the state of the system far exceeds that of the observations. Additionally, while the 3DVar-based identification step incorporates purely epistemic uncertainty in the estimation of the state $\boldsymbol{x}_j$ at time t_j, the additive forecasting uncertainty V_j brings an *irreducible* component of uncertainty at each step. Unlike the purely 3DVar formulation discussed in Section 6.4.1, the covariance matrix P^f_{n+1} of the *KF* forecast does not tend to zero even after a very large number of observational steps ($n -> \infty$): simply because of the forecasting variance term Q_{n+1}. From the time of analysis t_j, the uncertainty in the future state at times (t_k) $_{k=j+1,...n}$ can thus be computed through the recursive *KF* variance equations. They represent in fact a linear or Taylor approximation for uncertainty propagation (see Chapter 7, Section 7.3.2) through the chained recursive linear forecast model taken as the system model for future times of interest, combining both epistemic estimation and aleatory forecasting uncertainty (see Figure 6.11).

4DVar incorporates at once all the observations from time t_1 to t_j (as well as an initial prior before t_1) in order to update the complete trajectory of the states of the system from time t_1 to t_j. The technique is more precisethan 3DVar for the *re-analysis* of a past trajectory. This involves the chaining of the forecast models from the initial state up to the states corresponding to each of the t_k observations. A simplified formulation ignoring forecasting error results in the following program:

$$
\begin{aligned}
& x_b(t_1) = x_b \\
& \forall k > 1, HM_k(x) = H_k o M_{k-1 \to k} o M_{k-2 \to k-1} o \ldots M_{1 \to 2}[x] \quad HM_1(x) = H_1[x] \\
& x_a(t_1) = Arg \min \begin{bmatrix} J(x) = \dfrac{1}{2} \sum\limits_{k=1...j} (y_m(t_k) - HM_k(x))' R_k^{-1} (y_m(t_k) - HM_k(x)) \\ + \dfrac{1}{2} (x - x_b) \prime B^{-1} (x - x_b) \end{bmatrix} \\
& x_f(t_{j+1}) = M_{j \to j+1} o M_{j-1 \to j} o \ldots M_{1 \to 2}[x_a(t_1)]
\end{aligned}
\tag{6.163}
$$

Both approaches coincide in the idealised case where the forecasting model is perfect (noiseless given the true state) and linear. Extensions have been developed for significantly non-linear models: *Extended KF* (*EKF*) includes adaptive linearization while *Ensemble KF* (*EnKF*) involves a Monte-Carlo estimation of the expectations, or in other words the combined simulation of multiple uncertain trajectories.

Even when linearised, key computational difficulties arise when considering highly numerous degrees of freedoms: in 2011, computations for the weather forecast undertaken by the ECMWF (European Centre for Medium-Range Weather Forecasts) involved $p = 10^9$ with $nq \sim 10^7$ scalar observations. Indeed, a considerable literature has been developed in order to tackle to high-dimensional non-linear cases that are prevalent, for instance, in meteorological forecasting, which is the original field of application of the algorithms (Talagrand, 1997).

Dynamic data assimilation assumes implicitly that model parameters remain constant. *Abrupt change of the parameters* corresponds to another kind of uncertainty, think of the example of global warming. This kind of situation can be managed through sequential detection of change, see Siegmund (1985) or

Basseville and Nikiforov (1993) and Shiryaev (2010). It can also be validated by *a posteriori* change point analysis, see Csörgo and Horvath (1997) or Brodsky and Darkhovsky (1993).

6.5 Intrinsic variability identification

This section is devoted to identifying the full variability or uncertainty of model inputs, reviewing the main existing algorithms and the important research challenges remaining.

6.5.1 A general formulation

The inverse problem thus leads to the following program:

$$\left.\begin{array}{l} \boldsymbol{Y}_{mj} = H(\boldsymbol{X}_j, \boldsymbol{d}_j) + \boldsymbol{U}_j \\ \boldsymbol{X}_j \sim f_X(.|\boldsymbol{\theta}_X), \quad \boldsymbol{U}_j \sim f_U(.|\boldsymbol{\theta}_{uj}) \\ identify \quad \boldsymbol{\theta}_X \quad knowing \quad \Xi_n = \left(\boldsymbol{y}_{mj}, \boldsymbol{d}_j\right)_j and \quad \left(\boldsymbol{\theta}_{uj}\right)_j \end{array}\right\} \tag{6.164}$$

Note that realisations of $\boldsymbol{x}$ are deliberately assumed to be changing physically from the j-th to the $j+1$-th experiment, taking the form of i.i.d. random variables $\boldsymbol{X}_j$ following the fixed distribution f_X. Think, for instance, of mechanical observations of different pieces following a given design, but manufactured with (unavoidably) slightly varying dimensions. Thus, unlike traditional model calibration or data assimilation techniques, the inverse probabilistic algorithms considered hereafter do not foster the estimation of one *fixed* (although unknown) input value $\boldsymbol{x}$ explaining the data sample. They will process the information represented by Ξ_n in order to retrieve the *parameters* $\boldsymbol{\theta}_X$ of the input *uncertainty distribution* f_X.

Recall that in the general case, Maximal Likelihood Estimation (MLE) consists of maximising Equation (6.165) through the adjustment of some or all components of the vector $\boldsymbol{\theta}_X$ (and/or of $\boldsymbol{\theta}_{uj}$):

$$L\left[\left(\boldsymbol{y}_{mj}\right)_{j=1\ldots n}\Big|\left(\boldsymbol{d}_j, \boldsymbol{\theta}_{uj}\right)_{j=1\ldots n}, \boldsymbol{\theta}_X\right] = \prod_j \int_{\boldsymbol{x}} f_U(\boldsymbol{y}_{mj} - H(\boldsymbol{x}, \boldsymbol{d}_j)|\boldsymbol{\theta}_{uj}) f_X(\boldsymbol{x}|\boldsymbol{\theta}_X) d\boldsymbol{x} \tag{6.165}$$

Note that such an estimation program involves *i.non-i.d.* (non-identically distributed though generally assumed as independent) samples, which calls for more theoretical work to establish convergence properties. The proof of convergence and asymptotic normality in the *i.i.d.* case is a special application of the Law of Large Numbers and Central Limit Theorem (Section 5.4.3). Suppose that experimental conditions $\boldsymbol{d}_j$ do not vary, and that error variances $\boldsymbol{\theta}_{uj} = R_j$ are stable. In such a case, $(Y_{mj})_{j=1\ldots n}$ is an *i.i.d.* sample. Although complex, the normalised log-likelihood of the sample still appears as the empiric mean of *i.i.d.* log-densities:

$$\frac{1}{n} LL\left(\left(\boldsymbol{y}_{mj}\right)_{j=1\ldots n}, \boldsymbol{\theta}_X\right) = \frac{1}{n} \sum_{j=1\ldots n} Log f_{Ym}\left[Y_{mj}|\boldsymbol{\theta}_X\right]$$

Given that the classical MLE conditions recalled in Section 5.4.3 are fulfilled, convergence is thus guaranteed. The *i.i.d.* character of the observed sample is still preserved if experimental conditions $\boldsymbol{d}_j$ vary at random (*i.i.d.* following a known distribution) and error variances R_j are stable.

Considering then that the independent non-identically distributed (*i.non-i.d.*) case is trickier, more complex conditions would need to be formulated in the non-identically distributed case: some 'stableness' of the variation of experimental conditions $\boldsymbol{d}_j$ and R_j are wanted so that there still exists an equivalent of the expectation of the likelihood of the observed sample as is stated in classical non-linear regression theory (Jennrich, 1969; Hoadley, 1971).

In all cases, regularity conditions supposedly require at least the density of X (easily-fulfilled, since standard pdfs are generally involved), and of the observational model H: the flood model, for instance, is clearly C^2. While a more formal proof would be needed, it is reasonable to expect that the expectations further smooth the dependence on θ_X. The real check should come again from the identifiability point of view: it is a compulsory preliminary condition to be checked before undertaking a solving algorithm though it may be difficult to prove in the general non-linear case beyond simple checks (see Section 6.3.6).

Then, while the Expectation-Maximisation (E-M) algorithm is a classical choice for such likelihood maximisation, any optimisation algorithm could be mobilised in theory. The number of iterations required for a proper convergence of E-M may however be unrealistic due to the CPU requirement for a complex H. The big issue is the number of iterations required to ensure a stable (and robust) result: each estimate of log-likelihood (Equation (6.165)) for a given value of $\boldsymbol{\theta}_X$ requiring typically a large number of Monte-Carlo simulations. Later sections will review the algorithms that have been developed for that purpose, according to the type of hypotheses or approximations that may be made in the system model.

6.5.2 Linearised Gaussian case

Linear and Gaussian assumptions simplify greatly the problem. In order to estimate $\boldsymbol{\theta}_X = (\boldsymbol{x}_m, V)$, as already mentioned, log-likelihood (*LL*) becomes:

$$\begin{aligned} LL(\Xi_n|(x_m, V)) = &-\frac{1}{2}\sum_j \log\left(\det\left(2\pi\left[R_j + \mathbf{H}_j V \mathbf{H}'_j\right]\right)\right) \\ &-\frac{1}{2}\sum_j (y_{mj} - \mathbf{H}_j(x_m - x_o) - H_j^o)'(R_j + \mathbf{H}_j V \mathbf{H}'_j)^{-1}(y_{mj} - \mathbf{H}_j(x_m - x_o) - H_j^o) \end{aligned} \tag{6.166}$$

LL cannot be maximised with a closed-form solution in the general case where V is unknown, since Equation (6.166) is a non-quadratic multi-dimensional function. Note the major mathematical difference with the data assimilation case: there, MLE is a classical quadratic program, whereby R_j represent known weights associated with a classical quadratic cost function. In the fully probabilistic inversion, $R_j + \boldsymbol{H}_j V \boldsymbol{H}_j'$ represent partially unknown weights, within which V, representing the *intrinsic* input uncertainty around $\boldsymbol{x}_m$, needs also to be identified. Consequently, Equation (6.166) is no more quadratic as a function of the unknowns and no longer enjoys a closed-form solution. An optimisation algorithm is needed.

de Crécy (1997) introduced the CIRCE algorithm in order to solve such a problem when observable outputs Y are scalar; the algorithm is in fact based on a variant of the general *Expectation-Maximisation* (*E-M*) statistical algorithm (Dempster, Laird, and Rubin, 1977) called ECME (Expectation Conditional Maximisation Either) (Liu and Rubin, 1994). The algorithm can easily be generalised to vector observable outputs as suggested in Mahe and de Rocquigny (2005) and Celeux *et al.* (2010). It comes as a sequential updating program enjoying closed-form matrix transformation steps. Note ($\boldsymbol{x}^{(k)}$, $V^{(k)}$) the k-th iteration value for the desired parameter vector as well as the following quantities:

$$W_j^{(k)} = R_j + \boldsymbol{H}_j V^{(k)} \boldsymbol{H}_j' \tag{6.167}$$

$$A_j^{(k)} = \boldsymbol{y}_{mj} - \boldsymbol{H}_j(\boldsymbol{x}^{(k)} - \boldsymbol{x}_o) - \boldsymbol{H}_j^{\circ} \tag{6.168}$$

These quantities may be interpreted respectively as the k-th iteration approximate for the full output variance of the j-th observation combining the image through $\boldsymbol{H}_j$ of the intrinsic variance estimate $V^{(k)}$ as well as the observation error variance R_j; and the k-th iteration approximate for the deviation of the j-th observed value to the expected value of the observable output. Thus, the $k+1$-th iteration involves updating the parameter vector through the following equations:

$$V^{(k+1)} = V^{(k)} + \frac{1}{n}\sum_{j=1}^{n} V^{(k)}\mathbf{H}'_j W_j^{(k)-1}\left(A_j^{(k)} A_j^{(k)\prime} W_j^{(k)-1} - Id_q\right).\mathbf{H}_j V^{(k)} \tag{6.169}$$

$$\begin{aligned} \boldsymbol{x}^{(k+1)} &= \boldsymbol{x}^{(k)} + \left[\sum_{j=1}^{n} \mathbf{H}'_j W_j^{(k+1)-1}\mathbf{H}_j\right]^{-1}\left(\sum_{j=1}^{n} \mathbf{H}'_j W_j^{(k+1)-1} A_j^{(k)}\right) \\ &= \boldsymbol{x}_o + \left[\sum_{j=1}^{n} \mathbf{H}'_j W_j^{(k+1)-1}\mathbf{H}_j\right]^{-1}\left(\sum_{j=1}^{n} \mathbf{H}'_j W_j^{(k+1)-1}\left(\boldsymbol{y}_j - \boldsymbol{H}_j^{o}\right)\right) \end{aligned} \tag{6.170}$$

Note that these updating equations may be interpreted as follows:

- Variance updating involves the deviation between one and the ratio of output empiric dispersion $A_j^{(k)} A_j^{(k)\prime}$ to the k-th iteration estimate of total output variance $W_j^{(k)}$.
- Expectation updating reads simply as the data assimilation (or calibration) estimator taking, instead of the observation error variance (R only in data assimilation), the total output variance $W_j^{(k+1)}$ after k-th correction of the variance estimation.

As a variant to the CIRCE/ECME algorithm, standard EM can also be used for the linearised Gaussian problem. In the present case, the traditional EM algorithm is quite close to CIRCE/ECME: it involves the slightly modified updating equation for variance of the expectation, replacing Equation (6.169) by the following:

$$\begin{aligned} V^{(k+1)} &= V^{(k)} + \frac{1}{n}\sum_{j=1}^{n} V^{(k)}\mathbf{H}'_j W_j^{(k)-1}\left(A_j^{(k)} A_j^{(k)\prime} W_j^{(k)-1} - Id_q\right).\mathbf{H}_j V^{(k)} \\ &- \frac{1}{n^2}\left(\sum_{j=1}^{n} V^{(k)}\mathbf{H}'_j W_j^{(k)-1} A_j^{(k)}\right)\left(\sum_{j=1}^{n} V^{(k)}\mathbf{H}'_j W_j^{(k)-1} A_j^{(k)}\right)' \end{aligned} \tag{6.171}$$

$$x^{(k+1)} = x^{(k)} + \frac{1}{n}\left(\sum_{j=1}^{n} V^{(k)}\mathbf{H}'_j W_j^{(k)-1} A_j^{(k)}\right) \tag{6.172}$$

ECME is generally assumed to offer better convergence properties in that case than EM (see Celeux *et al.*, 2010) for an application to flood). Remember the key advantage of those algorithms (either Circé/ECME or EM): only a limited number of runs of the potentially high-CPU time consuming model H are necessary, those which are necessary to compute the n values of H_j° and of the partial derivatives $\boldsymbol{H}_j$. Either they are coded inside the numerical model or they may always be estimated through finite differences, involving at least $p.n$ additional computations of $\{H_j(\boldsymbol{x}+\varepsilon_i)\}_{i=1,\ldots p,\ j=1,\ldots n}$ where $\varepsilon_i = (\varepsilon\delta_{ii'})_{i'=1,\ldots p}$ and $\delta_{ii'} = 1$ if $i = i'$, or 0 otherwise.

6.5.3 Non-linear Gaussian extensions

If the uncertainties are too large to keep the linearisation acceptable (with respect to $\boldsymbol{x}$), while $\boldsymbol{X}$, $\boldsymbol{Y}$ and $\boldsymbol{U}$ might still be Gaussian, then two strategies could be considered (Barbillon *et al.*, 2011):

(i) adaptative linearisation;
(ii) non-linear Gaussian likelihood maximisation.

In fact, (i) consists of undertaking the same maximisation program such as that of Equation (6.166), albeit sequentially with multiple linearisations. Start with derivatives on a given linearisation point, then maximise the resulting Equation (6.166) (with, for instance, the CIRCE algorithm) in order to infer an expectation $\boldsymbol{x}_m^{(1)}$. Then linearise H around this new point, and launch again CIRCE to infer another expectation $\boldsymbol{x}_m^{(2)}$. Continue the process hoping that $\boldsymbol{x}_m^{(1)}$ will stabilise.

Under strategy (ii), still assuming that $\boldsymbol{Y}$ can be modelled as Gaussian (which is not automatic in this case where $\boldsymbol{X}$ is Gaussian but H is non-linear), the log-likelihood maximisation problem is written under its non-linear formulation:

$$LL(\Xi_n|(x_m,V)) = -\frac{1}{2}\sum_j \log\left(\det 2\pi\left[R_j + \mathrm{var}\left(H(X,\boldsymbol{d}_j)|(x_m,V)\right)\right]\right)$$
$$-\frac{1}{2}\sum_j \left(y_{mj} - E\left(H(X,\boldsymbol{d}_j)|(x_m,V)\right)\right)'\left(R_j + \mathrm{var}\left(H(X,\boldsymbol{d}_j)|(x_m,V)\right)\right)^{-1}\left(y_{mj} - E\left(H(X,\boldsymbol{d}_j)|(x_m,V)\right)\right) \tag{6.173}$$

Under the significantly non-linear behaviour of $H(\boldsymbol{x},\boldsymbol{d}_j)$, expectations $\left(E\left[H(X,\boldsymbol{d}_j)|(x_m,V)\right]\right)_{j=1\ldots n}$ and variances $\left(\mathrm{var}\left[H(X,\boldsymbol{d}_j)|(x_m,V)\right]\right)_{j=1\ldots n}$ can no longer be computed in a closed-form. They are simulated under Monte-Carlo sampling for each possible value of $\boldsymbol{\theta}_X$, thus making the likelihood maximisation quite CPU-consuming.

Note that those two strategies may become much more CPU-consuming, since multiple runs of H are needed at each step of the maximisation algorithm, either in (i) to compute the partial derivatives in the new expectation estimate or in (ii) to simulate the expectation and variance of the observable for the new expectation and covariance estimates: meta-modelling becomes quite attractive in that case. Remember nevertheless that, once the model is non-linear, it is more difficult to guarantee *a priori* the Gaussian nature of $\boldsymbol{Y}$ even if $\boldsymbol{X}$ and $\boldsymbol{U}$ are Gaussian. This may reduce the relevance of Equation (6.173) as an excessively approximate log-likelihood.

6.5.4 Moment methods

In the general (non-Gaussian, non-linear) case, the moments of Y_m can still be simulated for a given value of $\boldsymbol{\theta}_X$. An estimation algorithm may then involve an iterative simulation of those moments and their comparison to the empiric moments of the sample measurements Y_{mj} for progressively-adjusted values of $\boldsymbol{\theta}_X$ until acceptable fit. Such a procedure is comparable to the classical moment methods in standard statistical estimation: it may require fewer simulations than the computation of the n multiple integrals required by MLE.

A priori, the same limitations should be expected from such a procedure as those encountered when using moment methods for simple statistical samples: moments should generate less efficient estimators than MLE, although possibly less CPU-intensive.

In the context of a large number of uncertain physical variables and reduced observable data, a rather original choice has been put forward by Soize (2000) and applied especially in industrial vibration mechanics with a large number of degrees of freedom (Ratier, 2004). This consists of re-setting the parameters of the sources of uncertainty so that the vector $\boldsymbol{X}$ no longer indicates the physical variables (Young's modulus, dimensions of structures, etc.) but the coefficients of the system of equations (matricials) expressing the physical knowledge of the industrial reality (e.g. matrices of rigidity or mass in vibratory mechanics). The law in $\boldsymbol{X}$ is chosen by application of the principle of maximum entropy in a space representing the fixed physical constraints (e.g. matrices defined positive for physical objects/systems), and the identification of $\boldsymbol{\theta}_X$ requires a subsequent step B′, the X^i being, by construction, unobservable. The method of moments was used in that context to identify $\boldsymbol{\theta}_X$: note that $\boldsymbol{\theta}_X$ is deliberately of a very low dimension as the dispersion of matrices is modelled parsimoniously under the maximum entropy principle with only one coefficient per class of matrix, resulting in *dim* $\boldsymbol{\theta}_X = 3$ for standard vibrationary mechanical models.

6.5.5 Recent algorithms and research fields

The more general setting where the model is not linear and/when the output distribution cannot be assumed satisfactorily to be Gaussian is an area open for research. A brief introduction will be given hereafter to some recent algorithms that involve:

- Non-parametric simulated likelihood: the idea is to simulate directly the output observable density and smooth it through a kernel technique in order to maximise likelihood without calling upon the system model too often.
- Meta-modelling and stochastic variants of EM: in the absence of linearity, the conditional expectations involved in EM cannot be formulated in closed-form so that they are being sampled at each iteration; then, instead of directly calling upon the system model for such repeated samples, meta-modelling is involved in order to reduce the explosive computational cost.

Non-parametric simulated likelihood Inspired by a particular vibration mechanics issue in the nuclear industry, the purpose of this recent algorithm (de Rocquigny and Cambier, 2009) is to offer a solution for the general likelihood maximisation formulation in the case of a complex physical model with *intrinsic variability, outside linear or Gaussian hypotheses*. It assumes also that model-measurement error distribution is well known, or only approximately known but remaining small.

The general idea is to simulate the likelihood function of the observed sample with a limited number of calls to the complex physical model for a given value of $\boldsymbol{\theta}_X$, including a smoothing of the empiric cdf with a kernel method: this smoothing is designed to facilitate the later optimisation by compensating the inherent roughness remaining at low number of model runs. Hence, the maximisation can be done through the following iterative process:

- (1st step) choose $\boldsymbol{\theta}_X{}^{(k)}$,
- (2nd step) for each $\boldsymbol{d}_j$, simulate the cdf of the j-th model output, appearing theoretically as:

$$F_{Y_j}\left(\boldsymbol{y}_j\middle|\boldsymbol{\theta}_X^{(k)}\right) = \int_x f_X\left(\boldsymbol{x}\middle|\boldsymbol{\theta}_X^{(k)}\right) 1_{[y_j > H_j(x)]} d\boldsymbol{x} \qquad (6.174)$$

by standard Monte-Carlo simulation. Estimating Equation (6.174) involves n_s random samples $(\boldsymbol{x}_v)_{v=1\ldots ns}$ of the $\boldsymbol{X}$ vector according to its distribution (known conditionally to $\boldsymbol{\theta}_X{}^{(k)}$), and hence n_s runs (for each $\boldsymbol{d}_j$) of the complex physical model $\boldsymbol{y}_{jv} = H(\boldsymbol{x}_v, \boldsymbol{d}_j)$,
- (3rd step) Smooth the empiric pdf obtained through a kernel method, hence generating smoothed estimated model output densities $\hat{f}_{Y_j}^{(k)}\text{ker}$ for $j = 1 \ldots n$,

$$\hat{f}_{Y_j}^{(k)}\text{ker}(\boldsymbol{y}_j) = \frac{1}{(2\pi)^{m/2} \cdot n_s \cdot \prod_{l=1}^{m} h_{n_s}^l} \sum_{v=1}^{n_s} \exp\left(-\sum_{l=1}^{m}\left(\frac{y_j^l - y_{jv}^l}{h_{n_s}^l}\right)^2\right) \qquad (6.175)$$

where h_{ns}^l stands for the convolution bandwidth for each component y^l of the vector of observable outputs $\boldsymbol{y}$ supposedly of dimension m. Classical kernel estimation results (Silverman, 1986) show that given a decreasing bandwidth of the type $h_{n_s}^l = \hat{\sigma}_{n_s}^l \cdot n_s^{-\frac{2}{4+m}}$ where $\hat{\sigma}_{n_s}^l$ is an estimator of the standard deviation of y^l, $\hat{f}_{Y_j}^{(k)}\text{ker}$ on top of Monte-Carlo sampling generates a consistent and asymptotically unbiased estimator of the density of $\boldsymbol{y}_j$ (conditional to $\boldsymbol{d}_j$ and $\boldsymbol{\theta}_X{}^{(k)}$).
- (4th step) Compute the approximate likelihood of the observed sample:

$$Lik\left[(\boldsymbol{y}_{mj})_{j=1\ldots n}\middle|\boldsymbol{\theta}_X^{(k)}\right] \approx \prod_j \int_{u_j^*} f_{U_j}(\boldsymbol{u}_j)\hat{f}_{Y_j}^{(k)}\text{ker}(\boldsymbol{y}_{mj} - \boldsymbol{u}_j) d\boldsymbol{u}_j \qquad (6.176)$$

through numerical integration or Monte-Carlo with high number of runs, the measurement error density supposedly being known and closed-form: it can be computed quickly without involving new expensive runs of the physical model.
- (5th step) decide whether it is maximised efficiently or go back to the first step of a new iteration for refinement.

As for any optimisation program, it first assumes that there is a global likelihood maximum without too many local optima pitfalls It all depends then on the efficiency of the descent algorithm, for which several options can be considered in theory: gradient-based ones (involving here finite-difference estimations) or non-gradient based, involving adaptive design of experiment. One of the key point is that $\boldsymbol{y}$ is the output of a complex physical model. The numerical cost of the algorithm is essentially dependant on the number N of model runs, which is here:

$$N = n_{it}.n_s.n \qquad (6.177)$$

where n_{it} stands for the number of iterations. Thus, maintaining a maximum budget for N involves a trade-off between n_s, the size of the sample necessary to stabilise the estimation of model output density, and the number of overall iterations n_{it} necessary to refine the optimal value of $\boldsymbol{\theta}_X$. This may obviously be adapted in the course of the algorithms with increasing n_s when the iterations become more refined. In any case, N inflates if the observed sample Ξ_n involves *variable* environmental conditions $\boldsymbol{d}_j$ – the factor n in N – as was already the case in the linearised Gaussian algorithm with finite differences requiring about $p.n$ runs.

The field of applications motivating this new algorithm has the following features:

- The physical model is potentially quite non-linear, and no physical consideration can justify legitimately any prior parametric (such as Gaussian) hypothesis for the distribution of $\boldsymbol{y}$.
- Controlled external conditions ($\boldsymbol{d}_j$) are constant throughout the observed sample.
- In spite of that, it is physically necessary to account for intrinsic variability of some internal physical phenomena (X^i).
- While the dimension of the observable outputs $\boldsymbol{Y}$ or of the inner uncertainty sources $\boldsymbol{X}$ can be large, the dimension of $\boldsymbol{\theta}_X$ can be very low.

The first and third features clearly motivate the use of a general maximum likelihood formulation with intrinsic variability instead of calibration/assimilation. The second feature ensures also that only n_s model runs are necessary for each iterative value of $\boldsymbol{\theta}_X^{(k)}$. Even more importantly, the fourth feature should clearly favour identifiability (thus the existence of a unique global maximiser) and simplifies considerably the search for the optimum. In the simplest case where the unknown uncertainty parameter is unique, a simple grid of a few $\boldsymbol{\theta}_X$ values (possibly locally refined a few times according to the target accuracy) proves sufficient. A fruitful application of this algorithm to vibration mechanics can be found in de Rocquigny and Cambier (2009).

Meta-modelling, stochastic EM and advanced bayesian algorithms The general EM statistical algorithm already introduced in the linear Gaussian case can be extended to a generalised setting. In the absence of linearity, the conditional expectations involved in EM cannot be formulated in closed-form but may still theoretically be formulated, and possibly be estimated through sampling. Stochastic variants of EM, such as SEM (Celeux and Diebolt, 1985) or SAEM (Delyon *et al.*, 1999) were developed precisely for that kind of application: they include an additional sampling step (e.g. through the MCMC techniques introduced in Chapter 5 and applied in Perrin *et al.* (2007) within the Expectation-Maximisation steps. They may require thousands or more runs at each iteration in order to stabilise the expectation estimates. Keeping in mind that iterations themselves could be $10^3 - 10^5$ for the overall EM procedure to converge, this means that direct calls to the system model $H(.)$ are computationally unaffordable.

MCMC is also an essential recipe for the implementation of the fully-Bayesian formulation, including non-Gaussian and non-linear hypotheses, possibly with kriging-based models of residuals, which have developed in various fields of application such as hydrology (see http://mucm.group.shef.ac.ik for the GLUE developments), probabilistic mechanics, environmental impact assessment and so on. Besides the use of High Performance Computing (see Chapter 8), *meta-modelling* (meaning the substitution of a quickly-computable surrogate function approximating the original model well – see Chapter 7) becomes attractive if not compulsory. It should ideally include a suitable adaptation process in order to enrich/correct the meta-model in the course of convergence of EM, with calls from time to time upon the original system model: see Barbillon *et al.* (2011) for more on this and an application to the flood case or Bect *et al.* (2011) and Picheny *et al.* (2010) for advanced techniques involving Gaussian processes.

6.6 Conclusion: The modelling process and open statistical and computing challenges

Indirect modelling of the uncertain inputs is thought to have great potential for risk and uncertainty modelling: industrial samples often embrace varied degrees of freedom over the structures or systems under risk. There is thus quite a wealth of information to be retrieved in order to circumvent the variability of the systems and thus better control uncertain deviations from normal design. Amongst the various algorithms, a step-by-step approach is generally recommended starting with the less data-consuming framework of parameter identification and checking the residuals of such elementary model calibration before inferring intrinsic variability that requires full inversion techniques and more observations in order to converge. Linearisation around the best-estimate inputs is generally also a good starting point before going into non-linear formulations.

In this high-potential domain, considerable research is still needed. The existing methods mostly ignore the intrinsic variability of the model inputs through traditional calibration/assimilation or parameter estimation procedures. Only recently have the true inverse probabilistic methods been developed, mostly in the linearised Gaussian case. Non-linear and/or non Gaussian situations call for significant challenges to be overcome. This is obviously so on the computational side, as the integrals involved prove to be even more challenging than those appearing in uncertainty propagation. However, this also includes theoretical work on the convergence and regularity properties of the estimator that are based on non-*i.i.d.* likelihood estimation.

The question of identifiability, or more generally perhaps that of the optimal level of parameterisation of the uncertain models, is an even deeper challenge. Full inverse identification clearly involves a richer parameter vector to be identified than that of calibration/assimilation or parameter estimation techniques, as the inner variability calls for the estimation of the distribution of the inputs, not only of their more plausible point value. This means larger (epistemic) estimation variance when the dataset is fixed, thus possibly a poorer score on model selection criteria, or even a looser control of the adequacy of observation-model residuals to confirm the validity of statistical assumptions. This chapter has shown that there are clear connections between such thinking and sensitivity analysis with respect to the inputs to be estimated. The benefits associated with such a more refined understanding of the input uncertainty have to be carefully weighed against the costs involved, be they computational, metrological or even in terms of confidence given to the accuracy of the system model $G(.)$ itself. This opens up a large field of research when coming into the wider decision-making process involved in uncertainty modelling.

Exercises

There follows a list of exercises referring to the concepts developed in this chapter:

6.1 Compute the heteroscedastic calibration in the flood example. How does it compare to intrinsic variability identification, either in a linearised or non-linear formulation?

6.2 Compute the flood model calibration under log-log variable transforms. Are there any benefits?

6.3 Challenge the flood model against alternative simpler models (e.g. linear or power) or more elaborate ones (e.g. including Z_m and Z_v).

6.4 Test the robustness of the various algorithms in the flood example against a reduction of the observed sample. How does the account for epistemic uncertainty help in keeping conservative bounds?

6.5 Consider the auto-regressive lognormal process introduced in Chapter 5 and Section 6.1.3: using a sample of limited size, study how the estimation process, and associated epistemic uncertainty, impacts on the comparison between a forecast model and a risk model. How would that be affected by a change in the definition of the risk measure, taken now as a return-level of the maximum over instead a period of a daily value?

6.6 Compute the Bayesian updating on the flood example and test the sensitivity to priors. Extend the Gaussian Bayesian setting of Chapter 5 to the full Bayesian calibration in Chapter 6. To what extent does it modify the classical calibration approach?

6.7 Discuss the identifiability criteria of the general problem of intrinsic variability identification under maximal likelihood: linear Gaussian assumptions, more general distributions of the exponential family, and so on.

6.8 Extend the calibration formulation to a kriging model, that is replacing a model of residuals independently distributed from $\boldsymbol{d}_j$ to $\boldsymbol{d}_j'$ by a Gaussian process with, for example, a given family of correlation function over d. How does it change the identifiability issues?

References

Aven, T. (2003) *Foundations of Risk Analysis*. John Wiley & Sons, Ltd.

Barbillon, P., Celeux, G., Grimaud, A., Lefebvre, Y. and de Rocquigny, E. (2011) – Non linear methods for inverse statistical problems, *Computational Statistics and Data Analysis*, vol. **55**(1), pp. 132–142.

Basseville, M. and Nikiforov, I.V. (1993) *Detection of Abrupt Changes: Theory and Application, Prentice Hall Information and System Sciences Series*, Prentice Hall, Inc., Englewood Cliffs, NJ, pp. xxviii+528.

Beck, M.B. (1987) Water quality modelling: a review of the analysis of uncertainty, *Wat. Res. Research*, **23**(8).

Beck, J.V. and Arnold, K.J. (1977) *Parameter Estimation in Engineering and Science*, Wiley.

Beck, J.L. and Katafygiotis, L.S. (1998) Updating models and their uncertainties. *Journal of Engineering Mechanics*, **124**(4).

Bect, J., Ginsbourger, D., Li, L., Picheny, V., and Vazquez, E. (2011) Sequential design of computer experiments for the estimation of a probability of failure, *Statistics and Computing*, **11**, pp. 1–21.

Beven, K.J. and Binley, A.M. (1992) The future of distributed model: model calibration and uncertainty prediction. *Hydrological Processes*, **6**, 279–298.

Brodsky, B.E. and Darkhovsky, B.S. (1993) *Nonparametric methods in change-point problems. Mathematics and its Applications, 243*, Kluwer Academic Publishers Group, Dordrecht, pp. xii+209.

Casarotti, C., Petrini, L. and Strobbia, C. (2007) Calibration of Non-linear model parameters via inversion of experimental data and propagation of uncertainties in seismic fibre element analysis. Proc. of Computational Methods in Structural Dynamics and Earthquake Engineering, Rethymno, June 2007.

Celeux, G. and Diebolt, J. (1985). The SEM algorithm: a probabilistic teacher algorithm derived from the EM algorithm for the mixture problem. *Computational Statistics Quaterly* **2**, 73–82.

Celeux, G., Grimaud, A., Lefebvre, Y. and de Rocquigny, E. (2010) – Identifying intrinsic variability in multivariate systems through linearised inverse methods, *Inverse Problems in Science and Engineering* vol. **18**(3), pp. 401–415.

Csörgo, M. and Horváth, L. (1997) *Limit Theorems in Change-Point Analysis, Wiley Series in Probability and Statistics*, John Wiley & Sons, Ltd., Chichester.

de Crécy, A. (1997) CIRCE: a tool for calculating the uncertainties of the constitutive relationships of Cathare2. 8th International Topical Meeting on Nuclear reactor Thermo-Hydraulics (NURETH8), Kyoto.

de Rocquigny, E. and Cambier, S. (2009) Inverse probabilistic modeling of the sources of uncertainty: a non-parametric simulated-likelihood method with application to an industrial turbine vibration assessment. *Inverse Problems in Science and Engineering*, **17**(7), 937–959.

Delyon, B., Lavielle, M. and Moulines, E. (1999) Convergence of a stochastic approximation version of the EM algorithm. *Annals of Statistics* **27**, 94–128.

Dempster, F., Laird, N.M. and Rubin, D.B. (1977) Maximum likelihood from incomplete data from the EM algorithm. *Journal of the Royal Statistical Society (Ser. B)*, **39**.

Droguett, E.L. and Mosleh, A. (2008) Bayesian methodology for model uncertainty using model performance data. *Risk Analysis*, **28**(5), 1457–1476.

Friswell, M.I., Mottershead, J.E. and Mares, C. (2007) Stochastic model updating in structural dynamics. Proc. of Computational Methods in Structural Dynamics and Earthquake Engineering, Rethymno, June 2007.

Hoadley, B. (1971) Asymptotic properties of maximum likelihood estimators for the independent not identically distributed case. *Annals of Mathematical Statistics*, **42**(6), 1977–1991.

Jennrich, R.I. (1969) Asymptotic properties of non-linear least squares estimators. *Annals of Mathematical Statistics*, **40**, 633–643.

Kennedy, M.C. and O'Hagan, A. (2002) Bayesian calibration of computer models. *Journal of the Royal Statistical Society: Series B (Statistical Methodology)*, **63**(3), 425–464.

Krzysztofowicz, R. (1999) Bayesian theory of probabilistic forecasting via deterministic hydrologic model. *Water Resources Research*, **35**(9), 2739–2750.

Kurowicka, D. and Cooke, R.M. (2002) *Techniques for generic probabilistic inversion, Probabilistic Safety Assessment and Management* (eds E.J. Bonano *et al.*) Elsevier, pp. 1543–1550.

Leroy, O., Junker F., Gavreau B., de Rocquigny E. and Bérengier M. (2010) Uncertainty assessment for outdoor sound propagation, Proc. of the 20th International Congress on Acoustics ICA, pages 192–198 Sydney, Australia.

Liu, C. and Rubin, D.B. (1994) The ECME algorithm: a simple extension of EM and ECM with faster monotone convergence. *Biometrika*, **81**, 633–648.

Mahé, P. and de Rocquigny, E. (2005) Incertitudes non observables en calcul scientifique industriel – Etude d'algorithmes simples pour intégrer dispersion intrinsèque et ébauche d'expert, 37èmes. *Journ. Franc. de Stat., Pau.*

Nelder, J.A. and Mead, R. (1965) A simplex algorithm for function minimization. *Computer Journal*, **7**, 308–313.

Perrin, F., Sudret, B., Pendola, M. and de Rocquigny, E. (2007) Comparison of Monte Carlo Markov Chain and a FORM-based approach for Bayesian updating of mechanical models. Proc. of 10th Int. Conf. on Appli. of Stat. Proba. in Civil Engineering (ICASP), Tokyo.

Picheny, V., Ginsbourger, D., Roustant, O. and Haftka, R.T. (2010) Adaptive designs of experiments for accurate approximation of a target region. *Journal of Mechanical Design*, **132**(7).

Ratier, L. (2004) Modelling a mistuned bladed disk by modal reduction. Proceedings of ISMA2004 Leuven 2004 September 20–22.

Shiryaev, A.N. (2010) Quickest detection problems: Fifty years later. *Sequential Analysis*, **29**(4), 345–385.

Siegmund, D. (1985) *Sequential Analysis. Tests and Confidence Intervals,* Springer Series in Statistics, Springer-Verlag, New York, pp. xi+272.

Silverman, B.W. (1986) *Density Estimation for Statistics and Data Analysis*. Chapman and Hall, London.

Soize, C. (2000) A non-parametric model of random uncertainties for reduced matrix models in structural dynamics. *Probabilistic Engineering Mechanics*, **15**(3), 277–294.

Soize, C. (2005) Random matrix theory for modelling uncertainties in computational mechanics. *Computer Methods in Applied Mechanics and Engineering*, **194**, 1333–1366.

Talagrand, O. (1997) Assimilation of observations, an introduction. *Journal of the Meteorological Society of Japan*, **75**, 191–201.

Talagrand, O. and Courtier, P. (1987) Variational assimilation of meteorological observations with the adjoint vorticity equation, I: theory. *Q. J. R. Meteorol. Soc.*, **113**, 1311–1328.

Tarantola, A. (1987) *Inverse Problem Theory and methods for data fitting and model parameter estimation*, Elsevier, Amsterdam.

Tarantola A. (2004), *Inverse Problem Theory and methods for model parameter estimation*, SIAM

Tay, A.S. and Wallis, K.F. (2000) Density forecasting: A survey. *Journal of Forecasting*, **19**, 235–254.

Walter, E. and Pronzato, L. (1997) *Identification of Parametric Models from Experimental Data*, Springer-Verlag, Heidelberg.

7

Computational methods for risk and uncertainty propagation

This chapter discusses the computation of the risk measures or quantities of interest as well as the sensitivity indices once the system and uncertainty models have been estimated and calibrated (following the procedures in Chapter 5 or 6): such computations are often referred to as uncertainty propagation, or probabilistic computation. A detailed overview of the simulation and alternative numerical techniques is provided: prior simplification of the integrals involved (Section 7.1), the universal approach of Monte-Carlo simulation (Section 7.2); its alternatives (Taylor approximation and numerical integration, variance-reduction techniques, FORM-SORM and structural reliability methods, meta-modelling); and extensions to the propagation of uncertainty in two-layer separated epistemic/aleatory settings (Section 7.3). Of particular importance is the compromise between computational cost and propagation uncertainty in large-CPU phenomenological models most notably when distribution tails are involved: monotony and other regularity assumptions are shown to provide strong robustness in that respect (Section 7.4). A short review of some relevant techniques for sensitivity and importance analysis is given in Section 7.5 as well as the general numerical perspectives for risk modeling, which end the chapter.

Once the combined uncertainty and system models have been properly estimated thanks either to direct (Chapter 5) or indirect (Chapter 6) observations, the general issue is to compute the risk measure $c_Z(\boldsymbol{d})$ for a given set of actions $\boldsymbol{d}$, as shown in Figure 7.1.

In other words, this consists of *propagating* the uncertainty model $\{f_X(x \mid \theta_X), \pi(\theta_X \mid IK, \Xi_n)\}$ through the predictive system model $G(.)$ in order to assess the functional $c_z(\boldsymbol{d})$. Closely associated with this is the question of *sensitivity analysis*, that is the ranking of the relative importance of the uncertain component X^i with respect to c_z. It proves to be totally dependent on the computation methods chosen for the risk measure, and the measure of importance depends on it, as discussed at the end of the chapter.

In general, computation of risk measures, quantities of interest or sensitivity indices means estimating complex integrals – involving the known (but non-closed form) functions $G(.)$, $f_X(.)$ and $\pi(.)$ – possibly mixed with optimisation in mixed deterministic-probabilistic settings. $G(.)$ often refers

Modelling Under Risk and Uncertainty: An Introduction to Statistical, Phenomenological and Computational Methods, First Edition. Etienne de Rocquigny.

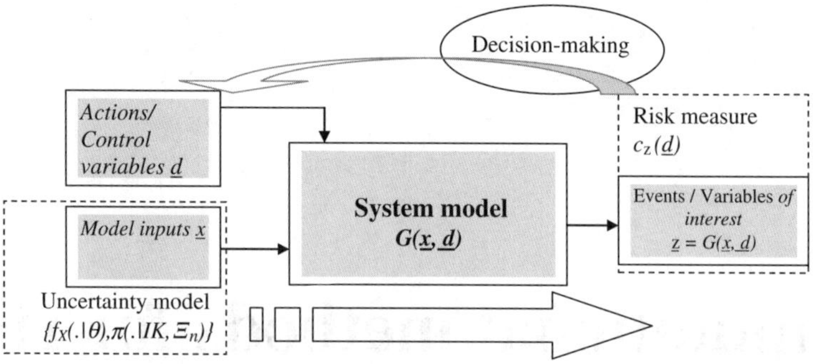

Figure 7.1 *Computing the risk measure from the predictive system model.*

to a complex phenomenological model with a non-negligible CPU requirement for each run, and multiple evaluations of the risk measure may be required in order to explore the impact of various actions $\boldsymbol{d}$. Thus, the numerical efficiency of the associated algorithms is a crucial point and requires compromising between the computing load and the control of residual propagation errors.

Note that the formal framework of the book covers a large range of differing situations regarding the complexity of $G(.)$ and that of the uncertain inputs $\boldsymbol{x}$, which can compound either discrete or continuous variables (denoted as $(\boldsymbol{x},\boldsymbol{e})$ in Chapters 1 and 2), or even temporal or spatial-depending random processes or fields $\boldsymbol{x}_t$ or $\boldsymbol{x}(\boldsymbol{r})$. The major focus of this chapter is given to the case of a *complex* $G(.)$ parameterised by a *vector* of *continuous* inputs. Also, discrete input events will be discussed as they allow for a simplified closed-form integration (Section 7.1.3); the case of a simple (or computationally *cheap*) $G(.)$ is generally easier as it allows for massive standard Monte-Carlo sampling, or rarely perhaps, direct numerical closed-form integration. Advanced stochastic techniques developed for the more challenging cases of space or time-dependent inputs are not discussed in this chapter, though they would often return to a vector case for $\boldsymbol{x}$ through an *ad hoc* parameterisation of the random fields or processes (see Table 7.9).

7.1 Classifying the risk measure computational issues

Chapter 2 listed the potential risk measures c_z as a series of functionals of the output uncertainty model, formally:

$$c_z = \mathcal{F}[f_Z(.), \pi(.)] \tag{7.1}$$

Such a functional may combine:

- for *purely probabilistic* risk measures:
  - integration involving the density functions $f_X(.)$, $\pi(.)$ as well as the predictive system model $G(.)$ and ancillary operations such as computing an indicator function $1_{G(\boldsymbol{x},\,\boldsymbol{d})<zs}$: it generally corresponds to computing an *expected utility*, that is expectation, exceedance probability and so on;
  - ancillary numerical operations may be needed for *non-expectation based probabilistic measures*: for example power or quotient when estimating the variance, a coefficient of variation or moments of the outputs $\mathbf{Z}$; inverting an integral function when computing a quantile; smoothing a series of threshold probabilities to estimate the distribution and so on.

- for mixed *deterministic-probabilistic risk measures*:
  - any of the previous operations;
  - mixed with optimisation of the similar functions or intermediate integration results with respect to some penalised variables $\boldsymbol{x}^{pn}$ or $\boldsymbol{\theta}_X^{pn}$ or of the raw outputs along a time interval or space field when Z is defined as a maximum output.

Note also that a decision criterion associated with a risk measure also appears as a derived functional of a logical (Boolean) nature:

$$c_Z = 1_{\{test(fZ)\}}$$
$$\text{where } \{test(f_Z)\} = \{z_{pn}(\boldsymbol{d}) < z_{pn}(\boldsymbol{d}^\circ)\} \text{ or } \{unc_z < i_s\}\ldots \tag{7.2}$$

which means that it is equally amenable to simulation as a straightforward post-treatment of the computation of the risk measure.

It is useful first to recall the double-probabilistic structure of uncertainty models involved with aleatory and epistemic components in order to understand their impact on the computation (Section 7.1.1). Section 7.1.2 will then discuss the essential core of *expectation-based* risk measures out of which most probabilistic risk measures derive through limited ancillary post-treatments. Section 7.1.3 introduces particular simplifications made possible by the discrete nature of input events. Eventually, Sections 7.1.4 and 7.1.5 will briefly discuss non-expectation based probabilistic or mixed risk measures.

7.1.1 Risk measures in relation to conditional and combined uncertainty distributions

The estimation phase results in the following predictive model:

$$\begin{aligned} &\boldsymbol{Z} = G(\boldsymbol{X}, \boldsymbol{d}) \\ &(\boldsymbol{X}, \boldsymbol{\Theta}_X) \sim f_X(.\,|\,\boldsymbol{\theta}_X).\pi(.\,|\,IK, \Xi_n) \end{aligned} \tag{7.3}$$

whereby the *joint input uncertainty model* is a distribution representing both aleatory and epistemic uncertainty components in extended vector $(\boldsymbol{X}, \boldsymbol{\Theta}_X)$. Focusing on the structure of the input uncertainty model, recall that the *conditional input distribution* f_X of input uncertainty provided a *fixed* value of parameters $\boldsymbol{\theta}_X$, that is that epistemic uncertainty is ignored:

$$\boldsymbol{X}\,|\,\boldsymbol{\theta}_X \sim f_X(.\,|\,\boldsymbol{\theta}_X) \tag{7.4}$$

The image distribution of f_X by $G(.)$ is the following *cdf* for **Z**:

$$F_z(z|\boldsymbol{\theta}_X) = P(Z < z|\boldsymbol{\theta}_X) = \int_x 1_{G(\boldsymbol{X},\boldsymbol{d})<z}\, f_X(\boldsymbol{x}|\boldsymbol{\theta}_X)d\boldsymbol{x} \tag{7.5}$$

Such an output distribution (with its associated density $f_Z(.\,|\,\boldsymbol{\theta}_X)$ obtained by derivation) represents the *conditional output distribution* of Z given a *fixed* value of parameters $\boldsymbol{\theta}_X$. It is sufficient to estimate the risk measures when epistemic uncertainty is neglected, that is in a single-level probabilistic uncertainty model. In that important practical case, the (conditional) risk measure appears through the propagation of the conditional distribution of $\boldsymbol{X}$ through the system model $G(.)$ as a standard *function* not only of the control variables $\boldsymbol{d}$ as was established in Chapter 2 but also *of the parameters* $\boldsymbol{\theta}_X$ of the uncertainty model:

$$c_z \mid \boldsymbol{\theta}_X = \mathcal{F}[f_Z(.\mid \boldsymbol{\theta}_X)] = C_z(\boldsymbol{\theta}_X, \boldsymbol{d}) \tag{7.6}$$

Take the example of an exceedance probability or more generally of an expected utility-based risk measure (once again assuming a fixed $\boldsymbol{\theta}_X$):

$$\begin{aligned} c_Z(\boldsymbol{d}) &= E[U(\boldsymbol{Z})|\boldsymbol{\theta}_X, \boldsymbol{d}] = \int_{\boldsymbol{Z}} U(z) f_Z(z|\boldsymbol{\theta}_X, \boldsymbol{d}).dz \\ &= \int_{\boldsymbol{x}} UoG(\boldsymbol{x}, \boldsymbol{d}) f_X(\boldsymbol{x}|\boldsymbol{\theta}_X, \boldsymbol{d}).d\boldsymbol{x} = C_Z(\boldsymbol{\theta}_X, \boldsymbol{d}) \end{aligned} \tag{7.7}$$

As a function of $\boldsymbol{\theta}_X$, C_z is generally far from being closed-form; yet such a functional dependence inspires some numerical techniques, such as a Taylor development around the best-estimate to approximate cheaply the impact of epistemic uncertainty on the risk measure (see Chapter 5, Section 5.4).

Indeed, the conditional output *pdf* represents only a *partial view* of the whole output uncertainty in the general case. Suppose now that the vector of parameters $\boldsymbol{\Theta}_X$ varies according to its epistemic uncertainty distribution π, then, as an image of $f_X(. \mid \boldsymbol{\theta}_X).\pi(. \mid IK, \Xi_n)$ by $G(.)$, the output distribution inherits the structure of joint input uncertainty:

$$(\boldsymbol{Z}, \boldsymbol{\Theta}_X) \sim f_Z(.\mid \boldsymbol{\theta}_X).\pi(.\mid IK, \Xi_n) \tag{7.8}$$

so that the image by $G(.)$ of the unconditional (or combined, predictive) input distribution:

$$\boldsymbol{X} \sim f_X^{\pi}(\boldsymbol{x}) = E_{\boldsymbol{\theta}_X}[f_X(\boldsymbol{x}|\boldsymbol{\theta}_X)|IK, \Xi_n] = \int_{\boldsymbol{\theta}_X} f_X(\boldsymbol{x}|\boldsymbol{\theta}_X)\pi(\boldsymbol{\theta}_X|IK, \Xi_n)d\boldsymbol{\theta}_X \tag{7.9}$$

yields the unconditional (or combined, predictive) output distribution defined by the following *cdf*:

$$\begin{aligned} F_Z^{\pi}(z) &= P(Z < z|IK, \Xi_n) = E_{\boldsymbol{\theta}_X}[F_Z(z|\boldsymbol{\Theta}_X)|IK, \Xi_n] \\ &= \int_{\theta} P(Z < z|\boldsymbol{\theta}_X)\pi(\boldsymbol{\theta}_X|IK, \Xi_n)d\boldsymbol{\theta}_X \\ &= \int_{\theta} \left[\int_{x} 1_{G(X,\boldsymbol{d})<z} f_X(\boldsymbol{x}|\boldsymbol{\theta}_X)d\boldsymbol{x} \right] \pi(\boldsymbol{\theta}_X|IK, \Xi_n)d\boldsymbol{\theta}_X \\ &= \int_{x} 1_{G(x,\boldsymbol{d})<z} f_X^{\pi}(\boldsymbol{x})d\boldsymbol{x} \end{aligned} \tag{7.10}$$

Taking into account epistemic uncertainty, the risk measure becomes in general a functional of the joint output uncertainty model:

$$c_z = \mathcal{F}[f_Z(.), \pi(.)] \tag{7.11}$$

Thus, it is no more a function of $\boldsymbol{\theta}_X$ but a function of the hyper-parameters (ζ) of distribution π, although that dependence may be intractable in elaborate Bayesian models yielding non-closed-form posterior distributions $\pi(\cdot \mid IK, \Xi_n)$.

7.1.2 Expectation-based single probabilistic risk measures

Expectation-based single-probabilistic risk measures are central in risk and uncertainty analysis, as shown in Chapter 2. Their formulation is in general as follows:

$$\begin{aligned} c_Z(\boldsymbol{d}) &= E_{Z,\boldsymbol{\Theta}}[U(\boldsymbol{Z})|IK, \Xi_n, \boldsymbol{d}] = \int_{z,\boldsymbol{\theta}} U(z) f_Z(z|\boldsymbol{\theta}_X, \boldsymbol{d}) \pi(\boldsymbol{\theta}_X|IK, \Xi_n) dz d\boldsymbol{\theta}_X \\ &= \int_{\boldsymbol{x},\boldsymbol{\theta}} UoG(\boldsymbol{x}, \boldsymbol{d}) f_X(\boldsymbol{x}|\boldsymbol{\theta}_X) \pi(\boldsymbol{\theta}_X|IK, \Xi_n) d\boldsymbol{x} d\boldsymbol{\theta}_X \end{aligned} \tag{7.12}$$

They include:

- expected variables of interest, such as expected dose or pollution over time as encountered in health or environmental impact analysis, or in IPRA;
- the expected utility model whereby a utility represents the preference of a decision-maker, a key model in cost-benefit analysis under risk;
- exceedance probabilities, frequency or probability of an undesired event, key quantities in SRA or QRA, which were shown to be associated formally with degenerate utility functions.

Expectation-based risk measures have key features facilitating their computation. They basically require the computation of an integral which will be seen as the standard application of Monte-Carlo sampling, that is generating random sequences of uncertain inputs and computing the empiric mean of the corresponding outputs through $UoG(.)$.

More specifically, consider next their epistemic-aleatory structure. When being defined as an expected utility, the complete (or *unconditional*) risk measure can be understood equivalently and computed with two interpretations. Within the complete uncertainty model, a *first interpretation* of c_z is the expected value of the *conditional* risk measure over the epistemic uncertainty model, as it has been seen as being a function of $\boldsymbol{\theta}_X$:

$$c_z = E_{\theta_X}[C_z(\boldsymbol{\Theta}_X, \boldsymbol{d})|IK, \Xi_n] = \int_{\theta_X} C_z(\boldsymbol{\theta}_X, \boldsymbol{d}) \pi(\boldsymbol{\theta}_X|IK, \Xi_n) d\boldsymbol{\theta}_X \tag{7.13}$$

Note that this may be seen formally as the computation of a risk measure (the output expectation) over a new model:

- the conditional risk measure, which was understood to be a deterministic function of the parameters and actions, becomes the new system model $C_z(\boldsymbol{\theta}_X, \boldsymbol{d})$,
- the epistemic distribution $\pi(\cdot)$ appears as the new single-probabilistic uncertainty model for the inputs of the new system model, that is the $\boldsymbol{\theta}_X$ parameters.

In other words, computing such an *unconditional* risk measure involves a double integration, meaning 'averaging the risk measure over the uncertainties of the risk components'. This first interpretation is very common in the area of system reliability when undertaking uncertainty analysis over a PRA or PSA. As reviewed in Chapter 1, the basic system model in that case may be a large combination of event/fault trees, that is a Boolean function; the conditional risk measure is the system failure probability which happens to have a closed-form. An unconditional risk measure is thus computed as the expected system failure probability when including uncertainties in the frequency of elementary events through Monte-Carlo sampling of the closed-form conditional risk measure.

The *second interpretation* of c_z is the expected value of the system model over the full unconditional input distribution:

$$\begin{aligned} c_Z(\boldsymbol{d}) &= E_{f_Z^\pi}[U(\boldsymbol{Z})|IK, \Xi_n, \boldsymbol{d}] = \int_z U(z) f_Z^\pi(z|IK, \Xi_n, \boldsymbol{d}).dz \\ &= \int_{\boldsymbol{x}} UoG(\boldsymbol{x}, \boldsymbol{d}) f_X^\pi(\boldsymbol{x}|IK, \Xi_n).d\boldsymbol{x} \end{aligned} \tag{7.14}$$

which is equivalent to the previous since when developing the definition of the unconditional input distribution:

$$\begin{aligned} c_Z(\boldsymbol{d}) &= \int_{\boldsymbol{x}} UoG(\boldsymbol{x}, \boldsymbol{d}) f_X^\pi(\boldsymbol{x}|IK, \Xi_n).d\boldsymbol{x} \\ &= \int_{\boldsymbol{x}} UoG(\boldsymbol{x}, \boldsymbol{d}) \int_{\boldsymbol{\theta}_X} f_X(\boldsymbol{x}|\boldsymbol{\theta}_X)\pi(\boldsymbol{\theta}_X|IK, \Xi_n) d\boldsymbol{\theta}_X.d\boldsymbol{x} \\ &= \int_{\boldsymbol{\theta}_X} \pi(\boldsymbol{\theta}_X|IK, \Xi_n) \int_{\boldsymbol{x}} UoG(\boldsymbol{x}, \boldsymbol{d}) f_X(\boldsymbol{x}|\boldsymbol{\theta}_X) d\boldsymbol{x} d\boldsymbol{\theta}_X \\ &= E_{\boldsymbol{\theta}_X}[C_z(\boldsymbol{\Theta}_X, \boldsymbol{d})|IK, \Xi_n] \end{aligned} \tag{7.15}$$

In other words, it is equivalent to compute it as a single-probabilistic risk measure replacing the conditional input distribution by the unconditional one, expected over the epistemic uncertainty distribution:

$$E_\theta(C_Z(\boldsymbol{\Theta}_X, \boldsymbol{d})) \Leftrightarrow c_Z \textit{ computed with the expected law } E_\theta(f_X(x \mid \boldsymbol{\Theta}_X))$$

Note, however, that this is different is general to computing the single-probabilistic risk measure using the original (aleatory) pdf shape taken with the mean (epistemic) values of its parameters θ_X:

$$E_\theta(c_Z(\boldsymbol{\Theta}_X, \boldsymbol{d})) \neq C_Z(E_\theta[\boldsymbol{\Theta}_X], \boldsymbol{d}) \tag{7.16}$$

A simple counter-example is as follows.

$$\begin{aligned} &X \sim U(0, \theta) \quad \theta \sim U(0, 2) \\ &Z = G(X, d) = X^2 \end{aligned} \tag{7.17}$$

Then:

$$E_\theta(c_Z(\boldsymbol{\Theta})) = 4/9 \tag{7.18}$$

while:

$$c_Z(E_\theta[\boldsymbol{\Theta}]) = E_{U(0,1)}[X^2] = 1/3 \tag{7.19}$$

That property has essential consequences for the computation of a risk measure in a full epistemic-aleatory uncertainty model as will be discussed below. When it is an expected utility, it means that it is not necessary to program a dedicated simulation for the estimation of the *conditional* risk measure $C_z(\boldsymbol{\theta}_X,\boldsymbol{d})$ for a given value of $\boldsymbol{\theta}_X$ on the one hand, and then sample the distributed values of $\boldsymbol{\theta}_{X\,j}$ representing epistemic uncertainty and run multiple computations of $C_z(\boldsymbol{\theta}_{X\,j},\boldsymbol{d})$ accordingly on the other. One can rely on a direct estimation of the unconditional risk measure through joint sampling of the joint distribution of $(\boldsymbol{X}, \boldsymbol{\theta}_X)$ for a much cheaper computational cost in general.

Note finally that while being of the nature of a similar formal expected-utility, a great computational difference lies between the two following variants; a central-based utility risk measure:

$$c_Z(\boldsymbol{d}) = E_{f_Z^\pi}[U(z)] = \int_z U(z) f_Z^\pi(z).dz = \int_{\boldsymbol{x}} UoG(\boldsymbol{x}, \boldsymbol{d}) f_X^\pi(\boldsymbol{x}).d\boldsymbol{x} \tag{7.20}$$

where utility function U is equal or close to $U(z) = z$, that is taking non-negative values over the entire distribution; and a rare exceedance probability:

$$c_Z(\boldsymbol{d}) = E_{f_Z^\pi}[U(Z)] = \int_z 1_{z>z_s} f_Z^\pi(z).dz = \int_{\boldsymbol{x}} 1_{G(\boldsymbol{x},\boldsymbol{d})>z_s} f_X^\pi(\boldsymbol{x}).d\boldsymbol{x} = P(Z > z_s) \tag{7.21}$$

whereby z_s represents a threshold that is rarely exceeded so that the utility is zero except in zones of low probability. The big difference is that a sampling method will take many more simulations to explore and yield a stable estimate of the risk measure, as will be obvious in the formulation of the corresponding Monte-Carlo estimator variances. This is of even greater concern since each evaluation of $z = G(\boldsymbol{x},\boldsymbol{d})$ involves in general a computationally-costly system model.

7.1.3 Simplified integration of sub-parts with discrete inputs

It is desirable to simplify the integrals involved in the risk measure in order to reduce the computational load. This concerns essentially the system model $G(.)$. Faster formulation of $G(.)$ or of some of its sub-parts, be it closed-form or simplified numerical equivalents such as response surfaces, is particularly interesting in that respect. A first opportunity arises when the system model incorporates discrete events for inputs (denoted as $\boldsymbol{e}$ inside the detailed notation of vector $\boldsymbol{x} = (\boldsymbol{x},\boldsymbol{e})$) and static risk models. Indeed, closed-form integration is conspicuous in classical fault tree approaches as was introduced in Chapter 1 for the PRA or PSA reliability studies.

Computing the risk measure of a logical model Consider a logical model as encountered in QRA, say a fault tree or event tree: it stands as a deterministic function between the indicator variables (denoted as $e\ldots = 1$ in the case of initiator realisation or component failure, $e\ldots = 0$ otherwise) representing the input events, that is each precise *elementary state* of the system components (e^{sy}) or initiators (e^{in}). Section 1.3.1 showed that the output event of interest is typically computed as a sum of products as follows:

$$z = e_f = \sum_i \left[e^{in^i} \cdot \prod_{j_i(\boldsymbol{d})} e^{sy^{j_i}} \right] = G\left(e^{in}, e^{sy}, \boldsymbol{d}\right) \tag{7.22}$$

Complex systems may require a recursive definition, with more sums of products embedded inside (e^{sy}) to represent the sub-systems, but the subsequent derivations could then be generalised easily. Consider, in a level-1 setting, the following risk measure:

$$c_z = P(E_f) = E[Z] = \int_e G(\boldsymbol{e}, \boldsymbol{d}) f_{XE}(\boldsymbol{e}) d\boldsymbol{e} \tag{7.23}$$

Assume that all those input events $\boldsymbol{E}$ are modelled by a discrete uncertainty model, that is a set of Bernoulli independent variables:

$$\begin{aligned} &P(E^{in\,i}) = P(1_{Ein\,i} = 1) = f^i P(E^{sy\,k}) = P(1_{Esy\,k} = 1) = p^k \\ &\boldsymbol{E} \sim f_E = \left\{ (f^i)^i, (p^k)^k \right\} \end{aligned} \tag{7.24}$$

The risk measure integral turns out to be a closed-form expression:

$$\begin{aligned} c_z &= \int_e G(\boldsymbol{e}, \boldsymbol{d}) f_{XE}(\boldsymbol{e}) d\boldsymbol{e} \\ &= \int_e \sum_i \left[e^{in^i} . \prod_{j_i(\boldsymbol{d})} e^{sy^{j_i}} \right] f_{XE}(\boldsymbol{e}) d\boldsymbol{e} \\ &= E\left[\sum_i \left[E^{in^i} . \prod_{j_i(\boldsymbol{d})} E^{sy^{j_i}} \right] \right] = \sum_i \left[E\left(E^{in^i}\right) . \prod_{j_i(\boldsymbol{d})} E\left(E^{sy^{j_i}}\right) \right] \\ &= \sum_i \left[f^i . \prod_{j_i(\boldsymbol{d})} p^{j_i}) \right] \end{aligned} \tag{7.25}$$

This classical result for systems reliability means that computing a risk measure associated with a fault tree is a simple combination of multiplications and additions. This also implies that level-2 uncertainty propagation in such risk models is practically identical to level-1 propagation on closed-form phenomenological models (e.g. metrological chains involving additions and multiplications).

Flood example

There are two initiators: limited overspill (E_{os}) and large overspill (E_{ol}) events. Thus, the risk measure over the protection system is the following simple closed form:

$$c_Z = P(E_p) = P(E_{os}).p_1.p_2.(p_{12} + p_{1i}.p_{2i}) + P(E_{ol}) \tag{7.26}$$

Computing the risk measure of a combination of a logical model and a phenomenological model Even more interesting is the fact that this property may still be used in the case of a more complex model, mixing a logical and a phenomenological (physical, economic ...) model. Consider the previous model and assume now that the initiator events $E^{in\,i}$ are the outputs of phenomenological sub-models $G_{in\,i}$:

$$\begin{aligned} P(E^{in\,i}) &= P(1_{Ein\,i} = 1) = P(z_{in\,i} = 1) \\ z_{in\,i} &= G_{in\,i}(\boldsymbol{x}_i, \boldsymbol{e}_i, \boldsymbol{d}_i) \end{aligned} \tag{7.27}$$

Think of Step Two or Four in Chapter 3, where the flood protection system is modelled through a logical function G_s, itself fed by the output of the hydro model G_h predicting the events of limited or large overspill. Assume additionally that the inputs of the phenomenological submodels $(\boldsymbol{x}_i,\boldsymbol{e}_i)$ are independent from those of the logical model: the risk measure then appears as a product of the two, and closed-form integration can be performed inside:

$$\begin{aligned} c_z &= \int_{x,e} G(\boldsymbol{x}, \boldsymbol{e}, \boldsymbol{d}) f_{XE}(\boldsymbol{x}, \boldsymbol{e}) d\boldsymbol{x} d\boldsymbol{e} = \int_{x,e} \sum_i \left[G_{in^i}(\boldsymbol{x}_{in}, \boldsymbol{e}_{in}, \boldsymbol{d}) . \prod_{j_i(\boldsymbol{d})} e^{sy^{j_i}} \right] f_{XE}(\boldsymbol{x}, \boldsymbol{e}) d\boldsymbol{x} d\boldsymbol{e} \\ &= \int_{x,e} \sum_i \left[G_{in^i}(\boldsymbol{x}_{in}, \boldsymbol{e}_{in}, \boldsymbol{d}) . \prod_{j_i(\boldsymbol{d})} e^{sy^{j_i}} \right] f_{XEin}(\boldsymbol{x}_{in}, \boldsymbol{e}_{in}) f_E(\boldsymbol{e}_{sy}) d\boldsymbol{x}_{in} d\boldsymbol{e}_{in} d\boldsymbol{e}_{sy} \\ &= \sum_i \left[\int_{x,e} G_{in^i}(\boldsymbol{x}_{in}, \boldsymbol{e}_{in}, \boldsymbol{d}) . \prod_{j_i(\boldsymbol{d})} e^{sy^{j_i}} f_{XEin}(\boldsymbol{x}_{in}, \boldsymbol{e}_{in}) f_E(\boldsymbol{e}_{sy}) d\boldsymbol{x}_{in} d\boldsymbol{e}_{in} d\boldsymbol{e}_{sy} \right] \\ &= \sum_i \left[\int_{x,e} G_{in^i}(., \boldsymbol{d}) f_{XEin}(.) d\boldsymbol{x}_{in} d\boldsymbol{e}_{in} \prod_{j_i(\boldsymbol{d})} p^{j_i} \right] \end{aligned} \tag{7.28}$$

The essential consequence is that the computation is brought back to that of the phenomenological models themselves, then combined through a closed-form formula so as to issue the overall model. Thus, Monte-Carlo Simulation need not be done on the input events of the logical model together with the inputs of the phenomenological ones: this simplifies the task and reduces variance.

Flood example
The risk measure combining the hydro and system models results as follows:

$$c_Z = P(E_p) = P(E_{os}).p_1.p_2.(p_{12} + p_{1i}.p_{2i}) + P(E_{ol}) = P(E_{os}).p_{nr} + P(E_{ol}) \tag{7.29}$$

Only $P_{os} = P(E_{os})$ and $P_{ol} = P(E_{ol})$ require Monte-Carlo Simulation (both produced by the same simulation generating the pdf of the overflow) while p_{nr}, the combined non-recovery probability of the system conditional to overflow, is closed-form. Indeed, when combining such restricted MCS and closed-forms, the variance appears as:

$$\begin{aligned}
\operatorname{var} \hat{c}_z &= \frac{1}{N}\operatorname{var}\left(p_{nr}.1_{2>G(X_j,d)>0} + 1_{G(X_j,d)>2}\right) \\
&= \frac{1}{N}\left[p_{nr}{}^2.P_{os}(1-P_{os}) + P_{ol}(1-P_{ol}) + 2p_{nr}Cov\left(1_{2>G(X_j,d)>0}, 1_{G(X_j,d)>2}\right)\right] \\
&= \frac{1}{N}\left[p_{nr}{}^2.P_{os}(1-P_{os}) + P_{ol}(1-P_{ol}) - 2p_{nr}P_{os}.P_{ol}\right] \\
&\approx \frac{1}{N}\left[p_{nr}{}^2.P_{os} + P_{ol}\right] \quad when \quad P_{ol} < P_{os} \ll 1
\end{aligned} \tag{7.30}$$

Conversely, pure MCS including the system failure events would have had the following variance:

$$\begin{aligned}
\operatorname{var} \hat{c}_z &= \frac{1}{N}P(E_p).(1-P(E_p)) = \frac{1}{N}[p_{nr}P_{os} + P_{ol}][1 - p_{nr}P_{os} - P_{ol}] \\
&\approx \frac{1}{N}[p_{nr}P_{os} + P_{ol}] \quad when \quad P_{ol} < P_{os} \ll 1
\end{aligned} \tag{7.31}$$

For a comparable number of Monte-Carlo runs, variance is certainly reduced using the closed-form integration. If P_{ol} is small in comparison to $p_{nr}P_{os}$, this means that variance is reduced approximately by a factor p_{nr}. In other words, if the branch that is impacted by the closed-form simplification is the dominating contributor to the overall risk measure, then the variance is deflated by the conditional probability that is withdrawn from simulation. As shown by the equations above where $1/N$ is a factor, such a reduction factor would stay the same whatever the number of simulations.

As a numerical example, the risk measure amounts 4.10^{-3} include a protection system but with the absence of a dike and an additional 1.3m-dike is enough to guarantee a 1/1000-yr return protection. Closed-form integration of the system component reduces the estimation variance from a $0.09\ 10^{-3}$ down to $0.06\ 10^{-3}$ after 500 000 simulations, or more generally by a factor 1.5.

7.1.4 Non-expectation based single probabilistic risk measures

Generally, single probabilistic risk measures are defined as a functional of the predictive output distribution (hence of the predictive input distribution and system model):

$$c_z = \mathcal{F}\left[f_Z^\pi(.)\right] = \mathsf{G}\left[f_X^\pi(.), G(., \boldsymbol{d})\right] = C_z(\boldsymbol{d}) \tag{7.32}$$

In fact, most such non-expectation based risk measures do appear as elementary post-treatments if not being expectation-based as such. This is the case of variance:

$$\begin{aligned}
\operatorname{var}(Z) &= E_{f_Z^\pi}(Z^2) - (E_{f_Z^\pi}Z)^2 = \int_z z^2 f_X^\pi(z).dz - \left(\int_z z f_X^\pi(z).dz\right)^2 \\
&= \int_{\underline{x}} G(\underline{x},\underline{d})^2 f_X^\pi(\underline{x})d\underline{x} - \left(\int_{\underline{x}} G(\underline{x},\underline{d}) f_X^\pi(\underline{x})d\underline{x}\right)^2
\end{aligned} \tag{7.33}$$

that is a simple post-treatment of two expectation-based risk measures, the computation of which immediately yields additionally the standard deviation and the coefficient of variation, two key metrological quantities of interest:

$$unc_z = \sqrt{E_{f_z^\pi} Z^2 - \left(E_{f_z^\pi} Z\right)^2} \tag{7.34}$$

$$\% \, unc_z = \frac{\sqrt{E_{f_z^\pi} Z^2 - \left(E_{f_z^\pi} Z\right)^2}}{E_{f_z^\pi} Z} \tag{7.35}$$

Output *moments* $E[Z^m]$ or $E[(Z\text{-}EZ)^m]$ for $m >= 3$ may sometimes be looked for in analytical studies although rarely in formal risk measures. By definition, they are similar elementary post-treatments of two expectations.

Computing an output *quantile* (e.g. the median of the output or any α-risk level, a Value-at-Risk) involves further post-treatment as it involves theoretically *inverting* the *cdf* of z, itself being an expectation:

$$\begin{aligned} F_z^\pi(z) &= P(Z < z) = E_{f_z^\pi}\left(1_{G(X,d)<z}\right) \\ z^\alpha &= F_z^{\pi^{-1}}(\alpha) \end{aligned} \tag{7.36}$$

which potentially requires computing $F_Z^\pi(z)$ for numerous different thresholds z in order to converge to z^α that is a number of different expectations. In practice, most quantile risk measures are associated with decision criteria evidence that generally targets an implicit exceedance probability (see Chapters 1 and 2). However, it will be seen that some propagation methods such as Monte-Carlo issue direct estimates of a quantile without requiring the inversion of an expectation.

Other risk measures may be defined, such as the expected shortfall or tail value-at-risk that are of particular importance in finance or insurance modelling (see Chapter 8):

$$TVaR_{\alpha\%} = E(Z \,|\, Z > VaR_{\alpha\%}) \tag{7.37}$$

If no specific algorithm can be thought of for any given risk measure, the generic solution involves the sampling of predictive output distributions and thus to compute the functional defining the risk measure.

$$\begin{aligned} \hat{f}_z^\pi &\approx f_z^\pi \\ \hat{c}_z &= F\left[\hat{f}_z^\pi\right] \approx F\left[f_z^\pi\right] \end{aligned} \tag{7.38}$$

Such an estimation of the density will be given a more mathematical definition in the section on Monte-Carlo.

7.1.5 Other risk measures (double probabilistic, mixed deterministic-probabilistic)

Double probabilistic risk measures stand as functionals not only of the predictive output distribution, but also of the full joint output distribution:

$$c_z = \mathcal{F}[f_Z(.), \pi(. \,|\, IK, \Xi_n)] \tag{7.39}$$

An important sub-class is that of iterated expectations of known functions $h_a(.)$ and $h_e(.)$:

$$c_z = E_{\Theta_X}[h_e(E_X[h_a(Z)|\Theta_X, \mathbf{d}])|IK, \Xi_n] \tag{7.40}$$

Think, for instance, of the (level-2 or epistemic) confidence that the (level-1 or aleatory) threshold exceedance probability – say dike overspill – remains below a risk target α (say $\alpha = 10^{-b}$), whereby:

$$\begin{aligned} h_a(z) &= 1_{z>zs} \\ h_e(z) &= 1_{p<\alpha} \end{aligned} \tag{7.41}$$

yielding:

$$\begin{aligned} c_z &= P_{\Theta|IK,\Xi_n}\left[P_{X|\theta_X}(Z > z_s|\Theta_X, \mathbf{d}) < \alpha\right] \\ &= \int_{\theta_X} 1_{\left[\int_x 1_{G(x,d)>z_s} f_X(\mathbf{x}|\theta_X)d\mathbf{x}\right]<\alpha} \pi(\boldsymbol{\theta}_X|IK, \Xi_n)d\boldsymbol{\theta}_X \end{aligned} \tag{7.42}$$

While still involving expectations, this kind of risk measure shows a double level of integration that cannot be solved by standard integration. As being the combination of two integrals, computation can be done through a two-tier Monte-Carlo sampling. As this results in a much costlier configuration than single-probabilistic risk measures (see Section 7.2.5), alternative shortcuts are often required.

Alternatively, one may look after the level-1 threshold exceedance probability for a given β-(level-2) confidence, that is a β-quantile of the threshold exceedance probability. Similar to single probabilistic cases, there is the same link between that double probabilistic quantile and the computation of the level-2 *cdf* of the threshold exceedance probability that theoretically needs to be inverted. However, two-tier Monte-Carlo sampling also yields direct estimation of such quantiles.

Mixed deterministic-probabilistic risk measures may be first understood in a *weak sense* as amounting to computing probabilistic risk measures conditionally upon given (penalised) assumptions. Think of a failure probability penalised on a given sub-vector $\mathbf{x}_{pn}$ of inputs or events:

$$c_z = P(Z > z_s|IK, \Xi_n, \mathbf{x}_{pn}, \mathbf{d}) = \int_x 1_{G(x,d)>z_s} f_X^{\pi}(\mathbf{x}|IK, \Xi_n, x_{pn}).d\mathbf{x} \tag{7.43}$$

which may alternatively be defined in any other (single or double) probabilistic risk measure, possibly also involving penalised epistemic parameters $\boldsymbol{\theta}_{pn}$. It does not alter the computing strategy by much, simply requiring selection of the appropriate penalised uncertainty model:

$$\{f_X(\mathbf{x}|\mathbf{x}_{pn}, \boldsymbol{\theta}_X), \pi(\boldsymbol{\theta}_X|IK, \Xi_n, \boldsymbol{\theta}_{pn})\} \tag{7.44}$$

Much more challenging is the *strong sense* case, that is where one looks for the maximal (or minimal) value of a conditional probabilistic risk measure over a range or set of deterministic inputs (or epistemic parameters):

$$c_z(\mathbf{d}) = Max_{\mathbf{x}_{pn}\in D_x \boldsymbol{\theta}_{pn}\in D_\theta}\left[c_z|\mathbf{x}_{pn}, \boldsymbol{\theta}_{pn}, \mathbf{d}\right] \tag{7.45}$$

This involves joining a simulation algorithm (for the conditional risk measure) and an optimisation algorithm (over the set of deterministic assumptions): a great numerical challenge, because $G(.)$ is generally neither linear nor convex where there may be a large number p of inputs. The accuracy of the result is often indiscernible in the case of the existence of the local minima. Its simplified version is often used, consisting of maximising an experimental design constructed on the limits of components of $\mathbf{X}$. There is then no guarantee of accuracy in the general non monotonous case.

7.2 The generic Monte-Carlo simulation method and associated error control

The section introduces the Monte-Carlo method which is a *universal solution* for the computation of *probabilistic* risk measures as well as for mixed deterministic-probabilistic ones to some extent. However, computational constraints generate a fundamental difficulty with complex phenomenological models in practice, the answer to which comes with, beyond the preliminary closed-form simplifications introduced in Section 7.1.3, a series of alternative computational methods that will be introduced in the rest of this chapter.

7.2.1 Undertaking Monte-Carlo simulation on a computer

Monte-Carlo simulation may be seen informally as a technique designed to simulate random variables on the input side of a numerical model, run the model for each sample and thus post-treat the resulting model outputs to approximate any desirable model output feature: either a probabilistic risk measure (expectation, variance or any moment; quantile, exceedance probability, etc.) or a deterministic one (e.g. an approximate maximal or minimal output) or any combination. It is more generally a quantity q defined as a functional of the distribution function of $\mathbf{Z}$:

$$q = \mathcal{H}[f_Z(.)] \tag{7.46}$$

that is another functional of the numerical model $G(.)$ and the input distribution of $\mathbf{X}$ assuming at this stage a single-probabilistic uncertainty model:

$$q = \mathcal{H}_X[f_X(.), G(.)] \tag{7.47}$$

While early trials have involved experimental settings (such as Buffon's needle), modern sampling is undertaken through pseudo-random numbers generated by a computer (Metropolis and Ulam, 1949), starting with that of *independent uniformly distributed* variables. Any language includes the sampling of a series of such normalised uniform random variables:

$$U \sim Unif[0, 1] \tag{7.48}$$

Bearing different names according to the software – runif (R software), ALEA() (MS Excel) and so on – they are all based on various techniques of pseudo-random number generation, a large field of research for which the reader is referred to the specialised literature, for example Rubinstein (1981) and Fishman (1996). Many of these generators are based partially at least on congruential generators involving quotients of large premier numbers, namely:

$$D(x) = (ax + b)mod(M + 1) \tag{7.49}$$

This generates a pseudo-random integer uniformly distributed in $\{0, 1 \ldots M\}$ where M stands for a very large integer, a and b large premier numbers.

Sampling a given *non-uniformly distributed* random variable X subsequently involves the inversion of its cdf F_X: indeed, $X = F_X^{-1}(U)$ transforms any uniformly distributed r.v. into a r.v. distributed along F_X. This basic method is not always selected as inversion may be either numerically costly or not practical in the absence of a closed-form expression of the cdf. Classical alternatives include, for instance, Box-Muller for the Gaussian distribution (whose cdf lacks a closed-form unlike its density) or the Accept-Reject technique (see Robert and Casella, 1999).

Sampling *linearly correlated variables* is very simple in the case of Gaussian vectors since ($\boldsymbol{X}$ and $\boldsymbol{V}$ being vectors and $\boldsymbol{A}$ being a matrix):

$$\begin{aligned} &\boldsymbol{X} = \boldsymbol{A}\boldsymbol{V} \\ &\Rightarrow \boldsymbol{varX} = \boldsymbol{A}.\boldsymbol{var}\boldsymbol{V}.\boldsymbol{A}' \end{aligned} \tag{7.50}$$

so that using the Cholesky decomposition of the target covariance matrix $\sum$ of vector $\boldsymbol{X}$

$$\sum = \boldsymbol{A}\boldsymbol{A}' \tag{7.51}$$

The simulation of a vector $\boldsymbol{V}$ of p i.i.d. standard normal r.v. enables generating the required correlated vector through $\boldsymbol{X} = \boldsymbol{A}\boldsymbol{V}$ as:

$$\begin{aligned} &\boldsymbol{var}\boldsymbol{V} = \boldsymbol{I} \\ &\boldsymbol{varX} = \boldsymbol{A}.\boldsymbol{var}\boldsymbol{V}.\boldsymbol{A}' = \sum \end{aligned} \tag{7.52}$$

Using matrix notation the algorithm goes as follows, $\boldsymbol{V}'$ being the p-dimensional transposed row vector, and $\underline{X}$ the N^*p matrix of sampled vector inputs:

$$\begin{aligned} \text{sample } \underline{V} &= \left(\boldsymbol{V}_j'\right)_{j=1\ldots N} \\ \text{compute } \underline{X} &= \left(\boldsymbol{X}'_j\right)_{j=1\ldots N} = \boldsymbol{V}.\boldsymbol{A}' \end{aligned}$$

Beyond Gaussian vectors, the sampling of *non-independent variables* becomes trickier as the variety of potential functional dependencies is very large.

A popular although approximate technique has been developed to sample *rank-correlated* (possibly non-Gaussian) random variables (McKay, Beckman, and Conover, 1979): it is the standard recipe of many commercial codes undertaking risk or uncertainty simulation that has been used in nuclear performance assessment studies (Helton and Davis, 2003). Iman-Conover algorithms all consist of rearranging the columns of an independently sampled matrix $\underline{X}°$ of input samples so that its rank correlation resembles that of its desired *rank* correlation matrix $\boldsymbol{R}$. To do so, the algorithms generate $\underline{T}$, a random N^*p-matrix of ranks the empiric (linear) correlation of which is $\boldsymbol{R}_N$, close to $\boldsymbol{R}$ (although not exactly because of sampling fluctuation). The elements of the independently sampled $\underline{X}°$ are then permuted inside each column so that the ranks follow exactly those of $\underline{T}$ so that the rank correlation of the resulting $\underline{X}$ is again $\boldsymbol{R}_N$. Variants differ typically on how to generate the rank matrices $\underline{T}$, typically through an initial matrix of independent and standardised variates, then transformed similarly into the above-mentioned simulation of correlated Gaussian vectors through Cholesky decomposition of $\boldsymbol{R}$.

In Chapter 5 it was said that dependence modelling involves more generally the specification of the *copula* function. Given the complete knowledge of the copula, sampling a vector of p dependent variables involves dedicated techniques designed to sample p uniformly-distributed random variables distributed along the copula function and then applying the inverse marginal *cdf* to the p components of such a uniformly-distributed dependant sample. More can be found in this field that is attracting a considerably growing interest in the specialised literature (Nelsen, 1999) as well as in the open source code www.openturns.org.

7.2.2 Dual interpretation and probabilistic properties of Monte-Carlo simulation

Monte-Carlo simulation may be formally viewed either:

- a numerical integration technique,
- or, more generally even, as a pseudo-empiric statistical estimation process.

Both interpretations will be discussed below.

Monte-carlo as a numerical integration technique According to the first interpretation, the purpose is to estimate a quantity q defined as the expectation of a function $Q(.)$ of the random vector $\boldsymbol{W}$ distributed over a sample space Ω according to a known pdf f_W:

$$q = E_W[Q(\boldsymbol{W})] \tag{7.53}$$

$$\boldsymbol{W} \sim f_W(\boldsymbol{w}) \tag{7.54}$$

The quantity q results in the following integral:

$$q = \int_{w \in \boldsymbol{\Omega}} Q(\boldsymbol{w}) f_W(\boldsymbol{w}) d\boldsymbol{w} \tag{7.55}$$

Various integration techniques may be mobilised to compute q, such as the Simpson method, Gauss points and so on mostly consisting of computing the integrand function on a discrete set of points $(\boldsymbol{w}_j)_{j=1...N}$ covering the integration domain and weighing those results through dedicated coefficients γ_j that are controlled by the integration method. Monte-Carlo does it through a random sampling of $(\boldsymbol{w}_j)_{j=1...N}$ evenly distributed according to the density $f_W(\boldsymbol{w})$ resulting in the following universal MCS estimator:

$$\hat{q}_N = \frac{1}{N} \sum_{j=1...N} Q(\boldsymbol{w}_j) \tag{7.56}$$

Interpreting it intuitively as an integration process that is distributed over an iso-probabilistic (random) N-partition of the entire sample space $\Omega = \bigcup_{j=1...N} E_j$, within each of which a random point $\boldsymbol{w}_j$ is selected:

$$\begin{aligned} q &= \int_{w \in \boldsymbol{\Omega}} Q(\boldsymbol{w}) f_W(\boldsymbol{w}) d\boldsymbol{w} = \sum_{j=1...N} \int_{w \in E_j} Q(\boldsymbol{w}) f_W(\boldsymbol{w}) d\boldsymbol{w} \\ &\approx \sum_{j=1...N} Q(\boldsymbol{w}_j) \int_{w \in E_j} f_W(\boldsymbol{w}) d\boldsymbol{w} = \sum_{j=1...N} Q(\boldsymbol{w}_j) \frac{1}{N} \end{aligned} \tag{7.57}$$

A more specific *stratification* of the sample space will appear as an extension of the Monte-Carlo so as method to explore space efficiently. The estimator MCS has considerable properties:

(1) it is unbiased:

$$E(\hat{q}_N) = \frac{1}{N}\sum_{j=1\ldots N} E[Q(\boldsymbol{W}_j)] = \frac{1}{N}\sum_{j=1\ldots N} q = q \tag{7.58}$$

(2) it converges *almost surely* to q (by virtue of the Strong Law of Large Numbers in Equation (7.57))

$$\hat{q}_N \underset{N\to\infty}{\longrightarrow} q \; a.s. \tag{7.59}$$

(3) it is asymptotically Gaussian distributed (by virtue of the Central-Limit Theorem in Equation (7.57)) with the following variance decreasing in 1/*N*:

$$Var(\hat{q}_N) = \frac{1}{N} Var(Q(\boldsymbol{W})) \tag{7.60}$$

Such variance can be estimated on the basis of an ancillary estimator (unbiased and almost surely convergent by another application of the Strong Law of Large Numbers), the simulated output variance denoted as s_Q^2:

$$Var(\hat{q}_N) \approx \frac{1}{N} s_Q^2 = \frac{1}{N(N-1)}\sum_{j=1\ldots N}\left[Q(\boldsymbol{w}_j) - \hat{q}_N\right]^2 \tag{7.61}$$

The required conditions being respectively: (i) that $Q(\boldsymbol{W})$ is a random vector, that is that $Q(.)$ is measurable; (ii) that the expectation value of $Q(\boldsymbol{W})$ is finite; (iii) that the variance of $Q(\boldsymbol{W})$ is also finite.

Such conditions would doubtlessly be met for the integrand functions encountered with most phenomenological system models. The typical integrand functions in single probabilistic expectation-based risk measures are *UoG*(.). The system model *G*(.) is generally continuous except for discontinuities in countable numbers or of null measure. Think of the potential phenomenological bifurcations within an accident scenario or the (finite) number of discrete events involved in IPRA. This is also the case with *UoG*(.) because the utility as such is generally a simple function that is also continuous except for discontinuities in finite numbers. Thus, the integrand can be integrated in Lebesgue terms, and is in fact integratable even in a Riemannian sense.

In theory, controversy arises with the *finiteness of the expectation and variance* of $Q(\boldsymbol{W})$. In practice, *G*(.) is always physically bounded. Which is also physically the case for the domains of $\boldsymbol{X}$: if unbounded distributions in $\boldsymbol{X}$ are used, this is only an approximation of real physically-limited domains, which real numerical simulation would in any case truncate: one is also obliged to do so in order to avoid both numerical overflows and non-physical input samples. The expected value and the variance of $Q(\boldsymbol{W})$, if taken as a bounded function of *G*(.), should thus also remain bounded. This is the case for the fundamental cases where $Q(\boldsymbol{W}) = G(\boldsymbol{X},\boldsymbol{d})$, $Q(\boldsymbol{W}) = G(\boldsymbol{X},\boldsymbol{d})^2$, $Q(\boldsymbol{W}) = 1_{G(\boldsymbol{X},\boldsymbol{d})>zs}$ or more generally *UoG*(.) with an increasing and bounded utility function *U* except in degenerate cases with infinite utility placed on one single event. Nevertheless, special care may be needed when handling, for instance, extreme value distributions for the input models, such as Fréchet/Pareto, as their moments may be infinite for large shape parameters (see Chapter 4). Even if a practical truncation would again be necessary and mandatory, such heavy-tailness would inevitably entail slow and unstable convergence: the flood example (Section 7.2.3) will show, for instance, how slowly an estimate of the expectation converges when variance is very large. As mentioned in Section 4.3.5 of Chapter 4, most puzzling rare and extreme events could indeed be *modelled* through such distributions with unbounded expectation or variance: assuming that the use of such extreme models remains credible regarding their estimation, the transformed output $Q(\boldsymbol{W}) = 1_{G(\boldsymbol{X},\boldsymbol{d})>zs}$ would then still converge normally and make more sense than an expected utility.

Conversely, these conditions for convergence are in no way limiting with regard to 'relative irregularities' encountered with some system models $G(.)$ at risk, for example non-linear, non-monotonic, or again non-differentiable characteristics or local discontinuities, for example in thermo-hydraulic models or accident analyses with phenomenological bifurcations. These irregularities will appear conversely as a limitation for many alternative propagation methods.

Note also that, contrary to most alternative methods, MCS propagation uncertainty *does not directly depend on the input uncertainty dimension p*. Indeed, the variance of the MCS estimator is as follows:

$$Var(\hat{q}_N) = \frac{1}{N} VarQ(\boldsymbol{W}) \tag{7.62}$$

As long as $VarQ(\boldsymbol{W})$does not depend on input dimension p, nor does the speed of MCS. Consider, for instance, the key case of an exceedance probability:

$$Q(\boldsymbol{w}) = 1_{G(\boldsymbol{w},\boldsymbol{d})>z_s} \tag{7.63}$$

Its variance appears as follows:

$$\begin{aligned} VarQ(\boldsymbol{W}) &= P(G(\boldsymbol{W},\boldsymbol{d}) > z_s)[1 - P(G(\boldsymbol{W},\boldsymbol{d}) > z_s)] \\ &= p_f(1 - p_f) \end{aligned} \tag{7.64}$$

so that the coefficient of variation, or relative computation error is:

$$C_V^N = \frac{\sqrt{Var(\hat{q}_N)}}{E(\hat{q}_N)} = \frac{\sqrt{\frac{1}{N}p_f(1-p_f)}}{p_f} = \sqrt{\frac{1}{N}\frac{(1-p_f)}{p_f}} \tag{7.65}$$

For a given system, there is no reason to believe that a more detailed system model requiring a much larger uncertainty input vector p (say finite-element model with hundreds of *df*) will result in a significantly different failure probability p_f than a much simpler one with simply a few random inputs. Or this becomes an issue of model-building, disconnected from the risk measure computation. The extreme example of system models $Q(\boldsymbol{W})$ being additions of widely-differing numbers (say $p = 10$ vs. $p = 1000$) of independent random variables of similar uncertainty variance – the most numerous one clearly having a smaller variance $varQ(\boldsymbol{W})$, simply describe two different systems at risk.

This point is often misunderstood by practitioners, perhaps because, contrary to propagation (i.e. 'uncertainty analysis'), the number of MCS calculations *increases with p for the step of importance ranking* (i.e. 'sensitivity analysis'). This is because the associated importance or sensitivity indices are typically based on estimators such as regression coefficients of Z as a function of X^i as will be discussed in Section 7.5.

Real difficulties associated with the use of MCS for the computation of risk measures thus do not appear with strong *p* but with the risk measures involving large number of runs of *costly system models*, such as *rare exceedance probabilities* (small p_f). Considering again the previous formula when $p_f \ll 1$, the relative computation error appears as:

$$C_V^N \approx \sqrt{\frac{1}{Np_f}} \tag{7.66}$$

so that the number of runs required to compute a rare threshold exceedance probability p_f increases inverse proportionally to p_f, say $100/p_f$ to $400/p_f$ in order of magnitude when looking for a coefficient of variation of 10 % to 5 %. Think of 10^5 system model runs when searching for a 1/1000-yr risk level of a system controlled in annual maximal values. Recall at this point the *rarity index* that was introduced in Chapter 5, Section 5.5.2 to control the amount of data required to estimate robustly a quantile. Switching now the number of degrees-of-freedom from the *data* sample size *n* to the *computational* sample size *N* yields:

$$\begin{aligned} RI &= Log_{10}\frac{1}{c_s} - Log_{10}N \\ &= -2Log_{10}C_V^N \end{aligned} \tag{7.67}$$

In other words, *RI* counts the number of orders of magnitude by which the risk measure $c_s = p_f = 10^{-x}$, is covered by the size *N* of the MCS sample: the previous comments indicate that a typical requirement would be for *RI* to be lower than -2, that is correspond to the area of 'quasi-asymptotics' suggested in Section 5.5.2. Smaller MCS samples turn the event of interest into a computationally *rare* to *very rare* event, thus fragilising the robustness of the estimate.

Note finally that all derivations made on a real-valued scalar *q* may obviously be extended to a vector quantity of interest $\boldsymbol{q}$. More importantly perhaps they may be extended to a two-tier process requiring the computation of a quantity *r*:

$$r = \int_{\theta} q(\boldsymbol{\theta}) f_{\boldsymbol{\Theta}}(\boldsymbol{\theta}) d\boldsymbol{\theta} \tag{7.68}$$

whereby the integrand function $q(\boldsymbol{\theta})$is itself estimated as an inner integral:

$$q(\boldsymbol{\theta}) = \int_{\boldsymbol{w}\in\Omega} Q(\boldsymbol{w},\boldsymbol{\theta}) f_{W}(\boldsymbol{w},\boldsymbol{\theta}) d\boldsymbol{w} \tag{7.69}$$

as will be exemplified by double-probabilistic risk measures (Section 7.2.5).

Monte-carlo as a pseudo-empiric statistical estimation process There is an even more general and perhaps intuitive interpretation of Monte-Carlo simulation.

This simply generates a random *i.i.d.* sample of any – however complex – random vector (or even time process) $\boldsymbol{Q}$ that can be modeled credibly enough by a model $Q(\boldsymbol{W})$ involving a potentially complex function (but amenable to computer simulation) $\boldsymbol{Q}$ and a random vector $\boldsymbol{W}$ distributed according to a known pdf f_W. Such a sample $(\boldsymbol{Q}_j)_{j=1...N}$ may just be taken as if it were a sample of observations, say *pseudo-empiric*, for being the output of computer simulation rather than empiric observation of nature.

Thus, it opens the way to any type of statistical estimate that would be contemplated with an empiric sample, that is:

- any probabilistic quantity of interest: expectation, variance or any moment; quantile, exceedance probability and so on;
- its empiric distribution and any post-treatment, for example histograms, empiric cdf and so on.Meaning that all of the estimation techniques discussed in Chapter 5 on the basis of an empiric sample could be applied to Monte-Carlo simulated output. Virtually any quantity q (real-valued, vector, function, graph ...) defined as a functional of the distribution function of $\boldsymbol{Q}$:

$$q = \mathcal{H}[f_Q(.)] \tag{7.70}$$

can be estimated on the basis of the pseudo-empiric sample. This goes well beyond the estimation of quantities q defined as integrals as in the previous section. In particular, an essential output of Monte-Carlo sampling is the empiric *cdf*:

$$\hat{F}_{Q\,N}(q) = \frac{1}{N}\sum_{j=1...N} 1_{Q_j<q} \tag{7.71}$$

which has considerable properties; whatever q, it converges almost surely to the true cdf $F_Q(q)$ and is asymptotically Gaussian as being the mean of i.i.d. random variates with finite expectation $E1_{Q_j<q} = F_Q(q)$ and variance $Var1_{Q_j<q} = F_Q(q)(1 - F_Q(q))$. Better even, later sections will even show that the *exact distribution* of $\hat{F}_{Q\,N}(q)$may be controlled, whatever the underlying system model function $Q(\boldsymbol{W})$.

Similar to Chapter 5, many statistical quantities of interest may be derived directly from such a Monte-Carlo estimator of the *cdf* of Q, such as:

- the empiric quantile or even any parametrically-modelled quantile, fitting a statistical distribution upon the output Q,
- the associated *pdf* of Q, possibly using kernel-smoothing techniques,

$$\hat{f}_N^{\,ker}(q) = \frac{1}{Nh_N\sqrt{2\pi}}\sum_{j=1}^{N}\exp\left(-\frac{1}{2}\left(\frac{q-q_j}{h_N}\right)^2\right) \tag{7.72}$$

non-parametric bootstrapping techniques to issue pseudo-confidence intervals,
- any more elaborate risk measure, for example the tail value at-risk, taking the empiric mean of the filtered sample above the empiric quantile.

> Remember that considering MCS as a credible pseudo-empiric observation process relies critically on trusting the combined system and uncertainty model as an appropriate representation of the likely behaviour of the variables of interest.

7.2.3 Control of propagation uncertainty: Asymptotic results

The great advantage of MCS is that it always provides an estimator of the *propagation uncertainty*, that is the error attached to the computation of $c_z(\theta,d)$ (even if θ is perfectly known, that is outside level-2 epistemic uncertainty). It greatly helps in anticipating the minimum number of calculations necessary to get reliable estimates of the risk measure, which proves particularly important in safety assessment.

In general, the above-mentioned probabilistic properties of the estimator (Equations (7.56)–(7.58)) ensure firstly an *asymptotic* control of the error variance, that is valid when N grows limitless. This was illustrated in the case of the threshold exceedance in Section 7.2.2 but it is true for most risk measures, see Table 7.1.

Note that the first two examples in Table 7.1 both stand as expectations estimated by empiric means. Thus, the Central Limit Theorem ensures that the estimators of the risk measure are both asymptotically Gaussian. The last example of risk measure is not directly an expectation, but comes as a closed-form compound of different expectations:

$$c_z = k.\sqrt{\text{var}(Z)} = k\sqrt{E(Z^2) - E(Z)^2}$$

Both $E(Z^2)$ and $E(Z)^2$ are estimated as standard expectations so that both enjoy the same property by virtue of the Central Limit Theorem. The compounding function $a,b -> k\sqrt{(b\text{-}a)}$ being continuous when $b > a$, c_z is again asymptotically Gaussian by virtue of the Slutsky theorem. Thus, an approximate minimum for the number of necessary calculations to guarantee at 95% confidence $+-10\%$ accuracy can be computed by fixing a maximum 5% for the coefficient of variation: this means

- $N > 200/p_f$ for the probability of exceeding a threshold;
- $N > 100/c_V$ for an expected utility where cv stands for the coefficient of variation of the random utility (but it is rarely known in advance);
- $N > 200$ for a risk measure derived from the standard deviation, whatever the value of such standard deviation.

Care should, however, always be taken with such asymptotic estimates as will be illustrated by the flood example.

Table 7.1 Asymptotic propagation uncertainty associated to MCS for various risk measures.

Risk measure c_Z	Estimator	Asymptotic variance	Asymptotic coefficient of variation
$c_z = P(Z > z_s) = p_f$	$Q(.) = 1_{G(.)>z_s}$ $\hat{c}_Z = \hat{q}_N = 1_{G(.)>z_s}$	$\text{var } \hat{c}_Z = \frac{1}{N}p_f(1-p_f)$ $\approx \frac{1}{N}\hat{c}_Z(1-\hat{c}_Z)$	$C_V^N \approx \sqrt{\frac{1}{N}\frac{(1-\hat{c}_Z)}{\hat{c}_Z}}$
$c_z = E[U(Z)]$	$Q(.) = UoG(.)$ $\hat{c}_Z = \hat{q}_N = UoG(.)$	$\text{var } \hat{c}_Z = \frac{1}{N}\text{var}[UoG(.)] \approx \frac{1}{N}\hat{s}_Q^2$	$C_V^N \approx \sqrt{\frac{1}{N}\frac{\hat{s}_Q}{\hat{c}_z}}$
$c_z = k.\sqrt{\text{var}(Z)}$	$Q(.) = G(.)$ $\hat{c}_z = k.\sqrt{\hat{s}_Q^2}$	$\text{var } \hat{c}_Z \approx \frac{1}{2N}k^2\hat{s}_Q^2 = \frac{1}{2N}\hat{c}_Z^2$	$C_V^N \approx \sqrt{\frac{1}{2N}}$

Flood example

Monte-Carlo sampling may be done firstly on the hydro component to simulate the flood level. Table 7.2 illustrates some examples of results with the raw probabilistic models (indicating in brackets twice the standard deviation when appropriate, that is the asymptotic 95 % confidence interval). Values of no or low significance are shown in italics.

Table 7.2 MCS sampling of the raw hydro-flood model.

Simulating Z_c	100 runs	10 000	1 000 000
Expectation	52.6 ± 0.24	52.6 ± 0.02	52.6 ± 0.01
standard deviation	1.2 ± 0.17	1.1 ± 0.02	6.7 ± 0.01
1/100-yr (99th quantile)	55.1	55.6	55.6
1/1000-yr (99th quantile)	58.9	58.1	57.8
$P(Z_c > 55.5)$ that is of overflow in absence of dike	$1\% \pm 2$	$1.1\% \pm 0.2$	$1.1\% \pm 0.02$

Beyond the fact that the number of runs is obviously too small to estimate some figures (such as the 1/1000-quantile or an exceedance probability of 1% with 100 runs), the examples illustrate both the orders of magnitude given by asymptotics and their limitations: undertaking 100 times more simulations should more or less increase the accuracy by a factor of 10. Which is true, for instance, for the exceedance probability but turns out to be only a rough approximate for mean and even more for standard deviation. The distribution of Z_c is so highly skewed and troubled by extremely high values when the number of runs becomes enough for extremes to be likely to happen that the standard deviation estimate is very slow to converge.

A closer look at the results reveals non-physical simulations as the maximum flood level exceeds ....1000mNGF (a mountain of water!) when more than 50–100 000 runs are undertaken. This means that the non-bounded input distributions have to be truncated in order to exclude non-physical values on the basis of phenomenological knowledge. For instance, most experts would exclude a friction Strickler coefficient $k_s < 3m^{1/3}/s$ (corresponding to a $\sim 10^{-4}$ quantile) even in heavily-obtruded rivers. Regarding flow, it is of little use to fix minimal values (as Z_c cannot go below Z_v in any case) and it is much more controversial to fix credible maximal values of non-negligible probability. Although 100 000m3/s (20 times the 1/1000-yr flood) could surely be excluded, such a level is exceeded for a truly-insignificant probability (far less than 10^{-15}). Table 7.3 gives the results including this truncation limited to K_s. After truncation, the maximum in Z_c result in 70m in 1 000 000 runs, still an extreme figure by physical standards (15m of flood above the riverbank) but low enough to reduce the skewness of the distribution so that standard deviation becomes much more stable. The asymptotic rate of convergence in $1/\sqrt{N}$ is now close to that observed. In any case, estimates of the most common risk measures used in safety or risk assessment such as quantile (or exceedance probability) are largely untroubled by such extreme

Table 7.3 MCS sampling of the hydro-flood model after truncation of the K_s distribution.

Simulating Z_c	100 runs	10 000	1 000 000
Expectation	52.6 ± 0.20	52.5 ± 0.02	52.54 ± 0.002
standard deviation	1.0 ± 0.14	1.08 ± 0.02	1.08 ± 0.001
1/100-yr (99th quantile)	55.0	55.6	55.6
1/1000-yr (99th quantile)	55.8	57.9	57.8
$P(Z_c > 55.5)$ that is of overflow in absence of dike	$1\% \pm 2$	1.1 ± 0.2	$1.1\% \pm 0.02$

values as only the ranks are important: thus the truncation is not always necessary but would be so when expectation or variance are looked for.

Eventually, this may also be done in the *economic component* to simulate cost with the results[1] in Table 7.4.

Table 7.4 MCS sampling of the economic flood model.

	No dike				Dike = 2m			
Simulating C_c(*in M€/yr*)	500 runs	5000	50 000	500 000	500 runs	5000	50 000	500 000
expectation	22 ± 74%	9.8 ± 38%	8.8 ±12%	8.8 ± 4%	0.2 ± 0%	0.9 ± 90%	1.6 ± 29%	1.7 ± 9%
standard deviation	184 ± 6%	132 ± 2%	122 ± 0.2%	124 ± 0.6%	0 ± 0%	30 ± 2%	50 ± 0.6%	53 ± 0.2%
P(C_c > *fixed cost*)	2%	0.9%	1.2%	1.2%	0%	0.1%	0.15%	0.15%

Note how slowly the expectation converges, and how the (asymptotic) error estimation of the estimates is again misleading though not because of the truncation effects as the cost function is bounded (a maximum yearly damage of 3MEUR). This is because of the very singular nature of the distribution of C_c, as shown in Figure 7.2. Most of the time, cost is limited to the small amount of *certain* investment cost (for $h_d = 0$, $C_c = 0$ *at* 99 %; for $h_d = 2$, $C_c < 0{,}3M€$ *at* 99,9 %), the rest of the distribution being distributed regularly (almost uniform) over [0,5-3 BEUR], representing the rare but large damage costs (*1 000 to 10 000) when overflow occurs. Note the enormous coefficient of variation of 1000–3000 % of the cost. Being very slow to converge, this expectation behaves just like an exceedance probability as the cost is essentially influenced by the rare occurrence of overflow: far from being Gaussian, the variance of the estimates converges slowly to the asymptotic figures. Convergence can be improved through accelerated techniques in that case (see Section 7.3.4)

7.2.4 Control of propagation uncertainty: Robust results for quantiles (Wilks formula)

Principle of the Wilks formula When looking for a α–quantile risk measure, there is a much greater robust control of the propagation error than the asymptotic result. Indeed, $c_z = z^\alpha$ is a by-product of the above-mentioned empiric *cdf*, formally:

$$c_z = \hat{Z}^\alpha = \hat{F}_{Q\,N}^{-1}(\alpha_N) \tag{7.73}$$

where:

$$\alpha_N = \frac{Int[\alpha \cdot N] + 1}{N} \tag{7.74}$$

[1] Unlike Chapter 8, the time basis is simplified into a *single* average year instead of simulating the whole period, for the sole purpose of illustrating convergence speed. This has no impact on the expected value but changes significantly the variance and exceedance probability, cf. Section 8.2.2. Correlations and the systems reliability component are also neglected.

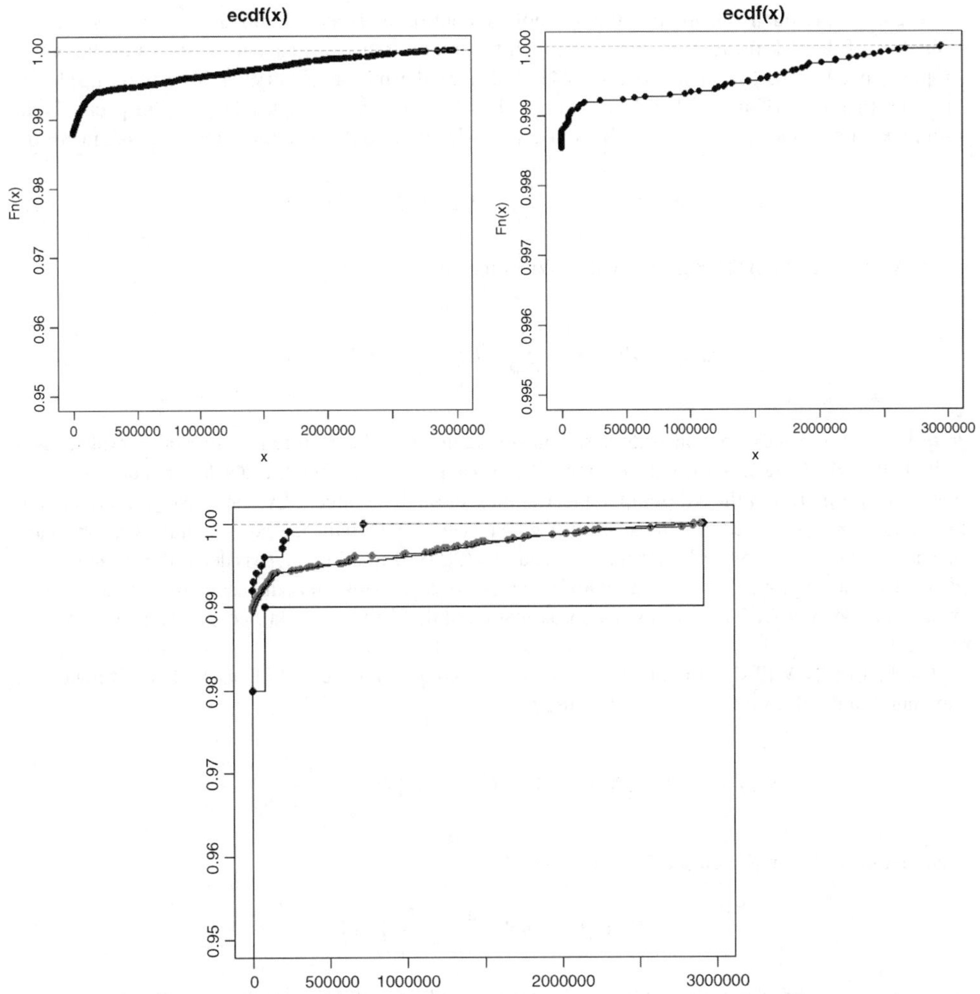

Figure 7.2 *Distribution of total yearly costs (in kEUR): left ($h_d=0$, $N=5.10^4$); right ($h_d=2$, $N=5.10^4$); down ($h_d=0$, convergence with $N=10^2$, 10^3, 10^4, 10^5).*

It reads simply as taking the closest empiric quantile, that is ranked simulated output to the wanted α. More can be said regarding the statistical fluctuation of such intuitive quantile estimation by Monte-Carlo, as it turns out remarkably that:

- the *exact distribution* of $\hat{F}_{Q\,N}$ enjoys a closed-form expression;
- such a distribution is *completely independent* from the underlying distribution of $Q(.)$, that is of the *system model* or its *input uncertainty distributions*;
- the result is true whatever the number of runs N, that is *even for small numbers of runs* for which asymptotic results may be questioned.

The key consequence being that it is possible to control the level of confidence associated with an estimation of the risk measure z^α simply through the total number of runs and the ranks of the output sample. Indeed, *Wilks' formula* (Wilks, 1941) indicates the minimum number N of trials needed to guarantee that the maximum value Z^N of the i.i.d. Monte-Carlo output sample $(Z_j)_{j=1...N}$ (respectively the m-th maximum value Z^{N-m+1}) exceeds for a given β-level of confidence the α-quantile risk measure:

$$for \quad N \geq N_m(\alpha, \beta), \quad P\left(z^\alpha < Z^{N-m+1}\right) > \beta \tag{7.75}$$

where $N_m(\alpha, \beta)$ is the integer given by the 'Wilks formula':

$$N_m(\alpha, \beta) = Min\left\{ N \left| \sum_{k=N-m+1}^{N} C_N^k \alpha^k (1-\alpha)^{N-k} < 1 - \beta \right.\right\} \tag{7.76}$$

Proof is given in Annex Section 10.5: based on simple order statistics, it requires no other probabilistic property than the mere existence of a *cdf* for Z. Note also that the number of trials does not depend in any way on the properties of the cdf of input uncertainties, or on the number of variates. Nor is it affected by correlations or dependence structures within $\boldsymbol{X}$ with respect to which the method is completely robust and unchanged. As recalled by Ardillon and de Rocquigny (2003), Equation (7.76) is derived from a binomial cdf which is asymptotically Gaussian with the variance $\alpha(1-\alpha)/N$: thus the asymptotic result already exhibited in Section 7.1.3 appears as an approximation of the Wilks formula, acceptable if $N(1-\alpha) \gg 1$ and $N\alpha \gg 1$.

The first-order Wilks' formula ($m = 1$) involves taking as the quantile conservative estimator the maximum value of a sample of the following size:

$$N_1(\alpha, \beta) = Min\left\{N \middle| \alpha^N < 1 - \beta\right\} = Min\left\{ N \middle| N > \frac{\log(1-\beta)}{\log(\alpha)} \right\} \tag{7.77}$$

Such first-order size has a closed-form expression:

$$N_1(\alpha, \beta) = Int\left(\frac{\log(1-\beta)}{\log(\alpha)}\right) + 1 \tag{7.78}$$

which can be simplified as follows when considering a small enough value for the exceedance probability $p_f = 1 - \alpha$, and the frequently-considered $\beta = 95\,\%$ confidence level:

$$N_1(\alpha, 0.95) = Int\left(\frac{\log(1-\beta)}{\log(\alpha)}\right) + 1 \approx \frac{\log(1-\beta)}{1-\alpha} \approx \frac{3}{p_f} \tag{7.79}$$

Thus, a first-order Wilks estimator at the 95 % confidence level reduces the minimal number of runs from the above-mentioned $100.p_f^{-1}$to $400.p_f^{-1}$ of standard MCS down to $3.p_f^{-1}$. For the higher-orders, the minimal size is not closed-form anymore but it may be tabulated easily, as illustrated in Table 7.5.

Using such a table is straightforward. The m-th maximum value resulting from Monte-Carlo sampling Z^{N-m+1} can be taken as a 'β-conservative' estimator of $c_z = z^\alpha$, denoted by Z^α_β. For instance, for $\alpha = \beta = 95\,\%$, $N_1(\alpha, \beta) = 59$ and $N_2(\alpha, \beta) = 93$ that is the maximum of a 59-sample or the second maximum of a 93-sample are 95 %-conservative estimators of the unknown true 95 %-quantile. Thus, if one prescribes in advance the desired quantile and level of confidence, Wilks' formula enables fixing the

Table 7.5 Minimum number of trials for given quantile and level of confidence and various orders Wilks methods.

α (quantile)	β (level of confidence)	Maximum $N_1(\alpha, \beta)$	2nd maximum $N_2(\alpha, \beta)$	3rd maximum $N_3(\alpha, \beta)$	4th maximum $N_4(\alpha, \beta)$
95 %	90 %	45	77	105	132
95 %	95 %	59	93	124	153
95 %	99 %	90	130	165	198

precise number of runs that are necessary: in fact a series of numbers of runs is available according to one's budget (e.g. 59, 93, 124, 153, and so on for $\alpha = \beta = 95\,\%$). Conversely, for a given number of runs, various estimators corresponding to various levels of confidence β can be taken as results for a risk measure defined as a given quantile α.

Wilks (1941) established this formula in the larger context of 'tolerance limits' in quality control, investigating the size needed for samples of manufactured pieces to predetermine, with a given confidence level, tolerance limits covering a prescribed probability. The formula gained a great deal of interest in the nuclear sector in the 1990s, particularly for accident analysis and uncertainty assessment where system models are computationally very expensive, with large potential irregularity of behaviours, such as non-continuous bifurcations and so on. It is indeed essential to stress that Wilks' formula is a simple post-treatment of standard Monte-Carlo sampling and therefore as robust regarding system model non-linearity or non-Gaussian input uncertainties as Monte-Carlo itself. Conversely, it is an exaggeration to speak about the 'Wilks method' as it is essentially a careful interpretation of Monte-Carlo outputs.

Conservatism of Wilks' formula for robust quantile computation Guaranteeing the confidence interval implies that the estimators Z^{α}_{β} can be quite conservative as compared to the real quantile z^{α}. There is a substantial probability that the Wilks estimator, that is the maximum (respectively the i-th maximum) corresponds to a real quantile far above the α-quantile: Table 7.6 illustrates the interest in doing, if possible, more simulations for a given (α, β) set as the higher-order Wilks estimators reduce gradually the risk of over-conservatism.

Note that this level of over-conservatism, tied to the non-parametric nature of the estimator, is not the 'worse'. Consider Kolmogorov-Smirnov statistic (KS) $d_N(\beta)$, such that:

$$P[F_N(z) - d_N(\beta) < F(z)] > P\left[\sup_z |F_N(z) - F(z)| < d_N(\beta)\right] = \beta \tag{7.80}$$

Table 7.6 Risk of overestimation using the i-th maximum-based Wilks estimator.

	Probability to exceed a higher quantile than 95 % using the i-th maximum-based Wilks estimator			
Real quantile α	Maximum (after 59 trials)	2nd maximum (after 93 trials)	3rd maximum (after 124 trials)	4th maximum (after 153 trials)
95 %	95 %	95 %	95 %	95 %
97 %	83 %	77 %	72 %	68 %
98 %	70 %	56 %	45 %	37 %
99 %	45 %	24 %	13 %	7 %

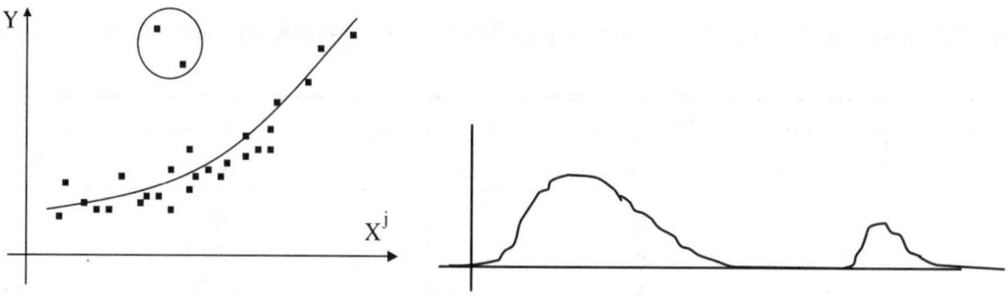

Figure 7.3 *(left) non-regular Z = G(X); (right) bimodal pdf for Z.*

This leads to an even more conservative estimate of the fluctuation of the sample maximum Z^N. It can be computed that after 153 trials, according to KS test, even the first maximum cannot be guaranteed at 95 % confidence to be higher than the real 95 % quantile whereas the Wilks formula guarantees that the 4th maximum is still conservative. Understandably, KS controls the maximum deviation of *all* empiric frequencies to the entire cdf.

The distinctive advantage of the Wilks method is that it does not assume any property for the model below the Z variate (beyond being a proper random variable, that is Lebesgue-measurable with a cdf). In particular, 'threshold effects' may occur in complex physical behaviour, that is the fact that for some values of the uncertain parameters (initial conditions, thermal exchange coefficients or material properties etc.), the phenomenology changes (bifurcation to new modes of failure, etc.), leading the output variables (such as temperature, stress, etc.) to jump to significantly higher or lower values. From a probabilistic point of view, this may imply non-regular dependence between output and inputs variates (Figure 7.3 – left) or a multi-modal support (as mixed heterogeneous populations) for the output pdf (Figure 7.3 – right).

The fully non-parametric Wilks estimators are robust to this type of behaviour. They simply take into account any part of the Z distribution to the extent of its probability weight, ignoring any form of connexity of the sets or system model functional properties. Conversely, it will be seen later that it is much more difficult to establish that alternative methods to Monte-Carlo relying on a regular approximation of the output variable based on a limited number of computations will cover the upper singularities. This is the case with structural reliability algorithms or response-surface based methods for which a regular output-input dependence is *a priori* necessary in order to control the conservatism of surface response. This is all the more difficult since parameter locations of the singularity regions (helpful in adapting multiple-surface response models) are *a priori* poorly known.

An alternative approach has been suggested (de Rocquigny, 2005) to the classical dilemma under computing constraints between robustness and over-conservatism of uncertainty propagation for a potentially non-regular physical system. It involves the use of statistical tests as a tool to infer the conservatism of a first series of calculations, to suggest checking physical relevance and finally to decide whether further calculations are worth making. The comparative advantage of this approach lies in the fact that while the robustness of the initial safety estimation is not affected, and no additional (high CPU) model trial is required, the operator has a tool to control and optimise the number of trials.

As a concluding remark, remember that propagation uncertainty should always be considered relative to other inaccuracies, in particular estimation uncertainty, that is to say an inaccurate specification of the sources of uncertainty, making mistakes in the choice of input distributions, generally turns out to be much more important than getting accurate numbers for a probability or variance. It may even be more important that propagation uncertainty is controlled in relative rather than absolute terms.

Flood example

Simulating the water level according to Monte-Carlo simulations resulted in the following Table 7.7 of estimators for 1/100 and 1/1000 return levels (i.e. $\alpha = 99$ % or 99,9 %) using the first three orders of Wilks and various confidence levels (β). They are compared not only to the 'true' quantiles, as obtained after 1 million runs, but also to the result for each particular run as if the empiric quantile was taken instead of the Wilks conservative estimate. Note that those results illustrate just one sampling experience for each, so that the reader will inevitably find different results for the same scenario.

A number of observations can be made:

(a) The β-50 % Wilks estimate coincides with the empiric quantile. There is no guarantee of conservatism for such an estimate (in fact only one in two), as illustrated in a result. This is a rather risky case in risk assessment.

(b) As imposed by the Wilks formula, the minimum number of trials grows with the Wilks order as well as the beta value. It should really be considered as a *minimal* figure. Getting a 50 %-1st order estimate requires only the intuitive $1/(1 - \alpha)$ minimum number of runs (99 runs to get a 1/100-yr estimate) but of course the former comment reminds one just how insecure it is. Getting a 95 %-1st order estimate means approximately three times more samples as a minimum: in other words, a *rarity index* of the computation pushed into a figure of $RI = -0.5$. This is largely higher compared to the above-recommended rarity index $RI < -2$ i.e. $100/p_f = 100/(1 - \alpha)$ and the associated cost is the overestimation of the level of dike design possibly by 2–3m!

(c) As expected, the Wilks estimates are conservative, all the more so when beta is high or the order is low – though of course this is only true in a statistical manner, not excluding some unlucky cases, for example $\alpha = 1/1000$yr, $\beta = 95$ % where the 2nd-order Wilks estimate turned out to be more conservative than the 1st-order. But note also that the 1st-order was not conservative in that particular case.

Getting close to the recommended $400/p_f = 400/(1 - \alpha)$ number of runs is necessary in order to issue more stable figures, while maintaining Wilks robustness: it can simply be done through raising the Wilks order. For instance, a 100th-order $\beta = 95$ % Wilks estimate for the 1/100-yr figure requires $N = 11691$ and issued in one sample 55.9 (55.7) which is much closer to the true quantile of 55.6.

Note of course that the β level of confidence has nothing to do with the level-2 epistemic uncertainty stemming from data limitations into the inputs.

Table 7.7 Wilks estimators for the hydro-flood model.

Simulating Z_c	z^α	$\beta = 50\%$ confidence		$\beta = 90\%$ confidence		$\beta = 95\%$ confidence	
1/100-yr*	55.6	99(1)	55.0	230(1)	57.6 (55.5)	299 (1)	58.6 (57.1)
		168(2)	55.8	388(2)	55.7 (55.4)	473 (2)	58.1 (55.9)
		268(3)	55.6	531(3)	56.8 (55.8)	628 (3)	56.6 (55.7)
1/1000-yr – id.	57.8	999(1)	57.7	2302(1)	57.6 (57.6)	2995 (1)	57.9 (57.6)
		1679(2)	56.5	3889(2)	60.7 (57.1)	4742 (2)	59.3 (58.4)
		2674(3)	58.2	5321(3)	59.2 (58.0)	6294 (3)	59.5 (58.2)

*Left/nb of runs required (order of Wilks estimate), right/result of Wilks estimate (result of empiric quantile)

7.2.5 Sampling double-probabilistic risk measures

The double-probabilistic structure of the uncertainty model may have a strong impact on the process and computational cost of Monte-Carlo simulation. In the most frequent case, double-probabilistic risk measures are based on iterated expectations:

$$c_z = E_{\Theta_X}[h_e(E_X[h_a(Z)|\Theta_X, \boldsymbol{d}])|IK, \Xi_n] \tag{7.81}$$

A two-tier sampling proves necessary to estimate, for example, confidence intervals in an exceedance probability:

- sample firstly $(\boldsymbol{\theta}_j)_{j=1...N_1}$,
- then estimate completely $E_X\left[h_a(Z)\middle|\boldsymbol{\theta}_X = \boldsymbol{\theta}_j, \boldsymbol{d}\right]$ conditionally to each of those $\boldsymbol{\theta}_j$ values through an inner conditional sampling strategy: $(\boldsymbol{X}_{j,k})_{k=1...N2} \sim f_X(|\ \boldsymbol{\theta}_j)$,
- eventually, average out the results:

$$c_z = \frac{1}{N_1}\sum\nolimits_{j=1...N_1} h_e\left[\frac{1}{N_2}\sum\nolimits_{k=1...N_2}\left[h_a(\boldsymbol{Z}_{j,k})\right]\right] \tag{7.82}$$

This has a considerable computational cost: think, for instance, of sampling $N_1 = 100$ times $N_2 = 10\,000$ samples when looking for 70 % confidence intervals in an exceedance probability close to 1 %. The method is the most computationally expensive, but has the advantage of retaining the separation details for intermediate analysis. Indeed, one can visualise the impact of various values for the epistemic parameters θ_X on the conditional risk measure (Helton, 1993). This is also pervasive in environmental or biostatistical studies where two-tier structures separate inter-subject or spatial variability from point uncertainty (Frey and Rhodes, 2005), as implemented for example in the `mc2d` package of R open statistical software.

Flood example

A two-tier sampling was undertaken (see Annex §10.4.5 for more detail) on the hydro model, resulting in the following results in Table 7.8. Graphically, the results generate the aleatory-epistemic ccdf given in Figure 7.4: left, the 50 first conditional ccdf are represented; right, the expected ccdf (central curve with °) as well as the 70 % confidence interval and associated lower and upper bounds (i.e. 85 % unilateral quantiles) are represented. Note that the confidence interval should be read vertically for a given threshold of water level z_s, as encompassing the (asymmetric) epistemic distribution of threshold exceedance probabilities $\{P(Z > z_s \mid \Theta), \Theta \sim \pi 1(.)\}$.

Table 7.8 Simulating water levels for given return periods with associated epistemic uncertainty, and the corresponding dike design.

	1/50	1/100	1/1000
Full probabilistic (correlation neglected; 100*50 000 samples)	*55.1-no dike (median)*	*55.6 – 0.1m (median)*	*57.8 – 2.3m (median)*
	55.2-no dike (mean)	*55.8 – 0.3m (mean)*	*58.0 – 2.5m (mean)*
	55.5-no dike (85%)	*56.1 – 0.6m (85%)*	*58.5 – 3.0m (85%)*

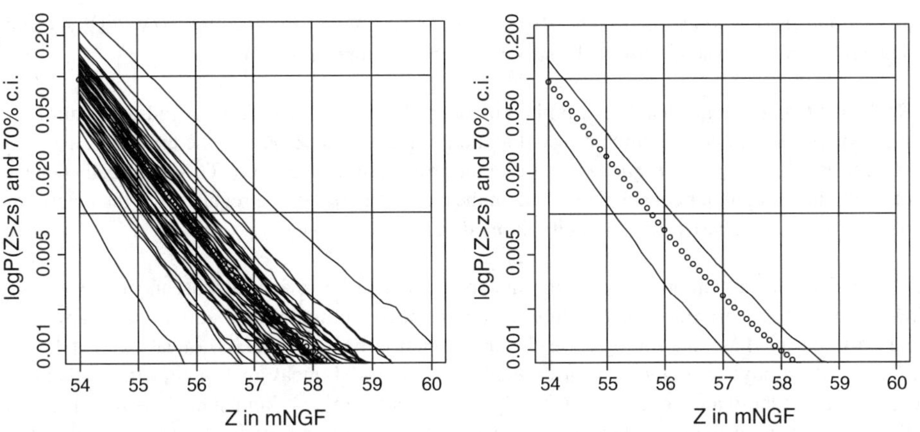

Figure 7.4 *Aleatory-epistemic ccdf for the water level (left) a 50-sample of conditional ccdf; (right) the combined expected ccdf and 70% bilateral confidence bounds (100* 50 000 runs).*

However, as already mentioned, remember that single-probabilistic risk measures which depend only on the predictive distribution f_Z^π may be sampled directly in a single-tier sampling of $E_\theta\left(f_X(x \mid \theta)\right)$, which is computationally much more economical.

7.2.6 Sampling mixed deterministic-probabilistic measures

Monte-Carlo simulation may also be used when optimisation is required in mixed deterministic-probabilistic settings, as stated in Section 7.1.5. This may even apply within the limit of a *fully deterministic* risk measure:

$$c_z(\boldsymbol{d}) = Max_{\boldsymbol{x}\in\boldsymbol{\Omega}_x}[G(\boldsymbol{x},\boldsymbol{d})] \tag{7.83}$$

which normally would require fully-fledged optimisation, a major challenge when input dimension is high and the system model is not simplistic or completely regular (e.g. convex).

Though convergence is much trickier to secure in that particular application, it may be seen as an interesting complement to deterministic (gradient or non-gradient based) optimisation algorithms as it is guaranteed to explore the whole space, avoiding trapping in local maxima.

7.3 Classical alternatives to direct Monte-Carlo sampling

7.3.1 Overview of the computation alternatives to MCS

When computational constraints prove to be substantially limiting, it is still possible to undertake alternatives to MCS. An abundant literature and a great deal of on-going research are devoted to that task, in particular concerning structural reliability. In other words, the methods aim at increasing the *computational rarity index* beyond the $RI < -2$ (standard MCS) or $RI < -0.5$ (Wilks). As will be discussed below, they should always be striking the best compromise between:

- Realistic *computational time* (i.e. a limited number of calls to $G(.)$). This depends, according to the method, on the number of variables (i.e. the dimension of $\boldsymbol{X}$ or $\boldsymbol{Z}$) and/or the criterion c_z.
- Appropriate *level of control of accuracy*, i.e. of the propagation uncertainty. This is critically dependent on the criterion c_z under consideration and on the characteristics of the regularity of the deterministic model $G(.)$ (e.g. linearity, monotonicity, limited variations etc.).

Table 7.9 provides a synoptic view of the methods that will be reviewed subsequently. The various alternatives to MCS are positioned within the compromise computational cost *vs.* accuracy control. Note particularly that this compromise depends essentially upon the risk measure that has been chosen for the study: some methods are specific to certain risk measures, such as the probability of exceeding a threshold and are generally highly inefficient for others. Remember also that these issues still mobilise a large field of research, with substantially-evolving methods and algorithms.

The response-surface approach is a complementary technique rather than a method of propagation as such: replace the initial system model $G(.)$ by an approximation $\tilde{G}(\boldsymbol{x})$ called 'the response surface' (or 'meta-model') that is numerically quicker to calculate, such as polynomials, neurone networks, stochastic developments, or even a simplified model of the pre-existing system (see (Myers, 1971; Box and Draper, 1987). It is therefore possible to apply MCS with its associated error control.

Although insufficiently-mentioned in the literature, the physical properties of the system model help greatly in selecting the best compromise and in controlling propagation error (de Rocquigny, 2009) as will be seen in the monotonous case in Section 7.4.

7.3.2 Taylor approximation (linear or polynomial system models)

Linearity and regularity of the system model An important class of system models displays the fruitful property of *linearity*. Linearity is quite reasonable physically if considering rather limited uncertainties. The linear property of a system model refers to the following intuitive physical property: increase one of the physical inputs x^i while all other components of $\boldsymbol{x}$ are fixed, and the system model will increase (or decrease) by a proportional factor, wherever the variation starts from. Mathematically, *linearity* of $h(\boldsymbol{x})$ with respect to a component x^i is defined as follows:

$$\begin{aligned}
&\forall(\lambda, i, dx^i), \forall \boldsymbol{x} = (x^1, \ldots x^i, \ldots x^p) \\
&h(x^1, \ldots, x^i + \lambda dx^i, \ldots x^p) - h(x^1, \ldots, x^i, \ldots x^p) = \lambda[h(x^1, \ldots, x^i + dx^i, \ldots x^p) \\
&\qquad - h(x^1, \ldots, x^i, \ldots x^p)]
\end{aligned} \tag{7.84}$$

This may be called more specifically a partial linearity with respect to that component. Global linearity meaning that it holds for any input component.

While some physical systems are truly globally linear, linearity should generally be seen as a *local* property, that is true at a given range or within limited uncertainty ranges. Mathematically, this is simply linked to the differential properties, embodied in the basic Taylor developments.

The Taylor approximation The Taylor approximation is a simplified method requiring very few runs of the system model under high regularity hypotheses. It is highly popular in metrology, for instance, where it has even been renamed the '*uncertainty propagation rule*' in the international standard Guide for

Table 7.9 Principal methods of uncertainty propagation.

Family of methods	Appropriate risk measure c_Z	Method of computation	Propagation uncertainty and level of control	Calculation time: number of calls to $G(.)$ depending on input dimension p	Applicability to various input structures $(\boldsymbol{x},\boldsymbol{e})$
Monte-Carlo sampling	All risk measures	Sampling	Full probabilistic control	Generally large, but does not depend on p	All input structures, continuous, discrete or time/space-dependent
Taylor quadratic approximation	Moments, especially $c_z = k.\sqrt{\mathrm{var}(Z)}$	Limited first order developments of moments of Z ($E(Z)$, $var(Z)$ or higher-order)	Very limited: outside linearity, requires additional differential estimates	Very low, increase linearly with p	Continuous inputs
Numerical integration	All probabilistic risk measures	Numerical computation of expectation integrals using a limited grid of runs (e.g. Gauss points).	Limited: numerical bounds require additional regularity hypotheses and differential estimates	Low if p very small; but grows very fast in n_d^p	Continuous and discrete inputs
Accelerated sampling	Especially $P[Z > z_s]$ but also moments	Sampling using modified sampling input density	Statistical error Statistical checks, though sometimes poor	Average in favourable cases	Continuous and discrete inputs. Extensions to time or space-dependent (particle-filtering)
Reliability methods (Form-Sorm, and derivatives)	$P[Z > z_s]$	Numerical approximation of the integral $\int_{\boldsymbol{x}} 1_{G(\boldsymbol{x})>z_s} f_X(\boldsymbol{x})d\boldsymbol{x}$ possibly associated with adaptive sampling	Functional approximation error checking is very difficult except in special cases though the sampling part may bring limited control	Low in favourable cases (regular system models $+ p$ not too big)	Mostly continuous inputs

(continued)

Table 7.9 (*Continued*)

Family of methods	Appropriate risk measure c_Z	Method of computation	Propagation uncertainty and level of control	Calculation time: number of calls to $G(.)$ depending on input dimension p	Applicability to various input structures $(\boldsymbol{x},\boldsymbol{e})$
Stochastic development and chaos polynomials	Especially moments	Development of $G(X)$ over stochastic functional bases through numerical integration	Truncation error (explicit control is seldom documented)	Average to high, but low for first moments and when $\boldsymbol{Z}$ vectorial with a high dimension (r)	Continuous and space-dependent inputs
Response surface (+MC sampling)	All risk measures	Build-up of a numerical approximate of the system model + intensive MCS	Approximation error associated to the response surface. Control requires regularity hypotheses.	Low in favourable cases (regular system models + p not too big)	Mostly continuous inputs but theoretically all input structures

Expression of Uncertainty in Measurement (GUM ISO, 1995). It consists of a limited order development of $z = G(\boldsymbol{x},\boldsymbol{d})$ around $\boldsymbol{x} \cong E(\boldsymbol{X})$:

$$\begin{aligned} z &= G(\boldsymbol{x}, \boldsymbol{d}) \\ &\approx G(E(\boldsymbol{X}), \boldsymbol{d}) + \sum_{i=1}^{p} \left(\frac{\partial G}{\partial x^i} \right) \left[x^i - E(X^i) \right] \\ &+ \frac{1}{2} \sum_{i_1, i_2} \left(\frac{\partial^2 G}{\partial x^{i_1} \partial x^{i_2}} \right) \left[x^{i_1} - E(X^{i_1}) \right] \left[x^{i_2} - E(X^{i_2}) \right] + o\left(\|\boldsymbol{x} - E(\boldsymbol{X})\|^2 \right) \end{aligned} \tag{7.85}$$

mostly limited in fact to its first order components, that is the local linearization of the system model if not linear itself:

$$z = G(\boldsymbol{x}, \boldsymbol{d}) \approx G(E(\boldsymbol{X}), \boldsymbol{d}) + \sum_{i=1}^{p} \left(\frac{\partial G}{\partial x^i} \right) \left[x^i - E(X^i) \right] + o(\|\boldsymbol{x} - E(\boldsymbol{X})\|) \tag{7.86}$$

Thus, at first-order the resulting approximations:

$$E(Z) = E[G(\boldsymbol{X}, \boldsymbol{d})] \approx G(E(\boldsymbol{X}), \boldsymbol{d}) \tag{7.87}$$

$$VarZ \approx \sum_{i=1}^{p} \left(\frac{\partial G}{\partial x^i} \right)^2 VarX^i + \sum_{i_1 \neq i_2} \left(\frac{\partial G}{\partial x^{i_1}} \right) \left(\frac{\partial G}{\partial x^{i_2}} \right) .Cov(X^{i_1}, X^{i_2}) \tag{7.88}$$

The second order development modifies the expected output into the following expression:

$$E(\boldsymbol{Z}) \approx E[G(\boldsymbol{X}, \boldsymbol{d})] + \frac{1}{2} \sum_{i_1, i_2} \left(\frac{\partial^2 G}{\partial x^{i_1} \partial x^{i_2}} \right) Cov(X^{i_1}, X^{i_2}) \tag{7.89}$$

Note that the basic hypothesis is that uncertainties are *small*. This is one of the reasons that possibly explain its relevance in the metrological field where uncertainty typically limits phenomenological deviations around perfect calibration. In fact, the key control is not the absolute (measured, e.g. by the input coefficients of variation) but the scale of the *input non-regularities of* $z = G(\boldsymbol{x})$. The first order, for instance, is valid so far as the second order derivative is negligible; for example the first-order derivative is constant (i.e. approximately linear) over the area of uncertainties.

Whatever the input distributions, the Taylor method can always be used to estimate risk measures using the first two moments. Using it for quantiles or exceendance probabilities requires in addition the Gaussian character of Z which is the case if, in addition to the weak non-linearity of $G(.)$, the X^i form a Gauss vector. Such a Gaussian interpretation of the method is close to the Delta method in statistics, whereby the image $G(\boldsymbol{U}_N)$ of a series of asymptotically Gaussian random vectors with decreasing variances is proved to be asymptotically Gaussian with a variance being the inflation of the input variance multiplied by its square derivative (or Jacobian matrix in multidimensional cases).

Equation (7.89) also illustrates the impact of the signs of correlation and of partial derivatives on variance. Rather intuitively, variance becomes inflated whenever correlations are in line with the derivative sign concordance, that is both inputs locally increasing or decreasing (signs of derivatives are equal) Z and positively correlated or one input locally increasing and the other locally decreasing (signs of derivatives are opposed) and they are negatively correlated. This generalises the comments made in Section 4.2.2 of Chapter 4: considering variance as measuring the amount of output uncertainty or 'error', the joint impact of correlations and derivatives can be interpreted as more or less compensation for errors brought by the inputs.

Computing the Taylor development requires the use of derivatives which may either be closed-form or numerically coded in some cases (see Section 7.6) but generally involve finite-differences. The simplest scheme for the first-order partial derivatives appears as follows:

$$\partial G/\partial x^i \cong \left[G\left(\mathbf{x}+\varepsilon u^i\right) - G(\mathbf{x})\right]/\varepsilon \tag{7.90}$$

thus requiring a minimum of $(p+1)$ runs of the system models when computing all first-order derivatives plus the central point (which are the only elements required by first-order Taylor). Such a number of runs remain very small in comparison to Monte-Carlo or other propagation alternatives; however, it grows with input dimension in a linear way (or polynomial way for higher orders) and may require more than the simplest scheme for more stable derivation.

Note that the Taylor method is also popular in a radically different industrial domain, that of scientific computing. There, $G(.)$ represents a major numerical code (e.g. a neutron calculation in a nuclear core); its input parameters are affected by all sorts of uncertainties, and what is required is an order of magnitude of precision for the results vector z. The numerical strategies deployed to obtain the least costly calculation of derivatives appearing in the formula lead to the use of a variety of names: by direct derivative = *Direct Sensitivity Method* (*DSP*), or by adjunct derivative = *Adjoint Sensitivity Method* (*ASM*), *see* (Cacuci *et al.*, 1980); or again, the 'method of disturbances' or the 'calculation of moments'.[2]

It also proves useful in level-2 studies when considering the impact of limited epistemic deviations of $\hat{\boldsymbol{\theta}}_X$ on the risk measure c_z. The Taylor method is applied to the function:

$$\boldsymbol{\theta}_X \rightarrow C_z(\boldsymbol{\theta}_X, \mathbf{d}) \tag{7.91}$$

If there are reasonable data and confidence so that the uncertainty in $\boldsymbol{\Theta}_X$ is limited, it is a very economical alternative to double Monte-Carlo sampling as each computation of $C_z(\boldsymbol{\theta}_X,\mathbf{d})$ may already require Monte-Carlo.

Care must, however, be taken to estimate the partial derivatives of $C_z(\boldsymbol{\theta}_X, \mathbf{d})$ in $\boldsymbol{\theta}_X$ by finite differences when such risk measure was estimated initially through Monte-Carlo sampling. Statistical fluctuations may seriously noise the $[C_z(\boldsymbol{\theta}_X + \varepsilon\boldsymbol{\theta}^k, \mathbf{d}) - C_z(\boldsymbol{\theta}_X, \mathbf{d})]/\varepsilon$ if ε is small. A convenient way is to fix the seed of pseudo-random computer sampling used for each simulation of the $(C_z(\boldsymbol{\theta}_X + \varepsilon\boldsymbol{\theta}^k, \mathbf{d}))_{k=1\ldots npar}$.

Flood example

Taylor can first be tried for simulating the flood water level, as a cheap alternative to Step 3 Monte-Carlo estimates of the design level. Taylor is used around the best-estimate (or mean input) $EX = (1335,30,50,55)$ and results in the approximate 1st-order Taylor $\{EZ = 52.5,\ s_Z = 1.0\}$ quite close to a Monte-Carlo estimate of $\{EZ = 52.6,\ s_Z = 1.1\}$ after 1 000 000 simulations.

However, it would be very naïve to deduce from those fair estimates a Gaussian-based estimate for quantiles: the resulting 1/100 yr at 54.9 m and 1/1000 yr at 55.8 m are underestimated by far as

[2] These appellations sometimes also include the deterministic propagation of uncertainties, which uses the same derivatives, or alternatively the estimation of other probabilistic moments though differential developments of order >1.

compared to the true 55.6 m and 57.8 m respectively. The Z_c distribution is just as asymmetric as that of Q so that a Gaussian fit grossly misses the upper quantile extension (see Chapter 5).

The results for the use of Taylor on level-2 uncertainty may be found in Section 7.4. It is even more erroneous in the case of cost figures, that is when $Z = Cc$: whatever the choice of dike height, the expected value of dike overflow is very negative $ES(h_d) = -3m - h_d$ so that the total cost taken at EX is constantly equal to the certain investment/maintenance cost, with null derivatives: the Taylor approximation for cost variance is thus null. This counter-example illustrates that Taylor is to be kept only for uncertain distributions moderately spread around the mean whereas the cost distribution is severely asymmetric with a large probability of being small and a small probability of reaching extremely large values (see Section 7.1.3)

Beyond the first order which fits well with approximately linear system models, Taylor expansion may be extended to higher orders enabling both:

- a more accurate approximation of expectation and variance when the system model has a non-linear but polynomial regularity (as illustrated above with the second-order development of expectation);
- an estimation of higher-order moments $E(Z^q)$ where $q > 2$.

Again, such an approximation relies on underlying regularity hypotheses.

7.3.3 Numerical integration

Numerical integration applies to the multi-dimensional integrals appearing in expectation-based risk measures such as:

$$p_f = P(Z > z_s) = \int_x 1_{G(x,d)>z_s} f_X(\mathbf{x}|\theta_X) d\mathbf{x} \quad (7.92)$$

$$E(Z^k) = \int_x G(\mathbf{x}, \mathbf{d})^k f_X(\mathbf{x}|\theta_X) d\mathbf{x} \quad (7.93)$$

Traditional numerical integration techniques may be mobilised, such as Simpson method, Gauss points and so on. They all consist of computing the integrand function on a discrete set of points $(x_j)_j$ covering the integration domain and weighing those results through dedicated coefficients γ_j that are controlled by the integration method.

$$c_Z = \int_x Q[G(\mathbf{x}, \mathbf{d})] f_X(\mathbf{x}|\theta_X) d\mathbf{x} \approx \sum_j \gamma_j Q[G(\mathbf{x}_j, \mathbf{d})] f_X(\mathbf{x}_j|\theta_X) \quad (7.94)$$

Note that this approach may equally be viewed as a choice of a design of experiment $(x_j)_j$ and integrating a response surface implicitly, which is the underlying interpolator between the results $Q[G(x_j, d)]$ that results, along with the density f_X, in the integration weights γ_j.

Traditional integration techniques prove competitive only when p is very small, as the number of calls to $G(.)$ grows exponentially in n_d^p where n_d denotes the number of points of integration for each input dimension X^i (typically $n_d \sim$ 3–6). The uncertainty of propagation (which decreases with n_d) can theoretically be limited, working from limits of integrand derivatives of increasing order; but in practice the estimation of these values, *a priori* unknown, is a delicate matter.

Stochastic developments through polynomial chaos (see subsequent sections) may be seen as a modern alternative to traditional numerical integration, whereby the choice of the design of experiment and the integration weights are deliberately adapted to the density of the output.

Additionally, a useful *mix* can be found between *MCS and partial numerical integration* of sub-system models of smaller parameterisation or of a subset of degrees of freedom/components of the input vector: see Conditional Monte-Carlo Sampling (Section 7.3.4).

7.3.4 Accelerated sampling (or variance reduction)

The methods called alternatively *accelerated sampling* or *variance reduction* methods are meant to reduce the number of simulations for a given accuracy, or, equivalently, to reduce the variance of the estimators for a given number of simulations (Rubinstein, 1981). Some of these are generic with respect to the array of risk measures, such as Latin Hypercube sampling or stratification; others are dedicated to outperform Monte-Carlo in computing a low exceedance probability p_f, such as directional sampling or conditional Monte-Carlo.

Importance sampling, LHS and stratification *Latin Hypercube Sampling* (LHS) and stratification (McKay, Beckman, and Conover, 1979; Rubinstein, 1981; Helton and Davis, 2003) both undertake sampling through a preliminary decomposition of the domain of definition aiming at better space-filling than crude Monte-Carlo sampling and increased sampling efficiency. The LHS estimator is formulated equivalent to that of MCS:

$$\hat{q}_N = N^{-1} \sum_{j=1...n} Q(\boldsymbol{x}_j) \tag{7.95}$$

Only the type of sample $(\boldsymbol{x}_j)_{j=1...n}$ differs: in crude MCS, $(\boldsymbol{x}_j)_{j=1...n}$ is completely random following f_X while for LHS it is sampled randomly inside sub-strata and reorganised component-wise. LHS is a simple and popular algorithm designed to 'stabilise' to some extent MCS as its space-filling deliberately covers the whole area of definition, which is only the case on average for MCS. In fact, the monotony of the function $Q(\boldsymbol{x})$ (which may embody the system model plus any post-treatment according to the risk measure, for example utility) guarantees the reduction of variance by the LHS estimator of $E[Q(\boldsymbol{X})]$ as compared to that of crude MCS. Yet LHS may not be the most appropriate method for the estimation of a rare p_f as its space-filling entails many useless simulations outside interesting regions.

Importance sampling involves the use of a deliberately-biased input distribution in order to intensify the number of runs in regions deemed more interesting. It is particularly appropriate for the case of threshold exceedance. Quite intuitively and similar to polls in classical statistics, the estimator needs to re-weigh each run in order to debias the results as follows:

$$\hat{P}_f^{IS} = \frac{1}{N} \sum_{j=1...N} 1_{G(\boldsymbol{X}_j, \boldsymbol{d}) > z_s} \frac{f_X(\boldsymbol{X}_j)}{\tilde{f}_X(\boldsymbol{X}_j)} \tag{7.96}$$

Remember that $\boldsymbol{X}_j$ are sampled *i.i.d* according to a distribution of importance $\tilde{f}_X$ (cf. Figure 7.5). Provided that such distribution of importance covers an area larger or equal to the intersection

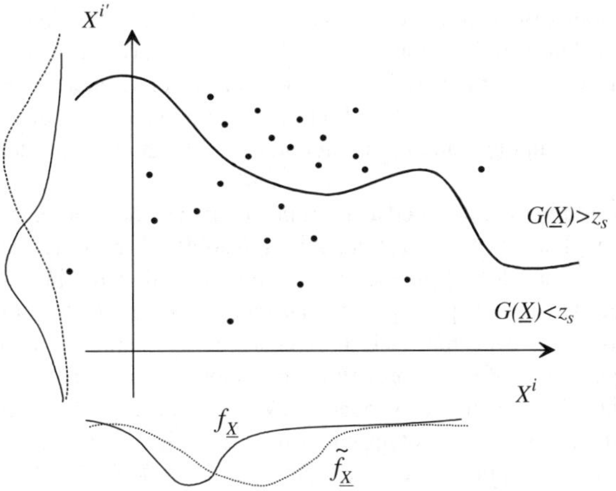

Figure 7.5 *Importance sampling and the estimation of a threshold exceedance probability.*

between that of the original density and that of failure, the importance sampling estimator proves to be unbiased:

$$\begin{aligned} E_{\tilde{f}_X}\hat{P}_f^{IS} &= E_{\tilde{f}_X}\left[1_{G(\boldsymbol{X},\boldsymbol{d})>z_s}\frac{f_X(\boldsymbol{X})}{\tilde{f}_X(\boldsymbol{X})}\right] \\ &= \int_x 1_{G(\boldsymbol{x},\boldsymbol{d})>z_s}\frac{f_X(\mathbf{x})}{\tilde{f}_X(\boldsymbol{x})}\tilde{f}_X(\boldsymbol{x})d\boldsymbol{x} = \int_x 1_{G(\boldsymbol{x},\boldsymbol{d})>z_s}f_X(\mathbf{x})d\boldsymbol{x} = p_f \end{aligned} \tag{7.97}$$

If prior physical informational helps in circumventing the zones where $G(\boldsymbol{x}, \boldsymbol{d}) > z_s$ is more likely, then craft a density $\tilde{f}_X$ ensuring that the trials are more densely distributed in those zones:

$$\{\boldsymbol{x}|G(\boldsymbol{x},\boldsymbol{d}) > z_s\} \subset \left\{\boldsymbol{x}\left|\frac{f_X(\boldsymbol{x})}{\tilde{f}_X(\boldsymbol{x})} \leq 1\right.\right\} \tag{7.98}$$

This guarantees that the variance of $\hat{P}_f^{IS}$ is lower than that of Monte-Carlo since:

$$\begin{aligned} \mathrm{var}_{\tilde{f}_X}\hat{P}_f^{IS} &= \frac{1}{N}\mathrm{var}_{\tilde{f}_X}\left[1_{G(\boldsymbol{X},\boldsymbol{d})>z_s}\frac{f_X(\boldsymbol{X})}{\tilde{f}_X(\boldsymbol{X})}\right] \\ &= \int_x \left[1_{G(\boldsymbol{x},\boldsymbol{d})>z_s}\frac{f_X(\boldsymbol{x})}{\tilde{f}_X(\boldsymbol{x})}\right]^2 \tilde{f}_X(\boldsymbol{x})d\boldsymbol{x} - p_f^2 \\ &= \int_x 1_{G(\boldsymbol{x},\boldsymbol{d})>z_s}\frac{f_X(\boldsymbol{x})}{\tilde{f}_X(\boldsymbol{x})}f_X(\boldsymbol{x})d\boldsymbol{x} - p_f^2 \leq \int_x 1_{G(\boldsymbol{x},\boldsymbol{d})>z_s}f_X(\boldsymbol{x})d\boldsymbol{x} - p_f^2 = \mathrm{var}\,\hat{P}_f^{MC} \end{aligned} \tag{7.99}$$

Ideally, the importance density should be distributed as closely as possible to the area of failure, so that the set inclusion Equation (7.98) is equality. Of course, it is challenging to craft in advance such an importance density as the whole rationale of random sampling is to explore the *a priori* unknown location of the region of failure. Thus, the distribution of importance $\tilde{f}_X$ is adapted gradually during the course of the sampling series in order to optimise the sampling as the past results allow for a better understanding of the system model. A considerable literature has developed on that very popular class of algorithms. A particular version of such adaptive sampling valuing the monotonocity of the system model will be outlined in Section 7.4.

Stratification may simply be understood as a variant of importance sampling whereby the importance density is designed through a collection of strata, that is disjoint subsets covering the area of definition, with specified weights that may differ from the original density, thereby forcing the sampling throughout. In the case where strata have equal probabilities and random selections by strata, McKay, Beckman, and Conover (1979) also established that such stratified sampling guarantees a lower estimator variance than MCS for the Equation (7.96) type of estimators although monotonous property for $Q(.)$ is not needed in that case. Thus, MCS estimated variance may also be taken as an upper bound. Due to space-filling, such stratification properties do not appear to be *directly* interesting for estimating a rare failure probability although dedicated applications for quantile estimation are being developed (see Cannamela *et al.*, 2008).

Directional sampling *Directional sampling* (DS) is a variant of sampling dedicated to estimating the integral p_f. Like the reliability methods, it is thought of in a transformed Gaussian space for the system model inputs, that is a space of i.i.d. standard Gaussian inputs $\boldsymbol{U}$ obtained by rescaling the distribution laws and de-correlating the original (so-called 'physical') vector of inputs $\boldsymbol{X}$ (for more detail see Section 7.3.5). The *DS* modified sampling procedure takes advantage of the *rotational symmetry* of the standard Gaussian space enabling a decomposition $\boldsymbol{U} = R\,\boldsymbol{A}$ where $\boldsymbol{A}$ denotes the uniformly distributed direction and R the p-degrees of freedom chi-square distributed radius. In other words, instead of viewing the sampling of a random Gaussian vector as that of its p Cartesian coordinates or vector components, consider it as sampling both a random direction or radius $\boldsymbol{A}_j$ (uniformly distributed over the p-dimensional sphere) and a random distance into such radius (distributed over a p chi-square law), that is its spherical coordinates (cf. Figure 7.6).

Note then that the final goal is to estimate the probability of the zone where the threshold is exceeded, the failure domain. As will be discussed in more detail in the section reliability and monotony section, a key additional hypothesis is that the failure domain is restricted to a certain part of the hyperspace and the

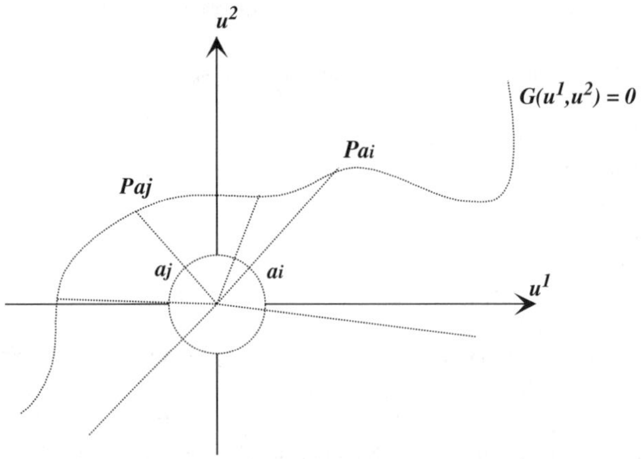

Figure 7.6 *Directional sampling.*

failure function has a somewhat regular behaviour along any given radius, so that $P(G(R\boldsymbol{A}_j) < 0 \mid \boldsymbol{A}_j)$ can be easily computed. Ideally it would possess one unique crossing of the limit state (for $R = r_j$), which would be identified by a quick zero search along the radius, thus generating the approximation of the conditional failure probability in that direction:

$$P(G(RA_j, \boldsymbol{d}) < 0 | A_j) \approx P(R > r_j) = 1 - \chi_p^2\left(r_j^2\right) \tag{7.100}$$

Thus, DS generates the following estimator:

$$\hat{P}_f^{dir} = \frac{1}{N} \sum_{j=1\ldots N} P(G(RA_j, \boldsymbol{d}) < 0 | A_j) \approx \frac{1}{N} \sum_{j=1\ldots N} \left(1 - \chi_p^2\left(r_j^2\right)\right) \tag{7.101}$$

Acceleration brought by the method lies in the supposedly lower calls of runs of $G(.)$ to perform this deterministic numerical zero search per radius replacing the random sampling along the radius. Unlike the latter, the former deterministic search is generally more or less sensitive to the fact that the conditional probability (i.e. a far distance from the origin) is very small when p_f is small. Directional Simulation is popular in structural reliability as those conditions prove true in practice.

However, such acceleration comes with a cost on the side of controlling propagation uncertainty. One may the question the residual error remaining when the search has been limited to a given maximum number of runs in one direction, which may not have identified all zeros. That source of error is generally not included inside the estimated variance of the estimator: in fact, relying on a deterministic search of zeros generally does not generate reliable error estimation, unless significant hypotheses are made (such as bounds on first or higher-order derivatives, tricky to estimate). A more elaborate formulation covers the case when multiple-zero (limit state crossings) occur: it involves standard probabilistic combinations of sub-events. In that case however, more runs become necessary in order to identify the various zeros. Alternatively, a Monte-Carlo sampling along the given direction would generate an estimation of variance; it would then cancel somehow the interest in the directional formulation by leading back to full sampling, albeit through a conditional step.

Monotony helps to some extent in controlling the residual error (cf. Section 7.4). More elaborate formulations also mix direction sampling with importance sampling and issue robust asymptotic estimators of the associated variance (Munoz-Zuniga *et al.*, 2011).

Conditional Monte-Carlo (or dimension reduction) *Conditional Monte-Carlo*, a generic variant of Monte-Carlo sampling (Rubinstein, 1981), provides an attractive variant dedicated to estimating a risk measure appearing as an exceedance probability also known as *dimension reduction.* To put it simply, assume first that the formulation of the system model $G(.)$ allows for algebraic extraction of one input X^{i_o}, that is that the threshold exceedance equation:

$$G(\boldsymbol{x}, \boldsymbol{d}) > z_s \tag{7.102}$$

can be rewritten equivalently as:

$$x^{i_o} > H\left(\left(x^i\right)_{i \neq i_o}, \boldsymbol{d}, z_s\right) \tag{7.103}$$

This is the case, for instance, when the system model involves an uncertain critical threshold with which one compares the performance or load of the system (e.g. toughness-stress intensity in fracture mechanics, acceptable dose-exposure in environmental impact, hazard-vulnerability in natural risk assessment etc.), connected to the initiator-frequency approach (Chapter 1):

$$G(\mathbf{x}, \mathbf{d}) = x^p - H\left((x^i)_{i=1\ldots p-1}, \mathbf{d}\right) \qquad z_s = 0 \tag{7.104}$$

MCS sampling produces a series of (p-1)-vectors $(X_j^i)_{i\neq i_o}$ generating the following new estimator $\hat{P}_{cmc}$ of p_f:

$$\hat{P}_{cmc} = \frac{1}{N}\sum_{j=1\ldots N}\left[1 - F_{X^{i_o}}\left(H\left(\left(X_j^i\right)_{i\neq i_o}, \mathbf{d}, z_s\right)\right)\right] \tag{7.105}$$

This new estimator appears also as empiric mean, so that its variance may be estimated simply by the resulting empiric variance. Note that the peculiar modified formulation involves recovering directly the cdf of the input that was extracted instead of sampling it. If such input has a large sensitivity index, it will decrease variance (see Rubinstein, 1981; Chia and Ayyub, 1994). To understand this simply, consider the estimator variance:

$$\operatorname{var}\hat{P}_{cmc} = \frac{1}{N}\operatorname{var}\left[1 - F_{X^{i_o}}\left(H\left(\left(X^i\right)_{i\neq i_o}, \mathbf{d}, z_s\right)\right)\right] \tag{7.106}$$

Such variance replaces the original Monte-Carlo estimator variance:

$$\operatorname{var}\hat{P}_f = \frac{1}{N}\operatorname{var}[G(\mathbf{X}, \mathbf{d})] \tag{7.107}$$

If all other inputs have a small impact on the system model output, $H\left((x^i)_{i\neq i_o}, \mathbf{d}, z_s\right)$should not vary much when $(x^i)_{i\neq i_o}$varies throughout the p-1-dimensional sample (nor $F_{X^{i_o}}\left(H\left((x^i)_{i\neq i_o}, \mathbf{d}, z_s\right)\right.$ if the cdf is regular enough). It should vary much less than $G(\mathbf{x}, \mathbf{d})$through the p-dimensional sample, so that the estimator variance is expected to decrease a lot.

Conditional Monte-Carlo (*CMC*) can be formulated more generally to accelerate the computation of any expectation-based risk measure, provided the input vector can be split into a part for which the behaviour of the system model is known or cheaply computable. Consider the following risk measure:

$$c_Z = \int_{X\in\Omega} Q(\mathbf{x})f_X(\mathbf{x})d\mathbf{x} = \int_{X\in\Omega} J[G(\mathbf{x}, \mathbf{d})]f_X(\mathbf{x})d\mathbf{x} \tag{7.108}$$

Assume that the particular decomposition $\mathbf{x} = (\mathbf{x}^I, \mathbf{x}^{II})$ of the uncertain input vector is such that the following *conditional* expectation of the integrand:

$$q(\mathbf{x}^{II}) = E\left(J[G(\mathbf{X}, \mathbf{d})]\middle|\mathbf{x}^{II}\right) \tag{7.109}$$

has a closed-form solution or can be computed cheaply and accurately for any value of $\mathbf{x}^{II}$, possibly through numerical integration being faster than MCS in low dimension (see Section 7.2.3). Note that this was the case for the previous simple formulation whereby:

$$\begin{aligned} E(J[G(\boldsymbol{X},\boldsymbol{d})]|\boldsymbol{x}^{II}) &= E\left(1_{[G(\boldsymbol{X},\boldsymbol{d})>z_s]}\middle|(x^i)_{i\neq i_o}\right) = P\left(G(\boldsymbol{X},\boldsymbol{d}) > z_s\middle|(x^i)_{i\neq i_o}\right) \\ &= P\left(X^{i_o} > H\left((x^i)_{i\neq i_o},\boldsymbol{d},z_s\right)\right) = 1 - F_{X^{i_o}}\left(H\left((x^i)_{i\neq i_o},\boldsymbol{d},z_s\right)\right) \end{aligned} \tag{7.110}$$

In general, the *CMC* estimator of c_z involves restricting the sampling process to the $\boldsymbol{x}^{II}$ part of the complete input vector $\boldsymbol{x}$:

$$\hat{C}_{cmc} = \frac{1}{N}\sum_{j=1\ldots N}\left[E\left(J[G(\boldsymbol{X},\boldsymbol{d})]\middle|\boldsymbol{X}_j^{II}\right)\right] \tag{7.111}$$

Its variance proves to be lower than that of crude Monte-Carlo Sampling, which goes:

$$Var\left[\frac{1}{N}\sum_{j=1\ldots N}Q(\boldsymbol{X}_j)\right] = \frac{1}{N}VarQ(\boldsymbol{X}) \tag{7.112}$$

Indeed, by virtue of the variance decomposition equation, the variance of MCS exceeds that of CMC by a positive factor:

$$\begin{aligned} \frac{1}{N}VarQ(\boldsymbol{X}) &= \frac{1}{N}E\left[Var\left(Q(\boldsymbol{X})|\boldsymbol{X}^{II}\right)\right] + \frac{1}{N}Var\left[E\left(Q(\boldsymbol{X})|\boldsymbol{X}^{II}\right)\right] \\ &= \frac{1}{N}E\left[Var\left(Q(\boldsymbol{X})|\boldsymbol{X}^{II}\right)\right] + \frac{1}{N}Var\hat{C}_{cmc} \end{aligned} \tag{7.113}$$

Additionally, the equation shows that the variance reduction is all the greater when the uncertain inputs $\boldsymbol{X}^{II}$ that are being sampled have only limited impact on $Q(\boldsymbol{X})$. This means that $VarQ(\boldsymbol{X})$ is little affected by fixing the inputs $\boldsymbol{x}^{II}$ so that it is close to $Var\ Q(\boldsymbol{X} \mid \boldsymbol{x}^{II})$ for most values of $\boldsymbol{x}^{II}$. In fact, Section 7.5 on sensitivity analysis will show that the following ratio of CMC variance to MCS variance:

$$\frac{\operatorname{var}\left(E\left(Q|\boldsymbol{X}^{II}\right)\right)}{\operatorname{var} Q} \tag{7.114}$$

corresponds exactly to the first-order Sobol sensitivity index for the inputs $\boldsymbol{X}^{II}$.

Flood example

Section 7.2.3 showed that the computation of the expected cost is slow to converge under standard Monte-Carlo because of the peculiar distribution of costs that includes a very long tail. Variance can be reduced with a combination of the above-mentioned techniques. Assuming at most one single yearly overspill, complete cost can be expressed as follows:

$$\begin{aligned} C_c &= c_i(\boldsymbol{d}) + c_d(Z_c, C_m, \boldsymbol{d}) \\ &= c_i(h_d) + c_{sm}(Z_c - h_d - z_b).C_m + c_g(Z_c - h_d - z_b).C_m \end{aligned} \tag{7.115}$$

Two simplifications can be brought to reduce variance: note first that the economic uncertain factor C_m is assumed to be independent from the uncertain water level Z_c so that the expectation of the product is the product of expectations; secondly, cost functions c_{sm} and c_g are *surely* null when their

Table 7.10 Flood model – comparing the simulations of total costs under standard MCS or accelerated sampling.

	Standard Monte-Carlo				Accelerated sampling		
Simulating C_c(*in M€/yr*)	500 runs	5000	50 000	500 000	500 runs	5000	50 000
Expectation	$22 \pm 74\%$	$9.8 \pm 38\%$	$8.8 \pm 12\%$	$8.8 \pm 4\%$	$8.7 \pm 11\%$	$8.9 \pm 3\%$	$8.8 \pm 1\%$
Standard deviation	$184 \pm 6\%$	$132 \pm 2\%$	$122 \pm 0.2\%$	$124 \pm 0.06\%$	$10.6 \pm 6\%$	$10.5 \pm 2\%$	$10.46 \pm 0.6\%$

input is less than $0.1\,m$ in the rare cases of overflow so that the expectation can be rewritten conditionally:

$$\begin{aligned} EC_c &= c_i(h_d) + E[c_{sm}(Z_c - h_d - z_b)].EC_m + E\left[c_g(Z_c - h_d - z_b)\right].EC_m = c_i(h_d) \\ &+ EC_m.E\left[c_{sm}(Z_c - h_d - z_b) + c_g(Z_c - h_d - z_b)|Z_c > h_d + z_b - 0.1\right]. \\ &P(Z_c > h_d + z_b - 0.1) \end{aligned} \tag{7.116}$$

Simulation variance is twice reduced: EC_m can be computed closed-form outside the simulation loop, thus reducing dimension; the conditional law of Z_c limited to overflow events that serves as an importance sampling density is much more concentrated in space. Compare the following numerical results with the standard Monte-Carlo sampling (no dike, no correlations, protection system excluded) in Table 7.10.

Superficially, the error estimates indicate a factor of 100 in the acceleration (indeed, the standard deviation is ~10 times less) but it is even more since the new random variable being sampled is much closer to a Gaussian variable than before so that the asymptotic error estimates are much more quickly reliable. This requires of course that the overflow probability is known elsewhere. Indeed, as the economic system model depends only on the scalar output of the hydro-model, sampling can be decoupled through a prior determination of the distribution of Z_c and then a sampling of the conditional cost expectation: this proves essential when testing different values for the decision variable h_d (Chapter 8) while avoiding the costly re-simulation of the hydromodel.

Others Quasi-random sampling has gained increasing interest (e.g. Bratley and Fox, 1988): it is a good example of the research field of *space-filling designs*. It consists of an experimental design of sequences, generally expressed in p-dimensional uniform spaces, which replace purely random samples through deterministic or pseudo-random procedures that aim to guarantee controlled space-filling properties (see also Iooss *et al.*, 2010).

7.3.5 Reliability methods (FORM-SORM and derived methods)

As introduced in Chapter 1 Section 1.2.1, FORM-SORM (Madsen, Kenk, and Lind, 1986) are the most famous methods developed in the field of structural reliability, targeting the efficient computation of the integral p_f (renamed the failure probability) for low probability and computationally-intensive system models (renamed failure function). Indeed, the methods prove to be practically very efficient, this being independent of the value of p_f and thus increasingly attractive as it becomes very low. The FORM

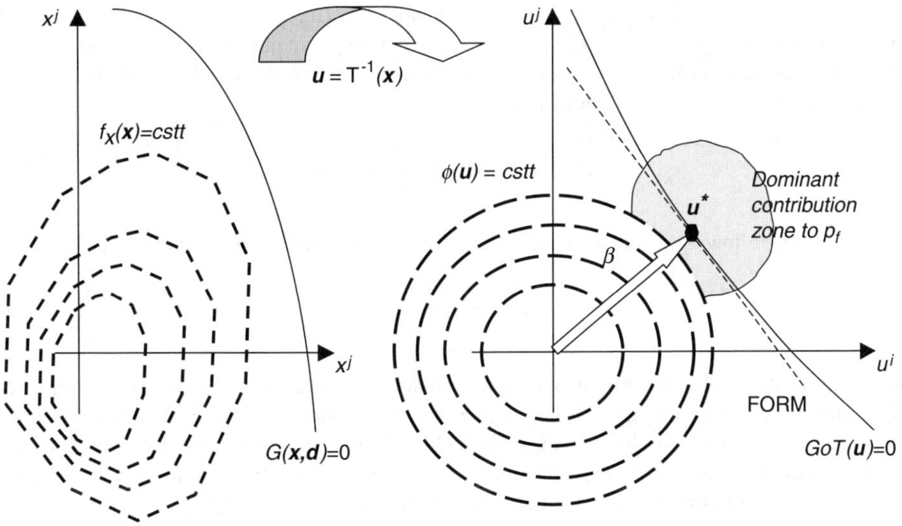

Figure 7.7 *The FORM approximation and steps of the algorithm.*

algorithm would typically involve some a few hundreds of runs or less to achieve reasonable convergence, regardless of the value of p_f which may be as low as 10^{-6}.

Indeed, the FORM-SORM methods approximate classically a threshold exceedance probability assuming that the region of failure for the various random variables is concentrated mainly around a design point, and that it can be estimated by a geometrically-approximated multi-normal region after transformation of the variables; validity and precision of the method cannot be controlled directly, although conservativeness of the method can be, up to some point, related to the monotonous behaviour of the failure function with regard to to the random variables (see Section 7.4).

FORM-SORM may be motivated by the following considerations (see Figure 7.7). Assume first that the inputs $\boldsymbol{X}$ follow an i.i.d. standard multivariate normal distribution:

- If the geometrical shape of the limit state hyper-surface $\{\boldsymbol{x} \mid G(\boldsymbol{x}) = z_s\}$ is simple that is hyperplane (or paraboloid, respectively), the calculation of p_f is of closed form and only depends on its distance from the origin, noted β, for FORM and on its curvatures at this point for SORM. The closest point, at distance β from the origin, is called the *design point*.
- The deviations from $\{\boldsymbol{x} \mid G(\boldsymbol{x}) = z_s\}$ relative to the hyperplane (or the second order surface respectively) tangental to the design point are negligible if they occur at a greater distance D than β, because of the very rapid decrease (in $exp(-D^2)$) of their likelihood.

In that case, a fair approximation of p_f, that is the volume of the failure domain, requires the mere knowledge of the design point, its distance from the origin by FORM. SORM adds a further refinement to the approximation using its curvatures. This is achieved through the use of a descent optimisation algorithm searching for the closest point of the failure domain to the origin (i.e. of the lowest norm in a standard Gaussian space):

$$\boldsymbol{x}^* = ArgMin_{\boldsymbol{x}}[\|\boldsymbol{x}\| \, |G(\boldsymbol{x}, \boldsymbol{d}) \geq z_s] \tag{7.117}$$

A common practice in reliability of structures is to denote $\{x \mid G(x) \geq z_s\}$ as $\{x \mid G(x) \leq 0\}$ which is equivalent to substituting the function $G(x) = z_s - G(x)$ for the system model $G(x)$; additionally, keeping implicit the dependence in d. When working on the Gaussian space, u is preferred to x to denote the inputs. Thus, the algorithm searches the following equivalently:

$$u^* = ArgMin_u \left[\|u\|^2 | G(u) \leq 0 \right] \tag{7.118}$$

It then issues the following estimate of the failure probability based on the Gaussian cdf Φ:

$$P_f^{FORM} = \Phi(-\beta) \tag{7.119}$$

When assuming that the limit state is an hyper-plane situated at a distance β from the origin, simply rotate the Gaussian space to find that p_f, that is the volume of the failure domain, is equivalent to the probability that one Gaussian input remains lower than $-\beta$. A number of variant algorithms are usually made available in software packages (e.g. conjugate gradient, BFGS, SQP or even genetic algorithms) and an extensive literature has discussed the convergence issues in searching for the design point.

Beyond standard Gaussian inputs, the method is easily generalised through a prior transformation of non-Gaussian possibly dependent inputs $\underline{X}$ into a reduced centred multivariate normal distribution. Change the variable $u = T^{-1}(x)$, henceforth considering the equivalent problem $P(GoT(u) > z_s)$ then search for the *design point* through the optimisation program. The exact transformation is simple in the case of independent sources using cumulative distribution functions (cdf):

$$u^i = \Phi^{-1} \circ F_{X^i}(x^i) \tag{7.120}$$

It becoms trickier for dependent inputs. The powerful Rosenblatt transformation proceeds by iteration of conditional transformations on the various inputs but it is often too complex in practice (Madsen, Kenk, and Lind, 1986). It is generally approached by the Nataf transformation which presupposes that the independently transformed variables U^i (according to Equation (7.120)) form a Gaussian vector, or equivalently that the dependence structure is that of a Gaussian copula along with the knowledge of the linear correlation coefficients. The independently transformed variables are then de-correlated using a Cholesky decomposition in the reverse manner to that which was discussed in Section 7.2.1 for the sampling of Gaussian correlated inputs. All in all, Figure 7.7 summarises the steps involved in FORM.

The real interest here lies in the fact that the number of calls to the physical code (although growing with p) becomes independent of the probability being sought, which is very efficient for rare quantiles. The residual *numerical* approximation of p_f is unfortunately not controllable in general. The literature developed to increase the sophistication of this method and check its conservatism is considerable; some strategies consist of checking on the existence of other secondary design points and even their inclusion in a 'multi-FORM'. Or again, the recovery of the (partial) control of accuracy associated with simulation methods, using hybrid approaches, for example the above-mentioned importance sampling centred around the design point; Axis Orthogonal Simulation which involves ordering the sampling along the design point radius mixed with a deterministic search for deviations to the tangent hyper-plane and so on (see Rackwitz, 2001).

Section 7.4 will also detail the intuitive and theoretical links between FORM and monotony of the system model, which help in controlling the conservatism of FORM to some extent.

Flood example

Computing the level-1 exceedance probabilities of overflow was undertaken with alternatives to Monte-Carlo as shown in Table 7.11.

FORM is groundbreaking is terms of acceleration, reducing by 4 decades the computing cost, yet lacking any control of error: in fact, FORM is not conservative in that particular case. Importance sampling (after FORM has found the design point) comes in that case with five times less sampling than MC for equivalent variance (the extra cost of FORM initial computations being quickly saved) although such variance is less robust (see Munoz-Zuniga *et al.*, 2011).

For FORM, some more detail is given in Table 7.12: the Abdo-Rackwitz descent algorithm is used, with varying starting points to assess the sensitivity of the design point. Evidencing a very limited sensitivity to starting the algorithm but a non-negligible (+20 %) upward SORM correction, hence a locally negative curvature. The number of runs comes as approximately five times the number of steps of the algorithms as each requires the gradient with respect to four inputs, thus at least five. SORM adds an additional ~10 to estimate the curvatures around the design point.

Table 7.11 Flood example – comparing alternative propagation methods.

	Monte-Carlo reference	FORM (SORM)	Conditional MC	Importance Sampling
$P(Z > 55.5)$ (overflow without dike)	$0.9\ 10^{-2}(10^4)$	$0.97\ 10^{-2}$	$1.16\ 10^{-2}(10^4)$	$1.19\ 10^{-2}(10^4)$
	$1.14\ 10^{-2}(10^5)$ $1.16\ 10^{-2}(10^6)$	$(1.2\ 10^{-2})$	$1.19\ 10^{-2}(10^5)$	$1.18\ 10^{-2}(10^5)$
Coefficient of variation (if any)	11% (10^4)	None	3.7% (10^4)	1.7% (10^4)
	3% (10^5) 0.9% (10^6)		1.2% (10^5)	0.6% (10^5)
Nb of runs	10^4 to 10^6	30-400		
Details and comments	—	See hereinbelow	With respect to flow Q	Undertaken around Form design point

Table 7.12 Details of FORM descent algorithms in the flood example.

Initial point in the U-space	Step of th algo.	Nb of iterations	U-space design-point (Q, Ks, Zv,Zm)	X-space design-point	FORM (SORM) proba
(0,0,0,0)	0,1	70	(1.656; −1.471; 0.724; −0.1719)	(2683; 19.0; 50.3; 54.9)	$0.973\ 10^{-2}(1\ldots 207\ 10^{-2})$
(2,0,1,0)	0,1	68	(1.657; −1.471; 0.723; −0.1720)	(2684; 19.0; 50.3; 54.9)	$0.972\ 10^{-2}\ (1\ldots 205\ 10^{-2})$
(0,0,0,0)	1	6	(1.650; −1.482; 0.719; −0.1708)	(2675; 18.9; 50.3; 54.9)	$0.971\ 10^{-2}\ (1\ldots 204\ 10^{-2})$

7.3.6 Polynomial chaos and stochastic developments

Chaos polynomials or stochastic development methods have seen considerable development since the originating work (Der Kiureghan and Ke, 1988; Sudret, 2008); they may be seen also as meta-models (Soize and Ghanem 2004; Sudret, 2008) but of a particular type as they also include the input distributions $\boldsymbol{X}$ inside the meta-model. In the originating work, these methods were inspired by problems involving random fields (in $\boldsymbol{X}$ and/or in Z) and not only random variables or vectors, and by modifying in an *intrusive* manner the numerical solvers of the underlying deterministic system model. However, their development, now frequently *non-intrusive* (i.e. without changing the deterministic code) is drawing closer to the technique developed for response surfaces (see Sudret, 2008). Indeed, chaos polynomials directly expand the output random vector of interest over the probability space:

$$\boldsymbol{Z} = G(\boldsymbol{X}, \boldsymbol{d}) = \sum_{l=1,\ldots+\infty} a^l(\boldsymbol{d}).\psi^l(\boldsymbol{X}) \tag{7.121}$$

Such a development relies on the fundamental results of functional analysis that guarantee that, given regularity properties over $G(.)$ such that $\boldsymbol{Z}$ is part of an Hilbert space, an infinitely-termed series fully develops the output of interest over a suitable Hilbertian basis. The basis of appropriate polynomials of uncertain inputs depends on the type of input cdf, for example Hermite polynomials if the uncertain inputs are (or can be transformed into) standard normal variables. However, practical estimation of the coefficients that require computing integrals involving the system model is always limited to a finite number P of terms (a number fast increasing with the input uncertainty dimension), leaving an error residual in the truncated development;

$$\boldsymbol{Z} \approx \tilde{G}(\boldsymbol{X}, \boldsymbol{d}) = \sum_{l=1,\ldots P} a^l(\boldsymbol{d}).\psi^l(\boldsymbol{X}) \tag{7.122}$$

The literature still needs to be developed so as to control such error residuals for which theoretical bounds themselves require the knowledge of higher-order derivatives or integral quantities involving the system model. Again, those derivatives could only be computed approximately through expensive additional runs of the system model. In general, the residual error cannot be estimated practically with a known probabilistic confidence.

7.3.7 Response surface or meta-models

As already mentioned, response surfaces or meta-models (Myers, 1971; Box and Draper, 1987; Kleijnen and Sargent, 2000) replace the model of the initial system $G(.)$ by an approximation:

$$\tilde{G}(\boldsymbol{x}, \boldsymbol{d}) \cong G(\boldsymbol{x}, \boldsymbol{d}) \tag{7.123}$$

that is numerically quicker to compute. Two great categories of meta-models can be distinguished:

- **Numerical methods**: either coming from the domain of classical numerical approximation (polynomials, neural networks, splines, ...) or dedicated to the fact that the input is a random vector and/or that a probabilistic control of the error is expected (chaos polynomials, Gaussian processes, ...).
- **Simplified physical**: (or phenomenological) model of the pre-existing system: using a 1D or 2D model instead of 3D, simplifying the physical-mathematical equations out of costly terms, or even integrating phenomenological sub-parts that are chained instead of fully coupled (as an extension to Section 7.1.3).

The literature has largely elaborated upon the first category though the second has great potential in practice, including its convincing virtue for regulators of having a domain-specific or physics background.

It is therefore possible to apply thereto a method of propagation that is greedy in terms of CPU but very robust: generally this is MCS, but closed-form results may be available for some risk measures and some types of meta-models. Chaos polynomials or stochastic developments may be seen also as meta-models that also include the input distributions $\boldsymbol{X}$ inside the meta-model. Meta-modelling is attractive in many cases: computing a risk measure involving a computationally-expensive system model, but also using inverse methods (Chapter 6) or optimising under uncertainty (Chapter 8) as both involve multiple propagation steps using the system model.

The key step is to optimise the size and type of experimental design $(\boldsymbol{x}_j)_j$ necessary for the construction of the surface while maintaining control of the residual $v(\boldsymbol{x}) = G(\boldsymbol{x}) - \tilde{G}(\boldsymbol{x})$ in particular in the zone of $\boldsymbol{x}$ that is most important for the risk measure c_Z. Quite similar to the inverse and identification techniques reviewed in Chapter 6, the robustness and accuracy of the meta-modelling method relies critically on the control of the approximation residual; it is generally formulated along a generalised linear regression (or kriging) approach with Gaussian processes, that is possibly heteroscedastic and/or self-correlated within a spatial process (Marrel *et al.*, 2008):

$$G(\boldsymbol{x}) = \tilde{G}(\boldsymbol{x}) + v(\boldsymbol{x}) = \sum_l \tilde{G}^l(\boldsymbol{x}).\beta^l + v \qquad (7.124)$$

β^l is typically estimated at least squares on $(\boldsymbol{x}_j)_j$ from whence the statistical check on v. Refined adaptive experimental design strategies can be built to increase the accuracy of the response surface in the vicinity of the limit state, using sometimes a preliminary FORM search (see, e.g. Kaymaz and McMahon, 2005). There is, however, no theoretical argument in general that ensures that the residuals are Gaussian or simply smoothly distributed. In fact, when the experimental design is deterministic, one could even question the validity of a random model for the residual as it is a difference between two deterministic functions (model and response surface) on a deterministic set of points. Error control of the residual of RSM proves quite heuristic (if ever mentioned), in spite of its importance when RSM is used to estimate rare failure probabilities. Gaussian characteristics have even more incidence in the use of meta-modelling for exceedance probabilities.

Once again, the monotonous property can in fact be valued inside a RSM, cf. Section 7.4.

7.4 Monotony, regularity and robust risk measure computation

Regularity of the system model is an important property in order to improve the risk measure computation: earlier sections have shown the simplifications brought by *linearity*. *Monotony* will prove to help greatly for robust accelerated computation of both probabilistic and mixed deterministic-probabilistic risk measures.

Beyond, monotony, it is expected that other forms of regularity such as boundedness of the system model or its derivatives could also help: a few perspectives will be outlined at the end of the section.

7.4.1 Simple examples of monotonous behaviours

Monotony was defined in Chapter 4 through a simple property relating to the monotonous behaviour of component-wise input-output functions constituting the system model. As commented upon in Annex Section 10.5.4, such a definition can easily be generalised to:

- models involving *discrete event* inputs, not only real-valued;
- *local* behaviour, whereby monotony is valid for known subsets of the domain of definition (or range of uncertainty) of the inputs.

An intuitive and quite useful property (also described in Annex Section 10.5.4) is the persistence of monotony through the chaining of sub-system models as well as through the iso-probabilistic transformation of input spaces, as was encountered with the FORM method (Section 7.3.5), or the copula functions (Chapter 5, Section 5.3.3).

Taking advantage of such model chaining considerations, consider the traditional basic structural safety examples (Melchers, 1999):

$$\begin{aligned} \boldsymbol{x} &= (r, s)' \\ G(\boldsymbol{x}, \boldsymbol{d}) &= r - s + d \end{aligned} \tag{7.125}$$

with r (respectively s) standing for a resistance (resp. load) variable and d a level of safety to be designed. Then, failure margin $G(.)$ is clearly globally increasing (resp. decreasing). It can be expressed as:

$$\begin{aligned} \boldsymbol{x} &= (K_{1c}, F, \sigma, a)' \\ G(\boldsymbol{x}, \boldsymbol{d}) &= K_{1c} - F\sigma\sqrt{\pi a} \end{aligned} \tag{7.126}$$

where K_{1c} stands for toughness, while F for the shape factor, σ for the stress and a for flaw dimension are the 'load' variables. $G(.)$ is still monotonous. Within the failure function of Equation (7.126) it is possible to further expand K_{1c} so that becomes a function of new input variables, such as within the ASME-RCC-M model (de Rocquigny *et al.*, 2004):

$$K_{1c}(u_{KIc}, T, RT_{NDT}) = \frac{1 + cu_{KIc}}{1 - 2c} Min\{a_1 + a_2 \exp[a_3(T - RT_{NDT} + a_4)], a_5\} \tag{7.127}$$

where a_i $(i = 1 \ldots 5)$ are positive scalars. T stands for the temperature at flaw, RT_{NDT} for the nil ductility temperature of the material and u_{KIc} parameterises the intrinsic toughness variability with c for its coefficient of variation. It is still monotonous within some physically reasonable domain (toughness being necessarily positive, u_{Kic} has to remain above $-1/c$), T and u_{KIc} playing the role of resistance variables while RT_{NDT} that of a load variable. The monotony property naturally remains true when this sub-model is plugged in the upper layer model of Equation (7.126).

Then a_i $(i = 1 \ldots 5)$ could also be modelled as uncertain (because of limitations of toughness testing or deviations between the structure and the test) and RT_{NDT} could also be predicted by another sub-model: again the model is monotonous with respect to those inputs. As a further complication of these basic examples, it will be seen that temperature T and σ can be the outputs of a finite-element model while still being physically clearly monotonous as a function of load sub-parameters (such as the limit conditions of an external cold shock).

The flood example is also globally monotonous, as described in Chapter 3, whatever its hydraulic, system reliability or economic components, and inputs and outputs considered.

On the other hand, if competition between several phenomena acting in opposite directions is plausible or if bifurcation, such as phase changes in fluid mechanics, is represented inside the failure model, monotony may be trickier to guarantee. It may even prove false, at least for some parts of the parameter space.

This is the case for instance in accidental nuclear fluid dynamics (e.g. the peak cladding temperature as a function of initial residual power) or dynamic seismic response (e.g. structural damage as a function of the peak ground acceleration (PGA), also depending on the dynamic shape of the accelerogram) though more elaborate conservative envelopes could still be derived and keep monotonous (see Annex Section 10.5.4).

7.4.2 Direct consequences of monotony for computing the risk measure

Monotony proves to be both naturally intuitive to risk analysts and quite efficient for a quick and conservative risk analysis. This is because it provides cheap and conservative bounds for the risk measures in an appreciable number of cases, as will be studied here and in Section 7.4.3.

Computing a risk measure when only one input is probabilised A classical result of probability calculus shows that quantiles are preserved through monotonous transformations. Indeed, if $dimX = 1$ and if $z = G(x)$ is a monotonous (e.g. increasing) system model mapping the area of definition D_x into D_z, then[3]:

$$\begin{aligned} &\forall x \in D_x, \forall z \in D_z \quad G(x) > z \Leftrightarrow x > H(z) \\ &P(Z = G(X) > z) = \int_x 1_{G(X)>z} f_x(X = x)dx \\ &\qquad = \int_x 1_{X>H(z)} f_x(X = x)dx = P(X > H(z)) \end{aligned} \tag{7.128}$$

So that, whatever $0 < \alpha < 1$.

$$z^\alpha = G(x^\alpha) \tag{7.129}$$

There is an essential consequence for the purpose of undertaking the *initiator-frequency approach* (as defined in Chapter 1), that is a mixed deterministic-probabilistic risk measure limited to one probabilised real-valued input. It only requires *one run* of the potentially-complex system model per design scenario $\boldsymbol{d}$. Indeed, consider now a p-input monotonous system model (supposedly increasing after appropriate sign changes) where penalised values are considered for the *p-1* first components and a *pdf* is chosen to model the last one, namely the initiator. Thus, twice benefiting from monotony:

$$\begin{aligned} &\forall \boldsymbol{x} = (x^1, \ldots x^p) \quad G(x^1, \ldots x^p, d) \le G\left(x^1_{pn}, \ldots x^{p-1}_{pn}, x^p, d\right) \\ &P\left[G(\boldsymbol{X}, d) \middle| (X^i = x^i)_{i=1\ldots p-1} > z\right] \le P\left[G\left(\boldsymbol{X}, d) \middle| \left(X^i = x^1_{pn}\right)_{i=1\ldots p-1}\right) > z\right] \end{aligned} \tag{7.130}$$

And:

$$P\left[G(\boldsymbol{X}, d) \middle| \left(X^i = x^i_{pn}\right)_{i=1\ldots p-1} > G\left(x^1_{pn}, \ldots x^{p-1}_{pn}, x^\alpha, d\right)\right] = P(x^p > x^\alpha) = \alpha \tag{7.131}$$

so that one requires only the computation of the following:

$$z^\alpha_{pn} = G\left(x^1_{pn}, \ldots x^{p-1}_{pn}, x^\alpha, d\right) \tag{7.132}$$

In order to design the z level that ensures a risk measure lower than α:

$$c_z = P\left[G(\boldsymbol{X}, d) > z^\alpha_{pn}\right] \le \alpha \tag{7.133}$$

[3] Considering *H(z)* as the generalised inverse. If *G* is supposed strictly monotonous, it is injective and thus constitutes a bijective mapping between the two domains. If *G* is only monotonous, then define *H(z)* as the lowest *x* fulfilling *G(x)* = *z*.

On the contrary, the probabilisation of $p \geq 2$ inputs discards the possibility of relying on just one run of the system model, even under monotony (cf. Chapter 4, Section 4.2.2). More general bounds can be developed as will be outlined below.

> *Flood example*
> Under an initiator-frequency approach, flow is the only variable enjoying a statistical model, while the other uncertain variables are given penalised values. The hydro-component being monotonous as a function of flow, designing up to a 1000-yr return protection level required only one computation of the system model for the following point:
>
> $$\boldsymbol{x} = \left(q^{1000}, k_{s\,pn}, z_{v\,pn}, z_{m\,pn}\right)$$
>
> Under a Gumbel model, the design level was found to be 60.7 m, that is a 5.2 m dike, see Step Three in Chapter 3. Note that this kind of initiator-frequency approach would generally use a level-2 confidence interval for q^{1000} as well as an additional margin on the final dike level, for example 0.2m (see Step Five and next section).

Computing a level-2 risk measure Considering a p-dimensional system model that is at least partially monotonous with respect to p-th component, a simplification occurs regarding the level-2 computations. In a number of cases, the highest risk measure is simply the image of the risk measure computed with the highest value of the parameter. The proofs are all given in Annex Section 10.5.

Note that EZ increases when any of the expectations of input uncertain components $\theta^p = E(X^p)$ increase. If composed with a monotonous function, the increase of the expectation of a random variable (with unchanged distribution otherwise) is preserved. In other words, the *output expectation* is *monotonous* as a function of the *input expectations.*

This applies notably to the case of an exceedance probability-based risk measure because the indicator function is monotonous when the system model is. Thus, it is monotonous as a function of some of the input pdf parameters $\boldsymbol{\theta} = (\theta^k)^k$:

$$c_Z(\boldsymbol{\theta}) = P\left[G\left(X^1, \ldots X^p\right) > z_s | \boldsymbol{\theta}\right] \tag{7.134}$$

The essential consequence is that a level-2 deterministic-probabilistic risk measure may be computed simply by a simple level-1 propagation using penalised parameters. Note, however, that this is not the case with all pdf parameters: dispersion parameters, such as the standard deviation in a Gaussian distribution, do not respect such property. Note also that it is very conservative to penalise all input parameters without taking advantage of error compensation.

Dedicated acceleration methods have been proposed by Limbourg *et al.* (2010) for the computation of level-2 risk measures with monotonous system models on the basis of the MRM methods that will be introduced in the following section.

> *Flood example*
> Risk measures computed in the hydro-component are monotonous with respect to input parameters such as Gumbel location for flow, normal mean for K_s, or triangular bounds for Z_m and Z_v. Thus, different forms of conservative (mixed deterministic-probabilistic) level-2 overflow probability could be computed within Step Five with simplified computations (Table 7.13). For the sake of comparison a

Table 7.13 Level-2 risk measures for the flood example.

Level 1 method	Level 2 method	1/100	1/1000	Number of runs	Comments
Probabilistic on flow + deterministic	Penalised values for each input parameter θ^k (70% quantile for flow, deterministically penalised for others)	59.2	60.9	2 (1 run for each return period)	A computationally cheap but highly penalised scenario
Fully probabilistic	Penalised values for each input parameter θ^k (70% quantiles)	56.8	60.7	10^{4-5} once the level-1 sampling	Less penalised on level-1 but still so on level-2; much more intense computationally
Fully probabilistic	Level-2 Taylor-based upper 70% quantile assuming Gaussian level-2 uncertainty$\boldsymbol{\Theta}$	55.9	58.1	9.10^{4-5} ($n_\theta + 1 = 9$ times the level-1 sampling (with the same seed))	A more realistic probabilistic measure, accelerated by the level-2 approx.
Fully probabilistic	Fully probabilistic (Double MCS)	56.0	58.1	10^6–10^7 100–1000 times the level-1 sampling	A reference scenario, with highest CPU.

pseudo-level of confidence of 70 % was taken for each case. In all cases, the Gumbel fit is chosen for flow and level-1 correlations are neglected.

Adding up the system component keeps it monotonous and adds parameters such as elementary component failures: the corresponding risk measure may be computed similarly.

7.4.3 Robust computation of exceedance probability in the monotonous case

Monotony turns out to be both intuitive and efficient in securing robust computation of exceedance probabilities. This is fundamentally due to elementary properties that will be explained below. Thus, monotony is valuable in controlling a number of reliability algorithms and even generated dedicated algorithms called the Monotonous Reliability Methods (MRM) (de Rocquigny, 2009).

Elementary properties of exceedance probabilities under monotony Suppose the value of a globally-monotonous system model $G(.)$ is known for a given point $\boldsymbol{x}_o$. Then the following basic *Property 1* will appear quite useful:

Property 7.1

(a) If $G(\boldsymbol{x}_o) < 0$ (respectively $G(\boldsymbol{x}_o) > 0$), then the 'upper hypercorner' $E^+_{x_o} = \{\boldsymbol{x} \in D_x | \forall i\, x^i \geq x^i_o\}$ (resp. 'lower hypercorner' $E^-_{x_o} = \{\boldsymbol{x} \in D_x | \forall i\, x^i \leq x^i_o\}$) constitutes a failure-dominated (resp. safety-dominated) subset, that is on any point of which the failure function $G(.)$ is guaranteed to be negative (resp. positive).

(b) Then the following bounds are certain: $P(E^+_{x_o}) \leq p_f$ (respectively $p_f \leq 1 - P(E^-_{x_o})$).

In other words, whenever a point in space is known to be safe, the chunk of space situated 'below' (the lower hypercorner, see Figure 7.8) is guaranteed to be included in the safe region, and thus remove the corresponding probability weight from the failure probability estimate. And conversely, once a failing point of space is known.

A number of extensions can be easily derived. A double bounding becomes available when $\boldsymbol{x}_o$ is part of the failure surface, that is $G(\boldsymbol{x}_o) = 0$. This is also the case when changing the threshold from 0 to $G(\boldsymbol{x}_o)$, namely:

$$P(E^+_{x_o}) \leq P[G(\boldsymbol{X}) < G(\boldsymbol{x}_o)] \leq 1 - P(E^-_{x_o}) \tag{7.135}$$

When $\boldsymbol{X} = (X^1, \ldots, X^p)$ is a set of independent variables, the bounds can be computed simply from the marginal cdfs:

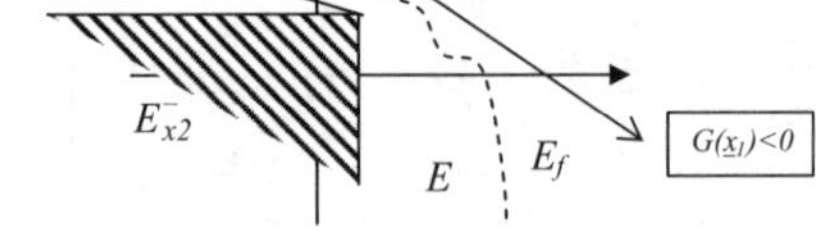

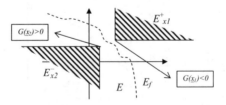

Figure 7.8 *Hypercorners associated to point values $\boldsymbol{x}_1$ and $\boldsymbol{x}_2$ – in dotted line, unknown limit state between safe and failure domains (de Rocquigny, 2009).*

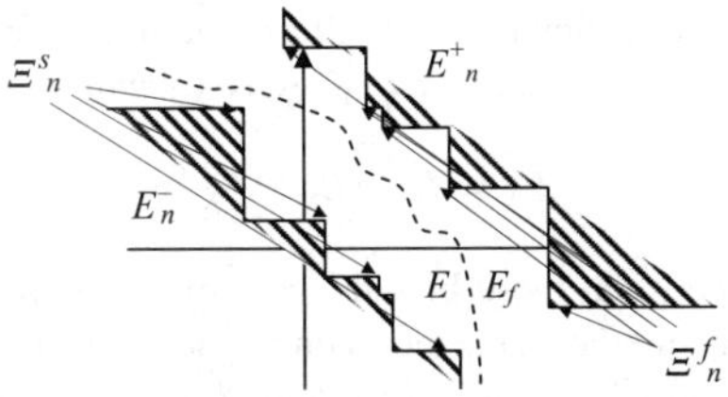

Figure 7.9 *Multiple hypercorners in the case of multiple results of a failure function (de Rocquigny, 2009).*

$$1 - P\left(E^-_{x_o}\right) = 1 - P\left\{X \middle| \forall i\, X^i \leq x^i_o\right\} = 1 - P\left[\bigcap_i \left\{X^i \leq x^i_o\right\}\right] = 1 - \prod_i F_{xi}\left(x^i_o\right) \tag{7.136}$$

$$P\left(E^+_{x_o}\right) = P\left\{X \middle| \forall i\, X^i \geq x^i_o\right\} = P\left[\bigcap_i \left\{X^i \geq x^i_o\right\}\right] = \prod_i \left[1 - F_{xi}\left(x^i_o\right)\right] \tag{7.137}$$

Note that the probability bounds hold whatever the space transformation, in particular into the uniform space or into the Gaussian space in the case of independence. This is because mapping regions through monotonous and injective transformations of variables do not change their probability measure. Note also that those bounds can be computed without any supplementary run of the model $G(.)$ beyond the mere knowledge of $G(\boldsymbol{x}_o)$. It is quite obvious and simple in the case of independent variables. In the general dependent case, computing $P\left(E^-_{xo}\right)$ or $P\left(E^+_{xo}\right)$ is a matter of multi-fold integration of a known function (the joint pdf of X) over a simple domain (the 'hypercorner') defined without reference to $G(.)$. They can be computed by numerical integration (in low dimension) or by Monte-Carlo simulation (in any dimension) either in the physical space or any transformed space. Although it may not be numerically immediate for complex dependence structures in high dimensions, it is always solved independently of the CPU-intense physical failure model $G(.)$. Later sections will show that computations are much easier in the Uniform-transformed space.

The hypercorner property can be applied successively (Figure 7.9) if n results $G(\boldsymbol{x}_j)$ are known for vectors of inputs $(\boldsymbol{x}_j)_{j=1\ldots n}$, whatever the nature of this sample (random, deterministic experimental design, etc.). Through elementary calculus of probabilities they may then be combined to refine the upper and lower bounds. Grouping those points into the failure and safe sub-samples:

$$\begin{aligned} \Xi^s_N &= \left\{\boldsymbol{x} \in \left(\boldsymbol{x}_j\right)_{j=1,\ldots N} \middle| G\left(\boldsymbol{x}_j\right) \geq 0\right\} \\ \Xi^f_N &= \left\{\boldsymbol{x} \in \left(\boldsymbol{x}_j\right)_{j=1,\ldots N} \middle| G\left(\boldsymbol{x}_j\right) \leq 0\right\} \end{aligned} \tag{7.138}$$

one can generate the upper and lower multiple-hypercorners:

$$\begin{aligned} E^-_N &= \left\{\boldsymbol{x} \in D_x \middle| \exists \boldsymbol{x}_j \in \Xi^s_N \forall i\, x^i \leq x^i_j\right\} \\ E^+_N &= \left\{\boldsymbol{x} \in D_x \middle| \exists \boldsymbol{x}_j \in \Xi^f_N \forall i\, x^i \geq x^i_j\right\} \end{aligned} \tag{7.139}$$

out of which generalised bounds can be derived (see Figure 7.9):

$$\int_{\bigcup_{x_j \in \Xi^f_N} \{x \geq x_j\}} f_x(\boldsymbol{x}) d\boldsymbol{x} = P\left(E^+_N\right) \leq p_f \leq 1 - P\left(E^-_N\right) = 1 - \int_{\bigcup_{x_j \in \Xi^s_N} \{x \leq x_j\}} f_x(\boldsymbol{x}) d\boldsymbol{x} \tag{7.140}$$

From the point of view of failure, the following restriction of the area of definition constitutes the only area remaining unknown once a first set of results $(\boldsymbol{x}_j)_{j=1\ldots N}$ is known:

$$E_N = D_x \backslash \left(E_N^+ \cup E_N^- \right) \tag{7.141}$$

An essential difference holds between these bounds, be they single-result or multiple-result, and the accuracy statements of simulation methods: the monotonous-derived bounds are probabilistically *certain*, whereas confidence intervals hold with a *limited likelihood* (e.g. 95 % confidence).

Form and monotony The monotonous property looks intuitively attractive for the FORM approximation validity, since the idea of monotony proves somewhat implicit in the genesis of FORM, as will be explained below. As mentioned in Chapter 1, the traditional partial safety factor approach obviously relies on the monotonous behaviour of the failure function. Consider the following classical expression:

$$G(\gamma.\boldsymbol{x}) = g_R\left(x_k^1/\gamma_R^1, \ldots x_k^r/\gamma_R^r\right) - g_S\left(x_k^{r+1}\gamma_S^{r+1}, \ldots x_k^p\gamma_S^p\right) \geq 0 \tag{7.142}$$

where γ denote partial safety factors and $\left(x_k^i\right)_{i=1\ldots p}$ denote the characteristic values for the uncertain inputs: the application of partial safety factors all fixed at generally more than 1 in effect assumes that the failure margin decreases when resistance variables $(x^1, \ldots x^r)$ decrease and load variables $(x^{r+1}, \ldots x^p)$ increase.

As a sophisticated evolution from the partial safety factor approach, recall that FORM/SORM involves first the search for the closest area to nominal values '*beyond which*' failure begins, encapsulated in the design point (DP), the reliability index being its distance to the failure surface. 'Closest' and 'beyond' assume that moving further away increases the risk. Mechanical common sense leads one to argue that stretching further down the design value of resistance variables or conversely up the value of load variables favours failure: it is equivalent to assuming monotony of the failure function, albeit with reverse signs according to whether resistance or load variables are considered. In Figure 7.10 – left, the region of failure is sketched traditionally as being located in the upper right, say increasing the load U^1 and decreasing the resistance $-U^2$ leads to failure (the typical FORM convention being that $G(.)$ decreases with U^i). Note that monotony could also apply to any search radius comprised in the 'North-East' quadrant (see Figure 7.10 – right), that is $\boldsymbol{u} = (u^1, u^2)' = (b\lambda^1, b\lambda^2)'$ where λ^1, λ^2 and b are positive scalars leading to a monotonous composite function $G(\boldsymbol{u}(b))$.

In the spirit of FORM/SORM this should be true at least in the vicinity of the limit state and particularly of the design point (local monotony) while other regions furthest from the origin would have negligible probabilistic weight due to the exponential decay of likelihood. It will be seen that this is not sufficient to establish FORM conservatism, although some limited results can be established in the case of independent variables, supposing also that $G(.)$ is positive at the nominal input values (the origin of the standard space), which is generally the case for safe structures. Consider henceforth the various quadrants generalising the 2-D map terminology of Figure 7.10 – right: these are the subsets within which the sign of each component x^i remains constant. It is straightforward to understand that there are 2^p such quadrants, all weighing a probability of $1/2^p$ because of rotational symmetry of the Gaussian-transformed standard space. We have then the following *Property 2*:

Property 7.2

(a) Here there cannot be any design point (i.e. a point of the limit state E_l lying at a distance from the origin lower or equal to that of any other E_l point) outside the North-East quadrant (or its boundary).

(b) Any design point (or more generally any point of the limit state E_l) noted $\mathbf{u}^* = (\beta^{1^*}, \ldots, \beta^{p^*})'$ generates the following lower and upper bounds:

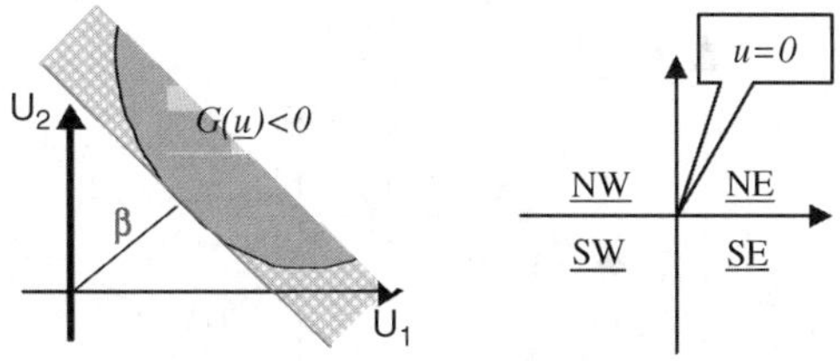

Figure 7.10 *left: a simple two-variable failure function with FORM limit-state approximation in the transformed Gaussian space – right: conventional names for quadrants around the origin of the standard space (de Rocquigny, 2009).*

$$\prod_{i=1\ldots p}\left[\Phi(-\beta^{i}*)\right] \leq p_f \leq 1-\prod_{i=1\ldots p}\left[1-\Phi(-\beta^{i}*)\right] \tag{7.143}$$

(c) Assuming additionally that no point of the limit state lies closer than β to the origin, an additional upper bound holds (χ_q^2 standing for the chi-square cdf with q degrees of freedom):

$$p_f \leq 1-\frac{1}{2^p}\left[1+\chi_p^2(\beta^2)+\sum_{i=1}^{p-1} C_p^i \chi_{p-i}^2(\beta^2)\right] \tag{7.144}$$

Obviously no design point can be located within the South-West quadrant $\{\boldsymbol{x}|\forall i\, x^i < 0\}$ since it is a safety-dominated hypercorner when the origin is safe. In fact neither can there be any design point in the NW or SE quadrants, that is those quadrants for which there are at least one negative component (and at most *p-1*). Indeed, if there was a u^* point in one of those quadrants both being on the limit state and closest to the origin, project it onto the NE quadrant boundary (i.e. putting to zero all negative components): thus find another point $p(u^*)$ lying closer to the origin, part of the NE quadrant and still in the failure zone because of monotony. Note also that Equation (7.143) is a straightforward consequence of the hypercorner bounds for a DP or any point of the limit state. Equation (7.144) is the result of simple probabilistic computation of the safety-dominated region using the directional properties of the standard Gaussian space already mentioned in Section 7.3.4 hereinabove): detailed proof may be found in de Rocquigny (2009).

Returning now to the validity of FORM, the fact that the DP must be situated inside the NE quadrant excludes in particular the possibility of multiple-design points. However, there cannot be excluded in spite of monotony: a negative curvature close to the design point or at the limit; and a multi-design point situation the limit state forming a hypersphere at the distance β in the NE quadrant and hyperplanes in the NW-SE quadrants. In that situation FORM is non-conservative although SORM could probably still be conservative. However, SORM computation might be trickier in situations still involving a monotonous failure function, but poorly distinguishable around the design points.

The upper bound 7.146 of *Property 2* offers a tighter bounding in those cases where one assumes that the reliability index β has been identified through an optimisation program robust enough to preclude any point of the limit state closer than β to the origin. There could obviously be multiple DP at distance β or farther but the limit state osculates the hypersphere of radius β in the NE quadrant, and stays constant in other NW-SE quadrants. It can be seen as the maximal envelope covering the kind of situations where, on top of monotony, one trusts the *inexistence of limit state points closer than* β.

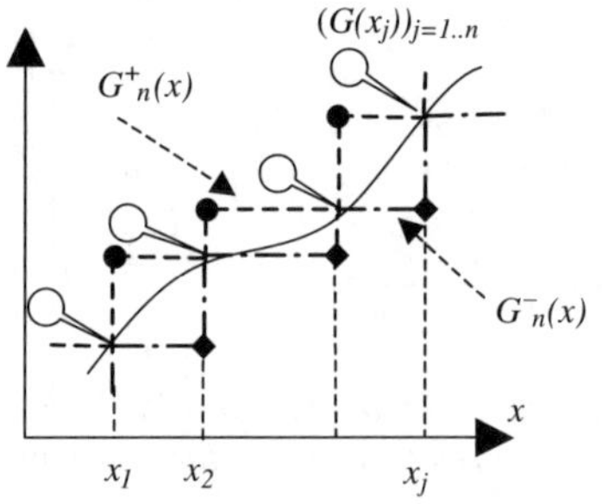

Figure 7.11 *Response surface under monotony in the 1D-case (de Rocquigny, 2009).*

Thus, although monotony is not enough to guarantee FORM conservatism, the bounds of *Property 2* are a starting point for identifying the robust bounds for p_f. They generally prove insufficient since this interval becomes very loose when the dimension of $\boldsymbol{X}$ increases or either when one of the component dominates. Thus, other mathematical properties than global monotony are needed to guarantee FORM (or SORM) conservativeness. Convexity of the failure domain in the standard space would be desirable, though it proves much less intuitive than monotony to infer on the basis of phenomenological expertise. Unless other properties are demonstrated, what can be said with certainty with FORM under monotony refers to the hypercorner bounds, possibly rather loose.

Other propagation methods under monotony *Directional sampling* (DS) enjoys some limited benefits from monotony. In the direction of the SW quadrant, considerations of the previous case when the origin of the standard space is safe show that a zero cannot occur and hence save some sampling. In the NE quadrant, it is straightforward that all directions evidence at most one zero: it guarantees a full error control for the NE quadrant conditional radial failure probabilities. It is, however, not the case for all other NW-SE directions, for which multiple zero cannot be excluded. In high dimensions, those uncontrollable quadrants represent (1-1/2^{p-1}), a majority of radial probability: for $p = 5$ they represent already 93 %.

Regarding *Latin Hypercube Sampling* (LHS), monotony is proved to guarantee the reduction of variance by the LHS estimator as compared to that of crude MCS which applies naturally to the estimation of the failure probability (McKay, Beckman, and Conover, 1979); though this is of little interest because of space-filling entailing many useless simulations.

In de Rocquigny (2005a) it is shown that a robust *response surface* is easily accessible under monotony. Whatever the design of experiment chosen, a piece-wise constant surface between the system model runs (see Figure 7.11 in dimension 1) is enough to build a fully-conservative response surface, and hence guarantee *with certainty* the control of the error.

In fact two response surface are built, an upper and lower version, guaranteeing that:

$$\forall \boldsymbol{x} \in D_x \quad G_N^-(\boldsymbol{x}) \leq G(\boldsymbol{x}) \leq G_N^+(\boldsymbol{x}) \tag{7.145}$$

or in other words that the residual is certainly negative or null (or respectively certainly positive or null).

Monte-Carlo and MRM dedicated algorithms for exceedance probability under monotony Important observations can be made when undertaking Monte-Carlo sampling under global monotony. First, whatever the simulation strategy, the multiple-result bounds given in Section 7.4.3 can always be computed besides the basic estimator during the simulation process: it provides *certain* bounds on p_f on top of the Monte-Carlo standard estimator and confidence interval. Those bounds being costless with respect to system model runs, they can help in stopping the simulation process when the bounding becomes satisfactory. The advantage of this technique is that the confidence that may be placed in the bounding will be 100 %, although the width of the bounds in large dimensions for $\boldsymbol{X}$ may limit its usefulness.

Secondly, as a consequence of the hypercorner property, every new model run for a sample $\boldsymbol{x}_j$ also generates a subset of the sample space that becomes useless to simulate in later steps. In fact, after N simulations, the growing subsets E_N^+ and E_N^- defined above encapsulate the zones where subsequent simulations are useless. This is the starting point for crafting an adaptive sampling strategy that suppresses the unnecessary bits of the sample space after each calculation.

A generalised formulation can be crafted in order to take full advantage of the monotonous property for any algorithm, not only Monte-Carlo sampling: iteratively update the hypercorner bounds and the remaining unknown sample space in order to adapt the design of experiment, whatever the specific technique favoured for that DoE. The meta-algorithm called MRM (Monotonous Reliability Methods (de Rocquigny, 2009), is formulated as follows:

(i) Choose a j-th set of new experiments (i.e. the j-th DoE) restricted to the remaining unknown domain E_{j-1} (at first iteration $E_0 = D_x$ i.e. the entire domain of definition).
(ii) Compute $G(.)$ on the j-th set.
(iii) Compute the updated bounds $1 - P(E_j^+)$ and $P(E_j^-)$, as well as the central estimators P_f if available (which is the case for MRM variants of MCS or DS, for instance).
(iv) Update the remaining unknown domain E_j into E_{j+1} and go to (i) of $j+1$-th iteration,

Stopping iteration if satisfied with upper and/or lower bounds, or if a prescribed maximum number of runs is reached. Many MRM variants can be developed according to how the design of experiment (DoE) is built in step (i) of the iterative process such as:

- *Fixed-type deterministic DoE* with a progressive refining of the meshing: think of the classical factorial or axial DoE associated with an iterative segmentation or meshing of E_j into levels along each input dimension (or factor). *Isoprobabilistic* meshings prove quite attractive as they simplify the handling of E_j and the computation of bounds.
- *Optimisation-based deterministic DoE*, in order to track quickly the failure surface. This includes gradient-steered algorithms in particular. Adapted *FORM* can be seen as an example.
- *Random-based DoE*: DoE is iteratively randomised in order to catch the space-filling properties of randomness which are attractive in higher dimensions. The above-mentioned adaptation of MCS fall into that category, but directional Sampling, AOS or importance sampling could also be adapted.

Obviously these different variants can be combined within a single iterative process. For instance, searching for a FORM design point can be tempting in early iterations of MRM in order to climb as efficiently as possible up to the closest limit state; switch to fixed deterministic DoE in later iterations to bound the rest of limit state. Alternatively, while fixed deterministic DoE may be considered in early iterations in order to map systematically the domain of definition, they may quickly lead to an explosion of the number of model runs when the limit state is identified approximately: launching random designs in remaining subsets may prove more efficient to improve the bounding accuracy, particularly in higher dimensions.

Important comments need to be made on the estimators generated and the convergence of the MRM. First, all MRM variants generate a bounding interval for p_f which can be seen as a first pair of estimators, although generally not in a statistical meaning of the word because when the design of experiment is not random, they cannot be seen as functions of random variables. This interval has two fair properties: it is *certain to include* p_f at any iteration; its width is *certainly decreasing*, that is any *additional model run always ameliorates* the estimation. This second point is due to the mere definition of subsets E_N^+ and E_N^-; at each iteration at least one of the two increases strictly because, in step (i), the j-th new design of experiment is built deliberately to exclude them.

Secondly, some variants of MRM do additionally offer another 'best-estimate' estimator: this is the case for the MRM sampling methods such as adaptive MCS or directional sampling and so on. That second estimator enjoys at least as good convergence properties as that of the standard corresponding one (without MRM); it often enjoys much improved ones, as shown below for the adaptive MCS. Moreover, it cannot stand outside the monotonous bounding interval by construction, which constitutes also a *probability one* error control of the estimation.

Variance reduction can be demonstrated for the central estimator of the *Sampling MRM* (S-MRM) variant, that is adaptive Monte-Carlo under monotony. Similar to importance sampling and using the fact the adaptive density is closed-form, the unbiased estimator appears as follows:

$$\hat{P}_f^{mis} = \frac{1}{N}\sum_{j=1\ldots N}\left[P\left(E_{j-1}^+\right) + 1_{G(x_j)\le 0}\left[1 - P\left(E_{j-1}^+\right) - P\left(E_{j-1}^-\right)\right]\right] \tag{7.146}$$

In other words, correct at each step the result of the raw sample $\frac{1}{N}1_{G(x_N)\le 0}$ by the probabilities of the dominated zones E_N^+ and E_N^- that were deliberately excluded from sampling. Note that because of adaptation $\boldsymbol{X}_1 \ldots \boldsymbol{X}_N$ are non-i.i.d. random samples. Variance can be proved to be as follows (proof in *Annex* Section 10.5.6):

$$\begin{aligned}\operatorname{var}\left[\hat{P}_f^{mis}\right] &= \frac{p_f(1-p_f)}{N}\left(1 - N^{-1}\sum_{j=1\ldots N-1} E_{X_1,\ldots,X_j}\left[\frac{P(E_j^-)}{1-p_f} + \frac{P(E_j^+)}{p_f} - \frac{P(E_j^-)P(E_j^+)}{(1-p_f)p_f}\right]\right) \\ &= \frac{p_f(1-p_f)}{N}\left(1 - N^{-1}\sum_{j=1\ldots N-1} c_j\right)\end{aligned} \tag{7.147}$$

The quantity c_N can be interpreted as the average reduction of the failure and safe region by the multiple-result hypercorners after N steps. It is not easy to compute in general. It is, however, certainly positive since $P(E_N^+)$ and $P(E_N^-)$ are lower than p_f and than 1-p_f respectively so that it is certainly lower than the variance of classical MCS.

Flood example

The following MRM variants were tested in the flood example (de Rocquigny, 2009). In each case, the central estimator is given as well as the *certain* bounds on the exceedance probability (assuming no dike, $p_f = P(Z_c > 55.5)$) in Table 7.14.

Note that while FORM was already mentioned as being groundbreaking in terms of acceleration, the figure is slightly under-conservative in the flood example, as already mentioned, and the associated *certain MRM* bounds are quite loose after 200 runs – completely useless for risk analysis in fact, as

Table 7.14 Comparing MRM to traditional propagation methods in the flood example.

Algorithm	Number of model runs	p_f estimate and MRM bounds
Form	203	$0.97\ 10^{-2}$ – $0.62 > p_f > 3.5\ 10^{-4}$
D-MRM	76	$0.24 > p_f$
	223	$0.11 > p_f$
	824	$5{,}2\ 10^{-2} > p_f > 6.6\ 10^{-4}$
S-MRM continuing D-MRM	2 063 (S-MRM has done 2063-824 = 1 239 runs)	$IC = 100\%$: $2.23\ 10^{-2} > p_f > 0.49\ 10^{-2}$
		or $1.14\ 10^{-2}$ - $IC = 95\%$: $1.19\ 10^{-2} > p_f > 1.105\ 10^{-2}$
Monte-Carlo	2 000	$0.9\ 10^{-2}$- $IC = 95\%$: $1.44\ 10^{-2} > p_f > 0.55\ 10^{-2}$
	200 000	$1.14\ 10^{-2}$- $IC = 95\%$: $1.19\ 10^{-2} > p_f > 1.09\ 10^{-2}$

they span over three decades of risk. FORM appears as a performing heuristic, but lacks robustness guarantees even under monotony. Dichotomic MRM (D-MRM) issues better bounds, but quickly becomes inefficient. The best algorithm comes with S-MRM (i.e. adaptive monotonous MC) on top of first dichotomic iterations: after 2000 runs, a robust bounding by a factor 5 is made available, 2000 runs being quite insecure for standard Monte-Carlo (see previous Section 7.2 illustrations, whereby 4 10^4 would be a minimum for a $\sim 10^{-2}$ exceedance probability).

Computing level-2 risk measures with accelerated monotonous algorithms Dedicated acceleration methods have been proposed by Limbourg *et al.* (2010) for the computation of level-2 risk measures such as level-2 quantiles or expected value of an exceedance probability assuming that the system model is monotonous, on the basis of the MRM methods. The principle is to undertake a double-Monte Carlo (cf. Section 7.2.) and take advantage of the gradual knowledge gained because of monotony in order to save some unnecessary simulations, much in the spirit of (level-1) MRM Monte-Carlo explained in previous sections. Such an approach proves to be much more economical than a full double Monte-Carlo or level-2 FORM-SORM: a few thousands of runs of the system models are enough to get a robust estimate with 95% confidence of a risk level around 10^{-3}, a figure that would require 10^7–10^8 runs of two-tier Monte-Carlo. See for more detail Limbourg *et al.* (2010).

7.4.4 Use of other forms of system model regularity

Beyond monotony, it is expected that other forms of regularity such as boundedness of the system model or its derivatives could also help. Consider, for instance, the Taylor approximation:

$$z = G(\boldsymbol{x}, \boldsymbol{d}) \approx G(E(\boldsymbol{X}), \boldsymbol{d}) + \sum_{i=1}^{p} \left(\frac{\partial G}{\partial x^i} \right) [x^i - E(X^i)] + o(\|\boldsymbol{x} - E(\boldsymbol{X})\|) \tag{7.148}$$

Classical functional analysis provides many alternatives to the formulation of the 1st-order differential development. Assuming that the system model $G(.)$ is continuously differentiable (C^1) over the area of uncertain inputs, the approximation error is not only asymptotically decreasing to zero as in

$o(\|X - E(X)\|)$ but can also be bounded on the basis of the upper bounds of the gradient variation itself. This is clearly the case where the area of uncertain inputs is compact (e.g. bounded distributions over IR^p), but can also be done in a probabilistic version through a bounded restriction of unbounded distributions for a given confidence level. Hence, upper bounds may be derived for the variance.

In some other cases, functional analysis may issue meta-models of a system model for which certain bounds can be established: Janon, Nodet, and Prieur (2011), for instance, suggest a partial-order decomposition (POD) of the complete numerical solution of a system model governed by certain partial-derivative equations, whereby the error is controlled for a *L-inf* norm over the definition of the function:

$$\forall \boldsymbol{x} \in \boldsymbol{\Omega}, \quad \left| \boldsymbol{Z} - \sum_{l=1\ldots P} a^l(\boldsymbol{d}).\psi^l(\boldsymbol{x}) \right| \leq \varepsilon |\boldsymbol{x}| \tag{7.149}$$

Equivalent to the monotonous response surface, the bounds suggested by this type of work should then be transfered easily into any risk measure. More research needs to be undertaken in that area beyond this simple example.

7.5 Sensitivity analysis and importance ranking

Chapter 2 showed how important it is for most types of risk or uncertainty assessment to undertake sensitivity analysis (or importance ranking) as part of the process. Sensitivity analysis involves also uncertainty propagation but the output requires dedicated post-treatment as one not only looks for the risk measure itself, but also requires input-output quantities called *sensitivity indices* (*or importance measures*) that enable a deeper understanding of the input-output mapping. Sensitivity analysis is a vast field of applied science and will only be briefly introduced hereafter in close connection with the propagation methods and regularity issues reviewed earlier. The reader should refer to the specialised textbooks or literature, such as Saltelli *et al.*, (2004).

7.5.1 Elementary indices and importance measures and their equivalence in linear system models

Before getting into numerical indices, the use of graphical methods is generally recommended as they give powerful insights through simple techniques. A simple approach to start with is to generate a sample (typically at random) and to use *scatterplots* visualising the results by couples $(Z_j, X_j^i)_{j=1\ldots N}$ for each *i*-th uncertain input. Even though very simple, such an approach enables the detection of more than linear or monotonous dependence although it is limited somewhat regarding the interactions between inputs with respect to the output response. More elaborate graphical methods can be found in Saltelli *et al.*, (2004).

Flood example

Scatterplots of the flood level *vs.* each of the four uncertain inputs have been plotted for a 1000-sample of the hydro models in Figure 7.12 (left) through their original values and (right) after a prior rank transformation, as will be explained below. Though the joint variation of all inputs results in variability around the main trend, an increasing (resp. decreasing) behaviour as a function of q or z_v (resp. of k_s) can be guessed at graphically while it is much less conspicuous for z_m which shows graphically a low impact on the output of the model; in ranks, it is even tempting to conclude that there is quasi-independence between z_m and the output.

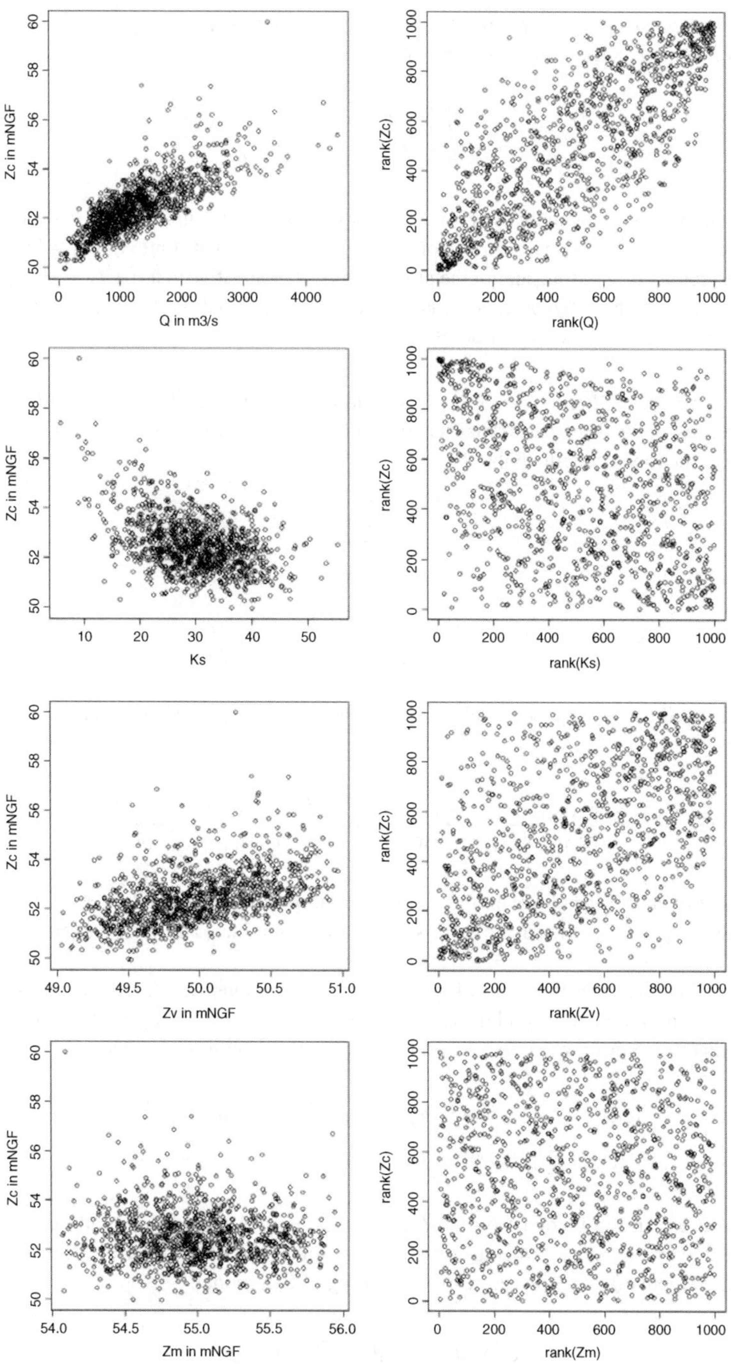

Figure 7.12 *Scatterplots in the flood example.*

Taylor-based linear sensitivity indices If the linear assumption is acceptable, the Taylor quadratic approximation generates the following elementary sensitivity indices:

$$s_i^{2\,Taylor} = \frac{\left(\frac{\partial G}{\partial x^i}\right)^2 \mathrm{var}(X^i)}{\sum_j \left(\frac{\partial G}{\partial x^j}\right)^2 \mathrm{var}(X^j)} = \frac{(G^i)^2 \mathrm{var}(X^i)}{\sum_j (G^j)^2 \mathrm{var}(X^j)} \tag{7.150}$$

Such elementary indices are widely used because they require few calculations of $G(.)$ and the formula is quite intuitive. A given uncertain input (also often called a *factor* in that context) is only important insofar as its own uncertainty $\sigma_{X^i}^2$ is strong and/or the physical model is sensitive to it (i.e. $G^i = \partial G/\partial x^i$ is high). As already mentioned, the base hypothesis is that uncertainties are small with respect to non-linearities of $z = G(x^1, \ldots x^p, \boldsymbol{d}) = G(\boldsymbol{x}, \boldsymbol{d})$ as a function of $\boldsymbol{x}$.

Monte-carlo sensitivity indices Monte-Carlo simulation also enables the determination of elementary sensitivity indices defined as the normalised ratio of input-output square linear (or Pearson) or rank correlation (or Spearman) coefficients:

$$s_i^{2\,Pearson} = \frac{\hat{Corr}(X^i, Z)^2}{\sum_{k=1\ldots p} \hat{Corr}(X^k, Z)^2} \tag{7.151}$$

$$s_i^{2\,Spearman} = \frac{\hat{Corr}(r_{X^i}, r_Z)^2}{\sum_{k=1\ldots p} \hat{Corr}(r_{X^k}, r_Z)^2} \tag{7.152}$$

Pearson normalised ratios are theoretically equal to the previous Taylor-based ones when the model is linear, since:

$$Corr(Z, X^i)^2 = \frac{\mathrm{cov}(X^i, Z)^2}{\mathrm{var}\,X^i \mathrm{var}\,Z} = \frac{(G^i)^2 \mathrm{var}\,X^i}{\mathrm{var}\,Z} = \frac{(\partial G/\partial x^i)^2 \mathrm{var}\,X^i}{\sum_k (\partial G/\partial x^k)^2 \mathrm{var}\,X^k} = s_i^{2\,Taylor} \tag{7.153}$$

Summing the previous formula over the $i = 1, \ldots p$ indices also shows that the square correlation coefficients sum up to 1 in the linear case, so that the denominator of the Pearson sensitivity index equals 1.

However, Monte-Carlo estimations of the correlation differ *empirically* from derivation-based ones due to sampling errors when the sampling size is low. Moving onto the cases when the model is not strictly linear, the indices also differ *theoretically* however large the sampling size since:

$$\begin{aligned} &\mathrm{cov}(X^i, Z) \neq \mathrm{cov}(X^i, \partial G/\partial x^i \,|_{x=EX} X^i) = (\partial G/\partial x^i)^2 \mathrm{var}\,X^i \\ &\mathrm{var}\,Z \neq \textstyle\sum_i (\partial G/\partial x^i)^2 \mathrm{var}\,X^i \end{aligned} \tag{7.154}$$

Thus, the linear expansion involved by the *Pearson coefficients* can only partially explain the output variance, a (random) residual U being left to incorporate the deviations from linearity. The linear correlation coefficients in that case correspond to a regression model:

$$Z = G(\boldsymbol{X}, \boldsymbol{d}) = b^o(\boldsymbol{d}) + \sum_{i=1\ldots p} b^i(\boldsymbol{d}) X^i + U \tag{7.155}$$

Slightly modifying the definition above of the sensitivity indices with a different normalisation, one obtains:

$$s_i^{2\,linreg} = corr\left(X^i, Z\right)^2 = \frac{b^i(\boldsymbol{d})^2 \operatorname{var} X^i}{\operatorname{var} Z} \tag{7.156}$$

One comes up with the following decomposition of variance which leaves a proportion of unexplained residual variance, notably incorporating the non-linear effects; the new input sensitivity indices sum up to a value of less than 1 and it is well-known as the R^2 coefficient or *correlation index*:

$$\begin{aligned} \operatorname{var}Z &= \sum_i b^i(\boldsymbol{d})^2 \operatorname{var} X^i + \operatorname{var} U = \operatorname{var} Z\left[\sum_i s_i^{2\,linreg} + \frac{\operatorname{var} U}{\operatorname{var} Z}\right] \\ 1 &= \sum_i s_i^{2\,linreg} + \left(1 - R^2\right) \end{aligned} \tag{7.157}$$

Note that the equality between correlation-defined and regression-coefficient-defined indices assumes that the residual is independent from the input factors; as it may no longer hold in the case of linearly-regressed non-linear models, it may become approximate. Additionally, the inclusion or not of an intercept (i.e. a bias in the residual) in the regression slightly modifies the variance decomposition in the non-linear case and thus the definition of the indices.

Spearman indices are more powerful as they do not to rely on system model linearity or on Gaussian distributions. As explained in Chapter 5, rank correlations rely instead on the monotonous dependence between variables, which means in the peculiar context of input-output correlations that the system model $G(\boldsymbol{x}, \boldsymbol{d})$ is required to be *monotonous* with respect to the $\boldsymbol{x}$ components instead of being linear. Moreover, rank correlation-based indices also require *additive* separation of input effects for the variance decomposition to work:

$$z = G(\boldsymbol{x}, \boldsymbol{d}) = G_0(\boldsymbol{d}) + \sum_{i=1}^{p} G^i\left(x^i, \boldsymbol{d}\right) \tag{7.158}$$

Thus, Spearman-based indices are appropriate for a much larger set of potential models than Taylor-based indices, though still excluding many non-monotonous or non-additive others (see below).

Coming onto sampling, the various definitions end up with slight variations between linear indices. Note also that the *number of MCS calculations* may now vary as a function of p for sensitivity analysis; this is due to the fact that the indices are based on $p + 1$ estimators of the coefficients of regression of Z as a function of X^i, plus that of the residual variance, on the common basis of only N vectors of *i.i.d.* samples $\left(Z_j, X_j^i\right)$. This important point is an introductory feature to the general area of *numerical design of experiment*. A (numerical) design of experiment may be seen indeed as a choice of a set of values for the input factors (here the matrix of sampled input vectors $\left(x_j^i\right)_{i=1\ldots p;j=1\ldots N}$) for which one wants the observation of the corresponding response (here the numerically-computed model output $(z_j)_{j=1\ldots N}$). As the costs generally increase with the size (N) of the design of experiment, it is desirable to make the most of it for understanding the response-factor relationship (here the input-output correlations). The challenge is to handle a large number of factors (p) as a given amount of robust information for response-factor relationship generally demands a rapidly increasing sample size as a function of the factors. For instance, deterministic designs of experiment involving the lower and upper ends of intervals of uncertainty increase the necessary sample size at a much faster *exponential* rate.

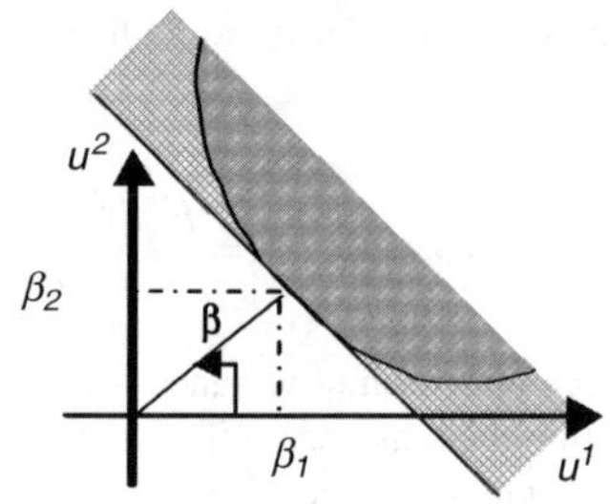

Figure 7.13 *FORM design point coordinates.*

The specialised literature (Kleijnen, 2007) should be referred to for in-depth presentation of that important domain, that involves recent development mixing deterministic and random designs (e.g. quasi-random sampling) or so-called 'supersaturated' designs for very high input dimensions.

Form sensitivity indices FORM also issues importance measures. Considering the coordinates of the design point $\boldsymbol{u}^* = (\beta^{1^*},\ldots,\beta^{p^*})$ it can be seen that their square values normalised by the square reliability index β^2 add up to 1 (Figure 7.13):

$$s_i^{2\,FORM} = \frac{(\beta^i*)^2}{\beta^2} = \cos(\alpha^i)^2$$

$$\|\boldsymbol{u}*\|^2 = \sum\nolimits_{i=1\ldots p}(\beta^i*)^2 = \beta^2 \Rightarrow \sum\nolimits_{i=1\ldots p} s_i^{2\,FORM} = 1 \tag{7.159}$$

Such quantities s_i^{2FORM} are the FORM indices: to understand this simply, consider the case where they are all close to zero except for the first input. This means that the design point is very close to the first axis as the quantity is also equal to the square cosine of the angle to such axis, hence close to zero. In such a case, this means that around the design point at least, the limit state function is parallel to all other axes, that is it does not depend much on the values of the other inputs. A rescaling of any of those would imply a displacement of the limit state along the corresponding axis (because of the transformation to standard space) with only little impact on β. FORM indices are interesting as they relate to behaviour close to the limit state, which should be more relevant when the risk measure is the exceedance probability. However, they assume the same limitations as the method itself.

Equivalence in the linearised Gaussian case Note that in the ideal linearised Gaussian case with independent uncertain inputs:

$$z = G(\boldsymbol{x},\boldsymbol{d}) = \sum_i G^i(\boldsymbol{d})x^i + G^0(\boldsymbol{d}) \tag{7.160}$$

these three categories of measures issue close results. We have already seen the equivalence of measures associated with linear Pearson correlations and with Taylor quadratic approximation. For Gaussian couples such as (Z,X^i), linear and rank correlations are linked theoretically through the following relation:

$$\rho_s = \frac{6}{\pi} Arc\sin\left(\frac{\rho}{2}\right) \tag{7.161}$$

It can be seen numerically that while the rank correlation of a Gaussian couple is always smaller or equal than its linear correlation, the difference never exceeds 5 %. Thus, sensitivity indices obtained by Taylor or Monte-Carlo (rank) correlation are similar up to that limited error (that is generally much less than non-linearities or statistical fluctuation of Monte-Carlo).

For FORM, transforming the physical inputs into standard inputs in such a case consists simply of normalising uncertainties through their standard variances. The associated limit state $G = z_s$ is plane, and its gradient $\partial_x G$ has constant components equal to $G^i\sqrt{\mathrm{var}(X^i)}$. As the design point is by definition the closest point of the limit state to the origin, the design point radius is also orthogonal to the limit state and hence the design point coordinates are proportional to those of the gradient:

$$\exists\mu\,\forall i = 1\ldots p, \quad \beta^{i*} = \mu G^i\sqrt{\mathrm{var}(X^i)} \tag{7.162}$$

Thus, the FORM sensitivity indices, given by those square components normalised to one, equal exactly the Taylor indices:

$$s_i^{2\,FORM} = \frac{(\beta^{i*})^2}{\sum_{i=1\ldots p}(\beta^{i*})^2} = \frac{(G^i)^2 \mathrm{var}\,X^i}{\sum_k (G^k)^2 \mathrm{var}\,X^k} = s_i^{2\,Taylor} \tag{7.163}$$

Flood example

Sensitivity indices for the variable of interest taken as the water level Z_c have been computed for the three methods, namely Taylor, Monte-Carlo Spearman indices (for 100, 1000 and 10^6 samples respectively – normalising by the sum of square correlations, and mentioning the correlation index) and FORM indices (with respect to the threshold $z_s = 55.5$, the 1/100-yr level), see Table 7.15.

Table 7.15 Sensitivity indices for the flood example, the case of water levels.

Inputs	Taylor	MC Spearman (100, 1000 and 10^6 samples)			FORM
Q	0.66	0.85	0.66	0.63	0.50
Ks	0.14	0.08	0.13	0.14	0.40
Zv	0.20	0.06	0.21	0.22	0.095
Zm	0.004	0.0005	0.0004	0.004	0.005
R^2	—	70 %	92 %	94 %	—

Note that Taylor required only five runs of the system model while MC requires obviously 1000 rather than only 100 samples to start stabilising although 100 samples are enough to understand the main contributors. MC Spearman also generates the correlation index of 94 % of the variance explained. While the rank order of sensitivities is similar, FORM indices increase greatly the importance of K_s to the detriment of both Q and Z_v: remember that FORM orders the sensitivities with respect to the exceedance probability, not to overall variance.

An alternative is to rank the inputs with respect to the total cost as being the variable of interest in the technical-economic system model, taking the 'no dike' scenario in Table 7.16.

Table 7.16 Sensitivity indices for the flood example, the case of total cost (in absence of dike).

Inputs	Taylor	MC Spearman (100, 10^3, 10^4, 10^5 samples)			
Q	0	0	0.33	0.47	0.49
K_s	0	0	0.48	0.42	0.41
Z_v	0	0	0.11	0.10	0.08
Z_m	0	0	0.03	0.005	0.007
C_m	0	0	0.04	0.001	$<10^{-4}$
R^2	—	0	5 %	5 %	5 %

For that 'pathological' variable of interest, which is quite asymmetric with zero mean and a fat tail, Taylor was already shown to be inappropriate; MC takes long to converge and results in a very low R^2 which is problemmatic; importance ranking is quite comparable to the 1/100-yr threshold-FORM results, not a surprising effect since the total cost behaves essentially as a threshold-exceedance index.

7.5.2 Sobol sensitivity indices

Sobol indices are the most powerful sensitivity tools with respect to a variance-based risk measure. Their limitations were shown in the previous section either regarding:

- The hypotheses: linearity, monotonicity or additivity of the system models were shown to be required in order to use Taylor-based or Spearman-simulated correlations properly.
- The coverage of the indices: so far the input-output relationship has only been investigated in a pairwise manner involved a single input, leaving aside the important question of interactions between inputs in explaining the output variance.

Both limitations can be resolved by the use of Sobol indices, which is very general though computationally challenging.

First-order indices Regarding the first concern, the hypotheses required, consider the following conditional expectation variance (also named correlation ratio or first-order Sobol index or importance measure):

$$S_i = \frac{\mathrm{var}[E(G(\boldsymbol{X}, \boldsymbol{d})|X^i)]}{\mathrm{var}\, G(\boldsymbol{X}, \boldsymbol{d})} \tag{7.164}$$

For any given value for the i-th uncertain input, $f_i(x^i) = E(G(\boldsymbol{X}, \boldsymbol{d})|x^i)$ represents the expected value for the uncertain output not knowing anything else about the system; such a function can equally be seen as manifesting the variation of the output depending only on the variation of the i-th input. Consequently, S_i represents the proportion of output variance explained by the i-th input variance. Indeed, the following elementary probabilistic theorem of total variance guarantees that such an index is comprised between 0 and 1:

$$\mathrm{var}\, Z = \mathrm{var}\, G(\boldsymbol{X}, \boldsymbol{d}) = \mathrm{var}\left[E\left(G(\boldsymbol{X}, \boldsymbol{d})\middle|X^i\right)\right] + E\left[\mathrm{var}\left(G(\boldsymbol{X}, \boldsymbol{d})\middle|X^i\right)\right] \tag{7.165}$$

Note the second term representing the expected value for the residual output variance when the i-th uncertain input is fixed $\mathrm{var}(G(\boldsymbol{X}, \boldsymbol{d})|x^i)$, that is the output variance not directly explained by that input. Such a definition does not involve anything about the model structure, presence of interactions with other variables and so on. It thus appears as an attractive measure of the (direct) impact of the i-th input on the output, provided that it can be computed efficiently. In the ideal linearised and Gaussian case with independent uncertain inputs, first-order Sobol indices simply coincide with the previous elementary indices:

$$s_i^{2\,Pearson} = \frac{(G^i)^2 \mathrm{var}\, X^i}{\mathrm{var}\, Z} = s_i^{2\,Taylor} = \frac{(\partial G/\partial X^i)^2 \mathrm{var}\, X^i}{\sum (\partial G/\partial X^k)^2 \mathrm{var}\, X^k} = \frac{\mathrm{var}(E(Z|X^i))}{\mathrm{var}\, Z} = S_i \tag{7.166}$$

In non-linear cases, the results are not closed-form anymore. Monte-Carlo sampling can be used through the following rearrangement:

$$\begin{aligned} S_i &= \frac{Var[E(Z \mid X_i)]}{\operatorname{var} Z} \\ &= \frac{E[E(Z \mid X_i)^2] - E[E(Z \mid X_i)]^2}{\operatorname{var} Z} = \frac{E\left[E(Z \mid X_i)^2\right] - E[Z]^2}{\operatorname{var} Z} \end{aligned} \tag{7.167}$$

Meaning that the estimation of the p first-order indices S_i involves that of EZ, of $var\,Z$ as well as of the p integrals denoted as $e_i = E\left[E(Z \mid X_i)^2\right]$. Brute-force Monte-Carlo could involve a double-loop Monte-Carlo costing $N_1{}^*N_2$ samples of the system model for each e_i:

$$\hat{e}_i = \frac{1}{N_1}\sum\nolimits_{j=1\ldots N_1}\left[\frac{1}{N_2}\sum\nolimits_{k=1\ldots N_2}\left[G\left(x^1_{j,k},\ldots,x^i_j,\ldots x^p_{j,k}\right)\right]\right]^2 \tag{7.168}$$

An elegant rearrangement suggested by Ishigami and Homma (1990) replaces such a computationally-explosive estimator by the following alternative costing only 2^*N samples of the system model for one index:

$$\hat{e}_i = \frac{1}{N-1}\sum_{k=1}^{N} G(\boldsymbol{x}_k, d)G(\boldsymbol{x}_k{}', d) \tag{7.169}$$

At each step k, two random and independent vectors $\boldsymbol{x}_k$ and $\boldsymbol{y}_k$ are sampled in the same pdf f_X, and $\boldsymbol{x}_k{}'$ is defined by reusing the i-th component of $\boldsymbol{x}_k$ as follows:

$$\boldsymbol{x}_k{}' = \left(y^1_k, \ldots y^{i-1}_k, x^i_k, y^{i+1}_k \ldots y^p_k\right) \tag{7.170}$$

Using the same twin samples with successive rearrangement of each input, the p first-order indices can be computed likewise at a total cost of $(p+1)^*N$ runs of the system model.

Higher-order Sobol indices First-order indices investigate a pairwise input-output relationship without looking at interactions. Consider the following quantity, called a second-order Sobol index:

$$\begin{aligned} S_{ij} &= \frac{V\left[E(Z|X_iX_j)\right]}{V(Z)} - Si - Sj \\ &= \frac{V\left[E(Z|X_iX_j)\right] - V[E(Z|X_i)] - V\left[E(Z|X_j)\right]}{V(Z)} \end{aligned} \tag{7.171}$$

The first term can be interpreted as the proportion of output variance depending only on (or explained by) the combined variation of the i-th and j-th inputs, while the second and third term were already interpreted as apportionating the proportion depending directly on each of the inputs considered separately. Interactions between those two factors are manifested in the fact that S_{ij} does not equal zero: the combined impact of both factors does not limit itself to that of each factor considered separately.

Indeed, a broader series of indices may be developed on the basis of a fundamental development of the system model. Sobol (1993) suggested that a system model (under certain integratability hypotheses) can be decomposed uniquely over a sum of properly-scaled and orthogonal functions of increasing levels of interaction. Similar to what was suggested for MRM methods, it is more convenient to work in a transformed uniform input space whereby each uncertain input $v^i = F_{xi}(x^i)$ is distributed according to a independent uniform distribution over [0,1] (for the sake of simplicity, inputs are assumed to be independent). Thus, the Sobol development goes as follows:

$$\begin{aligned} z &= G(\boldsymbol{x}, \boldsymbol{d}) = H(\boldsymbol{v}, \boldsymbol{d}) \\ &= H_0(\boldsymbol{d}) + \sum_{i=1}^{p} H_i(v^i, \boldsymbol{d}) + \sum_i \sum_{j>i} H_{ij}(v^i, v^j, \boldsymbol{d}) + \ldots + H_{1,2,\ldots,p}(v^1, v^2, \ldots, v^p, \boldsymbol{d}) \end{aligned} \quad (7.172)$$

Without further specification, there are infinite different ways to build such a decomposition of the system model into functions depending on growing sets of input components. However, this development is defined uniquely assuming additionally that the first constant term incorporates the expectation $H_o(\boldsymbol{d}) = EG(\boldsymbol{X},\boldsymbol{d})$ and that all other terms have null expectation over any of the input variables:

$$\begin{aligned} &\forall i_1, i_2 \ldots, i_k \quad \forall j \mid 1 \le j \le i_k \\ &E_{V^i}[H_{i_1\, i_2 \ldots i_k}(V^{i_1}, V^{i_2}, \ldots V^{i_k}, \boldsymbol{d})] = \int H_{i_1\, i_2 \ldots i_k}(v^{i_1}, v^{i_2}, \ldots v^{i_k}, \boldsymbol{d}) dv^{i_j} = 0 \end{aligned} \quad (7.173)$$

Thus, there is orthogonality of each term, that is any cross-product between different terms is of integral zero:

$$\begin{aligned} &E\left[H_{i_1 i_2 \ldots i_k}(V^{i_1}, V^{i_2}, \ldots V^{i_k}, \boldsymbol{d}) H_{j_1 j_2 \ldots j_k}(V^{j_1}, V^{j_2}, \ldots V^{j_k}, \boldsymbol{d})\right] \\ &= \int H_{i_1 i_2 \ldots i_k}(v^{i_1}, v^{i_2}, \ldots v^{i_k}, \boldsymbol{d}) H_{j_1 j_2 \ldots j_k}(v^{j_1}, v^{j_2}, \ldots v^{j_k}, \boldsymbol{d}) d\boldsymbol{v} = 0 \end{aligned} \quad (7.174)$$

The essential consequence being the following decomposition of total variance:

$$\begin{aligned} Var(Z) &= \operatorname{var} H(\boldsymbol{V}, \boldsymbol{d}) = \int \left(H(\boldsymbol{v}, \boldsymbol{d})^2 - H_o(\boldsymbol{d})^2\right) d\boldsymbol{v} \\ &= \int \left(\sum_{i=1}^{p} H_i(v^i, \boldsymbol{d}) + \sum_i \sum_{j>i} H_{ij}(v^i, v^j, \boldsymbol{d}) + \ldots + H_{1,2,\ldots,p}(v^1, v^2, \ldots, v^p, \boldsymbol{d})\right)^2 d\boldsymbol{v} \\ &= \int \left(\sum_{i=1}^{p} H_i(v^i, \boldsymbol{d})^2 + \sum_i \sum_{j>i} H_{ij}(v^i, v^j, \boldsymbol{d})^2 + \ldots + H_{1,2,\ldots,p}(v^1, v^2, \ldots, v^p, \boldsymbol{d})^2\right) d\boldsymbol{v} \\ &= \sum_{i=1}^{p} V_i + \sum_i \sum_j V_{ij} + \sum_i \sum_j \sum_k V_{ijk} \ldots + V_{1,2,\ldots,p} \end{aligned} \quad (7.175)$$

Whereby:

$$V_{i_1 i_2 \ldots i_k} = E\left[H_{i_1 i_2 \ldots i_k}(V^{i_1}, V^{i_2}, \ldots V^{i_k}, \boldsymbol{d})^2\right] = \operatorname{var} H_{i_1 i_2 \ldots i_k}(V^{i_1}, V^{i_2}, \ldots V^{i_k}, \boldsymbol{d}) \quad (7.176)$$

Each of these terms is closely related to the variance of different conditional expectations:

$$\begin{aligned} &Var[E(Z\,|\,X^i)] \\ &= Var[E(Z\,|\,V^i)] = Var[H_i(V^i, \underline{d})] = V_i \end{aligned} \quad (7.177)$$

Or:

$$\begin{aligned} &Var[E(Z\,|\,X^i, X^j)] \\ &= Var[E(Z\,|\,V^i, V^j)] = Var\left[H_i(V^i, \boldsymbol{d}) + H_j(V^j, \boldsymbol{d}) + H_{ij}(V^i, V^j, \boldsymbol{d})\right] = V_i + V_j + V_{ij} \end{aligned} \quad (7.178)$$

Thus, when defining the series of increasing-order Sobol indices defined as a generalisation of the above S_i and S_{ij}, the sum of all such indices appears to decompose likewise the total output variance:

$$Var(Z) = \mathrm{var}\, H(\boldsymbol{V}, \boldsymbol{d}) = \mathrm{var}\, H(\boldsymbol{V}, \boldsymbol{d}) \left[\frac{\sum_{i=1}^{p} V_i + \sum_i \sum_j V_{ij} + \sum_i \sum_j \sum_k V_{ijk} \ldots + V_{1,2,\ldots p}}{\mathrm{var}\, H(\boldsymbol{V}, \boldsymbol{d})} \right]$$

$$= \mathrm{var}\, H(\boldsymbol{V}, \boldsymbol{d}) \left[\sum_{i=1}^{p} S_i + \sum_i \sum_j S_{ij} + \sum_i \sum_j \sum_k S_{ijk} \ldots + S_{1,2,\ldots,p} \right] \quad (7.179)$$

Computing all such Sobol indices for each order results in a computationally-challenging task, though rearrangement such as that mentioned for first-order indices can be derived. Moreover, a rapidly-developing literature has suggested accelerated techniques, for example the FAST algorithm (Cukier *et al.*, 1978), meta-modelling and also chaos polynomials (Saltelli *et al.*, 2004).

7.5.3 Specificities of Boolean input/output events – importance measures in risk assessment

Implicitly, variables $\boldsymbol{X}$ and Z were assumed to be continuous: some specificity comes up when considering that either some of the uncertain inputs or outputs are uncertain events, hence Boolean (or more generally discretely-distributed) variables.

Assuming firstly that only $\boldsymbol{X}$ involves *discrete inputs* with a *continuous output* variable of interest Z, most of the measures introduced above still apply, save possibly FORM. Suppose x_i be Boolean with $P(X_i = 1) = p_i$. Graphical methods switch from bivariate scatterplots to a comparison of the distribution of outputs $Z \mid X_i = 1$ and $Z \mid X_i = 0$. Taylor, correlation or Sobol indices can be extended into an estimate of i-th input contribution to output variance, simplified as follows:

$$\left(\frac{\partial G}{\partial x^i}\right)^2 \mathrm{var}(X^i) = \left(G(EX^1, \ldots x^i = 1, \ldots EX^p) - G(EX^1, \ldots x^i = 0, \ldots EX^p)\right)^2 p^i(1 - p^i) \quad (7.180)$$

$$S_i = \frac{\mathrm{var}[E(G(\boldsymbol{X}, \boldsymbol{d})|X^i)]}{\mathrm{var}\, G(\boldsymbol{X}, \boldsymbol{d})} = \frac{(E(G(\boldsymbol{X}, \boldsymbol{d})|X^i = 1) - E(G(\boldsymbol{X}, \boldsymbol{d})|X^i = 0))^2 p^i(1 - p^i)}{\mathrm{var}\, G(\boldsymbol{X}, \boldsymbol{d})} \quad (7.181)$$

Taylor indices are improperly-defined for values of input events other than the i-th since expectation is meaningless: instead take either the *safe* state $x^j = 0$ or *failure* state for these. Sobol indices are more appropriate.

A more specific case is where the *output* is a (Boolean) *event of interest E*, typically the undesired system failure. Taylor indices then lose much of their meaning. Sobol indices remain appropriate, as follows:

$$S_i = \frac{\mathrm{var}[E(G(\boldsymbol{X}, \boldsymbol{d})|X^i)]}{\mathrm{var}G(\boldsymbol{X}, \boldsymbol{d})} = \frac{\mathrm{var}[P(E|X^i))]}{P(E)(1 - P(E))} \tag{7.182}$$

Such a case is particularly interesting when looking for *sensitivity indices* exploring the tail-associated risk measures as initiated by recent publications looking for the *exceedance probabilities* (Munoz-Zuniga *et al.*, 2011; Borgonovo, Tarantola, and de Rocquigny, 2011). It suffices to switch to the transformed output of interest $E = 1_{G(\boldsymbol{X},\boldsymbol{d}) > zs}$. The Sobol first-order index then intuitively compares the relative importance of the i-th input on the value of the exceedance probability, as already shown when studying the conditional Monte-Carlo method (Section 7.3.4.3).

When the i-th input is also Boolean with $P(X_i = 1) = p_i$ and when additionally assuming that failures are rare, the index becomes:

$$\begin{aligned} S_i &= \frac{\mathrm{var}[P(E|X^i))]}{P(E)(1 - P(E))} = \frac{p^i(1 - p^i)}{P(E)(1 - P(E))}[P(E|X^i = 1) - P(E|X^i = 0)]^2 \\ &\approx \frac{p^i}{P(E)}[P(E|X^i = 1) - P(E|X^i = 0)]^2 \end{aligned} \tag{7.183}$$

In other words, it compares the impact of the state of the i-th input component (or initiating event) to the output system failure probability, a *squared difference* normalised by the ratio of i-th component to system failure probability.

The risk analysis literature has developed a number of importance measures involving the same conditional output failures with various normalisations, as in the following brief review with strong reference to Zio (2009). Considering a Boolean system model $E = G(\boldsymbol{X},\boldsymbol{d})$ with (independent) input components and failure probabilities $(P(X_k = 1) = p_k)_{k=1\ldots p}$ over a given time of operation, the Birnbaum measure I^B can be expressed equivalently as the partial derivative of system failure probability with respect to component failure probability or the *difference* between conditional system failure probabilities:

$$I_j^B = \frac{\partial P\left(E; (p^k)_{k=1\ldots p}\right)}{\partial p^i} = \left[P(E|X^i = 1) - P(E|X^i = 0)\right] \tag{7.184}$$

In other words, assume that the i-th component fails, how much likelier becomes the system failure? The criticality importance measure I^{cr} normalises this difference with the ratio of input/output failure probabilities:

$$I_j^{cr} = \frac{p^i}{P(E)}\left[P(E|X^i = 1) - P(E|X^i = 0)\right] \tag{7.185}$$

It can also be proved to correspond to the probability that i-th component failure 'caused' system failure. Eventually, a normalisation using ratios – % higher or lower risk given that the i-th component fails or is kept safe – comes up with the *risk achievement worth* or *risk reduction worth*, respectively:

$$RAW_j = \frac{P(E|X^i = 1)}{P(E)}; \quad RRW_j = \frac{P(E)}{P(E|X^i = 0)} \tag{7.186}$$

Note that these measures have their limitations; they do not include explicitly the group behaviour of input components or design changes; they do not consider the accessible extent of failure probability change (generally much smaller than 0 % to 100 %), nor level-2 uncertainty and so on.

7.5.4 Concluding remarks and further research

Note that the Sobol development as well as all sensitivity indices, whatever the method employed, are directly dependent both on:

- the fixed variables $\boldsymbol{d}$ (as stated explicitly in the notation $H_{ijk}(v^i,v^j,v^k,\boldsymbol{d})$ in the Sobol case);
- and the choice of input distributions (less explicit, as being hidden in the prior transformation into the uniform variables v^i).

In other words, all computations need to be resumed whenever changing one of these. It is generally recommended to start with computationally-simple sensitivity methods in upstream steps where information regarding distributions or design choices is loose.

A classical distinction is made regularly in the literature on sensitivity analysis between:

- *local* sensitivity in the sense that only part of the input-output distribution is examined, for example in the neighbourhood of the origin, or in relation to the neighbourhood of a threshold exceedance (Taylor, FORM importance factors);
- *global* sensitivity in relation to the entire distribution (e.g. Sobol indices, or rank correlations in some cases).

Such a distinction, effective outside the *linear* Gaussian case, does not exactly cover the distinction that has to be made according to the quantity of interest with respect to which the sensitivity indices are defined; or in other words the risk measure with respect to which the uncertain inputs are ranked. All of the above-mentioned indices save the FORM indices were defined with respect to output variance: in fact, most of the sensitivity analysis literature concentrates, although often implicitly, on variance-based quantities of interest (de Rocquigny *et al.*, 2009). Risk assessment or industrial design commands a greater development of importance measures dedicated to distribution tails and threshold-exceedance events representing compliance with risk budgets or technical specifications.

Indices and measures are more delicate to specify in the case of *dependent* uncertain inputs; a meaningful sensitivity analysis may require that all the variables are partitioned into sub-groups that are independent of each other, or a re-parameterisation of the input vector through an inner model expressing it as a phenomenological function of hidden independent factors that could be guided by phenomenological knowledge (as suggested in Chapter 5, Section 5.3.5 for uncertainty modelling of physical dependence). The reader is referred to Saltelli *et al.* (2004) for further reading on modern sensitivity analysis.

7.6 Numerical challenges, distributed computing and use of direct or adjoint differentiation of codes

Risk and uncertainty assessments will inevitably lead to a number of runs of the system model that is much larger than for the traditional 'best-estimate' study (a single 'penalised' calculation). We have seen how, in a strong deterministic paradigm, the maximisation of the response for uncertain areas implies numerous optimisation calculations; in a simple probabilistic paradigm, even with accelerated methods, several dozens or hundreds of calculations are necessary at least. Even if the response surfaces become vital in these last three cases, the following are even greedier ($>10^3$ to 10^5 calculations):

- the 'mixed deterministic-probabilistic' paradigm, which nests a maximisation by intervals for the deterministic components with, for each point, a conditional probabilistic calculation for the probabilised variables;
- optimisation under uncertainty;
- inverse modelling of sources of uncertainty, likewise nesting optimisation and propagation in the general case.

Aside from the optimisation of the code solvers themselves and their parallelisation, the numerical challenge, depending on the propagation methods adopted, can benefit from:

- *High Performance Computing* (see Chapter 8, Section 8.4), and particularly *massive distributed computing*, which is immediate in certain methods such as MCS, but which probably requires optimisation for other methods (Berthou *et al.*, 2009).
- The *use of differentiated codes*, when propagation is applied to gradients (in the case of deterministic methods, Taylor quadratic approximation, FORM-SORM or derivations). While it remains always possible to use finite differences, one can substitute the coded derivatives $(\partial G/\partial x^i)_{i=1,\dots p}$ which are faster and/or more stable according to circumstances. However, that presupposes a numerical development effort specific to the coding of the physical system being studied, which will be more or less automatable depending on the complexity of its architecture (see Faure, 1996).

Note that the use of direct or adjunct modes of differentiation (excluding problems of numerical stability) theoretically provides the same final result, but each of the modes is more or less optimal as regards calculation time, depending on the ratio between the dimension of the inputs $\boldsymbol{X}$ compared to that of the intermediate outputs $\boldsymbol{Y}$ of an implied physical code for calculating the output variable of interest Z. Furthermore, the derivatives, which are new responses from the code, require a specific additional resolution: the algorithm may be coded 'by hand', or may rely on automatic differentiation algorithms (e.g. Tapenade, see Hascouet and Pascual, 2004) being restricted to sufficiently simple sub-sections of the major industrial codes.

Exercises

There follows a list of exercises referring to the concepts developed in this chapter:

7.1 Compute the 2nd-order and higher order corrections of Taylor approximation on the mean and variance of water level or costs. How significant are they? Why?

7.2 Study numerically the Wilks number as a function of α, β and of the Wilks order. Can there be closed-form approximations for large N at low orders? What about the case of large N at high orders?

7.3 Compute the Wilks estimates for the correlated flood case and discuss the changes with respect to the independent case.

7.4 Use the Wilks method for the level-2 quantiles of an exceedance probability. How does its overestimation compare with the additional margin introduced by a level-2 confidence interval (e.g. 70 %) on epistemic uncertainty?

7.5 Compute the level-2 quantiles with level-1 correlation included.

7.6 Compute the level-2 quantiles including the system reliability component.

7.7 Develop bounds for variance or quantiles in the case of monotonous system models with bounded-dervatives using the Taylor approximation. How conservative is it in the flood example?

7.8 Study the crossing between the type of input correlation structures and the signs of input-output monotony that keep the model monotonous after MRM transformation into a uniform space (see also Exercise 4.1). What about the case of linearly correlated inputs with all positive coefficients and input-output relationships all increasing with equal signs? In such a case, can we prove that increasing the correlation coefficients always increases (or decreases) some quantities of interest such as output variance, quantile or exceedance probability?

7.9 Study the evolution of FORM sensitivity indices with higher thresholds (i.e. lower exceedance probabilities): how does this compare with regression-based indices on the transformed output of interest $Z^* = 1_{G(\boldsymbol{X},\boldsymbol{d})>zs}$?

7.10 Is there a relationship between such importance measures recalled for the Boolean cases and the level-2 sensitivity indices defined when considering the system model as a function relating the input/output failure frequencies with input uncertainty in the elementary event frequencies?

7.11 Compute the sensitivity indices of level-2 uncertainty in the flood example. What is the most appropriate level-2 propagation method for that purpose? How do those sensitivity indices get impacted by the reduction of level-2 uncertainty, such as 10 times more data on flows (bootstrap it to check)?

7.12 Study numerically the speed of convergence of the Monte-Carlo sampled linear or rank sensitivity indices as a function of *p*, the dimension of input factors. Can a theoretical speed be established for the Gaussian linear case through mathematical results of regression theory? What would most influence the growth of the minimal number of runs *N* for a given input dimension *p* (non-linearity, interactions ...)?

References

Andrianov, G., Burriel, S., Cambier, S. *et al.* (2007) Open TURNS: An open source initiative to treat risk, uncertainties and statistics in a structured industrial approach. Proc. of ESREL'07, Stavanger, Norway.

Ardillon, E. and de Rocquigny, E. (2003) Probabilistic estimation of safety quantiles for complex non-regular physical systems with uncertain parameters. Proc. 25th ESReDA Seminar. Paris.

Berthou, J.-Y., Hamelin, J.-F. and de Rocquigny, E. (2009), XXL Simulation for XXIst Century Power Systems Operation, *International Journal of High Performance Computing Applications* vol. **23**(4), pp. 361–365.

Box, G.E.P. and Draper, N.R. (1987) *Empirical Model Building and Response Surface*, Wiley Series in Probability and Mathematical statistics, J. Wiley & Sons.

Borgonovo, E., Tarantola, S. and de Rocquigny, E. (2011) Importance Measures for Seismic Probabilistic Fragility Assessment. Submitted to *Risk Analysis*.

Boyack, B.E. (1990) Quantifying reactor safety margins-part I: An overview of the code scaling, applicability and uncertainty evaluation methodology. *Nuclear Engineering and Design*, **119**, 1–15.

Bratley, P. and Fox, B.L. (1988) Algorithm 659 implementing sobol's quasirandom sequence generator. *ACM Transactions on Mathematical Software*, **14**, 88–100.

Cacuci, D.G. *et al.* (1980) Sensitivity theory for general systems of nonlinear equations. *Nuclear Science and Engineering*, **75**.

Cannamela, C., Garnier, J. and Iooss, B. (2008) Controlled stratification for quantile estimation. *Annals of Applied Statistics*, **2**, 1554–1580.

Chia, C.-Y. and Ayyub, B.-M. (1994) *Conditional Sampling for Simulation-Based Structural Reliability Assessment, Structural Safety & Reliability*, Schueller, Schonzuka and Yao (eds) Balkema, Rotterdam.

Chick, S.E. (2000) Bayesian methods for simulation. Proc. of the 2000 Winter Simulation Conference, Washington D.C.

Cukier, R.I., Levine, H.B. and Shuler, K.E. (1978). Nonlinear sensitivity analysis of multiparameter model systems, *Journal of Computational Physics*, **26**, 1–42.

de Rocquigny, E., Chevalier, Y., Turato, S. and Meister, E. (2004) Probabilistic assessments of the reactor pressure vessel structural integrity: Direct coupling between probabilistic and finite-element codes to model sensitivity to key thermo-hydraulic variability. Proc. of ASME ICONE12.

de Rocquigny, E. (2005a) A statistical approach to control conservatism of robust uncertainty propagation methods; application to accidental thermal hydraulics calculations. *Proceedings of ESREL-05*, Tri City, Poland.

de Rocquigny, E. (2005b) Couplage mécano-probabiliste pour la fiabilité des structures – un cas industriel où la robustesse d'une surface de réponse est démontrable. Actes du 17ème Congrès Français de Mécanique, Troyes.

de Rocquigny E. (2009), Structural reliability under Monotony: Properties of Form, simulation or response surface methods and a new class of monotonous reliability methods (MRM), *Structural Safety* vol. **31**, pp. 363–374.

Der Kiureghian, A. and Ke, J.-B. (1988), The stochastic finite element method in structural reliability, *Prob. Eng. Mech.*, **3**(2), 83–91.

Ditlevsen, O. and Madsen, H.O. (1996) *Structural reliability Methods*, John Wiley & Sons.

Faure, C. (1996) Splitting of algebraic expressions for automatic differentiation. Proc. of the 2nd International Workshop on Computational Differentiation, 12–15 fév. 1996, Santa Fe.

Fishman G.S. (1996) *Monte Carlo: Concepts, Algorithms, and Applications*. New York, NY: Springer-Verlag.

Frey, H.C. and Rhodes, D.S. (2005) Quantitative analysis of variability and uncertainty in with known measurement error: Methodology and case study. *Risk Analysis*, **25**(3).

Helton, J.C. (1993) Uncertainty and sensitivity analysis techniques for use in performance assessment for radioactive waste disposal. *Reliability Engineering and System Safety*, **42**, 327–367.

Helton, J.C. and Davis, F.J. (2003) Latin hypercube sampling and the propagation of uncertainty in analyses of complex systems. *Reliability Engineering and System Safety*, **81**, 23–69.

Hascouet, L. and Pascual, V., TAPENADE 2.1 user's guide, INRIA report (2004) www.inria.fr/rrrt/rt-0300.html.

Homma, T. and Saltelli, A. (1996) Importance measures in global sensitivity analysis of non-linear models. *Reliability Engineering and System Safety*, **52**, 1–17.

Ishigami, T. and Homma, T. (1990) An importance quantication technique in uncertainty analysis for computer models. In *Proceedings of the ISUMA '90 First International Symposium on Uncertianty Modelling and Analysis*, 398–403.

Iooss, I., Boussouf, L., Feuillard, F. and Marrel, A. (2010) Numerical studies of the metamodel fitting and validation processes. *International Journal of Advances in Systems and Measurements*, **3**, 11–21.

ISO (1995) Guide to the expression of uncertainty in measurement (G.U.M.), EUROPEAN PRESTANDARD ENV 13005.

Janon, A., Nodet, M. and Prieur, C. (2011) Certified reduced-basis solutions of viscous Burgers equation parametrized by initial and boundary values, arXiv:1010.1761v1.

Kaymaz, I. and McMahon, CA. (2005) A response surface based on weighted regression for structural reliability analysis. *Problems in Mechanical Engineering*, **20**, 11–17.

Kleijnen, JPC. and Sargent, RG. (2000) A methodology for fitting and validating metamodels in simulation. *European Journal of Operational Research*, **120**, 14–29.

Kleijnen, J.P.C. (2007) *Design and Analysis of Simulation Experiments*, International Series in Operations Research & Management Science, Springer.

Limbourg, P., de Rocquigny E. and Andrianov G. (2010) Accelerated uncertainty propagation in two-level probabilistic studies under monotony. *Reliability Engineering and System Safety*, **95**, 998–1010.

Madsen, H.O., Kenk, S. and Lind, N.C. (1986) *Methods of Structural Safety*, Prentice-Hall Inc.

Marrel, A., Iooss, B., Van Dorpe, F. and Volkova, E. (2008) An efficient methodology for modeling complex computer codes with Gaussian processes. *Computational Statistics and Data Analysis*, **52**, 4731–4744.

Metropolis, N. and Ulam, S. (1949) The Monte-Carlo method. *Journal of the American Statistical Association*, **44**, 335–341.

McKay, M.D., Beckman, R.J. and Conover, WJ. (1979) A comparison of three methods for selecting values of input variables in the analysis of output from a computer code. *Technometrics*, **21**(2), 239–245.

Melchers, RE. (1999) *Structural Reliability Analysis and Prediction*, 2nd edn, J. Wiley & Sons.

Munoz-Zuniga, M., Garnier, J., Remy, E. and de Rocquigny, E. (2011) Adaptive directional stratification for controlled estimation of the probability of a rare event. *Reliability Engineering and System Safety*, **96**(12), 1691–1712.

Myers, RH. (1971) *Response Surface Methodology*, Allyn and Bacon, Inc., Boston.

Nelsen, RB. (1999) *An Introduction to Copulas*, Springer.

Nutt, WT. and Wallis, GB. (2004) Evaluation of safety from the outputs of computer codes in the presence of uncertainties. *Reliability Engineering and System Safety*, **83**, 57–77.

Olivi, L. (1984) *Response Surface Methodology*, Handbook for Nuclear Safety Analysis, J.R.C. European Commission.

Rackwitz, R. (2001) Reliability analysis – a review and some perspectives. *Structural Safety*, **23**, 365–395.

Robert, C. and Casella, G. (1999) *Monte-Carlo Statistical Methods*, Springer-Verlag.

Rubinstein, RY. (1981) *Simulation and the Monte-Carlo Method*, John Wiley & Sons.

Saltelli, A., Tarantola, S., Campalongo, F., Ratto, M. (2004) *Sensitivity Analysis in Practice: A Guide to Assessing Scientific Models*, Wiley.

Saltelli, A., Ratto, M., Andres, T. *et al.* (2008) *Global Sensitivity Analysis: The Primer*, Wiley.

Sobol, I.M. (1993) Sensitivity estimates for non-linear mathematical models, Mathematical Modelling and Computational Experiments.

Soize, C. and Ghanem, R. (2004) Physical systems with random uncertainties: chaos representations with arbitrary probability measure. *SIAM J. Sci. Comput.* **26**(2), 395–410.

Sudret, B. (2008) Global sensitivity analysis using polynomial chaos expansions. *Reliab. Eng. Syst. Safe.*, **93**, pp. 964–979.

Takeuchi, K. and Nissley, M.E. (2000a) Uncertainty evaluation of global model combined with local hot spot. Response Surface in the WCOBRA/TRAC BE Method & 2000b Validation of the WCOBRA/TRAC BE LOCA Method. Proc. of the Intern. Meeting on 'Best-Estimate' Methods in Nuclear Installation Safety Analysis, Washington D.C.

Wilks, S.S. (1941) Determination of sample sizes for setting tolerance limits. *Annals of Mathematical Statistics*, 12.

Zio, E. (2009) Computational methods for reliability and risk analysis. Series on Quality, Reliability and Engineering Statistics, Vol. **14**, World Scientific.

8

Optimising under uncertainty: Economics and computational challenges

Bringing costs inside the engineering models is essential from a decision-making perspective, particularly when the final goal is 'Select'. As will be discussed later in this chapter, engineering economics share mathematical similarities with phenomenological or physical modelling, thus opening up an interesting continuum between engineering risk and reliability considerations on the one hand, and economic, financial modelling or actuarial science on the other. Adding then an optimisation layer on top of uncertainty models brings another dimension of CPU complexity to real physical models: design reliability algorithms or adaptive response surface are being developed, but robustness of estimation remains difficult. This chapter does not introduce in depth the economics behind utility theory or non-linear rank transformed utilities, but rather offers some thoughts about the implementation of the computational challenges associated with the embedding of large physical models inside technico-economics. The role of time inside the formulation of the risk measure is discussed, as well as an extension to dynamic decision-making with the real options approach. The chapter concludes with a perspective on the promise of High Performance Computing for modelling in the context of risk and uncertainty.

8.1 Getting the costs inside risk modelling – from engineering economics to financial modelling

8.1.1 Moving to costs as output variables of interest – elementary engineering economics

Taking costs into account in risk assessment In risk assessment stages, system models are generally studied through variables of interest characterising their technical performance such as, for the flood

Modelling Under Risk and Uncertainty: An Introduction to Statistical, Phenomenological and Computational Methods, First Edition. Etienne de Rocquigny.

example, the overflow or plant failure. In decision stages, it is often useful to add a cost component to that type assessment through a comprehensive investment + detriment cost figure in order to help answer a fundamental question:

> 'What actions (system reliable design, protection design, maintenance) should be taken to best balance costs (& benefits) in a risky/uncertain future?' (Question 1)

Mathematically, with actions still being represented by vector $\boldsymbol{d}$, suppose that *technical* performance at the future time of interest t_f is embedded inside the uncertain variables of interest (noted $\boldsymbol{Z}^t$) being outputs of the technical system model. As was introduced in the flood example in Chapter 3, the engineering cost components generally comprise:

- *Controlled* (investment) costs $c_i(\boldsymbol{d})$: typically the cost of preventive actions, that depend on $\boldsymbol{d}$ while not (or not much) on the performance v.i. $\boldsymbol{Z}^t$. Think of the dike investment, or of the plant protection system.
- *Uncertain* (detrimental) costs $c_d(\boldsymbol{d}, z)$: typically the costs implied by a detrimental technical performance, such as damage costs associated with undesired events, repair, and so on. They depend on both $\boldsymbol{d}$ and $\boldsymbol{Z}^t$. They are often also intrinsically uncertain, even for a known technical variable of interest z^t. Think of the lack of knowledge of the costs that will be involved for a given flood event in an industrial area. These purely economic uncertain factors will be denoted as $\boldsymbol{x}^e$, or c_m in the flood example.

Note that the previous classification is somewhat schematic as the investment costs associated with engineering actions may also often be partially uncertain, due to delays or construction failures, for instance. Note also that those *costs* could be complemented by a view on the potential *benefits*, be they controlled or uncertain. While the inventory of the latter is quite essential and sometimes too much ignored in the 'pessimistic' context of risk analysis, benefits can simply be considered as 'negative costs' from a mathematical perspective. Thus, the word 'costs' will mostly be retained below.

From the point of view of the analyst, the consideration of these new variables may be seen as through the plugging in of another economic model downstream of the technical system model, as sketched in Figure 8.1.

Thus, a combined system model describes the impact of the choice of actions $\boldsymbol{d}$ on the economic variable of interest:

$$c_c = c_i(\boldsymbol{d}) + c_d(z^t, \boldsymbol{x}^e, \boldsymbol{d}) = G_e(z^t, \boldsymbol{x}^e, \boldsymbol{d}) = G_e o G_t(\boldsymbol{x}^t, \boldsymbol{x}^e, \boldsymbol{d}) \tag{8.1}$$

Minimising the expected costs Answering *Question 1* consists of choosing a value for vector $\boldsymbol{d}$ that will minimise *in a certain sense* the total cost c_c. The total cost figure is an uncertain variable of interest: there is a need to choose a certain risk measure or quantity of interest in order to give sense to that minimisation. The simplest choice is to minimise the total cost expectation, an elementary risk measure in engineering economics, as follows:

$$EC_c = E[c_i(\boldsymbol{d}) + c_d(\boldsymbol{Z}^t, \boldsymbol{X}^e, \boldsymbol{d})] \tag{8.2}$$

$$\boldsymbol{d}_{opt} = Arg\ \min_{\boldsymbol{d}} EC_c \tag{8.3}$$

As compared to traditional engineering economics neglecting risk or uncertainty, the cost expectation can be simply interpreted as follows. Uncertainty leads to considering a variety of technical performance

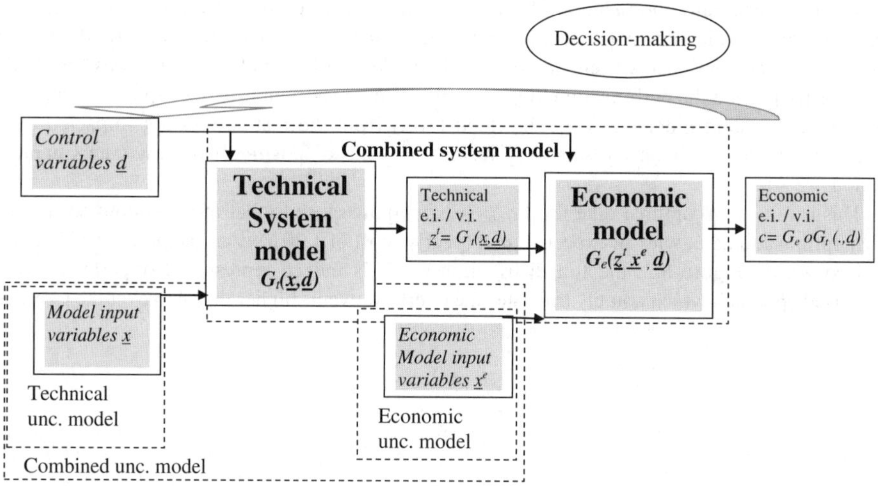

Figure 8.1 *Compounding economic and technical system models in engineering economics under risk.*

outcomes ($\boldsymbol{Z}^t$ follows a discrete distribution), the likelihood of each of which varies according to design actions. Instead of minimising a single-point computation of the total cost $c_c = c_i(\boldsymbol{d}) + c_d(z^t, \boldsymbol{x}^e, \boldsymbol{d})$ corresponding to supposedly well-known vectors of performance z^t and economic factors $\boldsymbol{x}^e$, the new figure averages costs over all possible scenarios of performance and economic factors $(z^t(k), \boldsymbol{x}^e(k))$ weighed by their likelihood:

$$EC_c = E[c_i(\boldsymbol{d}) + c_d(\boldsymbol{Z}^t, \boldsymbol{X}^e, \boldsymbol{d})] = c_i(\boldsymbol{d}) + \sum_k c_d(z^t(k), \boldsymbol{x}^e(k), \boldsymbol{d})P(z^t(k), \boldsymbol{x}^e(k)|\boldsymbol{d}) \tag{8.4}$$

Flood example

In the flood example, technical uncertain performance refers to the water level and the potential overflow above the dike, while the damage costs include additional purely economic uncertainty: a coefficient uniformly distributed over [50–150 %]. A deterministic cost optimisation is rather uneasy as the building of a dike does make sense only to cover rare and large damage. On the basis of best-estimate values for performance, that is the average water level, the answer is obviously 'no dike'; while assuming the worst case for the technical input parameters and an average economic coefficient results in a dike of 4.2 m (with a fixed cost of 0.8 M€/yr), just 0.1 m above the 'worst-case based technical design' (see Step Two in Chapter 4). A deterministic design is completely imposed by the technical assumptions, without much room for technical-economic trade-off.

On the contrary, the probabilistic uncertainty can be schematised over a limited number of scenarios. Section 7.2.3 showed that the distribution of water levels over riverbank Z_c-z_b is approximately (*0m at ~99 %; 1m at ~99,7 %; 2m at ~99,9 %; 3m at 99,95 %*). This can be schematised through five discrete events crossed independently with three equiprobable modalities for the economic factor (1/3 chance to be either at 50 %, 100 % or 150 %). Optimising the discretised expectation results in a dike of 3.5 m with an expected total (yearly) cost of 0,68 M€/yr (including 0,60 M€/yr of fixed costs). Note that this approach results in a lower dike than the deterministic design, the corresponding fixed costs of which were already higher than the expected total cost in the probabilistic approach, even without adding the reisudal expected damage costs.

The more accurate optimisation of the expected costs based on the thorough distribution of technical outcomes is more expensive to compute. At first hand, it requires as many Monte-Carlo samplings of the combined system and economic model as of a number of iterations (i.e. of dike values tested) inside the optimisation loop, though a simple shortcut is to decouple the sampling of the technical system model ($Z^t = Z_c$, the flood level) from the rest as it is not impacted by the design choice and to intensify sampling inside the zone of positive overflow (see Chapter 7, Section 7.3.4).

This results in an optimal dike level $d_{opt} \sim 3.5m$, quite close to that of the simplified scenario-based optimisation, but with an expected total (yearly) cost of 1.20 M€/yr (including 0,60 M€/yr of fixed costs), cf. Figure 8.2: the full distribution of Z_c is indeed represented imperfectly by the discretised approach which notably truncates the overflow events higher than 4 m over the riverbank.

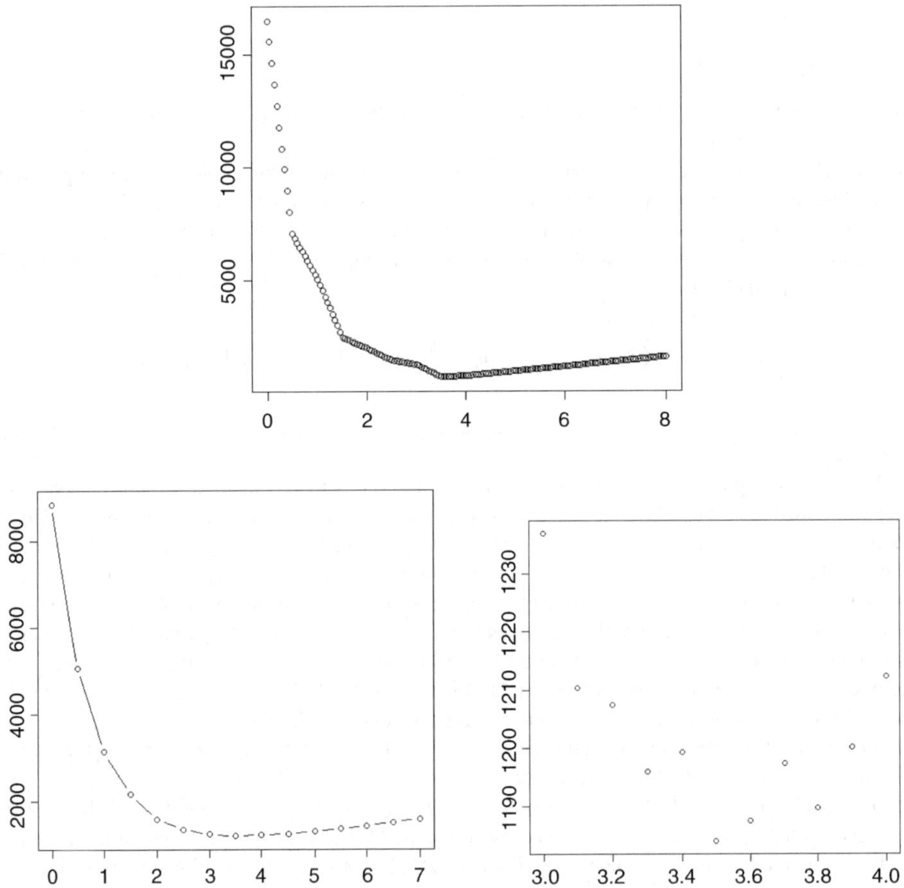

Figure 8.2 *The total expected yearly costs (in kEUR) as a function of dike height in a discretised probabilistic scenario approach (up) or full probabilistic expectation (down left and zoom on the right) after 10 000 samples for each (±0.3 % residual error).*

In spite of the preliminary variance reduction techniques applied to the simulation, dedicated techniques of stochastic optimisation would be needed so as to optimise finely because of the residual sampling noise of ±0.3 % (see Figure 8.2 – bottom right).

The limitations of expected costs as a risk measure Note the limitations of the expected cost optimisation. While it looks rather intuitive when dealing with moderate variations around the average cost, it may be seen as a somewhat risky attitude when rare and serious events are included. Consider the flood example and the distribution of cost.

Flood example
The raw cost expectation is minimal for ~3.5 m of dike. However, this leaves a yearly frequency of $0.3\ 10^{-4}$/yr (~1/3000-yr return-period that is ~1 % chance to occur over 30 years) to incur 0.5 m overflow with the associated consequences of half-ruining the dike and ~150 M€ damage. An additional investment of say 10 M€ (resulting in a yearly fixed cost of 1 M€/yr instead of 0.6 M€/yr) would have allowed for a dike 2 m higher, reducing the likelihood of the same damage by half, down to ~1/6000-yr return period. Though not optimal in the strict sense of expected value, many a decision maker would possibly prefer to add an extra 30 % investment to reduce such a catastrophic and costly event, although rare enough.

Beyond the simplest risk measure of expected cost and benefits, the economic theory of risk and uncertainty has introduced two types of sophistication:

(i) expected utility, replacing expected cost,
(ii) expected preference, including non-linear transformation of the utility distribution.

The abundant references in the literature should be consulted in order to obtain the details and rationale for choosing one of these concepts, starting with the seminal papers of Van Neumann and Morgenstern (1944) and Quiggin (1982). This book being centred on the statistical and numerical modelling aspects, the subsequent comments will be limited to a brief summary in subsequent Sections.

8.1.2 Costs of uncertainty and the value of information

Once cost functions have been made available so that the variable of interest becomes a total cost, fruitful additional quantities of interest may be computed besides the minimal expected cost in order to measure the 'cost of uncertainty' and/or the decision-making value of different attitudes towards uncertainty. Henrion (1982) and Henrion (1989) discussed two measures that compare the minimal expected costs under uncertainty $EZ(\boldsymbol{d}_{opt})$ to alternative informational quantities:

- The Expected Value of Perfect Information (EVPI) valuing the costs that would be saved in average if actions were optimised under perfect information on the system state. In other words, how much could be saved through a thorough reduction of uncertainty?
- The Expected Value of Ignoring Uncertainty (EVIU) valuing the costs incurred in choosing actions $\boldsymbol{d}_{be}$ that minimise the costs for the best-estimate guess of the system state $\boldsymbol{x}_{be}$. In other words, how much can be saved through the undertaking of uncertainty modelling in the present state of knowledge?

Each of these measures involves the consideration of all uncertain inputs involved in the overall cost function. We will consider formally the extended system model and uncertain input vector as follows:

$$\begin{aligned} \boldsymbol{X} &= (\boldsymbol{X}^t, \boldsymbol{X}^e) \\ C_c &= G_e o G_t(\boldsymbol{X}^t, \boldsymbol{X}^e, \boldsymbol{d}) = C(\boldsymbol{X}, \boldsymbol{d}) \end{aligned} \tag{8.5}$$

Mathematically, they appear as follows:

$$\begin{aligned} & EVPI(f_X) \\ & = Min_{\boldsymbol{d}} E_{f_X}[C(\boldsymbol{X}, \boldsymbol{d})] - E_{f_X}[Min_{\boldsymbol{d}}(C(\boldsymbol{X}, \boldsymbol{d})|\boldsymbol{X})] \\ & = E_{f_X}[C(\boldsymbol{X}, \boldsymbol{d}_{opt})] - E_{f_X}[Min_{\boldsymbol{d}}(C(\boldsymbol{X}, \boldsymbol{d})|\boldsymbol{X})] \end{aligned} \tag{8.6}$$

and:

$$\begin{aligned} & EVIU(f_X, \boldsymbol{x}_{be}) \\ & = E_{f_X}[C(\boldsymbol{X}, \boldsymbol{d}_{be})] - E_{f_X}[C(\boldsymbol{X}, \boldsymbol{d}_{opt})] \\ & = E_{f_X}[C(\boldsymbol{X}, Arg \min_{\boldsymbol{d}} [C(\boldsymbol{x}_{be}, \boldsymbol{d})])] - Min_{\boldsymbol{d}} E_{f_X}[C(\boldsymbol{X}, \boldsymbol{d})] \end{aligned} \tag{8.7}$$

Note that a correct figure for EVIU *cannot* be taken as the difference between the minimal expected costs under uncertainty $EZ(\boldsymbol{d}_{opt})$ and the approximative costs that would be computed $C(\boldsymbol{x}_{be},\boldsymbol{d}_{be})$: the latter figure may prove incorrect because of uncertainty. Note also that EVPI is much costlier to assess as it involves *stochastic optimisation* in order to compute the mix of an expectation step and an inner minimisation of the complex system model (see the conclusion of this chapter).

Both measures can also be computed *component-wise* as an alternative sensitivity analysis approach: to investigate the relative interest of investing so as to reduce uncertainty (for EVPI); or to assess the impact of simplifying the modelling process through the fixing of some inputs (for EVIU). EVPI matters essentially in the *epistemic* or *reducible* part of uncertainty. Those measures depend of course directly on the uncertainty model f_X used by the analyst: a fair model is required to honestly best represent his view of the potential outcomes and associated likelihoods.

Granger Morgan and Henrion (1990) also discussed the expected interest of using the measures as a function of the regularity properties of the system model. For instance, a *linear* dependence on the *uncertain inputs* leads to a zero EVIU when the best-estimate is taken at the expected value of the inputs because of the basic property that the expected output of a linear model is the image of the expected input. As a consequence, it could seem useless to model uncertainty on linear models: however, this is only true from the point of view of expected costs and later comments will show that decision makers would rather use an expected utility, utility being by essence a non-linear function.

Flood example

The EVIU can be computed either assuming that the 'best-estimate' is taken at the average water level (no flood) – then $\boldsymbol{d}_{be} = 0m$ – or at the 'worst case' – then $\boldsymbol{d}_{be} = 4.2m$. EVIU is worth 7,6 M€ and 0,04 M€ respectively: the latter is a sizeable figure as the yearly investment considered ranges over [0,03–0,5 M€]. Conversely, the decision maker is lucky enough that the 'worst-case' design (corresponding to a 1/10 000-yr return period) would not be much worse than a full probabilistic study because of the steepness of the damage cost functions: in fact, the worst-case does not 'ignore uncertainty' but schematises it in a simplistic way. The cost function being linear as a function of the uncertain economic factor $\boldsymbol{x}^e = c_m$, note that the EVIU for that uncertain input is zero: *from the expected-cost perspective*, it may be ignored. For the other uncertain inputs, Table 8.1 illustrates the component-wise EVIU.

Table 8.1 Comparing expected values of ignoring uncertainty in the flood example.

	Keep all X	Ignore Q	Ignore K_s	Ignore Z_v	Ignore Z_m
D	3.5 m	2 m	1 m	2.8 m	3.2 m
EVIU (M€)	—	0,4	2	0,07	0,01

Note that ignoring uncertainty through a partially probabilistic description while keeping best-estimates for the others results in a downsizing of the dike with respect to the optimum (which would be different if worst-case values were kept), all the more so since the input is important. EVIU offers an alternative ranking to the sensitivity indices given in Chapter 7, though close to the rankings identified in Section 7.5.1 except for the fact that K_s proves to be more important than Q.

The EVPI would result in close to 1,2 M€/yr (the minimal expected cost under uncertainty) as the expected cost under perfect information falls as low as 0,003 M€/yr: adjusting theoretically the dike close to the level of the yearly flood if any (which happens only with 1 % chance every year) costs eventually a very small amount. The figure is rather useless as such as there is a lot of irreducible uncertainty embedded as long as the short-term flood forecast leaves insufficient anticipation for the building of a dike.

8.1.3 The expected utility approach for risk aversion

Section 8.1.1.3 introduced the limitations to the expected cost approach. Indeed, most people would not consider as equivalent the two following outcomes of alternative decisions:

- (d_A) invest 0,1 M€ but accept a 10^{-3} residual risk of additional 1000 M€ of damage;
- (d_B) invest 1,1 M€ being (almost) sure of the absence of residual risk.

However, both decisions result in an equivalent total expected cost. A 'risk-averse' (respectively 'risk-prone') decision maker would prefer the second option (resp. the first), while a 'risk-neutral' person would consider them equivalent. The majority of human beings would rather be risk-adverse, particularly when facing rare and serious risk.

Expected utility is the subject of a huge branch of the econometric literature starting from the founding works of Van Neumann and Morgenstern (1944) (VNM)). Expected utility may be understood as a first way to represent the risk aversion affecting the perception of varying total costs, as follows:

$$EU = E[U(-c_i(\boldsymbol{d}) + -c_d(\boldsymbol{Z}^t, \boldsymbol{X}^e, \boldsymbol{d}))] \tag{8.8}$$

Expected utility means simply replacing the minimisation of the raw cost balance by the maximisation of the raw outcomes (the cost of each scenario) transformed through a function called *utility* which weighs the relative importance of the raw outcomes to the decision maker. In the terminology of economics, the distribution of outcomes $(z, f_Z(z|\boldsymbol{d}))$ (or $(c_c, f_c(c_c|\boldsymbol{d}))$ when considering costs in particular) over which utility is averaged is often referred to as a *lottery*: this is particularly the case in a discretised form $(c_k, p_k = P(c_k))_{k=1\ldots l}$ where there is a finite possible set of consequences with probabilities and costs c_k depending on $\boldsymbol{d}$. The question is thus to arbitrate between lotteries corresponding to various actions (values for $\boldsymbol{d}$), such as the twofold ('>' meaning 'preferred to'):

$$(\boldsymbol{d} = \boldsymbol{d}_B)\text{'}>\text{'}(\boldsymbol{d} = \boldsymbol{d}_A) \Leftrightarrow EU(\boldsymbol{d}_B) > EU(\boldsymbol{d}_A)$$

Returning to the example above, this means that a 'risk-averse' decision maker involves a *concave* utility function with respect to costs:

$$EU(d_B) = 1 \times U(-1.1) = U(-1000 \times 10^{-3} + -0.1001 \times (1 - 10^{-3}))$$
$$> 10^{-3} \times U(-1000) + (1 - 10^{-3}) \times U(-0.1001) = EU(d_A)$$

that is more generally (cf. Figure 8.3):

$$U(-pc_1 - (1-p)c_2) > pU(-c_1) + (1-p)U(-c_2)$$

A utility function is also necessarily an increasing function, the preference growing with the lesser costs (or the higher benefits). Eventually, utility being used only to weigh relative preferences between outcomes inside the maximisation program of Equation (8.8), the function used to model it makes sense only regardless of its location and scale normalisation:

$$\boldsymbol{d}_{opt} = Arg \max EU(Z(\boldsymbol{d})) = Arg \max\,(aEU(Z(\boldsymbol{d})) + b) = Arg \max\,(E[aU(Z(\boldsymbol{d})) + b]) \quad (8.9)$$

When considering a bounded interval of possible outcomes (as is the case for the total cost in the flood examples), it can thus be assumed that the function maps into *[0,1]*.

A useful concept associated with the utility function representing the preferences of the decision-making is the *certainty equivalent* c_c* defined so that (Figure 8.3):

$$U(-c^*) = E[U(-C_c)] \quad (8.10)$$

If $U(z) = az + b$ is linear, then c_c* coincides with the average cost EC_c: this is no longer so in the most common case where risk aversion leads to involving a *non-linear* utility. This entails the following definition of the risk premium:

$$RP(-C_c) = E(-C_c) - c_c^* \quad (8.11)$$

An essential interpretation of the *risk premium* is that it represents, for the decision maker, the *cost of uncertainty* or more precisely the willingness to pay an additional cost to get rid of the uncertain outcomes beyond a similar average result, for example through an insurance policy that refunds the uncertain outcomes. Most people are subsumed risk-averse, so that the risk premium would be positive.

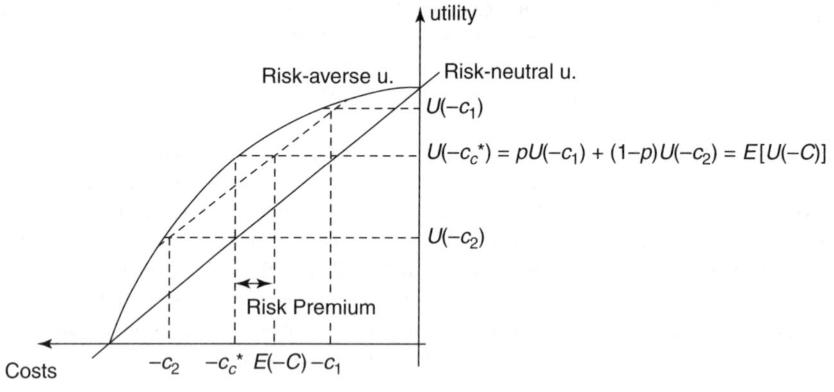

Figure 8.3 *Utility and complete cost of flood protection.*

Flood example

It is useful to compare the various distributions of cost associated with design choices in order to understand how a risk-averse utility may shift the choice towards more expensive investment as compared to expected cost-based, as in Table 8.2.

Table 8.2 Expected values and quantiles of the total costs (in M€/yr equivalent) for various choices of dike levels (performed through 10 000 time-based samples of the 30-year period).

	Fixed costs (*1/5 000*)	$P(C_c > \text{fixed costs})$	EC_c (% error)	$\sigma(C_c)$	$c_c^{55\%}$ (1/50)	$c_c^{74\%}$ (1/100)	$c_c^{97\%}$ (1/1 000)	$c_c^{99.4\%}$ (1/5 000)
No dike	0	32 %	9.7 (15 %)	23	0	2	83	110
d = *1m*	0.04	12 %	3.5 (8 %)	14	0.04	0.04	53	90
d = *2m*	0.2	4 %	1.6 (11 %)	9.6	0.2	0.2	4.6	78
d = 3.5m	0.6	1.6 %	1.2 (10 %)	6.4	0.6	0.6	0.6	54
d = *5m*	0.9	0.5 %	1.2 (6 %)	3.6	0.9	0.9	0.9	0.9
d = *8m*	1.6	0.1 %	1.6 (2 %)	2	1.6	1.6	1.6	1.6
d = *9m*	1.8	0.04 %	2.8 (1 %)	1.3	1.8	1.8	1.8	1.8

Since the complete cost is a yearly equivalent of a total cost over 30 years of operation, the definition of quantiles is related to the equivalent return period of the yearly flood through $\alpha = (1-1/T_r)^T$ (see Chapter 4). Obviously, increasing investment in the dike pushes to lesser probabilities the occurrence of uncertain damages above the fixed costs, themselves rising. Beyond the expected-cost optimum at $d = 3{,}5m$, risk-averse decision makers would probably still prefer increasing the fixed costs to reduce the risk of observing a residual damage, though it is unclear to what extent, $d = 5m$ (less than *1 %* chance to deviate from fixed costs) or even $d = 8m$ (less than *1 %*). As a matter of fact, the last two cost quantiles are imprecise figures due to the limitations of computational sampling, but even more of data, as would be reflected in a level-2 study of the total cost. To the best knowledge available at time of analysis, a *9 m* dike should be trusted to prevent the 1/10 000-yr flood (assumed to be *63,5–64mNGF*) while it is a bit unsure for a *8 m* dike.

VNM demonstrated that, given a number of axioms stating a given type of rationality of choices, a decision maker orders his preferences according to a unique utility function (though often unconsciously). Calibrating the utility function remains, however, a difficult task: this is especially true when addressing low-probability and high-loss risks, because the evidence about decision makers' preferences involves the consideration of more frequent uncertainty outcomes. A moderate degree of risk aversion might thus severely bias the optimisation towards protection against outcomes of quasi-negligible probability. The issue has been discussed a great deal in the literature as to whether the expected utility theory can still be applied and several alternatives were proposed.

8.1.4 Non-linear transformations

Expected utility suffers from the paradoxes first publicised by Allais (1953) and the concepts of expected preference have been developed as a remedy. Recall the expected utility as being the result of averaging the utility of each output level over the uncertain output distribution:

$$EU = \int_z U(z|\mathbf{d}) f_z(z) dz = \int_z U(z|\mathbf{d}) dF_z(z) \tag{8.12}$$

Rank-dependent utility consists of averaging the same function over a modified output distribution which represents the decision maker's as follows:

$$ERDU = \int_z U(z|\boldsymbol{d})d(w \circ F_z(z)) = \int_z w' \circ F_z(z)U(z|\boldsymbol{d})f_z(z)dz \tag{8.13}$$

As function $w(p)$ needs to map *[0,1]* into itself, it has to be non-linear so as to non-trivially transform the F_z cdf of z into a distinct distribution woF_z representing the subjective perception of probabilities attached to a given decision maker. Rewrite it with application to total cost as follows:

$$ERDU = E_{w[Fz]}[U(-c_i(\boldsymbol{d}) + -c_d(\boldsymbol{Z}^t, \boldsymbol{X}^e, \boldsymbol{d}))] \tag{8.14}$$

Implementing such an approach as an alternative to expected utility requires both the elicitation of the $w(.)$ function and a modified risk computation procedure. A simple way may be to first sample the complete system model in order to approximate F_z, for instance, through a kernel-based approximation (see Chapter 7, Section 7.2.2.2); then compute the new integrand of Equation (8.14) using either a numerical integration technique in low dimensions in Z or a sampling procedure of Z according to the modified density woF_z^{ker}.

8.1.5 Robust design and alternatives mixing cost expectation and variance inside the optimisation procedure

Robust design approaches in quality control Optimising the choice in $\boldsymbol{d}$ with respect to the uncertain cost of a system can be related to the *robust design* approach, pioneered by Taguchi in the early 1980s in the context of quality control of manufacturing (Taguchi, 1987; Phadke, 1989). In such a context, the output of interest z appears as a technical performance parameter of a product. Such a performance is affected by the so-called *control factors*, corresponding to $\boldsymbol{d}$, and *noise factors*, corresponding to $\boldsymbol{x}$: internal manufacturing variability, external disturbances in the later use of the product, ageing of the product and so on. Instead of simply controlling that such performance keeps within a deterministic tolerance interval (quality control seen as conformance to specifications and the associated proportion of scrap), Taguchi introduced a quadratic cost function penalising any deviation from the specification. The cost function involves either a nominal target z_n, or a minimal (or maximal) value for the performance parameter, as respectively:

$$c_Z = kE[(Z - z_n)^2] = k((EZ - z_n)^2 + \text{var}\, Z) \tag{8.15}$$

$$c_Z = kE[Z^2] = k(EZ^2 + \text{var}\, Z) \tag{8.16}$$

$$c_Z = kE[1/Z^2] \tag{8.17}$$

The optimisation in $\boldsymbol{d}$ is therefore looked for with respect to minisming any of those quality criteria called the *quality loss function* (all of which can be viewed as expected utilities) or equivalently maximising a log-modified criterion called the *signal-noise ratio* (denoted as η in the quality control literature) and normalised in dB as follows:

$$c_Z = -10 \log_{10} kE[(Z - z_n)^2] \tag{8.18}$$

$$c_Z = -10 \log_{10} kE[Z^2] \tag{8.19}$$

$$c_Z = -10 \log_{10} kE[1/Z^2] \tag{8.20}$$

Another signal/noise is also often considered, involving the coefficient of variation of the output:

$$c_Z = -10\log_{10}\left[\operatorname{var} Z/EZ^2\right] = -20\log_{10}\left[c_V(Z)\right] \tag{8.21}$$

From the perspective of this book, *robust design* can be seen as an approach of *optimisation under uncertainty* using the above-listed risk measures involving expectations of squared output costs or derived quantities. In that case, the uncertainty model is intended not so much to reflect the data variability of modelling imprecision but the likely ranges within which the key variable will stand because of changes in the detailed design steps, manufacturing deviations or operational changes with respect to the initial specifications. This is an essential approach for the safe and robust design of large industrial processes or systems, for instance that of airplanes that are being developed over more than 5–10 years of design and prototyping and possibly 30–50 years of operational cycle.

Robust design may be also understood in the wider context of the design process itself. In order to perform such an optimisation, Taguchi developed the associated designs of experiment that are required in the system between the control and noise factors using two experimental designs: the *inner array*, for the exploration of various values of $\boldsymbol{d}$, and the *outer array*, for the exploration of the impact of various 'noise' values of $\boldsymbol{x}$ for a fixed $\boldsymbol{d}$. The resulting design of experiment using both inner and outer arrays is referred to as a *cross array* and usually involves partial combinations of discrete levels for each factor with respect to a complete factorial design of all combinations. Its size grows exponentially with the dimensions of $\boldsymbol{x}$ and $\boldsymbol{d}$.

Historically, the method was developed with physical (real-world) designs of experiment though gradually extended to numerical (model-based) DoE. A large debate has questioned the limitations of the original approach either regarding the appropriateness of the quadratic loss function or derived S/N ratios (e.g. when cost implications of deviations are not symmetrical, not sensitive in a quadratic manner, etc.) or the relevance of the accelerated designs of experiment suggested to achieve the optimal design in an efficient way (e.g. when the impact of $\boldsymbol{x}$ and $\boldsymbol{d}$ involve complex interactions, see the comments in Chapter 7, Section 7.5 on sensitivity analysis and DoE). In general, the approach still means minimising a risk measure embodying the costs of uncertain deviations from specifications.

Other approaches to robust design (expectation-variance criteria, RBDO) A number of alternative risk measures can be handled so as to better control the variability around the optimum (see, e.g. Schuëller and Jensen, 2008). First, the quantity to be optimised can be transformed. Think, for instance, of the maximisation of a *quantile of utility* replacing that of the expected utility:

$$Max_{\boldsymbol{d}}(U_{\alpha}(\boldsymbol{d})) \tag{8.22}$$

which may be seen as a refinement of a deterministic robust optimisation approach

$$Max_{\boldsymbol{d}}[Min_{x}UoG(\boldsymbol{x},\boldsymbol{d})] \tag{8.23}$$

whereby the design $\boldsymbol{d}$ is chosen so that the utility in the worst-case in $\boldsymbol{x}$ is maximised. Similar to the comments made in Chapters 2 and 4 on the limitations of a deterministic approach that may undermine the quality of the optimum for implausible scenarios, the quantile-based risk measure may be seen as offsetting more the excessively-improbable if α is not too high. It is less used than expectation-variance criteria that appear generally as a multi-objective program:

$$\{Max_{\boldsymbol{d}}[EU(\boldsymbol{d})]; Min_{\boldsymbol{d}}[\operatorname{var}U(\boldsymbol{d})]\} \tag{8.24}$$

This generates a Pareto front of alternative compromises between maximising the expected utility and minimising its associated variance (see Figure 8.5 for the flood example). A single-criteria formulation can also be used:

$$Max_{\boldsymbol{d}}[EU(\boldsymbol{d}) - k\sqrt{\mathrm{var}}U(\boldsymbol{d})] \tag{8.25}$$

When the utility (or cost) can be approximated as a Gaussian-distributed output, such a variance-penalised expected utility criterion is equivalent to the quantile of utility. Expectation-variance optimisation can be less costly to compute than the quantile approach.

Alternatively, compound risk measures can also be handled through a constrained optimisation approach: optimise the expected cost or utility under constraints involving other risk measures, such as a maximal probability of exceeding a cost threshold or a minimum solvability threshold:

$$Min_{\boldsymbol{d}} EC_c(\boldsymbol{d}) | c_Z(\boldsymbol{d}) < c_s \tag{8.26}$$

where c_Z may be typically a maximal probability to exceed a given threshold (or, equivalently a VaR compared to a given threshold, see below). This is the case in reliability-based design optimisation (RBDO), whereby c_Z refers to the reliability of the system or more generally to a set of K reliability-like constraints on various components or system performance outputs:

$$Min_{\boldsymbol{d}} EC_c(\boldsymbol{d}) | \{P(Z^k > z_s^k | \boldsymbol{d}) < p_f^k\}_{k=1..K} \tag{8.27}$$

Dedicated extensions of the FORM-SORM reliability approaches (Chapter 7) have been developed to solve such a computationally-challenging compound optimisation problem (Thanedar and Kodiyalam, 1992; Chen, Hasselman, and Neill, 1997; Lemaire, Chateauneuf, and Mitteau, 2005).

8.2 The role of time – cash flows and associated risk measures

8.2.1 Costs over a time period – the cash flow model

In Section 8.1, costs were introduced in a 'lumped' view as output variables of interest occurring *once* at the *future time of interest*. Most economic models would more generally consider a variable of interest that accumulates costs occurring over a number of future time periods. Considering the standard engineering economics total cost, this would result in the following:

$$c_c = \sum_{k=1..K} c(t_k) = c_i(\boldsymbol{d}) + \sum_{k=1..K} c_d(z_t(t_k), \boldsymbol{x}_e(t_k), \boldsymbol{d}) \tag{8.28}$$

Investment cost is often concentrated at the beginning of the period, while the uncertain costs occur at the varying time periods, depending typically on the time realisations of the technical performance $z_t(t)$. For instance, the annual floods may generate costs or may not (when overspill is negligible).

In finance or insurance, one would consider more generally an income statement or cash flow, with events distributed over the successive time periods generating either positive or negative terms (meaning cost vs. revenues or the reverse), cf. Figure 8.4. A generalised cash flow may be viewed as a vector $(t_k, c_k)_{k=1...K}$ representing the times t_k and amounts c_k in- or out-flowing, where any of the components

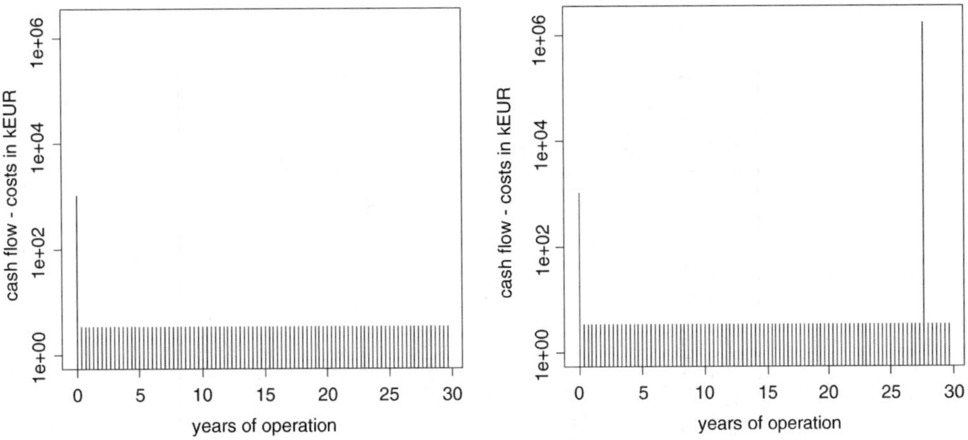

Figure 8.4 *Two random realisations of cash flows (costs only, over a log-scale) over the period of operation – (left) no flood occurred, costs are limited to the initial investment and yearly maintenance; (right) a flood occurs by the end of the period, entailing high damage costs.*

may be random: fixed times but partially uncertain payments, or known payments at uncertain times, random time lengths (as in life insurance) and so on. Similar to a natural alea (riverflow, windspeed . . .), a generalised cashflow in an all-random case may, for example, be a compounded Poisson process (Section 4.3.5), with random flow/cashflow events, and random conditional amounts. Thus, a number of possible variables of interest may be considered according to the nature of the underlying economic phenomena considered (net revenue or profit, free cash flow, etc.) as well as associated risk measures taken at expected values or quantiles (Table 8.3.).

Note that in the insurance field with regard to the control of the overall solvency of a firm, it is usual to denote the risk process as follows:

$$U(t_k) = s + ES + b(t_k) - S(t_k) \tag{8.29}$$

Reserve (or surplus) = initial reserve + net premium + mean surplus (safety function) – total claim

Thus, a key quantity of interest for the regulation of insurance is the *ruin probability* defined as follows:

$$c_z(\boldsymbol{d}) = P(Min_{[0,\Delta T]} U(t) < 0) \tag{8.30}$$

Take it over a given time period ΔT for the 'ultimate ruin probability'. Dependence assumptions regarding the elementary premia and claims involved in large sums of individual random variables representing elements of cashflow at fixed or random times are obviously essential in controlling the overall distribution of margin and the associated ruin probability. An increasing interest has been devoted to dependence modelling involving, for example, copula models and tail dependence issues (Chapter 5, Section 5.3), though the dimension of the vector of uncertain inputs may be extremely large. Note how close such ruin probability results are to the risk measures studied in natural risk assessment or reliability: logically it also mobilises the extreme value theory.

Table 8.3 Practices and terminologies from engineering economics to financial or decision theory.

	Engineering economics	Actuarial science	Financial modelling	Decision theory
System model $G(.)$	Physical + Economic	Risk generation + Accounting (cash flow + contractual provisions)	Financial structure (cash flow)	Technical + Decision (utility/probability transformation)
Event/Variable of interest z	TC - Total cost (s)	Total claim amount (SN)	Net revenue	Utility
	NPV - Net present value (d)	Net revenue (s)	Discounted cash flow	
		Cash flow (d)	Profit/loss of portfolio over a period	
		Reserve variable or solvency margin	Market value of a portfolio	
Risk measure $c_Z(d)$	$E(Z)$ total cost expectation	$E(Z)$ – 'fair premium'	$E(Z)$	$E(Z)$: expected utility, or expected rank-transformed preference
	$P(Z>z_s)$	Ruin or default probability, VaR	$\sigma(Z)$: volatility	
			z^{α}: Value at Risk	

8.2.2 The issue of discounting

When time becomes significant, some discounting or interest valuing becomes necessary, generating new variables of interest such as:

- discounted (cash flow) value;
- net accumulated value;
- net present value.

$$C = \sum\nolimits_{k=1..K} C_k v(T_k) \tag{8.31}$$

where $v(t) = (1\text{-}i)^t$ (*resp.* $v(t) = (1\text{-}i)^{-(T-t)}$) when considering the discounted value (resp. accumulated value) for an effective discount rate of i (net of inflation and taxes). The risk measures become then the *expected* (discounted) cash flow:

$$EC = \sum\nolimits_{k=1..K} E[C_k v(T_k)] \tag{8.32}$$

also called the 'fair premium' when considering the discounted value. The risk measure can alternatively involve any of the associated *VaR* or *TVaR*, that is quantile or conditional expectation of the variable of interest as defined in Section 4.4.3.

Note that while discounting is standard practice in standard investment choices or engineering decisions, it becomes controversial to some extent when considering public health, environmental benefits or severe societal risks. This is all the more so when applied to very *long-term decisions*, such as nuclear waste repositories where harmful doses are modelled up to the casualties 10 000–100 000 year-ahead. It is easy to understand that even a low discount rate completely discards such long-term consequences: 1 % yearly discounts 63 % (respectively 95 % or 99.995 %) of the consequences after 100 years (resp. 300 or 1000 years). A formula summing up total costs whatever the date of risk realisation – without any discounting – also has its shortcomings as recounted by Lind (2007), who advocates taking non-constant discount rates. On the subject of equivalent life-years[1] saved by risk mitigation, this neglects, for instance, the fact that money may be invested in low-risk securities (issuing a positive return rate, net of inflation and taxes) and mobilised in greater amounts later in competing public health investments in order to save more equivalent life-years at the identical time horizon. Lind (2007) advocates maintaining the discounting up to the financing horizon (such as commissioning time of the plant, amortisation period of the infrastructure, closure time for a waste repository ...), and then considering that the mitigation of any risk occurring in the future be counted as if occurring on the last year of the financing horizon.

Flood example

The variable of interest is the total cost over the time period of interest, typically 30 years, expressed in yearly equivalent. While the description has been focused so far on its expected value, the consideration of its detailed cash flow becomes necessary when moving on to the study of the distribution of the total (discounted) cost. It is therefore necessary to sample the following function:

$$\frac{1}{T}C_c = \frac{1}{T}C_i(h_d) + \frac{1}{T}\sum\nolimits_{k=1..K} C_d(Z_c^k, h_d)v(T_k) \tag{8.33}$$

[1] Take, for instance, the QALY, quality-adjusted life years, advocated by the U.S. Public Health Service Panel on Cost-effectiveness in Health and Medicine to measure life and health benefits.

over N random histories of flood conditions over 30 years of operation. In theory, this could involve not just the 30 yearly maximal flood events but even the potentiality of multiple floods $k = 1 \ldots K$ occurring more than once a year over the continuous history of daily water levels. For the purpose of cost study, this is limited, however, to those water levels that entail a non-zero damage cost, that is $Z_c\text{-}h_d\text{-}z_b>0$ which happens only $\sim 10^{-2}$ per year for the worst case of the absence of dike; which means a probability $\sim 5.10^{-5}$ (resp. $\sim 2.10^{-7}$) to observe two (resp. three) such floods per year. It is therefore useless to spend time sampling in too much detail as this would restrict the number N of 30-yr histories to be sampled and lead to less reliable estimates of rare quantiles. The sampling has been limited to three independant sub-periods per year through appropriate rescaling of Gumbel distributions (see Chapter 5). Table 8.4 illustrates the results with asymptotic estimates of $+-$ twice the coefficient of variation, which are then plotted in Figure 8.5.

Note how the variance has been reduced by a factor $\sqrt{30} \sim 5.5$ as compared to the previous computation of the standard year of cost, whereas here the average of *30* years i.i.d. is being taken; conversely, the probability for cost to be beyond the certain investment/maintenance cost rises to $1\text{-}(1\text{-}0.012)^{30} \sim 31\,\%$ chance to observe at least a $\sim 1.2\,\%$ yearly flood over 30 years. Note that the slight increase in the expected value is due to allowing for the possibility of more than one flood to occur every year. Besides the expectation figures, it is essential to study the expectation/variance diagram in a robust design approach. Aiming at a double-criterion optimisation (minimise expected cost *and* variance), the Pareto front appears to pass between *3,5m* to *8m* of dike (NB: the figures for *8m* are unstable and would require more than 10 000 samples), all leading to acceptable designs according to the relative importance given to expectation or variance minimisation.

Moreover, Table 8.5 shows that discounting reduces the optimal level of dike when reasoning with the expected cost only. While the optimal under no discounting rate is around *3.5m* (see above), the optimal level falls to $\sim$*2,5m* or $\sim$*2m* under *3 %* to *10 %* discounting. Understandably, a higher discount rate reduces the importance of rare costly damage occurring in later years over the period, so that the benefit for the initial investment is lower.

8.2.3 Valuing time flexibility of decision-making and stochastic optimisation

Although the risk measures considered so far already embodied a description of time, as illustrated in Section 8.2, the approach may still be seen as *static* in the following sense: optimisation is carried out assuming that all future events, albeit uncertain for external reasons, depend on design decisions taken once and for ever at the initial date of analysis. It is essential to note that this does not preclude the fact that those actions decided upon once may then spread over multiple future dates, as represented in the above-mentioned cashflows; however, once the decision has been taken, there is no feedback.

Real Options Analysis (ROA) or Valuation (Dixit and Pindyck, 1994; Mun, 2002) is a popular tool designed to help take optimising decisions in a more *dynamic* way in the sense that account can be made for flexible decisions that may change over time. Think, for instance, of waiting for the uncertain event to be better characterised – hopefully less uncertain – before taking all technical decisions. Conveying additional flexibility, this optional decision-making can be further modelled under ROA. Stochastic dynamic programming (SDP) is a stochastic refinement of the original dynamic programming approaches developed by Richard Bellman (Bellman, 1957). These aim to optimise the choice of control variables or actions over the time-based (partly-uncertain) evolution of a system as a dynamic function of the observed states. SDP has historically been developed on a discretised time-basis while ROA can be traced more into the area of continuous time-based financial option evaluation.

Table 8.4 Results of q.o.i. of the total costs for various dike levels with associated sampling accuracy.

Simulating C_c/T (*in M€/yr*)	No dike				Dike (10 000 samples)			
(no discounting)	10^6 runs (1 yr equivalent)	100 runs	1000	10 000	1 m	3,5 m	5 m	8 m (approx.)
Expectation	9.1 ± 3 %	9.2 ± 50 %	10.1 ± 16 %	9.9 ± 5 %	3.4 ± 8 %	1.1 ± 10 %	1.2 ± 6 %	1.6 ± 3 %
standard deviation	126 ± 0.1 %	20 ± 14 %	25 ± 4 %	24 ± 1 %	14 ± 1 %	6 ± 1 %	4 ± 1 %	2 ± *1* %
$P(C_c > \textit{fixed cost})$	1.2 %	29 %	31 %	32 %	12 %	1.6 %	0.5 %	0.1 %
VaR/ESa – 90 %	n.a.	44/67	45/77	45/74	0.8/34	0.6/5	0.9/3	1.6/2
VaR/ESa – 99 %	n.a.	131/131	110/133	98/124	81/94	3/48	0.9/22	1.6/8
Rarity index (RI)	−4	−1.5	−2.5	−3.5	−3.1	−2.1	−1.7	−1.1

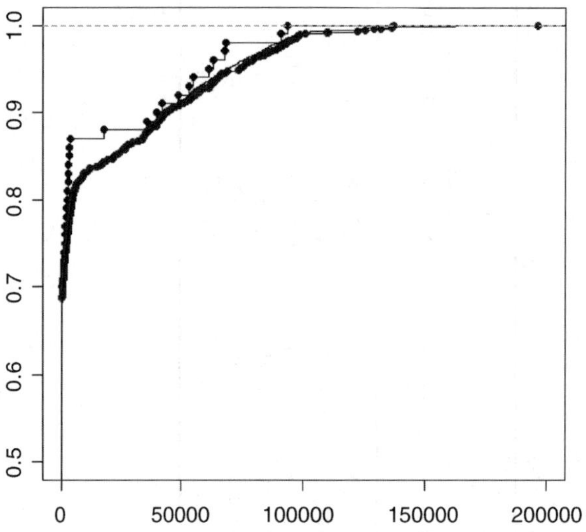

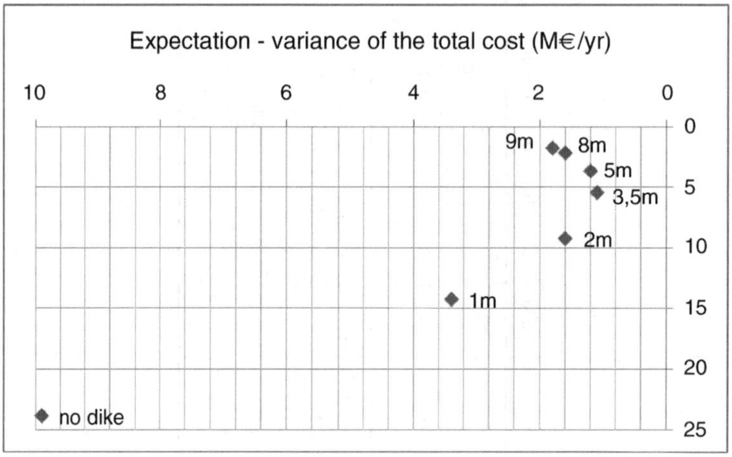

Figure 8.5 *(up) simulation of total costs without dike after 100 or 1000 samples – (down) Pareto front of expectation/variance of total cost for various dike levels.*

To help in understanding, consider the following simplified setting:

- The future period of interest involves K time steps at dates $(t_k)_{k=1...K}$ at which actions $\boldsymbol{d}_k$ which have been decided in earlier time steps, at the latest at time t_{k-1}, come into place.
- Costs $C_k(z_{k-1}, \boldsymbol{d}_k, X_k)$ are incurred as the result of the state of the system z_{k-1} at the end of the previous period as well as of the actions $\boldsymbol{d}_k$, and of the realisation of uncertain events X_k.
- The uncertain state of the system z_k becomes known to the analyst: it depends on its previous one z_{k-1} and on the actions $\boldsymbol{d}_k$ that are perturbated by the uncertain input events X_k according to a known (forecast) system model as follows: $Z_k = G_k(z_{k-1}, \boldsymbol{d}_k, X_k)$.

Table 8.5 The effect of discounting on the total cost quantities of interest.

Simulating C_c/T *(in M€/yr)*	Dike + discounting (10 000 samples)				
	No dike		2 m		3 m
EC_c $sd\ C_c$ $P(C_c > \textit{fixed cost})$					
0 % p.a.	9.9 ± 5 %		1.6 ± 11 %		1.3 ± 10 %
	24 ± 1 %		9.3 ± 1 %		7.2 ± 1 %
	32 %		4 %		2 %
3 % p.a.	6.6 ± 5 %		1 ± 12 %	(2.5 m)	1 ± 10 %
	16 ± 1 %		6.2 ± 1 %	0.9 ± 11 %	5 ± 1 %
	32 %		4 %	5.3 ± 1 %	2 %
				2 %	
10 % p.a.	3.2 ± 6 %	(1 m)	0.57 ± 12 %		0.63 ± 9 %
	9.7 ± 1 %	1.1 ± 11 %	3.5 ± 1 %		2.8 ± 1 %
	33 %	5.8 ± 1 %	4 %		1.8 %
		11%			

The initial state of the system z_o is known; the question is then to minimise the overall cost $\Sigma_{k=1\ldots K} C(z_k, d_k)$ over the finite future period of interest through an appropriate series of actions $(d_k)_{k=1\ldots K}$. A first strategy is that of classical expected cost minimisation:

$$Min_{(d_k)_k} E\left[\begin{array}{l} C_{K+1}(Z_K) + \\ \sum_{k=1..K} C_k(Z_{k-1}, d_k, X_k) \end{array} \middle| \begin{array}{l} Z_o = z_o \\ Z_k = G_k(Z_{k-1}, d_k, X_k) \\ X_k \sim f(.|z_{k-1}, \theta_k) \end{array} \right] \tag{8.34}$$

At the time t_0 of analysis, the whole program of actions $(d_k)_{k=1\ldots K}$ is optimised at once on the basis of the present view of the uncertainty in the future states $(Z_k)_{k=1\ldots K}$. This means optimising a single given expected cost function over the space of actions $(d_k)_{k=1\ldots K}$, simulating the temporal trajectories of $(X_k)_{k=1\ldots K}$ and $(Z_k)_{k=1\ldots K}$.

An alternative strategy can consist of making at t_0 the decision on action d_1 but waiting, say, for the knowledge of z_1 so as to make further decisions on action d_2, and so on. This makes sense, for instance, if the uncertainty in the future states $(Z_k)_{k=1\ldots K}$ changes conditional upon the realisation of z_1, that is the uncertainty in the next states is not independent from the previous state. Think of Z_k as being monthly stocks of water in a reservoir that depend not only on the uncertain monthly rainfall (a factor of uncertainty X_k), but also on the value of the previous stock as well as the action to withdraw an amount of water d_k.

In that case, the optimisation of the program of actions might benefit from being more flexible: instead of fixing the whole program in advance, decision-making consists of a strategy to fix, at each step k, actions $d_{k+1} = D_{k+1}(z_k)$ conditional upon the knowledge of z_k. This means a much more sophisticated analysis to be undertaken as one looks now not only for a *fixed vector* $(d_k)_{k=1,...K}$ containing the optimal program but for a *family of functions* $(D_k(y))_{k=1...K}$. Classical assumptions are generally required, for instance the Markovian property: actions d_{k+1} depend only on the knowledge of the present state z_k while not on the previous history of actions d_{k-u} or system states; uncertain events X_{k+1} may not be simply *i.i.d.*, but their distribution should only depend on the previous state and actions, not further in the past. Under a number of regularity hypotheses, it can then be proved that the optimal strategy (or 'policy'), that is the family of functions $(D_k(z))_{k=1...K}$ that minimises the expected total cost over the period as follows:

$$J^*{}_1(z_o) = Min_{(D_k(.))_k}\left(J_{D,1}(z_o) = E\left[\begin{array}{l} C_{K+1}(Z_K) + \\ \sum_{k=1..K} C_k(Z_{k-1}, d_k, X_k) \end{array} \left| \begin{array}{l} Z_o = z_o \\ d_k = D_k(Z_{k-1}) \\ X_k \sim f(.|z_{k-1}, \theta_k) \\ Z_k = G_k(Z_{k-1}, d_k, X_k) \end{array} \right. \right] \right) \tag{8.35}$$

is that which is built through the reverse optimisation as follows:

$$\begin{array}{l} J^*{}_{K+1}(z) = C_{K+1}(z) \\ \\ for\, k = 1..K, \\ J^*{}_k(z) = Min_d E[C_k(z, d, X_k) + J^*{}_{k+1}(G_k(z, d, X_k))|X_k \sim f(.|z, \theta_k)] \\ D^*{}_k(d) = Arg\, Min_d E[C_k(z, d, X_k) + J^*{}_{k+1}(G_k(z, d, X_k))|X_k \sim f(.|z, \theta_k)] \end{array} \tag{8.36}$$

Since the distribution of uncertain future states depend iteratively on the successive decisions, it would be impossible to start optimising from the past and move forward. The determination of the optimal strategy involves a reverse recursive optimisation (or *backward induction*) process also refered to as a variant of the *Bellman equations*. The minimal expected future residual costs $(J^*{}_{k+1}(z))_{k=0,...K-1}$ (or equivalently the maximal expected future residual benefits) being then refered to as the *Bellman values*.

Computational consequences, a key cost of the methods, are reviewed briefly in Section 8.3. Note that the essential recipe for these kinds of *methods valuing flexibility* is twofold:

- Uncertainty over the future variables of interest should have a substantial *epistemic* part so that the observation of the system state at step k and the availability of an associated forecast model greatly reduces the uncertainty in the states at step $k + 1$ as compared to the level of information at step t_0.
- The *expected value of perfect information* (EVPI) associated with the variables that may be forecast should be high enough for the adaptation of the actions to the expected improved state of knowledge to be worthwhile: keep in mind that only a limited part of EVPI would be recovered through adaptive strategies.

8.3 Computational challenges associated to optimisation

The previous paragraphs have illustrated the various formulations of optimisation under uncertainty that are required to be solved in order to make a decision on a technico-economic basis. More will be said regarding the computational implications. Indeed, stochastic optimisation generates tough computational challenges (Beyer and Sendhoff, 2007) both regarding the computational cost and the difficulties in guaranteeing robustness through theoretical results for the existence and unicity of optima and the convergence of the algorithms.

8.3.1 Static optimisation (utility-based)

An initial idea is to optimise the choice of $\boldsymbol{d}$ from the point of view of a given risk measure $c_Z(\boldsymbol{dd})$, which may be:

- maximise expectation of net present value, that is minimise expected net present costs of a loss scheme such as flood control;
- maximise the expected utility of the net present value or more complex expressions in time: formally, this includes also the case of minimising probability of undesired event, such as insolvency;
- optimise other combinations between risk measures of the output such as expected utility penalised by variance, utility quantiles and so on.

In general, a computational difficulty arises with the subsequent optimisation programs:

$$\boldsymbol{d}_{opt} = Arg\min_{\boldsymbol{d}} c_Z(\boldsymbol{d}) = Arg\min_{\boldsymbol{d}} E(U[c_i(\boldsymbol{d}) + c_d(Z_t, \boldsymbol{d})]) \tag{8.37}$$

as the distribution of the variable of interest z (total cost, utility, etc.) may depend on $\boldsymbol{d}$. That kind of problem proves to be similar to other numerical problems encountered in this book. Think firstly of the model calibration algorithms developed in Chapter 6, where one was optimising the observable likelihood with respect to a vector of parameters to be estimated. The associated cost function also requires the propagation of distributions through the system model:

$$\begin{aligned} &Arg\,Max_{\theta_X,\theta_u} L\left[\left(y_{mj}\right)_{j=1..n} \middle| \boldsymbol{d}_j, \theta_X, \theta_u\right] \\ &= Arg\,Max_{\theta_X,\theta_u} \prod_j E_{f_X|\boldsymbol{d}_j,\theta_X,\theta_u}\left[f_U(y_{mj} - H(\boldsymbol{X}_j, \boldsymbol{d}_j)|\theta_u)\right] \end{aligned} \tag{8.38}$$

It is also similar to the computation of a mixed deterministic-probabilistic risk measure as seen in Chapters 2 and 7:

$$\begin{aligned} c_z(\boldsymbol{d}) &= Max_{\boldsymbol{x}_{pn}\in D_x,\theta_{pn}\in D_\theta}\left[c_z \middle| \boldsymbol{x}_{pn}, \theta_{pn}, \boldsymbol{d}\right] \\ &= Max_{\boldsymbol{x}_{pn},\theta_{pn}}\left[EU(G(\boldsymbol{X}, \boldsymbol{d}) \middle| \theta_X, \boldsymbol{x}_{pn}, \theta_{pn})\right] \end{aligned} \tag{8.39}$$

although, in the latter case, finding the maximal output is more relevant than finding the parameter value that leads that maximum. Indeed, both $c_Z(.)$ and $L(.)$ are deterministic functions of $\boldsymbol{d}$ and θ_X respectively through the expectation of complex random distributions (that of $Z = G(\boldsymbol{X},\boldsymbol{d})$ or of $Y = H(\boldsymbol{X},\boldsymbol{d}) + U$ respectively).

Previous sections have also mentioned the even more challenging case of *probabilistically-constrained* utility optimisation such as those occuring in RBDO:

$$Min_{\boldsymbol{d}} EC_c(\boldsymbol{d}) | \{P(Z^k > z_s^k | \boldsymbol{d}) < p_f^k\}_{k=1..K} \tag{8.40}$$

Optimising may in fact involve not only a given risk measure as an objective function – typically an expected utility c_Z – but also a number of constraints $\{c_k(\boldsymbol{d})\}_{k=1\ldots K}$ that may either be of:

- a probabilistic type, $c_k(\boldsymbol{d}) = C_k[f_Z(Z|\boldsymbol{d}), \boldsymbol{d}]$: quantities of interest in the output distribution with pending requirements, for example a maximum budget for the coefficient of variation or some exceedance probabilities guaranteeing robustness or safety;
- or a deterministic type, $c_k(\boldsymbol{d}) = C_k(\boldsymbol{d})$: limitations on the feasible choices of $\boldsymbol{d}$, such as bounds on the characteristics or combinations of design variables.

$$Min_d c_Z(\boldsymbol{d}) | \{c_k(\boldsymbol{d}) < c_f^k\}_{k=1..K} \tag{8.41}$$

All those categories of problems belong to the area of *stochastic optimisation* for which dedicated algorithms are required.

8.3.2 Stochastic dynamic programming

Even stiffer computational challenges come with the use of *real options analysis* or *stochastic dynamic programming*. Real options analysis was introduced in Section as a means to value flexible decision-making through an optimisation of the risk measure that still allows for various options to be taken at future times of interest. It means a highly costly computational process as there is not just *one expected function* to be optimised over a space of *actions* $(d_k)_k$ but a series of K expectations each of which have to be optimised sequentially over a space crossing *actions and states* (d_k, z_k).

The corresponding areas of (stochastic) dynamic programming or optimal control involve many developments designed to tackle the heavy curse of dimensionality threatening such combined stochastic optimisation programs. Dedicated computational techniques tend to resolve the issues of the most appropriate discretisation of actions and states or conversely the extensions to continuous time basis and discounting the cash flows over time (traditional in the valuation of financial options) or sometimes non-Markovian dependence of uncertain events. Operational versions of this kind of algorithms in the power supply, logistics or financial industries usually rely on simplified system models (or meta-models); the plugging in of large physical-numerical system models either for the system evolution $(G_k(.))_k$ or cost functions $(C_k(.))_k$ should in principle bring considerable rewards in the shape of a finer anticipation and decision-making under uncertainty, but this remains a major computational challenge.

8.3.3 Computation and robustness challenges

Optimising the system model (either utility-based at the initial date with potential probabilistic constraints or through more sophisticated real-options analysis) is of a *stochastic* nature: it involves functionals incorporating random distributions being propagated through the system model that depend on the parameters to be optimised. This is generally a severe computational challenge as optimisation would involve as many propagation steps as the number of tentative values $\boldsymbol{d}^k$ required in the path to finding $\boldsymbol{d}_{opt}$; or, even more challenging, as many *trajectories* required to find the optimal set of strategies $(\boldsymbol{D}_{tk}(.))$.

The elementary methods that can be considered for any of those challenges are the following:

- Elementary design of experiment, such as a fixed-step grid of values discretising the space of control variables (as is done in Tagushi's approaches for robust design). As mentioned in Chapter 7, phenomenological knowledge and regularity properties are essential in order to choose the design properly and limit the interpolation errors.

- Gradient-based search algorithms: conjugate-gradient or any more sophisticated choice that adapts the type of uncertainty propagation undertaken to compute approximately-enough the expectations and gradients of expectations according to the stage of the optimisation loop (e.g. grossly-estimated at first iterations and later more finely sampled around the optimum).
- Adaptive response-surface: (i) polynomial, BBN or neural networks replacing the deterministic system model or its computationally-intensive sub-parts, or (ii) probabilistic response surfaces, such as kernel-smoothened distributions.

Optimisation may be much simplified in some privileged cases:

- If $\boldsymbol{d}$ intervenes only in the post-processing of a fixed system model output: think of the cases where there is a spatial or temporal field of quantities of interest $c_Z(\boldsymbol{x},t)$, representing, for instance, the spatial field of probabilities to find more oil than a given threshold, or of expected net benefits for a well location in a drilling project: optimisation may simply mean selecting the most interesting location $\boldsymbol{x}$ (or instant t) to drill. In those cases, no additional computation is needed.
- If dependence to $\boldsymbol{d}$ is quadratic: then a closed-form solution may be found.

Dedicated methods have been developed by the specialists in stochastic optimisation (e.g. Infanger, 1994). Further research is still needed in that field, particularly because of the heavy *curse of dimensionality*. A key requirement is that the numerical simplifications and approximations involved in the system model, uncertainty model or loops of uncertainty propagation to sample the expectations should not introduce more propagation uncertainty than the 'true' epistemic and aleatory uncertainty in the variables of interest for decision-making, thus calling for more *robust methods* in the sense developed in Chapter 7. This is all the more so when moving into real-options analysis or stochastic dynamic programming.

8.4 The promise of high performance computing

8.4.1 The computational load of risk and uncertainty modelling

Risk and uncertainty studies lead inevitably to a number of calls to the code for the phenomenological system models that is much larger than for the traditional 'best-estimate' study (a single 'penalised' calculation). Before adding the optimisation layer of this chapter, standard probabilistic uncertainty propagation involves at least several dozen or hundreds of calculations even when using the accelerated methods reviewed in Chapter 7, and much more when considering the robust computation of low-probability risk measures. Table 8.6 provides some orders of magnitude of the increase multiple in the computational budget with respect to a traditional engineering computation that does not represent uncertainty while already representing a large CPU cost because of the size of meshes and complexity of phenomenological equations.

Indeed, besides the traditional bottlenecks of structural reliability of highly-reliable systems, yet more exacting *computational challenges* come with the following:

- two-level uncertainty studies, typically involving epistemic quantiles on top of rare aleatory quantiles or low exceedance probabilities (Limbourg, de Rocquigny, and Andrianov, 2010);
- Bayesian modelling of input uncertainty, and the issue of *computing the posterior* with more economical algorithms than MCMC (cf. an example in Perrin *et al.*, 2007)) when involving large physical models for which 10 000 or more iterations for convergence are completely out of reach;
- stochastic optimisation.

Table 8.6 Computing challenges: number of code runs per type of analysis.

Task	Number of code runs	Comments
Risk computation – uncertainty propagation		
Level 1 – *EZ*, *Var Z*	$N = 10^1 - 10^2$	
Level 1 – quantile z^α or $P(Z > z_s)$	$N = 10^1 p$ (or 10^2) to $>10^4$ according to α	Linearly increases with $1/(1-\alpha)$ for robust methods (indep. of *p*), or less quickly for accelerated ones (+ linear with *p*)
Level 2	$N = 10^1 - 10^2 p$	As an additional factor multiplying the level-1 increase multiple
Importance ranking – sensitivity analysis		
For *var Z*	$N = 3p$ to $>10^3 p$ (or even a^p)	Linear to exponential increase depends on model regularity and cross-input interactions
For quantile/exceedance proba	largely unknown	Reserach needed; should cost at least as much as the level-1 quantile risk computation.
Model calibration – data assimilation		
«classical» assimilation/ calibration	$N = 10^0 p$ to $>10^1 p$	Depending on linear-Gaussian or non-linear/non-Gaussian hypotheses
full inverse probabilistic identification	$N = 10^1 p$ to $>10^4 p$	Idem

The three contexts requiring stochastic optimisation and identified above are much greedier: (i) the 'mixed deterministic-probabilistic' paradigm, (ii) optimisation under uncertainty, also known as stochastic optimisation, or (iii) inverse probabilistic modelling of the sources of uncertainty. In the former case, computational greediness is associated with the need for nesting a maximisation by intervals for the deterministic components with, for each point, a conditional probabilistic calculation for the probabilised variables. In the latter two cases, nesting optimisation algorithms with propagation by sampling is required in the general (non-linear, non-Gaussian) case, generating a large computational cost. It is harder to get reliable orders of magnitude as this depends broadly on the 'smoothness' of the functions to be optimised, although the optimisation layer would typically multiply the basic multiplicative factors already mentioned (typically $N = 10^1 – 10^2$ for uncertainty propagation required to compute *EZ*) by an factor generally reckoned to grow exponentially with the dimension of the control variables (***d***) or possibly of the time discretisation, hence at least 10^3 to 10^5 runs.

As a tentative synthetic figure, *a few thousands of simulations* could generally be considered as a recommendable average budget for risk and uncertainty studies in industrial models. Even if propagation may sometimes be optimised to require one order of magnitude less (given enough regularity and/or moderately-exacting risk measure), one or a few loops of sensitivity analyses would be desirable in order to understand the impact and importance of the input structure. In turn, this means that no more than a few tens of independent inputs could be reasonably studied within the input uncertainty structure (or even less if a high level of interaction between input components requires going into elaborate sensitivity methods). Conversely, going into more than a few thousand simulations should be left for quick and simple models or duly validated reference codes, or otherwise generally prefer to develop a meta-model or allow for the complex model to be further qualified and input data be completed in order not to overspend computing time.

8.4.2 The potential of high-performance computing

Handling uncertainty is therefore a great client of high-performance computing (HPC), an area into which industrial players are gradually joining the traditional academic champions, as shown by the halls of fame of supercomputing (www.top500.org). Aside from the internal optimisation of the code solvers themselves and their parallelisation, the numerical challenges posed by large-scale uncertainty treatment, depending on the propagation methods adopted, may indeed benefit from massively distributed computing. HPC is firstly useful in order to accelerate uncertainty propagation: in particular the computation of low probabilities of exceedance or rare quantiles. Monte-Carlo Sampling is a trivial recipient for computer distribution, and may be viewed as the very historical origin of computing remembering that Von Neumann's ENIAC machine in 1946 was designed essentially for Monte-Carlo neutronics. Although being a straightforward 'embarrassingly-parallel' scheme through its *i.i.d.* sampling algorithm, technological issues as well as a question of computing rights may still upraise. The lower layers of big machines need to be configurated in order to accommodate as many (identical and stable) executables of the physical-numerical codes as the number of processors instead of merely contributing to the execution of a single big code through, for example, domain decomposition of a mechanical solver.

Beyond simple Monte-Carlo, accelerated sampling (e.g. through LHS, stratification, importance sampling and so on, see Rubinstein and Kroese, 2007) is already a largely-disseminated practice in the nuclear industry. Like the other advanced uncertainty propagation algorithms or stochastic optimisation, advanced numerical development is required in order to take full advantage of parallel computing (as the algorithms are originally partially sequential) or accelerate through more or less automated code differentiation to benefit from the gradients, all of which represents an area with great research potential for computer science.

As a starting example, pioneering computations were undertaken by the author through Monte-Carlo sampling in a thermal-hydraulic computational fluid dynamics model with a partially phenomenological irregular response. Each *single* run involved a computing time of 5–10 days on a standard single-core CPU (as performed in 2007) according to the region of space, the numerical convergence of the underlying thermal hydraulics model depending partly upon the parameter values. Using high performance computing (Berthou, Hamelin, and de Rocquigny, 2009) with an IBM Blue Gene/L 8000-core machine (*23 TeraFlops*), the sampling size could be increased up to $N \sim 8000$ (in fact $N = 7733$ due to system computing failures) so as to investigate the decrease in the propagation uncertainty reflected in the Wilks estimator as compared to the final *95 %*-quantile. The result confirmed the expected $1/\sqrt{N}$ rate of decrease of propagation uncertainty that is added in the $z^{\alpha}{}_{\beta}$ estimator (i.e. in the width of the confidence interval). In particular, if the risk measure is taken as the upper $\beta = 95$ % bound of the confidence interval in assessing the $\alpha = 95$ % quantile of peak clad temperature, an increase of the computational power from 59 runs (59 being the lowest number of runs required to estimate a 95–95 % Wilks quantile, see Chapter 7) up to 7733 runs appeared to reduce by 30–80°C on average the overestimation of the quantile. The same machine was tested meanwhile for another computational challenge that involved stochastic dynamic programming requiring about one day of computing to optimise over 500 time steps, 35 commands and a state discretisation involving a few thousands of points in the design of the experiment.

This was a key demonstration of the value brought by combined high performance computing and robust uncertainty computation in such a case, as a similar 7733-set of Monte-Carlo sampling would require 20–30 years of computing time with standard single-core CPU run sequentially. Note also that such a massive computer experiment involving random sampling over a very-high dimensional space

(70 inputs) in irregular models also resulted in a non-negligible rate of computing failures, either due to low-layer computational failures or to slow convergence of the numerical algorithms involved in the underlying system model. An interesting by-product of such an uncertainty study was the proper re-calibration and improvement of the physical system model itself.

Exercises

There follows a list of exercises referring to the concepts developed in this chapter:

8.1 Perform sensitivity analysis of the total cost over a time period as a function of the uncertain inputs, grouping them into homogeneous subsets such as the yearly flood flows and so on.

8.2 Study the impact of sampling $G'(X, d_1, d_2) = G(X, d_1) - G(X, d_2)$ at once instead of two successive independent samplings of $G(X, d_1)$ and $G(X, d_2)$ when comparing the two designs d_1 and d_2 for different types of risk measures. Discuss the pros and cons in design comparison; more generally, what about the case of optimisation over a grid $\{d_k\}_k$ of competing designs?

8.3 Study the impact of modelling when switching aleatory uncertainty (independently sampled every year) into epistemic uncertainty (sampled once and constant over the time period) on the variance of the total cost over time, for instance for the Strickler friction coefficient or the economic uncertainty factor.

8.4 Study the interest of a real options approach for the flood risk problem. Considering first the risk model only, does a delayed decision to build a tall dike bring any benefit on average (or increase the risk?)? Assume then that a forecast model becomes available, taking the example of the log-autoregressive pseudo-hydrological model in Section 5.5. What should be the orders of magnitude of the times of interest and building delays, in comparison with the length of correlation, for option-taking to become beneficial?

References

Allais, M. (1953) Le comportement de l'homme rationnel devant le risque: critique des postulats et axiomes de l'école américaine. *Economica*, **21**(4), 503–546.

Barberà, S., Hammond, P.J. and Seidl, S. (eds) (1998a) *Handbook of Utility Theory, Volume 1 Principles*, Kluwer Academic Publisher.

Barberà, S., Hammond, P.J. and Seidl, S. (eds) (1998b) *Handbook of Utility Theory, Volume 2 Extensions*, Kluwer Academic Publisher.

Bedford, T. and Cooke, R. (2001) *Probabilistic Risk Analysis – Foundations and Methods*, Cambridge University Press.

Bellman, R.E. (1957) *Dynamic Programming*, Princeton University Press, Princeton, N.J.

Berthou, J.-Y., Hamelin, J.-F. and de Rocquigny, E. (2009) XXL simulation for XXIst century power systems operation. *International Journal of High Performance Computing Applications*, **23**(4), 361–365.

Beyer, H.-G. and Sendhoff, B. (2007) Robust optimization – A comprehensive survey. *Comput. Methods Appl. Mech. Engrg.*, **196**, 3190–3218.

Charles, T., Guéméné, J.M., Corre, B., Vincent, G. and Dubrule, O. (2001) Experience with the quantification of subsurface uncertainties. Proc. of the Soc. of Petrol. Eng. Asia Pacific Oil and Gas Conference, Jakarta, April 2001.

Chen, X.C., Hasselman, T.K. and Neill, D.J. (1997) Reliability based structural design optimization for practical applications. In Proceedings of the 38th AIAA/ASME/ASCE/AHS/ASC Structures, Structural Dynamics, and Materials Conference, number AIAA-97-1403, pp. 2724–2732.

Dixit, A.K. and Pindyck, R.S. (1994) *Investment under Uncertainty*, Princeton University Press, Princeton, N.J.

Granger Morgan M. and Henrion M. (1990), Uncertainty – A Guide to Dealing with Uncertainty in Quantitative Risk and Policy Analysis, Cambridge University Press.

Henrion, M. (1982), OT The Value of Knowing How Little You Know: The Advantages of a Probabilistic Treament of Uncertainty in Policy Analysis. PhD diss., Carnegie Mellon University, Pittsburgh.

Henrion, M. (1989) *The Value of Knowing How Little You Know: Part I, manuscript, Department of Engineering and Public Policy*, Carnegie Mellon University, Pittsburgh.

Infanger, G. (1994) *Planning under uncertainty*, Boyd & Fraser Publ.

Lemaire, M., Chateauneuf, A. and Mitteau, J.C. (2005) *Fiabilité des Structures: Couplage Mécano-Fiabiliste Statique*, Hermes Science Publication.

Limbourg, P., de Rocquigny, E. and Andrianov, G. (2010) Accelerated uncertainty propagation in two-level probabilistic studies under monotony. *Reliability Engineering and System Safety*, **95**, 998–1010.

Lind, N. (2007) Discounting risks in the far future. *Reliability Engineering and System Safety*, **92**, 1328–1332.

Mun, J. (2002) *Real Options Analysis: Tools and Techniques*, Wiley.

Perrin, F., Sudret, B., Pendola, M. and de Rocquigny, E. (2007) Comparison of Monte Carlo Markov Chain and a FORM-based approach for Bayesian updating of mechanical models. Proc. of 10th Int. Conf. on Appli. of Stat. Proba. in Civil Engineering (ICASP), Tokyo.

Phadke, MS. (1989) *Quality Engineering Using Robust Design*, Prentice Hall, N.J.

Quiggin, J. (1982) A theory of anticipated utility, *Journal of Economic Behaviour and Organization*, **3**, 323–343.

Rubinstein, R.Y. and Kroese, D.P. (2007) *Simulation and the Monte Carlo Method*, 2nd edition, Wiley & Sons.

Schuëller, G.I. and Jensen, H.A. (2008) Computational methods in optimization considering uncertainties – an overview. *Computational Methods and Applications in Mechanical Engineering*.

Taguchi, G. (1987) System of experimental design, in *Don Clausing*, vol. **1 & 2**, UNIPUB/Krass International Publications, New York.

Thanedar, P.B. and Kodiyalam, S. (1992) Structural optimization using probabilistic constraints. *Structural Optimization*, **4**, 236–240.

Uncertainty in Industrial Practice, Chapter Hydrocarbon Exploration

Van Neumann, J. and Morgenstern, O. (1944) *Theory of Games and Economic Behaviour*, Princeton University Press, Princeton, N.J.

9

Conclusion: Perspectives of modelling in the context of risk and uncertainty and further research

The book has introduced a variety of methodologies, techniques and algorithms now available for the analyst wishing to model consistently in the context of risk and uncertainty and to best value all quantitative information available to support decision-making. However, important scientific challenges lie ahead. The whole rationale of this book was intended to show that the coupling of probabilistic approaches with large-scale physical-numerical models generates new requirements from applied mathematics, statistics and numerical analysis: it also renews old epistemological debates and decision theory issues. The scientific challenges will be summarised in Section 9.1. More importantly perhaps, practical dissemination of the techniques and methodologies also raises a number of challenges that will be recalled briefly in Section 9.2.

9.1 Open scientific challenges

Once probabilistic computation as well as sensitivity analysis science and techniques start disseminating, the need for careful handling of data and expertise designed to perform accountable *input uncertainty modelling* becomes all the more important. Indeed, it appears as the typical top priority for industrial applications. One is likely to ask what type of input probabilistic model should be fed into uncertainty and sensitivity analysis when considering largely missing data sets. In that respect, *Bayesian settings* still have much further to go, regarding notably the issue of choosing accountable *priors* in the risk context where *distribution tails* are important. However, Bayesian approaches should not disregard the attention paid to

Modelling Under Risk and Uncertainty: An Introduction to Statistical, Phenomenological and Computational Methods, First Edition. Etienne de Rocquigny.

the underlying *identifiability* issues that are essential in the risk context. The development of sophisticated physical-numerical models spurred on by computational availability tends to enrich input parameterisation much quicker than the true amount of data available to calibrate the input distributions: an elementary requirement, however, is that *modelling should not increase risk.*

There is a delicate issue regarding the modelling of input distribution tails when quantities of interest involved in the study are *extreme quantiles* or exceedance probabilities. While the extreme value theory is quite abundant and consensual when dealing with scalar random variables (cf. the review by Coles, 2001), the modelling of multivariate uncertain inputs is still a large field of research, especially in the real cases where multivariate samples are rare or impractical to acquire. In the view of the author, a promising area is to couple physical model simulation and scalar extreme value statistics, as was initiated within the current research in the case of extreme low sea events (Lafon and de Rocquigny, 2006). Phenomenological models of extreme events could, in principle, provide for missing multivariate data sets, the conjunction of multiple hazards of uncertain dependence or regional dependence which proves essential for spatially-distributed industrial assets. Think of large ocean/estuarine models run over decades under randomised climatic forcing that can provide multivariate samples of maximal coastal waves in conjunction with barometric tides and possibly estuarine floods, all of which are difficult to observe directly through empiric records. Extreme value modelling also faces challenges regarding non-stationarity, abrupt phenomenological changes or the issue of long-range dependence within complex systems that may characterise the most disruptive high-loss events involved in 'systemic risk'.

As was explained in Chapter 6, the associated approach is to calibrate such phenomenological models under uncertainty through advanced inverse techniques. While robust two-level algorithms were pioneered recently, other algorithms would obviously be needed in order to further address the need for *inverse probabilistic modelling* for a larger range of situations regarding physical model regularity and input/output dimensions. Those could be: the use of mixed moment methods and later local likelihood maximisation in order to optimise the number of model runs; the use of adaptive response surfaces, be it through classical polynomials, neural networks or even chaos polynomial expansion which would approximate the physical model itself, while relying on an 'exact' likelihood maximisation process, and so on. Once again, important scientific computing and statistical estimation challenges arise with these application-rich research perspectives.

High-performance computing has only very recently been pioneered so as to support the associated risk and uncertainty simulations. A number of new scientific issues are likely to be raised such as parallel or distributed computing in a manner quite different to that envisioned by the solving of one big job, that is a massive mesh associated with a deterministic set of coupled PDE. Finance already routinely relies on Monte-Carlo sampling; some companies use very large machines to distribute the computations for that purpose, but this is much rarer in other industries and is generally limited to brute-force Monte-Carlo. Alternative propagation methods (as discussed in Chapter 7) do pose trickier parallelisation issues as the *i.i.d.* sampling gives way to adaptive sampling, the density or design of experiment depending on previous results.

In all such cases, the particular context of risk analysis for serious events demands an increased control of the propagation error or uncertainty introduced by the limitation in the number of samples, if not the increasing numerical rounding errors or computational failures that are likely to grow in many-core processors: an extension of the type of research discussed in Chapter 7 in relation to robust propagation is needed.

At the end of the day, large-scale industrial applications involve not only the modelling and simulation results, but also the whole issue of simulation-based *decision-making*. While extended modelling and simulation make for a much richer feedback for the analyst, this obviously raises both practical and epistemological/organisational/regulatory issues in handling the mass and complexity of analytical data and risk metrics in risk management and industrial certification processes. Advanced *visualisation* techniques could greatly assist both uncertainty and sensitivity analysis when handling high-dimensional models and data sets, or when trying to disseminate easy-to-grasp decision-making under uncertainty. While scientific and information visualisation have largely permeated the field of model-based design, and graphical methods have become a must in sensitivity analysis (Saltelli *et al.*, 2008; Kurowicka and Cooke, 2006), visualisation under uncertainty should still offer broad scientific perspectives. This involves typically a more user-friendly depiction of the limitations of model limitations or simulation inaccuracies beyond confidence curves around point predictions, as well as more interactive settings helping the modeller to further refine model results on the basis of large-screen results and more deeply the human-computer interaction issues in handling uncertainty.

Eventually, *epistemological* as well as *organisational/regulatory issues* are likely to arise with the dissemination of extended risk metrics involving, for example, double probabilistic settings for risk and uncertainty modelling. The old debate on separating epistemic/aleatory uncertainty that included a move in the 2000s into the alternative use of non-probabilistic settings – such as Demspter/Shafer or evidence theory – involves not only deep epistemological and computational issues but also the organisational/regulatory issues when moving into large-scale industrial applications. The regulation of industrial facilities took a long time to open up traditionally-deterministic rules into probabilistic arrangement – if it did so at all, which was not the case in every industrial country nor in all branches of industry (see de Rocquigny, Devictor and Tarantola, 2008). It is known to present a number of organisational difficulties in internal engineering organisations, as experienced by the author in launching internal training and industrial computing development designed to disseminate the approaches inside EDF or large industrial companies associated with the programme. Further major challenges appear with the introduction of double probabilistic or non-probabilistic risk measures in some cases, especially regarding the issue of decision-making and model validation in the context of rare risk or deep uncertainty. More research on the epistemological and organisational issues attached to those developments is anticipated.

In the *long term* (7–10 years), a number of stronger challenges could be considered, as described briefly here:

- Generalise *physical uncertainty modelling* through *fully probabilised micro-macro statistical physics.* Although being statistical at the molecular/crystal network scale, multi-scale physical modelling (e.g. *ab initio* computation in material science) is pioneered presently in a macroscopically-deterministic approach (i.e. structural characteristics are deterministic at a macro-scale) with already-gigantic computational obstacles, and modelling difficulties (e.g. how should scale interfaces be homogeneized, both physically and probabilistically?). This could change in the long term with multi-scale modelling probabilised at all scales.
- Generalise time-based modelling in large physical models through *fully uncertain complex physical dynamics.* At the present time, uncertainty description in physics is either well-detailed in a static models (complex physical models, and elaborated non-Gaussian parametric uncertainty, but limited to static parameters) or grossly-approximated in dynamical models (such as stochastic PDE, based on brownian-derived uncertainty factors that are limited to Gaussian-like pdfs and highly-simplified physical models). This is because of severe computational difficulties, as well as challenging statistical non-identifiability or data insufficiencies to calibrate such a model under uncertainty. In the long term, it could become possible to integrate stochastic dynamics inside sophisticated non-Gaussian uncertainty description as well as fully developed coupled multi-physical models.

Both could become fascinating scientific challenges with pioneering applications . . . given not only the continuous rise of computing power (HPC), but also the greater availability of empirical data at all scales (‘big data’).

9.2 Challenges involved by the dissemination of advanced modelling in the context of risk and uncertainty

Beyond the open scientific issues, a number of generic challenges may be encountered in applying industrial-wide modelling for decision-support. Here are a few key factors for success linked to changes in industrial and engineering practices as well as the regulation of risk:

- Cultural and organisational: the consideration of uncertainty disrupts to some extent the traditional engineering habits and regulatory settings and often faces claims of costly, confusing or insufficiently guided sophistication, as risk analysis proves generally difficult to communicate to clients or furthermore to the general public.
- Policy: decision-making under uncertainty requires the tricky choice of quantities of interest or risk measures (e.g. expected utilities) that should properly represent risk aversion or behaviour facing uncertainty.
- Technical and computational: regarding, for instance, the need to tackle *large-scale physical-numerical models* as most design or operation control studies rely on sophisticated engineering processes involving complex numerical codes, all the more so since high performance computing unlashes ever-increasing meshing sizes or micro-macro equation formulations. They require large CPU budget to run and are fed by quite heterogeneous sources of information (noised data on various physical variables, incomplete expertise), far away from the closed-forme examples originally studied in the literature.

'True uncertainty accounts for the peculiar income of the entrepreneur' (Knight, 1921). Eventually, honesty, parsimony and a great deal of common sense should always guide the analyst and the decision maker ambitious to model reality for the purpose of controlling it; which would always remain unknown to some extent, and often surprise our prior opinion, though hopefully not exempting the courage to act.

References

Coles, S. (2001) An introduction to statistical modelling of extreme values, in *Springer Series in Statistics*, Springer-Verlag.

de Rocquigny, E., Devictor, N. and Tarantola, S. (eds) (2008) *Uncertainty in Industrial Practice, A Guide to Quantitative Uncertainty Management*, John Wiley & Sons, Ltd.

Knight, F.H. (1921) *Risk, Uncertainty & Profit.*

Kurowicka, D. and Cooke, R. (2006) *Uncertainty Analysis with High Dimensional Dependence Modeling*, Wiley Series in Probability and Statistics, John Wiley & Sons.

Saltelli, A., Ratto, M., Andres, T. *et al.* (2008) *Global Sensitivity Analysis: The Primer*, Wiley.

Lafon, F. and de Rocquigny, E. (2006) Niveaux marins extrêmes: contrôle d'une méthode analytique par simulation d'un processus couplé, 38èmes. *Journ. Franc. de Stat., Clamart.*

10

Annexes

10.1 Annex 1 – refresher on probabilities and statistical modelling of uncertainty

This section provides a short and elementary refresher on the basics of probabilistic and statistical modelling as a means of representing *uncertain* variables.

10.1.1 Modelling through a random variable

Suppose that a given property of the system denoted as x (temperature, toughness, oil price, water flow...) has been observed at several past times – or at a given date for several spatial locations, indivuals of a population, components of a system – issuing the following data sample or series of $N = 20$ observations $x(t_j)$ (Figure 10.1).

Two types of questions could be asked:

- What is the next value for the property (short-term)?
- What value would the property take in the future (possibly mean-term)?

A statistical (or probabilistic) representation of the variable X reckons that it is too hard to understand, causally explain and/or predict what happened in the observed data. No obvious regular pattern could be discerned so that it is quite puzzling to anticipate (i) what will happen at the next time step and even worse (ii) what will happen later on. The phenomenon is reckoned to be '*erratic*' or '*random-like*'.

Instead of predicting precise values, a statistical approach limits itself to *counting the frequency of occurrence* of typical values in the observations assuming that they *should happen likewise* in the future, for example 35 % of the time it falls in the medium interval of the six intervals splitting the observed range (see the *histogram* in Figure 10.2). The uncertainty model (denoted f_X) is the following discrete

Modelling Under Risk and Uncertainty: An Introduction to Statistical, Phenomenological and Computational Methods, First Edition. Etienne de Rocquigny.

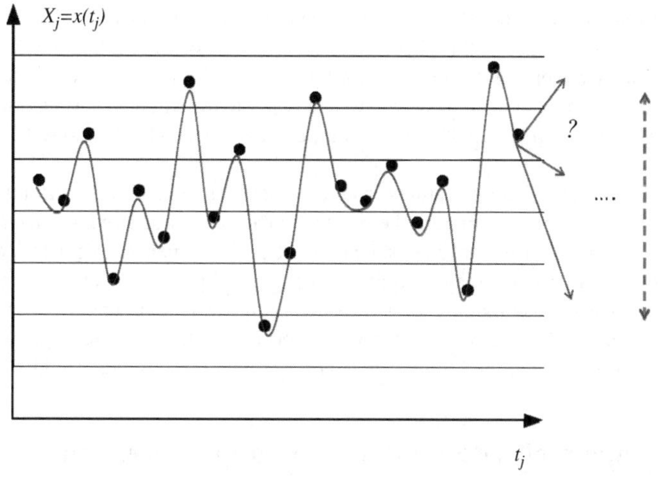

Figure 10.1 *Erratic values of a property over time.*

probabilistic distribution, which contains the specification of the successive intervals of possible values with the corresponding proportions (per definition = *probabilities*) of occurrence:

$$\forall j, k \quad P(X_j \in [x^k, x^{k+1}]) = p_k \tag{10.1}$$

$$X_j \sim f_X(.|\boldsymbol{\theta}) = \{[x^k, x^{k+1}], p_k\}_{k=1,\ldots K} \tag{10.2}$$

In that case, the *parameter vector* of the uncertainty distribution of *X* contains the discrete probabilities of each modality or interval of variation: $\boldsymbol{\theta}_X = (p_k)_{k=1\ldots K}$.

Note therefore the essential features of a statistical model:

- It *will not* predict precise values either for the short or for the longer-term.
- It does, however, predict a bit *more than* a simple (min-max) deterministic *range* observed in the past: relative frequencies or likelihoods that should occur in average over the mean-term.

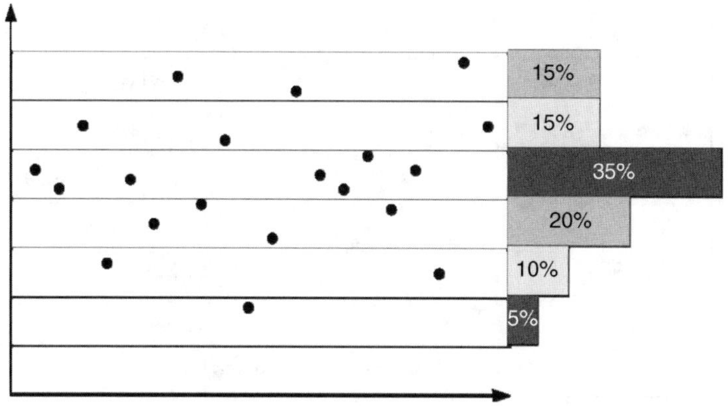

Figure 10.2 *An empirical histogram of 20 observations.*

- It fundamentally assumes that the phenomenon is analogous in the future (in terms of range of variation and proportions of occurrence within that range) to what was observed, that is *stationary*.
- It will always remain a mere *mathematical model* of an uncertain reality: whether the physics or phenomenologies of the real system do truly behave according to the Kolmogorov's axioms of probability theory would generally remain unclear or even un-decidable (see also Sections 4.1 and 10.3)

It is essential to keep each of those features in mind whenever modelling uncertainty (or variability, or error, or lack of knowledge . . .) through the mathematical model of a *random variable*. The first one may not fit the expectations or mental representations of decision makers, regulators or of the general public; in turn, the third property may be quite challenging to prove in practice (see Section 10.1.2. More elaborate statistical or combined statistical-deterministic models could partially relieve those features, as introduced in Chapters 5 and 6, through deterministic trends in the mean value or the range (or standard deviation) although the '*erratic*' *behaviour* around those trends would still keep the same fundamental features.

10.1.2 The impact of data and the estimation uncertainty

Suppose then that 20 new observations have been made available (Figure 10.3). Most likely, those new observations would resemble to some extent to the 20 previous ones, but could lead to slightly different proportions of occurrence of the observed intervals of variations and even possibly extend the observed domain.

Such new data (denoted formally as *IK'* as a distinction with the original 20-sample data *IK*) would thus lead to updating the probability distribution best describing the plausible variability (or uncertainty) about the phenomenon, with a slightly modified vector of parameters $\boldsymbol{\theta}'$.

$$X_j \sim f_X(.|IK') = \{[x_k, x_{k+1}], p'_k\}_{k=1,\dots K'} \tag{10.3}$$

$$\hat{p}'_k = \frac{1}{n}\sum_{j=1}^{n} 1_{x_j \in [x_k, x_{k+1}]} \tag{10.4}$$

This essential process thus illustrates that the estimation of a probabilistic model ($\hat{p}'_k$, the hat figuring the *estimator*) for an uncertain variable inevitably suffers from fluctuation due to the level of data available.

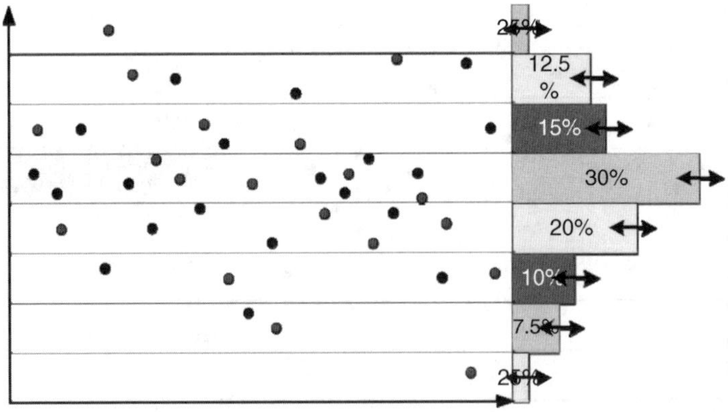

Figure 10.3 *An empirical histogram of 40 observations.*

When projecting into the future, there is thus *estimation uncertainty* in the proportions that will be observed over the range of variation. This affects similarly any of the derived characteristics of the variable that would be estimated – an *estimator* being by definition a quantity computed on the basis of the observations – such as its *empiric* mean or variance (square of standard deviation), two fundamental figures of the average value and order of magnitude of variation around that value:

$$\begin{aligned} \bar{x} &= \frac{1}{n}\sum_{j=1}^{n} x_j \\ s^2 &= \frac{1}{n}\sum_{j=1}^{n}(x_j - \bar{x})^2 \end{aligned} \tag{10.5}$$

Consequently, the empiric mean or variance computed on a sample should only be viewed as 'approximate' values which would slightly change if more data would become available. A fundamental result of the *Law of Large Numbers* and *Central-Limit Theorem* guarantees that such fluctuation of $\hat{p}_k^l$, $\bar{x}$ or s^2 decreases at speed $1/\sqrt{n}$ when the size of the sample grows given the following fundamental and intuitive conditions:

- Observations are *independent* (see Section 10.1.3) and *identically distributed* (*i.i.d.*); to put it simply, the distribution remains constant over the time period of observations.
- (For the mean and standard deviation only) the phenomenon *varies* at a *finite* (or bounded) '*rate*'; mathematically, the expectation, variance (and possibly higher-order moments) are bounded.

The *former* condition typically requires one to consider observations over a limited time period at the scale of, *for* example, climate change, or operational evolutions; although, conversely, time steps between observations are required to be long enough to keep them independent, see Section 5.5. Though normal phenomenology would fit the *latter* condition, this may not be the case with some kinds of *extreme events* – typically modelled through so called *fat-tail* distributions, cf. Sections 4.3.5, 5.2.1 and 7.2.2 – for which mean or variance become inappropriate though statistical estimates of interval probability (and more importantly for risk assessment, threshold exceedance) remain relevant.

In more formal terms, the *Law of Large Numbers* and *Central-Limit Theorem* can be expressed as follows (with many extensions beyond these elementary formulations). Given *i.i.d.* observations with finite expectation, the empiric mean converges (almost surely with the *Strong Law of Large Numbers*) to the expectation:

$$\begin{aligned} &(X_j)_j i.i.d., \quad |EX| < +\infty \\ &\overline{X} = \frac{1}{n}\sum_{j=1}^{n} X_j \xrightarrow[N\to\infty]{} EX \quad a.s. \end{aligned} \tag{10.6}$$

Given additionally that the variance is finite, the empiric mean is distributed asymptotically around the expectation with a variance decreasing in $1/\sqrt{n}$ (*Central-Limit Theorem*):

$$\begin{aligned} &(X_j)_j i.i.d., \quad \operatorname{var} X < +\infty \\ &\sqrt{n}\left(\overline{X} - EX\right) \xrightarrow[N\to\infty]{} N(0, \operatorname{var} X) \quad \textit{in law} \end{aligned} \tag{10.7}$$

Supposing n to be very large (i.e. the so-called 'asymptotic' conditions), the proportions would thus only deviate imperceptibly from some (unknown) stable proportions, or from the (unknown) fixed mean or standard deviation of the phenomenon. In the context of uncertainty modelling, the uncertainty (or variability) in X would be called aleatory/irreducible (or random or level-1 ...) while that in the proportions $\boldsymbol{\theta}$ would be called epistemic/reducible (or estimation or level-2 ...) in the sense that the latter decreases with the size of the data sample. However, except for textbook theoretical examples, those 'true' or asymptotic values will *always remain unknown* to the analyst as the duration of real-life observations of systems for decision-making is always limited in time. Given a fixed sample made available at a given present time, both kinds of uncertainty combine into a single view of the likely values describing the uncertainty in the future value of the system property at a given future time.

10.1.3 Continuous probabilistic distributions

It is quite useful to consider *continuous probabilistic distributions* as they are generally much more convenient to manipulate than collections of discrete proportions.

A continuous probability distribution model replaces the discrete distribution (or discrete density or *pdf*):

$$X_j \sim f_X(.|\boldsymbol{\theta}) = \{[x_k, x_{k+1}], p_k\}_{k=1,\ldots K} \tag{10.8}$$

by the following continuous distribution function (or continuous density or *pdf*):

$$X \sim f_X(x|\boldsymbol{\theta}_X) \tag{10.9}$$

so that the *cumulative distribution function* (*cdf*) denoted as F_X, defined as the probability for X to stay under a given threshold, switches from a standard sum:

$$P(X \leq x_k) = F_X(x_k) = \sum_{l<k} p_l \tag{10.10}$$

to an integral:

$$P(X \leq x) = F_X(x) = \int_{u<x} f_X(u|\boldsymbol{\theta})du \tag{10.11}$$

Empirically, a continuous pdf may be seen as the limit of a discrete pdf with interval widths $[x_k, x_{k+1}]$ that become infinitely small $[x, x+dx]$. Such an approximation refers to the asymptotic refinement of the x-scale which is quite different to the asymptotic behaviour regarding the size of the data sample. Similar to discrete models, the value of $\boldsymbol{\theta}_X$ suffers the impact of the sampling variability when trying to estimate it on a set of data. It becomes itself a random variable, as being computed on the basis of random observations. It will be called a second probabilistic level of *epistemic* estimation uncertainty in the sense of being reducible, in theory at least, with access to more data at speed $1/\sqrt{n}$ (see Figure 10.4).

10.1.4 Dependence and stationarity

Consider more than one uncertain property characterising a system, say, for instance, the temperature and windspeed X_j^1 and X_j^2 at a given place for similar dates t_j, $j = 1\ldots n$. A fundamental feature for model-based decision-making or risk assessment concerns the *dependence* or *joint behaviour*

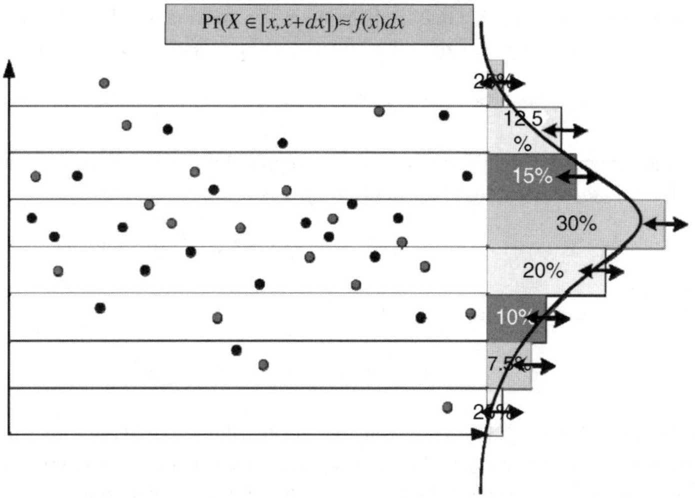

Figure 10.4 *Discrete and continuous distribution functions.*

of the two properties: assuming $X_j^1 \leq x^1$, for example temperature lower (or higher) than a prescribed threshold, how does that impact upon the (uncertain) value of the other property X_j^2, for instance being higher (or lower) than a prescribed threshold x^2? Is it likelier for the wind to be stronger when the temperature is lower, do the likely values of windspeed remain similar to the overall average, whatever the temperature? Figure 10.5 illustrates the analysis of a bivariate joint sample: counting the empiric frequencies of the sample shows that the contrary tends to be true. Assuming that the temperature is lower than x^l (which corresponds to the 16 observations standing to the left of the vertical line out of the total 40) proportionally, there seem to be fewer cases (4 out of 16, i.e. 25%) for windspeed X^2 being higher than threshold x^2 than on average over the entire population (24 out of 40, i.e. 60%).

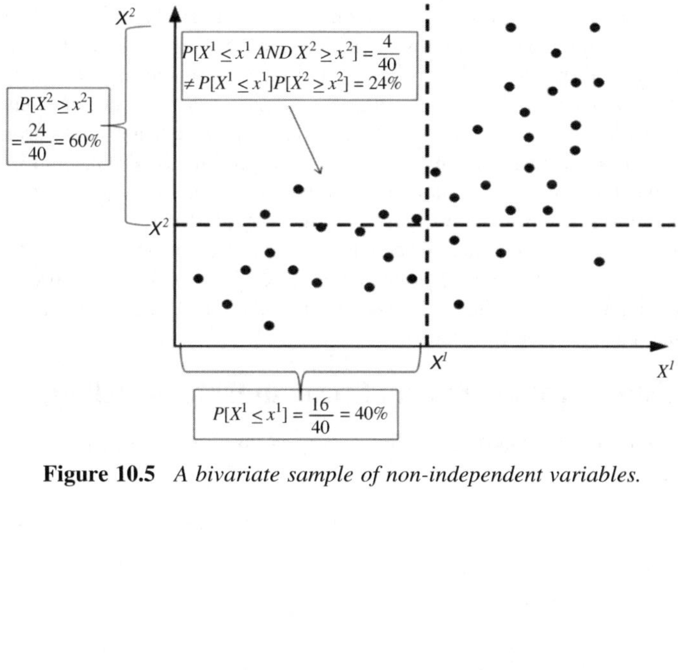

Figure 10.5 *A bivariate sample of non-independent variables.*

Independence characterises the fact that the behaviour – or information about the behaviour – of one property *does not influence in any way* the other, formulated mathematically as follows:

$$\forall x^1, x^2; \quad P[X_j^1 \leq x^1 \; AND \; X_j^2 \leq x^2] = P[X_j^1 \leq x^1].P[X_j^2 \leq x^2] \tag{10.12}$$

Or equivalently, introducing the definition of a *conditional* probability which generates the concept of the 'likelier values for X_j^1, *knowing something* about the values of X_j^2':

$$P[X_j^1 \leq x^1 | X_j^2 = x^2] = P[X_j^1 \leq x^1 \; AND \; X_j^2 \leq x^2]/P[X_j^2 \leq x^2] \tag{10.13}$$

Then independence can be reformulated as follows:

$$\forall x^1, x^2; \quad P[X_j^1 \leq x^1 | X_j^2 \leq x^2] = P[X_j^1 \leq x^1] \tag{10.14}$$

In other words, knowing that $X_j^2 \leq x^2$ does not change the distribution of likely values for X_j^1 and this is true whatever the threshold x^2 considered. Such a property is quite strong and requires in theory a large sample and many tests for the varying values of both thresholds to be checked.

There are many forms of *dependence*, that is cases where such a strong property cannot be established plausibly. This includes the very popular but much weaker property of *linear correlation*, meaning that the bivariate cloud of observations tends to be massed along a linear trend of increase or decrease (as is roughly the case in Figure 10.5), that is one variable is likelier to increase when the other does so, for positive correlation (or to decrease while the other increases, for negative correlation). Correlation is much easier to check through only one simple formula, but while independent variables are surely *un*correlated, the contrary is wrong: see Chapter 5 for more.

Time dependence can also be considered as a property characterising the inter-relationship within a series of observations made at successive times for a single property, called a time series. Returning to the example $x_j = x(t_j)$, hypotheses so far were that the observations are independent from each other, requiring not only that x_j and $x_{j'}$ be independent for any $j \neq j'$, but more extensively that any subset (x_{j1}, x_{j2}, ... x_{jk}) be independent. Such a process is also referred to as *white noise*. Time series could indeed involve time dependence to a large extent, such as linear correlation between observations at different times, called *auto-correlation* (see Chapter 5).

Stationarity is an essential property of a time series generalising the *i.i.d.* assumption for time-dependent cases. Stationarity requires the series to behave similarly, that is: (i) each $x_j = x(t_j)$ is identically distributed, whatever the observed time t_j; (ii) any subset ($x(t_{j1})$, $x(t_{j2})$, ... $x(t_{jk})$) is distributed identically to the subset $(x(t_{j1}+h), x(t_{j2}+h), \ldots x(t_{jk}+h))$ whatever the *time shift h*. Weaker stationarity is usually considered, through the requirement of constant auto-correlation. Though *stationarity* is a basic *requirement* for model-based inference into *uncertain inputs*, system models can help model non-stationary outputs of interest: think, for instance, of a future system configuration differing from past observations of the system or of elementary phenomena. Additionally, various cases of non-stationary uncertain inputs can still be handled easily through elementary statistical modelling such as seasonality for mixed periodic phenomenology (e.g. in winds according to seasons), or a given drift in time for example of the mean, and variance, or location and scale of the distribution for slow phenomenological changes (see Chapter 5).

10.1.5 Non-statistical approach of probabilistic modelling

Probabilistic modelling of uncertainty can usefully be applied to cases where there is no data regarding the variable of interest, either:

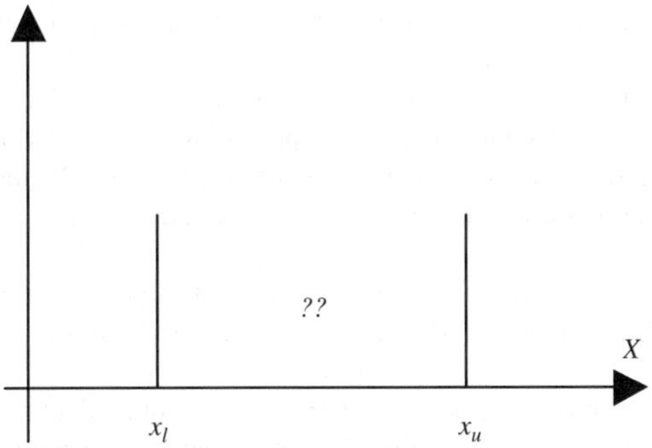

Figure 10.6 *Uncertainty over an interval of values.*

- because of the absence (or excessive cost) of data collection procedures for a truly variable feature (over time, space, a population of systems);
- or because the feature characterises a system that would be realised only once, or even not at all (in upstream design stages, where some designs may be tested and then not developed anymore because they have a lesser performance than others).

In those cases, Figure 10.1 does not fit the purpose. Consider rather the situation where the only piece of information available on a given feature or property of the system of interest X places its plausible range between a lower x_l and upper bound x_u (see Figure 10.6).

A deterministic model of the uncertainty would simply state that $x \in [x_l, x_u]$. A probabilistic model may be seen as a *richer* alternative model, taking the form of a density $X \sim f_X(x \mid \boldsymbol{\theta}_X)$ – a positively-valued and integratable function defined over the interval $x \in [x_l, x_u]$ with sum 1 – which should represent the relative *likelihoods* for x of being anywhere within the interval. Such a density should be tuned in order, for instance, to represent the likelihood for x to be within the sub-part $[x_v, x_w]$ as computed as follows:

$$P(X \in [x_v, x_w]) = \int_{u \in [x_w, x_w]} f_X(u \mid \boldsymbol{\theta})du \tag{10.15}$$

The essential difference with the previous approach lies in the fact that such likelihood need not be understood as representing a frequency over, for example, time, space, population of components under which the property takes a value in $[x_v, x_w]$. It may merely represent the view of the analyst, given any piece of professional intuition, analogous information or lack of knowledge, of what the 'chances' are for the feature of interest to be within $[x_v, x_w]$ when (or if) it realises at the future time of interest. Some authors would therefore qualify that second type of probabilistic as a 'subjective' approach as opposed to the 'objective' approach based on frequency data, though such a distinction can be somewhat simplistic in real systems, where both tend to be mixed to some extent.

In the absence of any more information that the mere bounds, a classical suggestion is to model through a *uniform density* $f_X(x \mid \boldsymbol{\theta}_X) = (x_u\text{-}x_l)^{-1}.1_{x\in[xl,\, xu]}$, stating that there is no reason for believing that the feature is likelier to take any of the lower or upper values (within sub-intervals of equal lengths). It should be noted that the statement 'feature x takes its values inside $[x_l, x_u]$' is not equivalent to the statement 'feature x is uniformly distributed over $[x_l, x_u]$', the latter containing more information about the system than the former, as is discussed in Chapter 4.

Alternatives arise when expertise or background information suggests either:

- that one of the bounds (respectively both) is difficult to establish plausibly: then an exponential (respectively a Gaussian) density may be more a appropriate representation;
- that the feature is likelier to be centred around the mean value, instead of being anywhere in the interval; then a triangular distribution may be more appropriate (see Annex on Z_m, Z_v flood examples below).

In each case, it is then necessary to elicit likely values for the ancillary parameters that are required (mean for the exponential, mean and standard deviation for the Gaussian model, most likely value for the triangular).

Though it may be more controversial than the data-based sampling fluctuation, *epistemic uncertainty* can be viewed, for instance, as a representation the debate between experts of the likely features of a system, thus entailing further lack of knowledge of the bounds or values of additional parameters.

10.2 Annex 2 – comments about the probabilistic foundations of the uncertainty models

The probabilistic setting is a central feature of the book. In this section, the formal foundations of the underlying model are given more detail although it does not aim to provide a full discussion of the foundations of probability and statistics as well as their interpretation in the risk and uncertainty litterature. The resulting setting is comparable with the well-publicised views of Kaplan and Garrick (1981) or Helton (1993) albeit with different notation and a few additional features.

10.2.1 The overall space of system states and the output space

Suppose that the whole state of the system (states of natural or internal initiating events, true values of the mesurand in metrology, existing design characteristics or actions taken, as well as all consequences of interest such as fatalities, costs, etc.) can be described in a point ω of a space Ω which may be very complex: ω may contain an infinite (even non-countable) collection of real values characterising all components of the systems, all material properties and so on, or even of real functions characterising the trajectories in time of flows, loadings, and so on. Containing virtually any possible state of the system (including maybe any state in any choice of design variables), Ω may be referred to as the underlying *overall sample space*, space of system state, or ω -space.

In the decision-making process, we generally wish to control essentially a very limited part of the system state linked to what we called 'stake indicators'. It is supposedly characterised by the vector $\boldsymbol{z}$ of variables or events of interest, that is itself a point of Ω_z, the z-space of all possible realisations of the v.i., or space of consequences, or else *output space*: it is the subset of Ω, limited to what is important for decision-making. Ω_z is typically a subset $\mathcal{R}^r \times \{0,1\}^s$ when we have r variables of interest and maybe s binary events of interest (typically system failure), but we could of course also have discrete v.i. (e.g. counts of fatalities for a given period) or multi-state events of interest.

Formally, $\boldsymbol{z}$ can be defined as the result of an 'observation function' of the whole state of the system ω, which may be simply understood as a projection of ω onto those components relevant for decision-making:

$$\begin{aligned} &\Omega \rightarrow \Omega_z \\ &\omega \rightarrow \boldsymbol{z} = \boldsymbol{Z}(\omega) \end{aligned} \tag{10.16}$$

Talking about risk and uncertainty in the system means of course that we do not have the ambition of predicting exactly the realisation of z (even less of ω) but merely having a measure controlling the parts of the space Ω_z where realisations are most likely. This is where a formal structure of a probability space can model the uncertainty in the system state at a given time of interest.

Probability theory involves developing over the sample space Ω a probability space structure $(\Omega, \mathcal{A}, P_o)$. The probability function P_o is defined in a suitably-restricted collection $\mathcal{A} = (A_\lambda)_\lambda$ of subsets (events) A_λ of Ω that enjoys the properties of a σ-algebra. This means $\mathcal{A}$ contains $\varnothing$ and Ω and is stable through the countable elementary operations of complementation, union and intersection as follows:

$$A_\lambda \subset \mathcal{A} \Rightarrow \Omega \backslash A_\lambda \subset \mathcal{A} \tag{10.17}$$

$$(A_\mu)_\mu \subset \mathcal{A} \Rightarrow \bigcup_\mu (A_\mu) \subset \mathcal{A} \tag{10.18}$$

Note that the combination of those two properties obviously also implies stability under countable intersections. Thus, *probability*, as a function mapping such a collection of events $\mathcal{A}$ into [0,1] is required to respect $P_o(\varnothing) = 0$ and $P_o(\Omega) = 1$ and to be sub-additive: provided that $A \cap B = \varnothing$ then $P_o(A \cup B) = P_o(A) + P_o(B)$, and such additivity holds for any countable collection of disjointed events $(A_\mu)_\mu$. The elementary consequences of such features for an arbitrary (non-disjointed) finite collection of events $(A_i)_i$ are the following basic formulae of probability computation, the second one also known as the Poincaré formula:

$$P_o(A_1 \cup A_2) = P_o(A_1) + P_o(A_2) - P_o(A_1 \cap A_2) \tag{10.19}$$

$$P_o\left(\bigcup_{i=1,\ldots p} A_i\right) = \sum_{k=1,\ldots p} \left((-1)^{k-1} \sum_{1 \le i_1 < i_2 < \ldots \le p} P_o(A_{i_1} \cap A_{i_2} \cap \ldots \cap A_{i_k})\right) \tag{10.20}$$

A third useful formula known as Boole's inequality guarantees that the sum of event probabilities is always a conservative (i.e. upper bound) estimate of the probability of the union of events:

$$P_o\left(\bigcup_{i=1,\ldots p} A_i\right) \le \sum_{i=1,\ldots p} P_o(A_i) \tag{10.21}$$

Events are sets of points of Ω such as the following: all the states that may characterise the system when undergoing an earthquake of a magnitude larger than a given reference; or when a length measurement gives a result comprised between *1 m* and *1.1 m* and so on. Yet, not all collections of events $(A_\lambda)_\lambda$ can be modelled properly through a probabilistic measure and thus given a probability of occurrence: only a suitably-restricted collection of measurable events. Though rarely made explicit in applied modelling, the overall sample space would be generally a subset of, for example, $\mathcal{R}^u$, assuming that an algebra of events be constructed on the elementary pieces of products of real-valued intervals. The Borel sets constitute classically the smallest σ-algebra containing those combinations of intervals, and the Lebesgue measure (uniform over products of real-valued intervals with finie endpoints) and associated Lebesgue integration over $\mathcal{R}^u$ are the fundamental basis for defining probability measures. This can be further generalised to an overall sample space mixing continuous and discrete events of interest, typically $\mathcal{R}^u$ x $\{0,1\}^v \times \mathcal{N}^w$.

More complex generalisations have been developed when dealing with the description of the state of the system ω through a spatial or temporal field as the probability space then requires to be defined over a functional space. This may be the case when moving into explicitly time-dependent reliability or risk modelling over random trajectories, as mentioned in Chapter 1, Section 1.3.3 or Chapter 4, Section 4.3.3. For the sake of simplicity, the comments in the remainder of Section 10.2 will be developed solely for the (time-implicit) static model where uncertain variables are standard random variables.

Supposing that Ω has been given the structure of a probability space $(\Omega, \mathcal{A}, P_o)$, Equation (10.16) defines $\boldsymbol{Z}$ as a random (vector) variable provided that $\boldsymbol{Z}(.)$ is a *measurable function* in the sense that, providing Ω_z with the σ-algebra of, for example, Borel sets, the pre-image $\boldsymbol{Z}^{-1}(\mathcal{A}_Z)$ of any measurable z-event $\mathcal{A}_Z$ is also a measurable ω-event, that is part of the σ-algebra of Ω. The definition of the P_o probability measure upon Ω entails the definition of the (image) probability measure P_Z of $\boldsymbol{Z}$ through $P_Z = Z \ o \ P_o$: when z is real-valued, a probability measure is uniquely and equivalently defined by its (cumulative) distribution function or cdf noted F_Z:

$$F_Z(z_s) = P_z[Z \leq z_s] = P_o\left[Z^{-1}\{z \leq z_s\}\right] = \int_{\Omega} 1_{Z(\omega) \leq z_s} dP_o \tag{10.22}$$

Note that such an integral, as all those which follow, should be understood in the Lebesgue sense. In most cases, it would be assumed that the distribution of Z additionally enjoys a *density*, that is that its distribution is differentiable so that:

$$F_Z(z_s) = P_z[Z \leq z_s] = \int_z 1_{z \leq z_s} dF_z = \int_z 1_{z \leq z_s} f_Z(z) dz = \int_{\{z \leq z_s\}} f_Z(z) dz \tag{10.23}$$

although such a density could possibly mixed with a Dirac density, that is non-zero masses over a (finite) set of points in some cases, and thus be differentiable except on that set of points. The definition of the cdf is easily extended to a vector $\boldsymbol{Z}$, using the partial order in $\mathcal{R}^r$, that is $\boldsymbol{z} \leq \boldsymbol{z}_s \hat{\Leftrightarrow} \bigcap_{k=1 \ldots r} z^k \leq z_s^k$ so that the probability measure will subsequently always be handled through its cdf F_Z assuming also the existence of a (joint) density:

$$F_Z(\boldsymbol{z}_s) = P_z\left[\bigcap_{k=1 \ldots r} \{Z^k \leq z_s^k\}\right] = \int_z 1_{\bigcap_{k=1 \ldots r} \{z^k \leq z_s^k\}} dF_z = \int_{\bigcap_{k=1 \ldots r} \{z^k \leq z_s^k\}} f_Z(\boldsymbol{z}) d\boldsymbol{z} \tag{10.24}$$

Note that a measure of probability (also referred to as the probabilistic *measure of uncertainty* in de Rocquigny *et al.*, 2008) will not generally be given over all variables included in ω, characterising the overall sample space: variables denoted as $\boldsymbol{d}$ will be fixed, though possibly varying in a controlled manner during later parametric risk/uncertainty studies. This is the case with design or decision variables but can be also a choice in fixing conditional studies (see Chapter 3). Similar to $\boldsymbol{z}$, one can define $\boldsymbol{d}$ as an 'observation function' $\boldsymbol{d} = D(\omega)$ mapping Ω into the set Ω_d of possible values for decision/design variables (a vector of real-value or discrete components). Formally then, we would define a collection of probability spaces $(\Omega(\boldsymbol{d}), \mathcal{A}_d, P_d)_{\boldsymbol{d}}$ indexed by $\boldsymbol{d}$, defined on the subsets $\Omega(\boldsymbol{d})$ of Ω where $\boldsymbol{d}$ keeps a given fixed value:

$$\Omega(\boldsymbol{d}) = \{\omega \in \Omega | D(\omega) = \boldsymbol{d}\} \tag{10.25}$$

This also entails the definition of a collection of probability measures upon $\boldsymbol{z}$, conditional upon $\boldsymbol{d}$:

$$P_Z(\boldsymbol{z}|\boldsymbol{d}) = ZoP_{\boldsymbol{d}} \tag{10.26}$$

representing the impact of design choices or other decision variables upon the distribution of consequences represented by the probability measure of v.i.. As this does not change the formal setting much, indexing by $\boldsymbol{d}$ may sometimes be omitted in subsequent comments in order to simplify notations.

Differing interpretations of this probabilistic structure are very classically given, whether frequentist (assuming virtually that an infinite number of observations can be made of the system state Ω, the proportions of which define the probabilities of events of $\mathcal{A}$) or Bayesian (a measure of credibility under decision-theory axioms): Chapters 2 and 4 recalled some of this. However, these interpretations do not change the formal structure at this stage.

10.2.2 Correspondence to the Kaplan/Garrick risk analysis triplets

Note that a correspondence can be made with the famous *triplets* defined by Kaplan and Garrick (1981) to better illustrate the fundamental meaning of *risk analysis*, which go as follows in their original notation as a finite series of triples ($i = 1, \ldots N$):

- s_i: the states of the system – or more precisely the part of the system state that correspond to the *hazard* (What can happen?);
- p_i: the *likelihood* of each of those states, that is a description of the associated *uncertainty* (How likely is that that will happen?);
- x_i: the *consequences* of each of those states for the decision-maker, encompassing the *vulnerability* (If it does happen, what are the consequences?).

In that respect, (s_i, x_i) can be viewed as a breakdown of the overall state of the system ω making a distinction between the variables characterising the hazard scenarios (e.g. flood, reservoir breakdown, extreme temperature, ...) and the variables representing the associated consequences (fatalities, costs, etc.). That formal presentation was introduced in a finite discretisation of the sample space – distinguishing N scenarios or more comprehensively categories of scenarios grouped through the listing of similar consequences – though it may be generalised into an infinite but countable collection. Reformatting the notation, it would here correspond to the following:

$$\{S_\lambda, P_o(S_\lambda), \mathbf{Z}(S_\lambda)\}_\lambda \tag{10.27}$$

where $(S_\lambda)_\lambda$ constitutes a subset of the σ-tribe $\mathcal{A}$ of possible events in the system that constitute a partition Ω and that are grouped by the fact that they lead to the same consequences. $\mathbf{Z}(S_\lambda)$, which are the images of the subsets (S_λ) through the function $\mathbf{Z}$ representing the variables of interest – valuing consequences of interest to the decision-maker – constitutes subsets of Ω_z supposedly disjointed and concentrated in small deviations around a given value of the consequences z_λ. Considering a scalar definition of the consequences (say a cumulated cost), they could represent small intervals around an increasing cost value z_λ.

10.2.3 The model and model input space

In simplistic risk situations where we would be able to directly observe the v.i., and thus estimate the future risk (for the same system), that would be enough: this would be the case when observing the frequency of failure for a given component over a past period and inferring the failure frequency in the future of a completely identical component under unchanged operation. In many cases this is not enough either because direct observations are not practical, or because some of the system operating conditions inside $\boldsymbol{d}$ would change, and so on. Then the modelling option, which is the purpose of this book, relies on the basic assumption that information can be retrieved from other characteristics of the system, from which a model can predict the v.i. Following the notations introduced in earlier Chapters, a vector ($\boldsymbol{x}$, $\boldsymbol{e}$, $\boldsymbol{d}$) is supposed to characterise the system sufficiently (resp. system characteristics, initiating events, and design variables or actions), so that, in combination with a supposedly available model $G(.)$, we are able to predict z through:

$$z = G(\boldsymbol{x}, \boldsymbol{e}, \boldsymbol{d}) \tag{10.28}$$

Like z and $\boldsymbol{d}$, ($\boldsymbol{x}$, $\boldsymbol{e}$) can be defined formally as the result of another 'observation function':

$$\begin{aligned} &\Omega \rightarrow \Omega_x \\ &\omega \rightarrow (\boldsymbol{x}, \boldsymbol{e}) = (\boldsymbol{X}(\omega), \boldsymbol{E}(\omega)) = (X, E)(\omega) \end{aligned} \tag{10.29}$$

where Ω_x, the *input* space, covers all possible realisations of the model input variables or events: note that subscript e will be skipped so as to simplify the notations. It is typically a subset $\mathcal{R}^p x\, \{0,1\}^s$ when we have p model input variables of interest and maybe s binary events (e.g. initiators) of interest. The probabilistic structures ($\Omega(d)$, $\mathcal{A}_d$, $P_d)_d$ also generate probabilistic structures upon Ω_x with a joint probability distribution for the random vector ($\boldsymbol{X}$, $\boldsymbol{E}$) conditional on $\boldsymbol{d}$:

$$F_{XE}(|\boldsymbol{d}) = (X, E) o P(|\boldsymbol{d}) \tag{10.30}$$

Assuming that $G(.)$ is a measurable function over its area of definition (a subset of $\Omega_x x\ \Omega_d$ comprising the image of Ω by (X,E,D)), it again defines a random variable $\boldsymbol{Z}_m$ with a joint distribution function conditional on $\boldsymbol{d}$:

$$\begin{aligned} &\Omega \rightarrow \Omega_z \\ &\omega \rightarrow \boldsymbol{Z}_m(\omega) = G(\boldsymbol{X}(\omega), \boldsymbol{E}(\omega), \boldsymbol{D}(\omega)) \\ &F_{Zm}(|\boldsymbol{d}) = Go(X, E) o P(|\boldsymbol{d}) \end{aligned} \tag{10.31}$$

The central assumption of the system modelling is that the random variable $\boldsymbol{Z}_m(\omega)$ does *reasonably* resemble $\mathbf{Z}(\omega)$ – the 'true' variables of interest characterising the system state from the decision-making point of view – so that we may infer relevant risk information using the model G instead of directly estimating $\mathbf{Z}$. When working in a probabilistic setting it is enough that they be *almost surely* equals (i.e. excluding possibly infinite numbers of ω but with nil probability). Weaker acceptable conditions could still be defined, such as within reasonable accuracy ($\boldsymbol{Z}_m(\omega)$-$\boldsymbol{Z}(\omega)$ being small), and/or only for a subset of design values $\boldsymbol{d}$.

Apart from Chapter 6 on model calibration, this (strong) basic modelling hypothesis can be assumed and thus not distinguish $\mathbf{Z}_m(\omega)$ from $\mathbf{Z}(\omega)$. Note also that upon that same hypothesis, it is no longer necessary to consider Ω: supposing we know the probability distributions F_{XE} $(\boldsymbol{x},\boldsymbol{e})\,|\,\boldsymbol{d})$ upon Ω_x conditional on $\boldsymbol{d}$ (the *uncertainty model*), the whole probability distribution F_Z $(z\,|\,\boldsymbol{d})$ associated with random variable $\mathbf{Z}$ is determined through $G(.)$.

$$F_Z(z_s) = P_Z[\boldsymbol{Z} \leq z_s] = P_{XE}\left[\boldsymbol{Z}_m^{-1}\{z \leq z_s\}\right] = \int_{\Omega_x} 1_{G(x,e,d) \leq z_s} dP_{XE} \tag{10.32}$$

Subsequently any risk measure or quantity of interest c_Z defined as a functional of F_Z $(z\,|\,\boldsymbol{d})$ can be computed, and becomes a function c_Z $(\boldsymbol{d})$ of the given design or decision variables $\boldsymbol{d}$.

Note that so far, there is only one 'probabilistic layer', whatever their interpretation, objective or subjective. This is satisfactory for risk measures that are defined according to traditional 'one-level' probabilistic criteria (e.g. expected fatalities, probability of undesired event or threshold exceedance); but in some cases, Chapter 1 showed that risk measure involves also a form of 'probability of a frequency (or probability)' or 'uncertainty about the risk': a formal interpretation of the latter will be given when coming to the estimation stage in the context of data limitations.

10.2.4 Estimating the uncertainty model through direct data

As mentioned above, it is therefore necessary to define an uncertainty model $F_{XE}(\boldsymbol{x},\boldsymbol{e})\,|\,\boldsymbol{d})$ on the inputs of the $G(.)$, that is a description of the likely variability of the system characteristics or initiator events that will explain in turn the distribution of consequences $F_Z(z\,|\,\boldsymbol{d})$. Indeed, though the sample space may have been proved to be *measurable* given its functional structure (e.g. $\mathcal{R}^u \times \{0,1\}^v$), for instance through the definition of the appropriate (uniform) Lebesgue measure, it still needs to be given a particular probability measure that plausibly models the relative likelihoods of all states of the system viewed from the $(\boldsymbol{x},\boldsymbol{e})\,|\,\boldsymbol{d}$, which is generally not uniform. Thus, an *estimation* of the uncertainty model is necessary and will generally be carried out on a mixture of both 'hard' and 'soft' data, that is observation samples Ξ_n and expertise associated with the system. In most cases, only limited confidence can be given to that estimation (due to limited sample sizes, observation errors, expertise limits, etc.) so that, a certain distance will remain between the 'true' probability distribution $F_{XE}(\boldsymbol{x},\boldsymbol{e})\,|\,\boldsymbol{d})$ that would be given unlimited information, at least in thought experiment, and its estimation given limited data. This distance, representing the estimation of uncertainty or error, constitutes a second level probabilistic which can be given a formal statistical meaning, either classical or Bayesian.

The *classical* context in which to estimate $F_{XE}(\boldsymbol{x},\boldsymbol{e})\,|\,\boldsymbol{d})$ is introduced in Chapter 5, Section 5.4.1: suppose we use a sample of 'hard data' with n direct *i.i.d.* observations of $(\boldsymbol{x},\boldsymbol{e})$ that are all supposed to be observed for the same $\boldsymbol{d}$ so that the notation may be simplified transitorily into $F_{XE}(\boldsymbol{x},\boldsymbol{e}))$.

$$\Xi_n = (\boldsymbol{x}_j, \boldsymbol{e}_j)_{j=1\ldots n} \tag{10.33}$$

The introduction of a more complex probabilistic model is necessary to account for these statistical observations: a product space Ω_x^{n+1}, representing the product of n i.i.d. systems (or sequential stationary and independent realisations of a given system over time) that have been observed at past dates as well as of the i.i.d. system at the time of interest. Such product space is given the following product probability distribution:

$$P_{XE}^{n+1} = \prod_{j=1\ldots n} [F_{XE}((\boldsymbol{x}_j, \boldsymbol{e}_j))].F_{XE}((\boldsymbol{x}, \boldsymbol{e})) \tag{10.34}$$

defining the random variable $((\boldsymbol{X}_j, \boldsymbol{E}_j)_{j=1..n}, (\boldsymbol{X}, \boldsymbol{E}))$ of which the sample Ξ_n is one realisation of the first n components and the state of the system at the (generally future) time of interest will be a realisation of the last component. Most of the time, expertise would lead to defining a family of probability distributions with a density $f_{XE}(\cdot\,|\,\theta_{XE})$ parameterised by a vector θ_{XE} which is common to all $n+1$ uncertain vectors, thus generating a family of product distributions over Ω_x^{n+1} with the following density:

$$f_{XE}^{n+1}(.|\theta_{XE}) = \Pi_{j=1\ldots n}[f_{XE}((\boldsymbol{x}_j, \boldsymbol{e}_j)|\theta_{XE})].f_{XE}((\boldsymbol{x}, \boldsymbol{e})|\theta_{XE}) \tag{10.35}$$

Size n_p of vector $\boldsymbol{\theta}$ may be very large, typically comprising one to three parameters per component x^i or e^i plus the parameters of the dependence structure (correlation coefficients, common mode probabilities, copula parameters etc.). Vector $\boldsymbol{\theta}_{XE}$ belongs to a given parameter space Θ, typically a subset of $\mathcal{R}^{np}$, within which the unknown 'true' parameters $\boldsymbol{\theta}_o$ characterising the real probability distribution that has generated the observations is postulated to be. Estimators, such as maximum likelihood, moments, and so on would then be set up as a function of the sample Ξ_n, and generate an approximation of $\boldsymbol{\theta}_o$ with suitable characteristics (unbiased, convergent, etc.). Estimators are defined as random variables, that is deterministic measurable functions mapping Ω_x^n into Θ:

$$\begin{aligned}&\Omega_x^n \rightarrow \Theta \in R^{n_p}\\&(\boldsymbol{X}_1, \boldsymbol{E}_1), \ldots (\boldsymbol{X}_n, \boldsymbol{E}_n) \rightarrow \hat{\Theta}_{XE} = Q((\boldsymbol{X}_1, \boldsymbol{E}_1), \ldots (\boldsymbol{X}_n, \boldsymbol{E}_n))\end{aligned} \tag{10.36}$$

Turning now to the inference of meaningful properties for the decision maker, the overall sample space should now also be considered as a product space Ω^{n+1} defined similarly to its subset Ω_x^{n+1} with respect to the $n+1$ i.i.d. realisations of the state of the system at a past observed or (future) time of interest, now including not only the model inputs $(\boldsymbol{X}_j, \boldsymbol{E}_j)_{j=1..n}, (\boldsymbol{X}, \boldsymbol{E}))$ but also the variables of interest $(\boldsymbol{Z}_j)_{j=1..n}, (\boldsymbol{Z}))$. Note that the random variable $\boldsymbol{Z}_j$ though defined in theory, could not be observed. The risk measure should characterise the probabilistic properties of the random variable $\boldsymbol{Z}$, and depends on $\boldsymbol{\theta}_o$ through a known (though computationally complex) deterministic function $c_Z(\boldsymbol{\theta}_{XE}{}^\circ, \boldsymbol{d})$: it can only be estimated through a new estimator, derived from that of $\boldsymbol{\theta}_o$:

$$\begin{aligned}&\Omega_x^n \rightarrow R\\&(\boldsymbol{X}_1, \boldsymbol{E}_1), \ldots (\boldsymbol{X}_n, \boldsymbol{E}_n) \rightarrow \hat{C}_Z = c_Z[Q((\boldsymbol{X}_1, \boldsymbol{E}_1), \ldots (\boldsymbol{X}_n, \boldsymbol{E}_n)), \boldsymbol{d}]\end{aligned} \tag{10.37}$$

Chapter 5 discusses the key question of the *estimation uncertainty* generated by this type of procedure, which is embodied by the probability distribution of the random $\hat{C}_Z$ around the fixed but unknown $c_Z(\boldsymbol{\theta}_{XE}{}^\circ, \boldsymbol{d})$.

Under a *Bayesian* estimation setting, defined in Chapter 5, Section 5.4.5, estimation uncertainty is given an even more explicit treatment. The appropriate sample space now becomes a product between Ω_x^{n+1} (or in fact Ω^{n+1}) and the parameter space Θ; it receives a probability measure defined as the product of the previously-defined P_{XE}^{n+1} and a (prior) parameter distribution $\pi_o(\boldsymbol{\theta})$, whereby experts represent their prior knowledge of the plausible values of $\boldsymbol{\theta}_{XE}$, as possibly very poor with non-informative priors, as follows:

$$\begin{aligned}&(\boldsymbol{X}, \boldsymbol{E})|\theta_{XE} \sim \Pi_{j=1...n}[f_{XE}((\boldsymbol{x}_j, \boldsymbol{e}_j)|\theta_{XE})].f_{XE}((\boldsymbol{x}, \boldsymbol{e})|\theta_{XE})\\&\Theta_{XE} \sim \pi_o(\theta_{XE}) = \pi(\theta_{XE}|IK, \zeta)\end{aligned} \tag{10.38}$$

As discussed in Chapter 5, in the spirit of Helton (1993), Θ can now be interpreted as the epistemic uncertainty (or level 2) sample space, while Ω_x^{n+1} incorporates the aleatory uncertainty (or level 1) structure conditional upon a given value of $\boldsymbol{\theta}_{XE}$. The risk measure characterising the probabilistic properties of the random variable $\boldsymbol{Z}$ is again a function of $\boldsymbol{\theta}_{XE}$ through a known (though computationally complex) deterministic function $c_Z(\boldsymbol{\theta}_{XE}, \boldsymbol{d})$, so that it becomes a random variable mapping Θ into $\mathcal{R}$. Its properties of interest (its expectation defining the *Bayesian estimator* of the risk measure, its quantiles generating level-2 risk measures and so on, see Chapter 5, Section 5.4.5) are to be studied conditionally on $(\mathbf{x}_j, \mathbf{e}_j)_{j=1...n}$ and $\mathbf{d}$.

10.2.5 Model calibration and estimation through indirect data and inversion techniques

As discussed in Chapter 6, a more general form of the data sample available to model the state of the system is $\Xi_{tn} = (\mathbf{y}_m(t_j), \mathbf{d}(t_j))_{j=1..n}$ where $\mathbf{y}_m$ denote observable features of the state of the system under known experimental conditions $\mathbf{d}$ that may relate to an observational model (assuming an additive residual structure for the sake of simplicity and simplifying the notation $(\mathbf{x},\mathbf{e})$ into $\mathbf{x}$):

$$y_m(t_j) = H(\mathbf{x}(t_j), \mathbf{d}(t_j)) + \mathbf{u}_j = \mathbf{y}_j + \mathbf{u}_j \tag{10.39}$$

where residuals $\mathbf{u}_j$ could encapsulate many things beyond the 'pure' measurement errors: modelling (H) inaccuracy in describing the phenomenological quantities measured by $\mathbf{y}_m$, or even the lack of characterisation by $\mathbf{x}$ and $\mathbf{d}$ of the true system state ω being observed. Assuming then that the observation system model $H(.)$ is a measurable function over its area of definition (a subset of $\Omega_x\, x\, \Omega_d$ comprising the image of Ω by (X,D)), it again defines a random variable $\mathbf{Y}$ with a joint distribution function conditional on $\mathbf{d}$:

$$\begin{gathered}\Omega \rightarrow \Omega_y \\ \omega \rightarrow \mathbf{Y}(\omega) = H(\mathbf{X}(\omega), \mathbf{D}(\omega))\end{gathered} \tag{10.40}$$

$$F_Y((.)|\mathbf{d}) = H(., \mathbf{d}) o F_X(.) \tag{10.41}$$

On the measurement side, observed variables $\mathbf{y}_m$ can be defined as the result of an 'observation function' of the whole state of the system ω, that incorporates, beyond simple projections, all the characteristics of the measurement device, 'errors' and artefacts included and so on:

$$\begin{gathered}\Omega \rightarrow \Omega_y \\ \omega \rightarrow \mathbf{y}_m = \mathbf{Y}_m(\omega)\end{gathered} \tag{10.42}$$

The basic modelling assumption will be that the observations $(\mathbf{y}_m(t_j), \mathbf{d}(t_j))_{j=1..n}$ reasonably resemble the model $H(\mathbf{x}(t_j), \mathbf{d}(t_j))$ so that the residuals $\mathbf{u}_j$ remain 'small enough' or, perhaps more importantly, that their distribution can be modelled through a known family of densities: henceforth, inference with controlled model error could be made on the basis of the model $H(.)$ for differing conditions $\mathbf{d}$ at the future time of interest from those at past observed dates. Similar to the previous paragraphs, a product probability probability space Ω^{n+1} (Ω now assumed to contain Ω_y) can be developed to make use of the sample $\Xi_{tn} = (\mathbf{y}_m(t_j), \mathbf{d}(t_j))_{j=1..n}$ in order to infer the characteristics of the distribution of $\mathbf{Y}$ or $\mathbf{Y}_m$ *conditional* on $\mathbf{d}$ at the (past) observed and (future) time of interest.

The inverse algorithms described in Chapter 6 can be viewed as procedures designed to estimate these *conditional* distributions with variants according to the model developed in $\mathbf{x}(t_j)$ and $\mathbf{u}_j$. When $\mathbf{x}(t_j) = \mathbf{x}_o$ is fixed in a *parameter identification* or *calibration* approach, the algorithms do in fact estimate the common distribution of residuals that are independent although possibly non-identically distributed, as being heteroscedastic and hence only conditionally *i.i.d.* in $\mathbf{d}$ (see Section 6.3.4):

$$\mathbf{u}_j = [\mathbf{y}_m(t_j) - H(\mathbf{x}_o, \mathbf{d}(t_j))] \sim F_U(.|\mathbf{x}_o, \theta_U, \mathbf{d}) \tag{10.43}$$

Estimation refers here to the vector $\boldsymbol{\theta} = (\boldsymbol{x}_o, \boldsymbol{\theta}_U)$ which formally plays the same role as the $\boldsymbol{\theta}_{XE}$ in the previous section.

In a *full inversion* approach, it is acknowledged that the features of the state of the system $\boldsymbol{X}(\omega_j) = \boldsymbol{x}(t_j)$ do vary significantly through the past dates and future time of interest though not being observable because ω_j is only *partially* observed through $\boldsymbol{Y}_m(\omega_j) = \boldsymbol{y}_m(t_j)$. In that second case, the residual $U(\omega_j) = \boldsymbol{Y}_m(\omega_j) - H(\boldsymbol{X}(\omega_j), \boldsymbol{d}(t_j))$ cannot be observed anymore than $\boldsymbol{X}(\omega_j)$ can be. Thus, instead of estimating the distribution of the residuals, the algorithms will directly estimate the distribution of $\boldsymbol{Y}_m$:

$$F_{Ym}(.|\boldsymbol{d}) \sim [H(., \boldsymbol{d})oF_X(.|\boldsymbol{\theta}_{XE})] \oplus F_U(.|\theta_U) \tag{10.44}$$

Such a distribution is a complex sum of the image through H (., $\boldsymbol{d}$) of F_X and of F_U, thus involving the convolution of the densities referred to in the sign $\oplus$ (see Section 6.3.3): it is parameterised by the vector $\boldsymbol{\theta} = (\boldsymbol{\theta}_{XE}, \boldsymbol{\theta}_U)$. Remember finally that one is not interested in the distributions of X and E as such. One could even accept that the accuracy in describing of X is only useful up to the point that the later prediction of Z (or more precisely of its risk measure or quantity of interest c_Z) by propagation is satisfactory.

10.3 Annex 3 – introductory reflections on the sources of macroscopic uncertainty

A large variety of causes or phenomena give rise to uncertainty in the state of a system at a macroscopic decision-making scale. Brief introductory reflections will be made on the sources of such uncertainty, zooming out from the smaller to the larger scales of reality. It should be seen as an incentive to study further the fruitful relationship of a macroscopic uncertainty modelling to quantum and statistical physics or complex system dynamics. A thorough theoretical discussion of the complex concepts involved, not exempt from on-going interpretational controvery, is completely beyond the scope of this book.

It is essential first to estimate the underlying *intrinsic variability* of reality.

Starting with the fundamental constituents of matter and energy, a set of associated phenomena generate intrinsic variability: the *atomic or particular nature* and the *quantum behaviour of reality*, epitomized, for instance, by the *discrete* nature of radioactive emissions or of photon fluxes, accompanied by energy fluxes varying random-like in time between discrete values; the fundamental indeterminacy in the states of a system (and its dependence to observation) as reflected in *Heisenberg principle of indeterminacy* (*or uncertainty principle*). The latter concept reflects the intertwined *particular and wave* natures of matter, energy or light; it involves a series of relations specifying an *irreducible* lower bound for the product of uncertainty (say standard deviation) of measured values of conjugate physical variables of interest which fluctuate empirically even with highly-refined metrological settings. For instance, $\Delta x.\Delta p \geq \hbar$ for the uncertainty in the observed location and quantity of movement of a particle or $\Delta E.\Delta t \geq \hbar$ for the uncertainty in the measured energy and particle lifetime or duration of the experiment: two conjugate variables of interest characterising a system can never be observed with infinite precision at the same time. Notwithstanding the complex controversies of interpretation, a probabilistic description of the fundamentals of physical reality proved to be a powerful model for such microscopic phenomena. They lead one to consider any system with some amount of elementary uncertainty as being distributed at random between a discrete array of micro-states – that is sets of discrete values for the variables of interest such as energy, location, quantity of movement, each of which is defined with a limited precision – the complexity of such quantum mechanics concepts also excluding a classical *movement* over time of the system between states, reflecting more deeply an indeterminacy with respect to *observable* variables of interest.

Consider now a meso-scopic system, typically a volume of homogeneous matter (e.g. a volume of gas), the field of description of *statistical physics*; micro-states are now characterised by the complex array of discrete values for microscopic variables of interest quantifying each particle of the system (possibly including the fluctuating number of particles included because of elementary disintegrations or inflow/outflow). Adding to the elementary behaviour of particles, dynamics now involves collisions or more complex interactions between particles – corresponding to a dynamic process of the system between successive micro-states in the phase space. This yields a complex motion that can now be computed through molecular dynamics models, though limited so far to short time periods and requiring a large set of initial conditions: it typically leads to a random-like behaviour for longer periods. In theory, such a system now combines multiple forms of uncertainty or variability: elementary quantum indeterminacy, 'granular' incremental variation of moving fluxes of discrete particles, time variability of the complex inter-particle dynamics, and possibly even more the lack of knowledge in the incredibly large sets of initial conditions required. Hopefully, they are mostly suppressed when handling large systems from the point of view of *macroscopically-defined* variables of interest given the key assumption of *equilibrium* as follows:

(i) Heisenberg indeterminacy rapidly becomes negligible above the atomic scale. For an individual particle in an ideal gas under normal atmospheric conditions, $\Delta x \sim 0.01$ Å to be compared with a molecular scale of a few Å or an inter-particle distance or free paths of a few tens/hundreds of Å; this is much more negligible when considering a macroscopic variable, such as the location of a mechanical piece. Yet, this may not be the case when considering micro-physical or nano-physical systems, or quantum engineering.

(ii) Macroscopic variables of interest do generally stand either as sums of elementary contributions of each particle or micro-state for extensive quantities (e.g. energy) or expectations for intensive quantities (e.g. density); both are additionally averaged over the macroscopic duration of measurement. Under the fundamental hypotheses of thermodynamic equilibrium and ergodicity, complexity evolves after enough time into a set of micro-variables of interest or micro-states of the system all distributed identically, close to independent and stationnarily over time, the system most likely occupying a distribution of maximal entropy. By a similar effect of averaging as shown in Chapters 4 or 7, the macroscopic variables of interest involve a coefficient of variation of $1/\sqrt{\alpha(N)}$ of an order of magnitude close to $1/\sqrt{N}$. Given the incredibly large number N of particles involved – Avogadro's number $6\ 10^{23}$ means that even in a few mm^3 of gas there are $N \sim 10^{17}$ particles, much more for solids – the associated fluctuation represents in most cases no more than 10^{-6} to 10^{-10}, a completely negligible amount of variability. Time-wise, the elementary Brownian discrete fluctuations can also become completely averaged out in a continuous and imperceptible flow: in an ideal gas under normal conditions, collisions leading to a change in the microscopic variables of interest occur every 10^{-10} s.

However, such macroscopic averaging of the microscopic complexity and variability is not operative in many cases: think of flows of rare elementary events such as radioactive disintegrations or, much more importantly, when considering dynamic systems *outside the equilibrium*, notably in physical-chemical systems with a high degree of interaction beyond pure collisions. Similar to the limitations to risk mutualisation introduced in Chapter 4 with highly-correlated variables, complex systems are often characterised as evidencing long-range correlation structures either in time or space; these rule out the averaging effects of equilibrium, thus leaving large variability and emerging patterns of behaviour different from averaged microscopic ones.

Moving on to *macroscopic* systems either of the size of a piece of iron, a mechanical sub-system, a river basin or the atmosphere over a region, the dominating source of variability could indeed be originated in the *instability* of the *system dynamics*. The well-known butterfly effect epitomises the extreme sensitivity to the initial conditions of a system in the classical example of *deterministic chaos*. This was first introduced by E.N. Lorenz in 1963 (Lorenz E.N., Deterministic nonperiodic flow, *J. Atmos. Sci.*, **20**: 130–141) in a schematic model of weather dynamics, although simplified with respect to a more realistic description of atmospheric dynamics in modern weather forecasting models. That context characterises many other dynamic systems fraught with non-linear and coupled phenomena, including classically even a system as simply-specified as the astronomic 3-body gravitational equilibrium studied by Poincaré at the beginning of the 20th century. Although dynamics could be understood as being perfectly determined for given initial conditions, the extreme sensitivity after sufficient time means in practical terms a certain range of unpredictability of the variables of interest as the initial conditions could never be infinitely precise. This can be qualified not only as an *epistemic* source of uncertainty, but also as *irreducible* to some extent because of the variability arising from the previously-mentioned lower layers. Instability also inter-relates with the spatial and temporal scales at which systems are considered – as, for instance, in the case of turbulence – featuring possibly a fractal reality, thus generating an essential interplay between variability and the definition and scale of the macroscopic variable of interest.

Atmospheric variability, coupled with other forms of geophysical or astronomical radiation or matter flows, can also be understood as the essential drivers of *environmental variability* in general, be it of hydrology and maritime deviations from predictable astronomic tides, topography through erosion/sedimentation and more generally soil and geophysical dynamics including the *spatial heterogeneity* of most samples of natural matter (heterogeneous physical-chemical micro-structures) which thus replaces variability in the macroscopic variables of interest of any system. The evolution of life itself – be it at the level of genetics, biological development of an individual or ecological dynamics of populations – can also be contemplated as consubstantial to variability for multiple reasons: intrinsically-instable dynamics, partially random genetic mutations or reproductive mechanisms, coupled dynamics with that of atmospheric or other environmental variability and so on.

Man-made systems also bear the consequences of environmental variability, starting with the sources of metrological noise and artifacts generating measurement uncertainty where it dominates though not always suppresses the above-mentioned quantum or Brownian variability of underlying layers. Moreover, in man-made or man-influenced systems, the variability of human behaviour (vulnerability, actions, opinions or values, etc.) often becomes the largest source of uncertainty. It can be involuntary deviations or 'human errors' from expected manufacturing or operating processes; or much more importantly the ontological freedom of mankind to experiment, invent unexpected products and processes or react innovatively to uncertain situations: Knight (1921), for instance, insisted on the essential link between uncertainty and innovation through entrepreneurship. Nonetheless, similar to mesoscopic physical systems, the complex variability of human behaviours can be considerably regularised in many cases through the large population of pseudo-independent behaviours: statistical regularity can characterise surprisingly well such variables of interest as political voting rates, car insurance claims or social-economic features (consuming habits, suicide rates, etc.).

Eventually, all forms of *intrinsic variability* of reality introduced above are amplified by the pervasive *lack of knowledge* affecting the analyst: shortcomings of data collection or modelling efforts in a cost-benefit or time-constraining context or even deeper limitations of scientific knowledge for complex systems, newly-designed or yet un-experimented products or processes.

Information comes at a cost, a strong driver of uncertainty.

10.4 Annex 4 – details about the pedagogical example

In this section, complementary details are provided for the example introduced in Chapter 3: numerical values, samples and cost functions.

10.4.1 Data samples

The following Table 10.1 provides the detailed flow recorded as $(Q_j)_{j=1...n}$: one maximal value per year.

The following Table 10.2 provides the simultaneous records $(Q_j, H_{mj})_{j=1...n}$ of yearly maxima for flow (in m3/s) and water depth (in m). Data is missing for water depth, evidencing the unreliability of local water depth measurement, thus blurring direct statistical estimation on that variable (see Chapter 3). Significant metrological noise generally affects water depth records (Figure 10.7): they have been simulated intentionally on the basis of the original flow records and random sampling of the flood model (other sources of uncertainties sampled along the pdf mentioned below), to which a multiplicative $+-10\,\%$ uniform noise has been added.

Table 10.1 Flow records.

Year	Flow	*Year*	Flow	*Year*	Flow	*Year*	Flow	*Year*	Flow	*Year*	Flow	*Year*	Flow	*Year*	Flow
1849	3854	*1869*	2417	*1889*	3330	*1909*	1307	*1929*	424	*1949*	635	*1969*	1837	*1989*	793
1850	1256	*1870*	1125	*1890*	1858	*1910*	1275	*1930*	2017	*1950*	733	*1970*	1629	*1990*	856
1851	1649	*1871*	903	*1891*	1359	*1911*	2706	*1931*	1958	*1951*	758	*1971*	1421	*1991*	1903
1852	1605	*1872*	1462	*1892*	714	*1912*	582	*1932*	3192	*1952*	1368	*1972*	2204	*1992*	1594
1853	341	*1873*	378	*1893*	1528	*1913*	1260	*1933*	1556	*1953*	935	*1973*	956	*1993*	740
1854	1149	*1874*	1230	*1894*	1035	*1914*	1331	*1934*	1169	*1954*	1173	*1974*	971	*1994*	3044
1855	868	*1875*	1149	*1895*	1026	*1915*	1283	*1935*	1511	*1955*	547	*1975*	1383	*1995*	1128
1856	1148	*1876*	1400	*1896*	1127	*1916*	1348	*1936*	1515	*1956*	669	*1976*	541	*1996*	522
1857	1227	*1877*	2078	*1897*	1839	*1917*	1048	*1937*	2491	*1957*	331	*1977*	703	*1997*	642
1858	1991	*1878*	1433	*1898*	771	*1918*	1348	*1938*	881	*1958*	227	*1978*	2090		
1859	1255	*1879*	917	*1899*	1730	*1919*	383	*1939*	846	*1959*	2037	*1979*	800		
1860	1366	*1880*	1530	*1900*	1889	*1920*	1526	*1940*	856	*1960*	3224	*1980*	651		
1861	1100	*1881*	2442	*1901*	3320	*1921*	789	*1941*	1036	*1961*	1525	*1981*	1153		
1862	1837	*1882*	2151	*1902*	352	*1922*	811	*1942*	1830	*1962*	766	*1982*	704		
1863	351	*1883*	1909	*1903*	885	*1923*	1073	*1943*	1391	*1963*	1575	*1983*	1771		
1864	1084	*1884*	630	*1904*	759	*1924*	965	*1944*	1334	*1964*	1695	*1984*	1433		
1865	1924	*1885*	2435	*1905*	731	*1925*	619	*1945*	1512	*1965*	1235	*1985*	238		
1866	843	*1886*	1920	*1906*	1711	*1926*	3361	*1946*	1792	*1966*	1454	*1986*	122		
1867	2647	*1887*	1512	*1907*	1906	*1927*	523	*1947*	136	*1967*	2595	*1987*	1306		
1868	1248	*1888*	1377	*1908*	1543	*1928*	493	*1948*	891	*1968*	706	*1988*	733		

Table 10.2 Flow–heights records.

Year	Q	H_m	*Year*	Q	H_m	*Year*	Q	H_m
1849	3854	missing	*1899*	1730	missing	*1949*	635	1.6
1850	1256	3.7	*1900*	1889	missing	*1950*	733	2.6
1851	1649	2.7	*1901*	3320	missing	*1951*	758	2.1
1852	1605	3.1	*1902*	352	missing	*1952*	1368	2.9
1853	341	1.3	*1903*	885	missing	*1953*	935	2.2
1854	1149	2.3	*1904*	759	missing	*1954*	1173	2.3
1855	868	1.6	*1905*	731	missing	*1955*	547	1.2
1856	1148	1.9	*1906*	1711	missing	*1956*	669	1.3
1857	1227	2.3	*1907*	1906	missing	*1957*	331	1.4
1858	1991	3.1	*1908*	1543	missing	*1958*	227	0.9
1859	1255	2.2	*1909*	1307	missing	*1959*	2037	4.6
1860	1366	2.0	*1910*	1275	missing	*1960*	3224	5.1
1861	1100	2.0	*1911*	2706	missing	*1961*	1525	2.5
1862	1837	2.4	*1912*	582	1.4	*1962*	766	2.1
1863	351	1.3	*1913*	1260	3.0	*1963*	1575	2.8
1864	1084	2.6	*1914*	1331	2.2	*1964*	1695	missing
1865	1924	3.6	*1915*	1283	2.6	*1965*	1235	missing
1866	843	2.0	*1916*	1348	2.8	*1966*	1454	missing
1867	2647	3.1	*1917*	1048	2.5	*1967*	2595	missing
1868	1248	2.8	*1918*	1348	2.4	*1968*	706	missing
1869	2417	3.6	*1919*	383	1.1	*1969*	1837	3.6
1870	1125	missing	*1920*	1526	3.0	*1970*	1629	3.2
1871	903	missing	*1921*	789	1.4	*1971*	1421	2.7
1872	1462	missing	*1922*	811	2.2	*1972*	2204	3.3
1873	378	missing	*1923*	1073	2.9	*1973*	956	2.0
1874	1230	missing	*1924*	965	2.4	*1974*	971	2.4
1875	1149	missing	*1925*	619	1.8	*1975*	1383	2.4
1876	1400	3.6	*1926*	3361	4.6	*1976*	541	1.7
1877	2078	3.4	*1927*	523	1.5	*1977*	703	1.8
1878	1433	2.7	*1928*	493	1.7	*1978*	2090	3.1
1879	917	1.5	*1929*	424	1.0	*1979*	800	2.8
1880	1530	4.1	*1930*	2017	3.2	*1980*	651	1.4
1881	2442	3.6	*1931*	1958	2.9	*1981*	1153	2.9
1882	2151	3.6	*1932*	3192	3.5	*1982*	704	1.7
1883	1909	2.4	*1933*	1556	3.1	*1983*	1771	3.7
1884	630	1.8	*1934*	1169	1.9	*1984*	1433	1.9
1885	2435	3.3	*1935*	1511	2.3	*1985*	238	1.1
1886	1920	2.7	*1936*	1515	2.7	*1986*	122	0.5
1887	1512	3.5	*1937*	2491	4.2	*1987*	1306	2.3
1888	1377	3.0	*1938*	881	1.5	*1988*	733	1.8
1889	3330	3.7	*1939*	846	2.0	*1989*	793	2.2
1890	1858	7.4	*1940*	856	1.4	*1990*	856	2.1
1891	1359	2.9	*1941*	1036	1.9	*1991*	1903	4.5

Table 10.2 (*Continued*)

Year	*Q*	H_m	*Year*	*Q*	H_m	*Year*	*Q*	H_m
1892	714	2.0	*1942*	1830	2.9	*1992*	1594	3.4
1893	1528	2.3	*1943*	1391	2.5	*1993*	740	2.2
1894	1035	5.5	*1944*	1334	3.1	*1994*	3044	6.0
1895	1026	2.2	*1945*	1512	2.6	*1995*	1128	2.0
1896	1127	2.3	*1946*	1792	4.1	*1996*	522	1.7
1897	1839	2.9	*1947*	136	0.6	*1997*	642	1.6
1898	771	missing	*1948*	891	2.1			

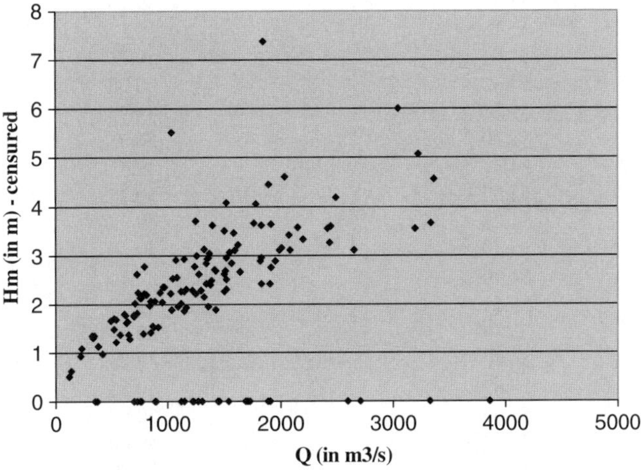

Figure 10.7 *Flow-height records.*

The following Table 10.3 provides the simultaneous records $(Z_{m\,k}, Z_{v\,k})_{k=1...l}$ of riverbed elevations respectively upstream and at the level of the facility studied (both in *m*). Dependence is apparent (Figure 10.8).

10.4.2 Reference probabilistic model for the hydro component

Tables 10.4 and 10.5 as well as Figure 10.9 introduce the fixed parameters and uncertain probabilistic model for hydro inputs. Note that the distributions of Q and K_s are in fact truncated in their lower tails at 0.1 in order to stay physically meaningful. This has a negligible impact on K_s (probability to be negative for a $N(30,7.5^2)$ being 3.10^{-5}), as well as on Q (probability to be negative is 2.10^{-3}) since lower values for Q are not important for the risk analyses illustrated in this book.

10.4.3 Systems reliability component – expert information on elementary failure probabilities

All events are given probabilities *conditional on* the event of limited overspill ($0 < S < 2m$), see Table 10.6.

Table 10.3 Bivariate sample data for upstream/downstream riverbed elevations.

Z_m	Z_v
55.09	50.39
55.00	50.28
54.87	50.23
54.28	49.92
54.74	50.51
55.48	50.42
55.36	50.16
55.39	50.16
54.80	49.76
55.18	50.17
54.94	50.71
54.42	50.08
55.34	49.95
55.30	50.63
54.31	49.51
55.57	50.77
54.28	49.98
55.49	50.30
54.49	50.10
55.11	50.12
55.15	50.54
54.40	49.21
55.87	50.55
55.63	50.67
54.93	50.00
55.61	50.70
54.95	50.27
55.38	50.06
54.57	49.49

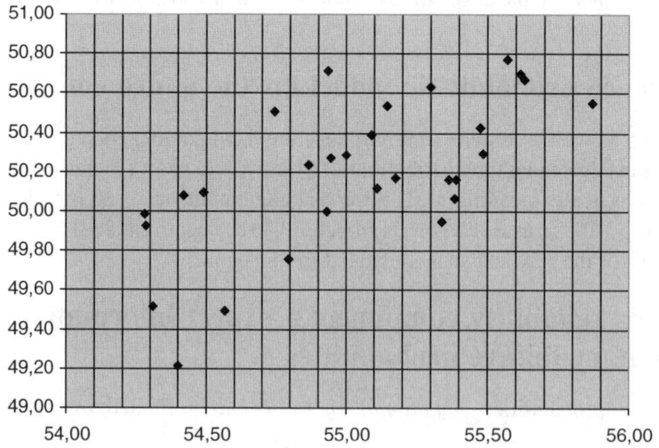

Figure 10.8 *Riverbed elevation data and correlations.*

Table 10.4 Deterministic inputs of the hydro component.

L	Length of the river reach (in m)	5000
B	River width (m)	300
z_b	Bank elevation (m NGF)	55.5

Table 10.5 Probabilistic parameters of the hydro component.

Uncertain input	Type of distribution	Position parameter	Scale parameter	Rank correlations (with other ***X***)
Q (m3/s)	Gumbel	1013 (mode)	557 (scale)	Independant
K_s ($m^{1/3}/s$)	Normal	30 (mean)	7.5 (standard deviation)	−0.3 with Z_m and −0.5 with Z_v
Z_v (m NGF)	Triangular	50 (mean)	1 (half-range)	+0.66 with Z_m and –0.5 with K_s
Z_m (mNGF)	Triangular	55 (mean)	1 (half-range)	+0.66 with Z_v and –0.3 with K_s

All in all, the non-recovery probability of the system conditional on limited overspill:

$$p_{nr} = p_1.p_2.(p_{12} + p_{1i}.p_{2i})$$

Modelling the expert ranges through uniform distribution, this overall epistemic distribution is as shown in Figure 10.10: the vertical line indicating the best-estimate value of 25 %, while the mean and standard deviation are 23 and 8 % respectively, and the range goes from [6–53 %].

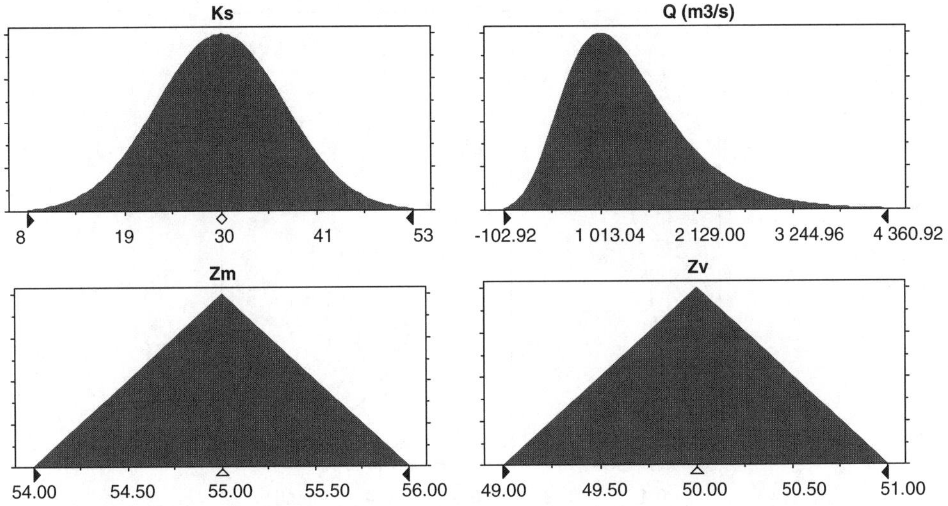

Figure 10.9 *Hydro Input distributions.*

Table 10.6 Probabilistic parameters of the systems reliability component.

Event	Probability	Expert best-estimate	Expert range	Comment
L_1 (*resp.* L_2): failure of passive system 1 (*resp. system 2*)	$p_1 = p_2$	90 %	80–95 % identical and completely dependent for the two epistemic	Assuming that both systems were identically produced, epistemic uncertainty is completely dependent
V_1 (*resp.* V_2): Total failure to activate protection system 1 (*resp. 2*)	$P(V_1) = P(V_2)$	50 %	25 %–60 %	Idem
V_{12}: Common mode failure to activate protection systems 1 & 2	p_{12}	25 %	25 %* $P(V_1)$ to 100 %* $P(V_1)$	The values are expressed *unconditionally* to the event so that $P(V_1) = p_{12} + p_{1i}$
V_{1i} (*resp.* V_{2i}): Failure to activate protection system 1 (*resp. system 2*) independently from system 2 (*resp. system 1*)	$p_{1i} = p_{2i}$	25 %	0 %* $P(V_1)$ to 75 %* $P(V_1)$ (but not independently from that of V_{12}; in fact, $p_{1i} = P(V_1)\text{-}p_{12}$)	

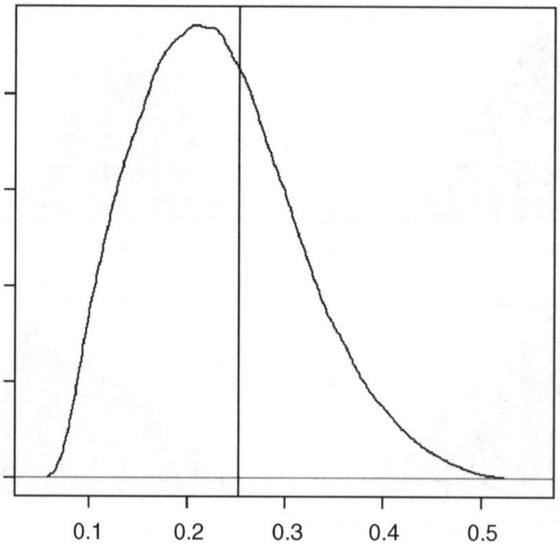

Figure 10.10 *Epistemic uncertainty in the non-recovery conditional probability of the protection system.*

10.4.4 Economic component – cost functions and probabilistic model

We note (Tables 10.7 and 10.8):

- s overspill (in m), z_b bank elevation at the foot of the dike and $h_d = z_d - z_b$ dike height (in m); s^i is the i-th year peak flood;
- l_d dike length (necessary to protect the installation at a given dike crest elevation);
- c_{ic} dike investment cost and c_{mn} annual maintenance cost (taken at 1 % of investment cost);
- c_i dike total cost including investment and maintenance over lifetime;
- c_s installation damage cost function, being the product of a function of overspill $c_{sm}(s)$ and an uncertain vulnerability coefficient c_m;
- c_g dike damage cost function;
- c_d^i damage cost for one event-year; c_d damage cost over lifetime;
- c_c complete cost over lifetime (including dike total cost, and damage cost);

Table 10.7 Damage cost function.

Overflow s in m	c_{sm} (s) in MEUR	Dike damage cost fraction c_g
−0.1	0	0 %
0	0	10 %
0.5	150	50 %
1	1500	100 %
1.5	2000	100 %
>2	2000	100 %

Table 10.8 Dike construction cost function.

Dike height h_d	Dike cost per metre of dike length EUR/m	Dike length l_d in m	Dike investment cost c_{ic} in kEUR
0	—	—	—
0.1	100	1000	100
0.5	150	1500	225
1	424	2437	1034
2	1200	3959	4750
3	2205	5000	11 023
4	3394	5000	16 971
5	4338	5000	21 692
6	5302	5000	26 509
7	6282	5000	31 408
8	7275	5000	36 377
9	8282	5000	41 409
10	9300	5000	46 498

Table 10.9 Probabilistic parameters of the cost component.

Uncertain input	Type of distribution	Position parameter	Scale parameter	Rank correlations
C_m (unitless coefficient)	Uniform	1 (mean)	0.5 (half-range)	Independent from all other inputs

- *T* lifetime (in years) over which the system is studied: 30 years. Dike lifetime may in fact be longer but 30 years is a reasonable lifetime for typical industrial investment studies.

$$c_i = c_{ic}(h_d) + c_{mn} \cdot T$$
$$c_{d^i} = c_s + c_g = c_{sm}(s^i, h_d).c_m + c_g(s^i).c_{ic}(h_d)$$
$$c_c = c_i + c_d = c_i + \Sigma_{i=1\ldots T} c_{d^i}$$

Detailed explanation:

- Dike cost per metre supposedly increases quicker than linearly with dike height h_d since when h_d increases, the minimal dike width required to sustain the structure increases as well. Cost growth model is typically $c_{ic} = (h_d/h_{ref})^c l_d$ with $c = 1.5$ at low heights and $c = 1.1$ above 4 metres of dike height: this represents a technological change moderating the impact of height onto cost.
- Dike length l_d increases less than linearily with h_d ($c = 0.7$) and reaches a threshold at 5 km since topographically the perimeter of the installation to be protected is limited.

Damage cost uncertainty is modelled to be within the range of [50 %, 150 %] of the prior cost estimate. A uniform distribution is assumed, in the absence of any particular information (Table 10.9).

10.4.5 Detailed results on various steps

Step one Overflow probability is estimated via water depth records (note that $z_{vpen} = 51.0$; $z_b = 55.5$; overflow starts when $H > 4.5\,m$) modelled alternatively: see Tables 10.10 and 10.11.

Table 10.10 Risk assessment results of Step One.

	Model parameters	No dike	$h_d = 1$ m	$h_d = 2$ m	$h_d = 3$ m
Bernoulli model	p – depending on dike level	4.9 %	2.4 %	0.81 %	??
Gaussian	$\mu = 2.6$ $\sigma = 1.1$ $K_S = 0.079$	3.4 %	0.3 %	0.01 %	$1.6\ 10^{-6}$
Lognormal	$l\mu = 0.86$ $l\sigma = 0.42$ $K_S = 0.062$	6.3 %	2.3 %	0.82 %	0.31 %
Gumbel	l = 2.1 s = 0.84 $K_S = 0.057$	5.6 %	1.8 %	0.54 %	0.16 %

Table 10.11 Design results of Step One.

	1/50	1/100	1/1000
Bernoulli model	1	1.5	?? (>2.9)
Gaussian	0.2	0.5	1.3
Lognormal	1.1	1.8	4.2
Gumbel	0.9	1.5	3.4

Note that large overflow probability is inaccessible to the Bernoulli model once the dike exceeds *2.5 m* because there is no observed flood the height of which exceeded *7 m*.

Step two Given the knowledge of the probabilities of limited overspill (E_{os}) and large overspill (E_{ol}) events, the protection system shows the following failure probability:

$$\begin{aligned} P(E_p) &= P(E_{os})p_1 \cdot p_2(p_{12} + p_{1i} \cdot p_{2i}) + P(E_{ol}) \\ &= P(E_{os})*0.253 + P(E_{ol}) \end{aligned}$$

Thus, the following results: see Tables 10.12 and 10.13.

Step three The fully deterministic approach involves taking penalised values ($q = 3854$ m3/s, $k_s = 15\ \mathrm{m}^{1/3}/\mathrm{s}$, $z_m = 54$ mNGF, $z_v = 51$ mNGF), which leads to a 'pseudo-worst case' design of 59.6 mNGF, that is 4.1 m above the natural riverbank. It is hard to assess the protection level associated

Table 10.12 Risk assessment results of Step Two.

Risk assessment				
	No dike	$h_d = 1$ m	$h_d = 2$ m	$h_d = 3$ m
Bernoulli model	1.8 %	?? > 0.6 %	?? > 0.21 %	???
Gaussian	0.87 %	$7.2\ 10^{-4}$	$2.6\ 10^{-5}$	$4.1\ 10^{-7}$
Lognormal	2.2 %	0.80 %	0.30 %	0.11 %
Gumbel	1.8 %	0.57 %	0.17 %	0.05 %

Table 10.13 Design results of Step Two.

Design (*dike heights*)			
	1/50	1/100	1/1000
Bernoulli model	no dike	?? 0.5	???
Gaussian	no dike	no dike	0.9
Lognormal	0.1	0.8	3.2
Gumbel	no dike	0.6	2.5

Table 10.14 Flow quantiles according to distribution fit.

	Model parameters	Quantiles (90 %)	Quantiles (99 %)	Quantiles (99.9 %)
Gaussian	$\mu = 1335$ $\sigma = 714$	2250	2996	3542
Lognormal	$1\ \mu = 7.041$ $1\ \sigma = 0.6045$	2478	4660	7394
Gumbel	$1 = 1013$ $s = 557$	2265	3573	4857

Table 10.15 Penalised water level as a function of flow.

Water level (dike)	55.5 – 0	56.5 – 1	57.5 – 2	58.5 – 3	59.5 – 4	60.5 – 5	61.5 – 6	62.5 – 7
Flow	1350	1900	2500	3200	3900	4700	5550	6450

with such a combination so that the following Table 10.16 of results mentions 'no answer' to the risk assessment issue: however most engineers would reckon that it is a bit higher than the historical period, that is a risk of ~ 1/150 yr.

When considering the probabilistic-deterministic approaches, the first step is to fit a distribution onto the flow, which is illustrated according to three types of prescribed parametric model (see Chapter 5): Normal, Gumbel or Lognormal. The results are as follows (Table 10.14).

To resolve the risk assessment issue, the system model considered as a function of the flow input $q \rightarrow z_{c\,pn} = G(x_{pn}, q)$ needs to be inverted in order to recover the flow beyond which a given dike level overflows. Then, according to the statistical model developed, the corresponding probability of exceedance may be computed. In the case of the regularity smoothness of such function, a restricted number of runs (say 20) may be sufficient to build an elementary (piece-wise linear) response surface and assess the flows for the variety of dike levels, as issued in the following Table 10.15 derived from the Figure 10.11[1].

Conversely, such inversion is not necessary for the design issue, whereby the level of protection translates into a flow using the appropriate quantile, and thus to a water and dike level through a single computation of the system model.

The fully probabilistic method involves a large number of Monte-Carlo runs of the system model issuing an approximate cdf of the water level, thus resolving the risk assessment or design issue directly. All of this allows for the following tables of results Tables 10.16 and 10.17.

The correlations bring a significant but only modest difference of +20–30 % of risk of 0.2–0.3 m of dike. Note that the correlations play a conflicting role because of respective signs of monotony: (K_s, Z_v) negative correlations tend to increase the risk as the monotony signs of K_s and Z_v are opposed. While (K_s, Z_m) negative and (Z_v, Z_m) positive correlations reduce the risk as K_s and Z_m monotony signs are similar while Z_v and Z_m monotony signs are opposed (see Section 7.3.2 of Chapter 7 for a justification in linearised models).

[1] Of course, with the simple flood model, closed-form inversion is possible. This would not be the case with a complex phenomenological model, so that such closed-form is not used for illustration.

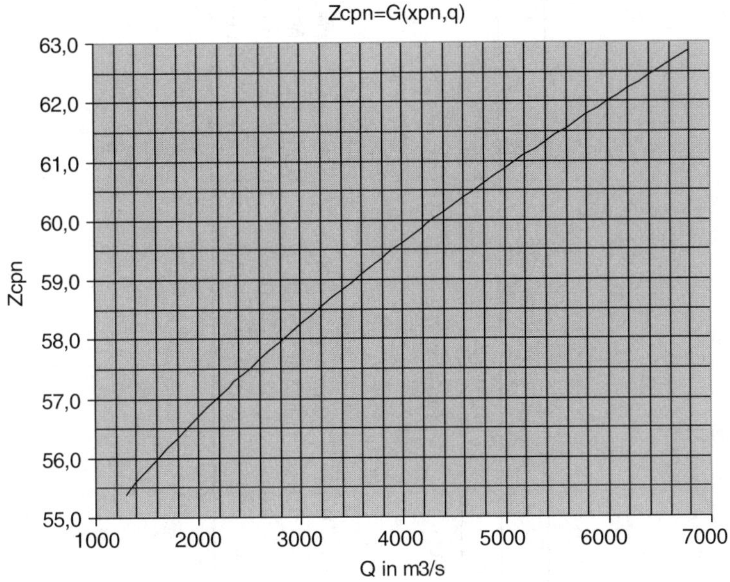

Figure 10.11 *Penalised water level computed as a function of flow and penalised values for the other uncertain inputs.*

Note the following results (see Table 10.18), in a hypothetical case where Z_m would be assumed non-correlated either with K_s or Z_v (which contradicts empiric evidence in the bivariate sample): this would add a further 0.2 m of dike.

More importantly perhaps, more elaborate dependence structures (e.g. Gumbel copula) would emphasise the tail dependence (which is limited when using linear or rank correlation models) with a larger impact on quantiles.

To finish with, Figure 10.12 represents the (empiric) cdf of Z_c in the standard full probabilistic Gumbel case without correlation sampled at 1000 000 samples, the upper values having been truncated at 65 mNGF in order to remain physically-sensible: resulting return levels of 1/5 000 and 1/10 000 are 60.7 mNGF and 63.7 mNGF respectively, they probably represent the rarest quantiles that can be modelled honestly owing to data limitations.

Step four From now on, all of the cases include Gumbel distribution for flow. Tables 10.19 and 10.20 illustrate the inclusion of the protection system in the figures.

As explained in Section 7.1.3, risk assessment is a closed-form post-processing of Step Three results for the full probabilistic case as follows:

$$\hat{c}_z = p_{nr} \cdot \frac{1}{N} \sum_{j=1..N} 1_{2>G(X_j,d)>0} + \frac{1}{N} \sum_{j=1..N} 1_{G(X_j,d)>2}$$

Step five A first step is to estimate the level-2 uncertainties. Regarding flow, different methods may be undertaken on the basis of the 149-sample: asymptotic approximation or several parametric Bootstrap approaches (Table 10.21).

Table 10.16 Risk assessment results of Step Three.

Risk assessment	No dike	$h_d = 1$ m	$h_d = 2$ m	$h_d = 3$ m
Fully deterministic (pseudo-worst case) + 0.2 m	(no answer)	(no answer)	(no answer)	(no answer)
Proba. on flow + deterministic - Normal	49 %	22 %	5 %	0.5 %
Proba. on flow + deterministic - Gumbel	42 %	19 %	6 %	2 %
Proba. on flow + deterministic – Lognormal	39 %	20 %	10 %	4 %
Full probabilistic[a] – Normal, no correlation	0.87 % (+− 0.02)	0.24 % (+ − 0.01)	0.10 % (+ − 0.006)	0.05 % (+ − 0.006)
Full probabilistic – Gumbel, no correlation	1.2 % (+ − 0.02)	0.35 % (+ − 0.01)	0.13 % (+ − 0.007)	0.06 % (+ − 0.005)
Full probabilistic – Lognormal, no correlation	2.2 % (+ − 0.03)	0.82 % (+ − 0.02)	0.33 % (+ − 0.01)	0.14 % (+ − 0.008)
Full probabilistic – Gumbel, correlation included	1.5 % (+ − 0.02)	0.45 % (+ − 0.01)	0.16 % (+ − 0.008)	0.07 % (+ − 0.005)

[a] all full probabilistic cases are obtained after 1000 000 runs and their accuracy expressed $(+-2\sigma)$.

Table 10.17 Design results for Step Three.

Design (*i.e. dike heights*)	1/50	1/100	1/1000	1/5000	1/10 000
Fully deterministic (pseudo-worst case) +0.2 m	(no answer)	~< 4 m	~>4 m		
Proba. on flow + deterministic – Normal	58.0 – 2.5	58.3 – 2.8	59.0 – 3.5		
Proba. on flow + deterministic – Gumbel	58.5 – 3	59.1 – 3.6	60.7 – 5.2		
Proba. on flow + deterministic – Lognormal	59.6 – 4.1	60.5 – 5	63.5 – 8		
Full probabilistic – normal, no correlation	54.9 – no dike	55.4 – no dike	57.5 – 2.0		
Full probabilistic – Gumbel, no correlation	55.1 – no dike	55.6 – 0.1	57.8 – 2.3	60.7 – 5.2	~ 62 – 7 m (approx.)
Full probabilistic – lognormal, no correlation	55.6 – 0.1	56.3 – 0.8	59.0 – 3.5		
Full probabilistic – Gumbel, Correlations included	55.3 – no dike	55.8 – 0.3	58.1 – 2.6		

Table 10.18 Design results for Step Three with alternative correlation structure.

	1/50	1/100	1/1000
Full probabilistic – Gumbel, Correlations included (except Z_m)	55.4 – no dike	56.0 – 0.2	58.3 – 3.3

Here the asymptotic ML are kept in the uncertainty model. Regarding Z_m and Z_v, we assumed $+-0.5$ uniformly distributed for the upper and lower bounds, neglecting so far correlations, either level-1 and level-2. Regarding the K_s, no data is available. Reasonable assumptions are the following:

- The mean is not known very well: there could be a typical 15 % standard error of appreciation, in connection with calibration issues: EK_s will supposedly be $N(30,4.5^2)$.
- The standard deviation represents the level of inter-annual variability typically due to sedimentation/erosion processes. It should stay at 15 % variation as a function of the mean. Thus, the coefficient of variation will be supposed to have negligible level-2 error.

In the context of very low data availability, the idea here is not to overparametrise the uncertainty model: only one level-2 random variable will be considered here.

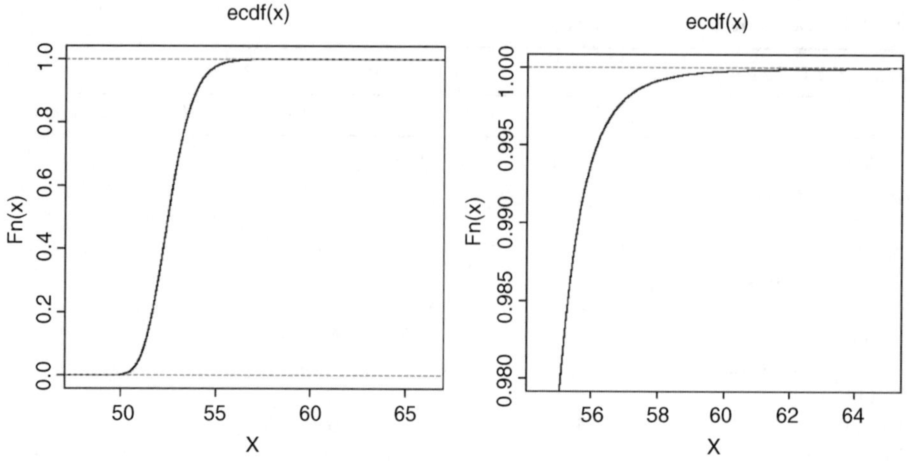

Figure 10.12 *Empiric cdf of the water level Z_c after 1 000 000 samples in the Gumbel non-correlated case (left, overall cdf; right zoomed at the upper tail).*

Thus, the full two-level propagation can be completed, as summarised in the following Table 10.22 firstly excluding the protection system in the level-2 based-design.

The level-2 based-design here refers to the upper bound of a 70 % c.i. (i.e. unilateral 85 %-quantile) of a return period, with an additional margin of 20 cm included to cover other non-quantified uncertainties, as suggested by the French nuclear regulations. When only flow is probabilised, the level-2 value is the

Table 10.19 Risk assessment results for Step Four.

Risk assessment (including protection system)				
	No dike	$h_d = 1$ m	$h_d = 2$ m	$h_d = 3$ m
Proba. on flow + deterministic	16 %	6 %	2 %	0,6 %
Full probabilistic – no correlation[a]	0.40 %	0.13 %	0.06 %	0.03 %
	±0.006 %	±0.004 %	±0.003 %	±0.002 %
Full probabilistic – correlation included[a]	0.51 %	0.17 %	0.07 %	0.04 %
	±0.007 %	±0.007 %	±0.003 %	±0.002 %

[a] After 500 000 simulations.

Table 10.20 Design results for Step Four.

Design (i.e. dike heights)			
	1/50	1/100	1/1000
Full probabilistic	*No dike*	*No dike*	*1.3 m*
Full probabilistic – correlation included	*No dike*	*No dike*	*1.6 m*

Table 10.21 Level-2 flow estimates.

	Asymptotic maximum likelihood	Non-param bootstrap of MM	Non-param bootstrap of ML	Param bootstrap of ML
Central estimate for m and associated stdd deviation	1013 ± 48 (5 %)	1016 ± 49 (5 %)	1014 ± 49 (5 %)	1014 ± 48 (5 %)
Central estimate for e and associated stdd deviation	557 ± 36 (6 %)	553 ± 41 (7 %)	555 ± 34 (6 %)	554 ± 35 (6 %)
Correlation of estimates	0.31	0.18	0.36	0.30
0.99 quantile and associated stdd deviation	3573 ± 184	3562 ± 203 (5.7 %)	3567 ± 182 (5.1 %)	3561 ± 183 (5.1 %)
95 % confidence interval for the 0.99 quantile	[3212 – 3934]	[3167 – 3960]	[3221 – 3930]	[3205 – 3927]
0.999 quantile and associated stdd deviation	4857 ± 264	4839 ± 295 (6.1 %)	4847 ± 260 (5.4 %)	4839 ± 263 (5.4 %)
95 % confidence interval for the 0.999 quantile	[4340 – 5375]	[4263 – 5417]	[4352 – 5367]	[4329 – 5365]

Table 10.22 Design results for Step Five, excluding the protection system.

Design (i.e. dike heights – without protection system)	1/50	1/100	1/1000
Proba. on flow + deterministic – Gumbel (with 20 cm margin)	59.0 – 3.5 m (85 %)	59.5 – 4.0 m (85 %)	61.2 – 5.7 m (85 %)
Full probabilistic-no correlation (100*50 000 runs)	55.1– no dike (median)	55.6 – 0.1 m (median)	57.8 – 2.3 m (median)
	55.2 – no dike (mean)	55.8 – 0.3 m (mean)	58.0 – 2.5 m (mean)
	55.5 – no dike (85 %)	56.1 – 0.6 m (85 %)	58.5 – 3.0 m (85 %)
Idem – computed through quadratic approximation + Gaussian approximate	55.1 mean	55.6 mean	57.6 mean
	55.4 (70 %)	55.9 (70 %)	58.1 (70 %)
	55.8 (90 %)	56.4 (90 %)	58.8 (90 %)
	56.0 (95 %)	56.6 (95 %)	59.1 (95 %)
	Stdev = 0.5	Stdev = 0.6	Stdev = 0.9

Table 10.23 Risk assesment results for Step Five.

Risk assessment (including protection system)				
	No dike	$h_d = 1$ m	$h_d = 2$ m	$h_d = 3$ m
Full probabilistic – no correlation	0.4 % (median) 0.5 % (mean) 0.8 % (85 %)	0.1 % (median) 0.2 % (mean) 0.3 % (85 %)	0.05 % (median) 0.07 % (mean) 0.1 % (85 %)	0.02 % (median) 0.03 % (mean) 0.05 % (85 %)

Table 10.24 Design results for Step Five including protection system.

Design (i.e. dike heights – including protection system)			
	1/50	1/100	1/1000
Full probabilistic – no correlation	No dike (mean) No dike (85 %)	No dike (mean) No dike (85 %)	1.7 m (mean) 2.2 m (85 %)

unilateral 85 % on flow, without going back to parameters. Other sources of uncertainties are maintained at their 'penalised values'. The 20 cm margin is not included in other methods.

The full probabilistic method is highly time-consuming: results are given here for 5 000 000 simulations, broken into 100*50 000. The accuracy is not yet complete however, especially regarding the level-2 quantiles of 1/1000-yr protection level. A boostrapping of level-1 quantiles can nevertheless be performed to simulate the variability of these level-2 quantiles: this issued about 0.1 m variation at most (including 1/1000-90 %).

The mean level-2 is 0.1–0.2 m higher in average than the median evidencing the asymmetry of distribution, while the median is close to the level-1 results. Level-2 uncertainty taken at the 85 % epistemic quantile adds a further of 0.4–0.7 m according to the return period considered.

Alternatively, the Taylor approximation is not bad at estimating those figures quantile-wise (the mean being noised by the neglected asymmetry), though the computation of derivatives remains somewhat instable, according to the differentiation increment (taken here at 1 % of standard deviation) and the number of level-1 (fixed-seed) samples: due to eight level-2 input parameters it requires about $(1+8)$*10 000 samples which is only one or two orders of magnitude less than double MC.

Adding the protection system (Tables 10.23 and 10.24) does not change the computing cost much. In fact, for any given value of θ_j, the computation of the conditional non-recovery probability is closed-form. Thus, it is simply a post-processing of the level-1 Monte-Carlo on a phenomenological model, to be included within the level-2 Monte-Carlo loop.

Step six The flow-water height data is represented in Figure 10.13 (left) as against the computation of the modelled water height for mean values of Z_m and Z_v and varying values of K_s.

Under a simple *parametric identification* approach, the non-linear cost function as a function of k_s is illustrated in the Figure 10.13 (right). The minimum is obtained for $k_s = 26.97$, with $Cost = 27.13$ after 24 gradient iterations with a hessian of 0.443. A linearisation can be done in $k_{so} = 30$ with results only slightly different after linearised identification with varying gradients ($k_s = 26.68$). Note indeed that H_{oj} varies from 0.6 to 4.4 m and the gradient H'_j from -0.01 to -0.08 according to flows taken as d_j, so that an assumption of constant gradients H'_j is grossly false.

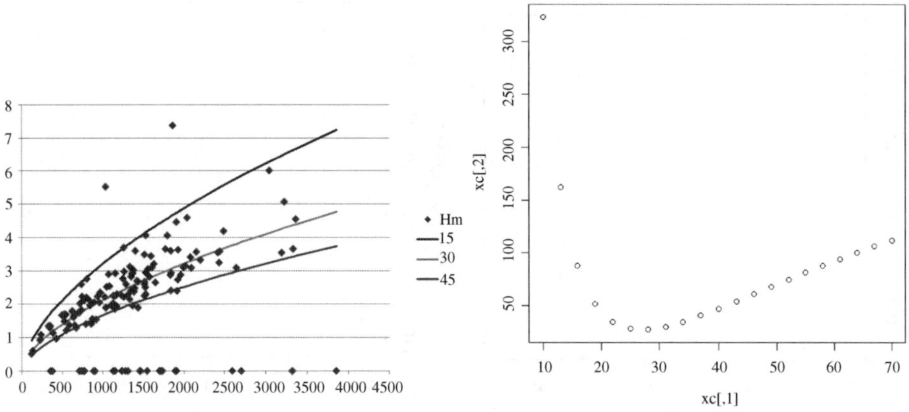

Figure 10.13 *Flow-water height data against modeled water height (left); non-linear parameter identification cost function against k_s (right).*

The *calibration* results are as follows:

- $X_m = 27$ with $VarX_m = 0.99$, thus a st. dev. for x_m of 1 unit of k_s, either through numerical hessian or through analytical formula, including if neglecting the second-order derivatives.
- Residual variance is estimated at 0.66 m, graphically obvious (see Figure 6.7-right in Chapter 6).

Linearised Intrinsic variability identification leads to the following likelihood plots (Figure 10.14). Convergence requires the fixing of measurement noise variance σ^2 for intrinsic variance v^2 to be identifiable, taken at $\sigma^2 = (0.66/2)^2$ as a fraction of total variance.

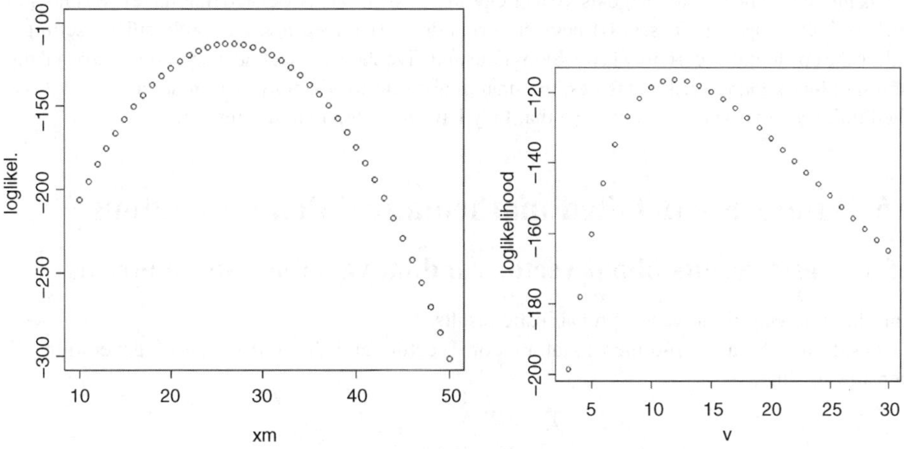

Figure 10.14 *Plot of linearised loglikelihood for $v^2 = 11^2$ and $x_m = k_s$ varying in abscissa (left) or $x_m = 30$ and v varying in abscissa (right); both assuming that $\sigma^2 = (0.66/2)^2$.*

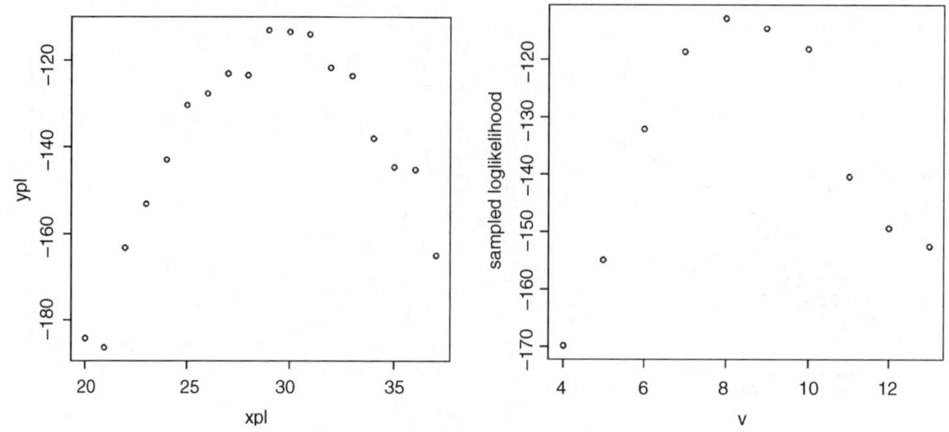

Figure 10.15 *Plot of sampled non-linear loglikelihood for $v^2 = 8.5^2$ and $x_m = k_s$ varying in abscissa (left) or $x_m = 30$ and v varying in abscissa (right); both assuming that $\sigma^2 = (0.66/2)^2$.*

The non-linear formulation involves random sampling so as to estimate the loglikelihood at each iteration of optimisation. Figure 10.15 provides somes results performed with 500 samples per iteration, again with $\sigma^2 = (0.66/2)^2$. The non-linear algorithm leads to a minimum of loglikelihood for a lower intrinsic variance for k_s, closer to 8^2–9^2, and a slightly higher value for the expectation of k_s, around to 29–31, though the underlying sampling is required to be larger so as to stabilise the estimates. Note that Figure 10.14 already cost $\sim 30^* 500^* 123 = 1.8\ 10^6$ calls to the system model (in fact, for each θ_k, a new 500-sample is drawn but the sample is then used throughout $(d_j)_{j=1..n}$ to reduce the estimation variance of $E[H(X \mid \theta_k, d_j)]$ and $var[H(X \mid \theta_k, d_j)]$ inside the loglikelihood function).

Step seven The details can be found in Chapter 8. Step Seven insists on the computation of expectation-type risk measures and the salient goals 'Select/Optimise' so that all associated runs have been undertaken under two following hypotheses: (i) neglect correlations, (ii) keep a single probabilistic setting, not detailing the epistemic uncertainty layer. Steps Four and Five showed that both phenomena have an impact on threshold-exceedance risk measures, the final goal being to 'Comply' with risk criteria, but only a limited influence on expected values, so that they have not been detailed further.

10.5 Annex 5 – detailed mathematical demonstrations

10.5.1 Basic results about vector random variables and matrices

Recall the following basic vector probabilistic results.

Considering X, as a p-dimensional random vector, and H, a q^*p matrix generating Y (of dimension q), then:

$$\begin{aligned} &Y = HX \\ &\Rightarrow EY = H.EX \\ &\Rightarrow \mathrm{var}Y = H.\mathrm{var}X.H' \end{aligned}$$

Considering any r^*s matrix denoted as D, then the symmetric matrix DD' is also *semi-positive definite*. Indeed, for any vector y of dimension s, then:

$$y'DD'y = (D'y)'(D'y) = z'z = \sum_{k=l...r} (z^k)^2 \geq 0$$

The following property is useful in the context of model identifiability (Chapter 6):

H is a full-rank (i.e. injective) q^*p matrix $\Leftrightarrow H'H$ is an invertible p^*p matrix.

Proof

Suppose H is injective, then:

$$\begin{aligned} &H'Hx = 0 \\ &\Rightarrow x'H'Hx = 0 \\ &\Rightarrow ||Hx||^2 = 0 \\ &\Rightarrow Hx = 0 \\ &\Rightarrow x = 0 \end{aligned}$$

which proves that $H'H$ is also injective. Being a p^*p matrix, it is invertible.

Conversely, if $H'H$ is invertible then:

$$Hx = 0 \Rightarrow H'Hx = 0 \Rightarrow x = 0$$

which proves that H has a void kernel, that is, it is injective.

This property still holds if $H'RH$ is invertible, R being a definite-positive q^*q matrix. Indeed, R may be decomposed under Cholesky as a product of a q^*q matrix and its transpose:

$$R = C'C$$

C being necessarily also full-rank, the previously-established property applies when replacing H by CH.

Because of the previous property ensuring that $H'H$ is semi-positive definite, we can add to the equivalence:

H is a full-rank (i.e. injective) qxp matrix
$\Leftrightarrow H'H$ is a (p^*p) invertible matrix
$\Leftrightarrow H'H$ is positive-definite matrix.
$\Leftrightarrow H'RH$ is positive-definite matrix (given any positive-definite matrix R).

10.5.2 Differentiation results and solutions of quadratic likelihood maximisation

Many results regarding statistical estimation depend on a generic quadratic optimisation program: this is the case especially with the considerartions in Chapter 6 on the inverse methods involved in parameter identification or data assimilation algorithms. Detailed algebric demonstrations will therefore be provided hereafter.

Differentiating quadratic likelihoods A function of the form (where x is a scalar and $R_j(x)$ twice-differentiable but potentially non-linear):

$$L(x,\sigma^2) = -\frac{n}{2}\log(2\pi\sigma^2) - \frac{1}{2}\frac{1}{\sigma^2}\sum_j (R_j(x))^2 \tag{10.45}$$

acquires the following first and second partial derivatives:

$$\begin{aligned}
\partial L(x,\sigma^2)/\partial x &= -\sum_j \frac{1}{\sigma^2} R_j(x)\nabla R_j(x) \\
\partial^2 L(x,\sigma^2)/\partial x^2 &= -\sum_j \frac{1}{\sigma^2}\left(\nabla R_j(x)^2 + \Delta R_j(x) R_j(x)\right) \\
\partial L(x,\sigma^2)/\partial(\sigma^2) &= -\frac{n}{2\sigma^2} + \frac{1}{2\sigma^4}\sum_j (R_j(x))^2 \\
\partial^2 L(x,\sigma^2)/\partial(\sigma^2)^2 &= \frac{n}{2\sigma^4} - \frac{1}{\sigma^6}\sum_j (R_j(x))^2
\end{aligned} \tag{10.46}$$

When considering the maximisation of L, the first-order condition leads to the following pair of equations:

$$\begin{aligned}
0 &= \sum_j R_j(\hat{x})\nabla R_j(\hat{x}) \\
\hat{\sigma}^2 &= \frac{1}{n}\sum_j R_j(\hat{x})^2
\end{aligned} \tag{10.47}$$

At the optimal point the hessian matrix of second-order derivatives appears as follows:

$$\begin{aligned}
\partial^2 L/\partial x^2|\hat{x} &= -\sum_j \frac{1}{\sigma^2}\left(\nabla R_j(\hat{x})^2 + \Delta R_j(\hat{x}) R_j(\hat{x})\right) \\
\partial^2 L/\partial(\sigma^2)^2|_{\hat{\sigma}^2} &= \frac{n}{2\hat{\sigma}^4} - \frac{1}{\hat{\sigma}^6}\sum_j R_j(\hat{x})^2 = \frac{n}{2\hat{\sigma}^4} - \frac{1}{\hat{\sigma}^6} n\hat{\sigma}^2 = -\frac{n}{2\hat{\sigma}^4} \\
\partial^2 L/\partial x\partial(\sigma^2)|_{\hat{x},\hat{\sigma}^2} &= \frac{1}{\hat{\sigma}^4}\sum_j R_j(\hat{x})\nabla R_j(\hat{x}) = 0
\end{aligned} \tag{10.48}$$

An extension can be made to the case where $\boldsymbol{x} = (x^i)_{i=1..p}$ is a vector:

$$\begin{aligned}
\partial L(\boldsymbol{x},\sigma^2)/\partial x^k &= -\sum_j \frac{1}{\sigma^2} R_j(\boldsymbol{x})\partial R_j/\partial x^k \\
\text{i.e.}\quad \nabla_{\boldsymbol{x}} L(\boldsymbol{x},\sigma^2) &= -\frac{1}{\sigma^2}\sum_j R_j \nabla_{\boldsymbol{x}} R_j \\
\partial^2 L(x,\sigma^2)/\partial x^k\partial x^l &= -\frac{1}{\sigma^2}\sum_j \left(\partial R_j/\partial x^k \partial R_j/\partial x^l + R_j(x)\partial^2 R_j/\partial x^k \partial x^l\right) \\
\text{i.e.}\quad \Delta_{\boldsymbol{x}} L(\boldsymbol{x},\sigma^2) &= -\frac{1}{\sigma^2}\sum_j \left(\nabla_{\boldsymbol{x}} R_j' \nabla_{\boldsymbol{x}} R_j + R_j \Delta_{\boldsymbol{x}} R_j\right)
\end{aligned} \tag{10.49}$$

taking the gradient $\nabla_x R_j$ as a row vector. There is no change to the derivatives with respect to σ^2. The first-order conditions lead to:

$$\begin{aligned} 0 &= \sum_j R_j(\hat{x}) \nabla \hat{x} R_j(\hat{x}) \\ \hat{\sigma}^2 &= \frac{1}{n} \sum_j R_j(\hat{x})^2 \end{aligned} \tag{10.50}$$

Note that nothing is changed regarding the estimation of σ^2: still worth the mean residuals, and uncorrelated with the estimator of $\boldsymbol{x}$, with an unchanged variance.

Suppose now that the quadratic function takes the following generalised 'least square' form:

$$L(\boldsymbol{x}, (Q_j)_{j=1\ldots n}) = -\frac{1}{2} \sum_j \log(\det(2\pi Q_j)) - \frac{1}{2} \sum_j (Y_j - H_j(\boldsymbol{x}))' Q_j^{-1} (Y_j - H_j(\boldsymbol{x})) \tag{10.51}$$

Its gradient with respect to $\boldsymbol{x}$ takes the following form:

$$\nabla_x L(\boldsymbol{x}, (Q_j)_{j=1\ldots n}) = -\sum_j \nabla_x H_j(\boldsymbol{x})' Q_j^{-1} (Y_j - H_j(\boldsymbol{x})) \tag{10.52}$$

When H_j is linear, the Hessian can be expressed simply as follows:

$$\Delta_x L(\boldsymbol{x}, (Q_j)_{j=1\ldots n}) = -\sum_j \nabla_x H_j(\boldsymbol{x})' Q_j^{-1} \nabla_x H_j(\boldsymbol{x}) = -H' Q^{-1} H \tag{10.53}$$

using the block matrix notation introduced in Section 6.3.6.

Solving quadratic likelihood optimisation programs The above-mentioned function may be seen as a general formulation of the loglikelihoods involved in many Gaussian models mentioned in Chapter 5 or 6 included in the generic model as follows (Y_j are independent q-dimensional multi-normal vectors but possibly non-identically distributed):

$$\begin{aligned} Y_j &= H_j(\boldsymbol{x}) + U_j \\ & \quad Y_j | \boldsymbol{x}, Q_j \sim N(H_j(\boldsymbol{x}), Q_j) i.ni.d. \\ LL(\boldsymbol{x}, (Q_j)_{j=1..n}) &= -\frac{1}{2} \sum_j \log(\det(2\pi Q_j)) - \frac{1}{2} \sum_j (Y_j - H_j(\boldsymbol{x}))' Q_j^{-1} (Y_j - H_j(\boldsymbol{x})) \end{aligned} \tag{10.54}$$

In the linear (or linearised case), given linearisation and appropriate rescaling of constant terms, change of notation and block grouping of matrices (see Section 6.3.6), this may be reformulated as a *Gaussian multi-linear regression problem.*

$$Y = X.\beta + U \quad \textit{where} \quad U \sim N(0, Q) \tag{10.55}$$

the likelihood of which is re-written as follows:

$$LL[Y|X, \beta] = -\frac{nq}{2} \log(2\pi) - \frac{1}{2} \log(\det Q) - \frac{1}{2} (Y - X\beta)' Q^{-1} (Y - X\beta) \tag{10.56}$$

In the cases where Q is known, the likelihood maximisation consists of finding the solution β for the following quadratic (least-square) program:

$$\hat{\beta} = Arg \max LL[Y|X, \beta] = Arg \min \left[\frac{1}{2}(Y - X\beta)' Q^{-1}(Y - X\beta)\right] = Arg \min \|Y - X\beta\|_{Q^{-1}} \quad (10.57)$$

Using the previously-recalled differentiation results, the first-order condition yields:

$$\nabla_{\beta} LL[Y|X, \beta] = -X' Q^{-1}(Y - X\beta) = 0 \quad (10.58)$$

Assuming that $X'Q^{-1}X$ is invertible (a necessary requirement, see Section 6.3.6) a classical result of least-square projection yields the following estimator:

$$\hat{\beta} = (X'Q^{-1}X)^{-1} X'Q^{-1}Y \quad (10.59)$$

It is easy to check that the estimator fulfils the first-order condition:

$$\begin{aligned} -X'Q^{-1}(Y - X\hat{\beta}) &= -X'Q^{-1}(Y - X(X'Q^{-1}X)^{-1}X'Q^{-1}Y) \\ &= -X'Q^{-1}Y - X'Q^{-1}X(X'Q^{-1}X)^{-1}X'Q^{-1}Y = 0 \end{aligned} \quad (10.60)$$

Coming to the second-order equation, it can be derived from the previous differentiation results that:

$$\Delta_{\beta} LL[Y|X, \beta] = -X'Q^{-1}X \quad (10.61)$$

It is easy to see that this matrix is closely related to the variance matrix of the estimator (Q is symmetric as being a variance matrix):

$$\begin{aligned} Var\, \hat{\beta} &= (X'Q^{-1}X)^{-1}X'Q^{-1} Var Y \left[(X'Q^{-1}X)^{-1}X'Q^{-1}\right]' \\ &= (X'Q^{-1}X)^{-1}X'Q^{-1}Q\left[(X'Q^{-1}X)^{-1}X'Q^{-1}\right]' \\ &= (X'Q^{-1}X)^{-1}X'Q^{-1}X(X'Q^{-1}X)^{-1} \\ &= (X'Q^{-1}X)^{-1} \end{aligned} \quad (10.62)$$

This result is also a consequence of the estimator being the maximal likelihood estimator (see Chapter 5).

As being an invertible variance matrix $(X'Q^{-1}X)^{-1}$ is positive-definite and thus also $X'Q^{-1}X$, satisfying the second-order condition of optimality: the estimator is the unique solution. Note that it appears as a linear function of the observables Y that is unbiased as an estimator:

$$\begin{aligned} E\hat{\beta} &= (X'Q^{-1}X)^{-1}X'Q^{-1}EY \\ &= (X'Q^{-1}X)^{-1}X'Q^{-1}X\beta = \beta \end{aligned} \quad (10.63)$$

It can also be proven that $\hat{\beta}$ is the linear unbiased estimator of least variance, thus being called BLUE (Best Linear Unbiased Estimator). Suppose there is another such unbiased linear estimator:

$$\hat{\beta}_2 = CY \quad (10.64)$$

Because of linearity, C cannot depend on Y. Because of unbiasedness,

$$\beta = E\hat{\beta}_2 = ECY = CEY = CX\beta \tag{10.65}$$

C cannot vary as a function of β either: the previous equation, being true for any β, implies that $CX = I_p$, the p-dimensional identity matrix and thus, considering the difference:

$$\begin{aligned} \hat{\beta}_2 - \hat{\beta} &= CY - (X'Q^{-1}X)^{-1}X'Q^{-1}Y = DY \quad \textit{where} \\ D &= C - (X'Q^{-1}X)^{-1}X'Q^{-1} \end{aligned} \tag{10.66}$$

Thus, we have $DX = 0_p$ and the variance of the second estimator re-expresses as follows:

$$\begin{aligned} Var\,\hat{\beta}_2 &= C\,\mathrm{var}\,YC' = \left[D + (X'Q^{-1}X)^{-1}X'Q^{-1}\right]Q\left[D + (X'Q^{-1}X)^{-1}X'Q^{-1}\right]' \\ &= DQD' + (X'Q^{-1}X)^{-1}X'D' + DX(X'Q^{-1}X)^{-1} + (X'Q^{-1}X)^{-1} \\ &= DQD' + (X'Q^{-1}X)^{-1} \end{aligned} \tag{10.67}$$

This states that the variance of the second estimator is always superior to that of the BLUE which is $(X'Q^{-1}X)^{-1}$. Indeed, Q being positive-definite as any variance matrix, DQD' is semi-positive definite and any diagonal term of DQD' is positive or null, i.e. a positive excess of variance of each component of the second estimator with respect to that of the corresponding component of the first estimator.

10.5.3 Proof of the Wilks formula

This section will detail the proof of the Wilks formula as shown in Chapter 7. $F(z)$ denoting the real (unknown) *cdf* of the output of interest Z, the cdf of the m-th maximum value of the i.i.d. N-sample Z^{N-m+1} appears classically as the following combinatorial expression:

$$P(Z^{N-m+1} < z) = \sum_{k=N-m+1}^{N} C_N^k F(z)^k (1 - F(z))^{N-k} \tag{10.68}$$

As being by definition the sum of N i.i.d indicator variables $1_{Zi<z}$ of common mean $p = F(z)$, the scaled empiric distribution $N.\hat{F}_{QN}(z)$ is, for any fixed z, distributed exactly according to a binomial distribution:

$$N.\hat{F}_{QN}(z) \sim B_{N,p}(k) \tag{10.69}$$

so that:

$$P\left(\hat{F}_{QN}(z) = \frac{k}{N}\right) = C_N^k F(z)^k (1 - F(z))^{N-k} \tag{10.70}$$

Thus, the probability of the event that out of N simulations the number exceeding z is less or equal to m is that given by Equation (10.68). Apply then such function to the (unknown) quantile $z = z^\alpha$:

$$P(Z^{N-m+1} < z^\alpha) = \sum_{k=N-m+1}^{N} C_N^k F(z^\alpha)^k (1 - F(z^\alpha))^{N-k} = \sum_{k=N-m+1}^{N} C_N^k \alpha^k (1 - \alpha)^{N-k} \tag{10.71}$$

Thus, if N is such that:

$$\sum_{k=N-m+1}^{N} C_N^k \alpha^k (1-\alpha)^{N-k} < 1-\beta \tag{10.72}$$

Then

$$P(Z^{N-m+1} < z^\alpha) < 1-\beta \Longleftrightarrow P(z^\alpha < Z^{N-m+1}) > \beta \tag{10.73}$$

which proves that Z^{N-m+1} is a Z^α_β-type conservative estimator of z^α.

It should be noted that this proof based on simple order statistics requires no other probabilistic property than the mere existence of a *cdf* for Z. Note also that the number of trials does not depend in any way on the properties of the cdf of input uncertainties, nor on the number of variates. Nor is affected by correlations or dependence structures within $\boldsymbol{X}$ with respect to which the method is completely robust and unchanged.

10.5.4 Complements on the definition and chaining of monotony

Generalising the definitions Monotony was introduced in Chapter 4 through a system model mapping D_x, a subset of $\mathcal{R}^p$ into either $\mathcal{R}$ or $\{0,1\}$ when considering respectively an output variable or event of interest: the *global monotony* of $Z = G(\boldsymbol{x},\boldsymbol{d})$ over D_x (or simply *monotony* when non-ambiguous) is defined as follows:

$$\begin{aligned} &\forall i, \exists s^i \in \{1,-1\}, \forall a \geq 0, \boldsymbol{d}, \boldsymbol{x} = (x^1, ..x^i, ..x^p) \in D_x \\ &G(x^1, .., x^i + s^i a, ..x^p, \boldsymbol{d}) \leq G(x^1, .., x^i, ..x^p, \boldsymbol{d}) \end{aligned} \tag{10.74}$$

where s^i represents the sign of monotonous dependence: $s^i = 1$ when $G(.)$ is decreasing with i-th component x^i, and $s^i = -1$ when it is increasing.

While the definition above encompasses cases where Z is a binary event of interest, it may also be extended to input *events*, a situation that occurs frequently in risk analysis.

$$\begin{aligned} &\forall i_1 = 1..p_1, \exists s^{i_1} \in \{1,-1\}, \forall a \geq 0, \forall \boldsymbol{d}, \forall \boldsymbol{xe} = (x^1, ..x^{i_1}, ..x^{p1}, e^1, ..e^{p2}) \in D_{xe} \\ &G(x^1, .., x^{i_1} + s^{i_1} a, ..x^{p1}, e^1, ..e^{p2}, \boldsymbol{d}) \leq G(x^1, .., x^{i_1}, ..x^{p1}, e^1, ..e^{p2}, \boldsymbol{d}) \\ &\forall i_2 = 1..p_2, \exists s^{i_2} \in \{1,-1\}, \forall \boldsymbol{d}, \forall \boldsymbol{xe} = (x^1, ..x^{p1}, e^1, ..e^{i_2}, ..e^{p2}) \in D_{xe} \\ &G(x^1, .., x^{p1}, e^1, ..e^{i_2} + s^{i_2}, ..e^{p2}, \boldsymbol{d}) \leq G(x^1, ..x^{p1}, e^1, ..e^{i_2}, ..e^{p2}, \boldsymbol{d}) \end{aligned} \tag{10.75}$$

All input events are gathered inside $(e^1, \ldots e^{p2})$ and since those events may only take two values, 0 or 1, that is may only decrease or increase once for the sake of formal consistence with the variables, the convention is that $e^{p2} = 2$ is similar to $e^{p2} = 1$ and $e^{p2} = -1$ is similar to $e^{p2} = 0$.

Global monotony is in fact twice global: (a) it is assumed to be true throughout D_x and (b) it is assumed to be true whatever the component considered. Physically, (a) may only be true for a reasonably-bound physical domain within which phenomenology does not change radically: this is in fact an implicit hypothesis in many models which describe a given phenomenology (e.g. elastoplastic flaw initiation).

Care is, however, required in properly bounding their domain of definition, particularly when considering mathematically-limitless pdf (e.g. normal, lognormal) in order to model uncertainty: the physical plausibility of the model and of the ranges of the physical variables represented by model inputs is rarely limitless. Truncating pdfs over a physically-sound D_x will be generally desirable and may solve some unphysical ruptures of monotony in some cases (e.g. negative toughness).

It is, however, quite possible that failure involves more than one phenomenon, with differing behaviours regarding monotony with respect to physical inputs, thereby contradicting (a). A less exacting property would then be *local monotony*, that is Equation (10.74) limited to subsets of D_x, for instance associated with distinct failure modes. The base hypothesis assumed later on will be that monotony is true over the entire physical domain of definition D_x. But the conclusions would still hold when supposing at the outset that D_x denotes a subset within which monotony is verified, although the probabilistic meaning would then differ, generating conditional probabilities instead of full probabilities. This hypothesis may also be challenged in some physical systems where competing physical effects make it hard to secure *a priori* monotony with respect to some components, while it may still be obvious for others. *Partial monotony* is the alternative hypothesis, assuming that Equation (10.74) is valid only for a subset of indices i although it proves much more difficult for the algorithms to handle.

Eventually, it may be interesting to consider a conservative monotonous envelope of the model $G(.)$. For example, a locally increasing model can be converted into a (conservative) globally increasing monotonous envelope through the following transformation using the partial order in a p-dimensional space:

$$\tilde{G}(\underline{x}) = Max\{G(\underline{x}')|\underline{x}' \leq \underline{x}\} \tag{10.76}$$

Permanence of monotony through model chaining and space transformation The elementary results of real-valued scalar functional analysis state that the composition of two monotonous functions generates a monotonous function and that the inverse of a monotonous function (which can always be defined if the function is injective) is again monotonous, with two important practical consequences for our purposes. First, given a monotonous failure function $G(\boldsymbol{x})$ over D_x, consider a more detailed description of a component developed through the chaining of a monotonous sub-model, that is a function $x^p = h(y^p, y^{p+1}, y^q)$ globally monotonous as a function of new variables y^p, y^{p+1}, y^q over D_y. This generates a new composed system model:

$$G[x^1, x^2, ..xp^{-1}, h(y^p, y^{p+1}, ..y^q)] = G_c(\boldsymbol{y}) \text{ of } \boldsymbol{y} = (x^1, ..x^{p-1}, y^p, ..y^q) \tag{10.77}$$

that has the same property. In practise this enables the progressive chaining of models in a study while keeping monotonous benefits: alternatively, for the purpose of demonstration, it enables the elaboration of a proof of global monotonous property *physical layer by physical layer* in a complex model. Note, however, that this property holds in general only if the sub-model involves variables $y^p, y^{p+1}, ..y^q$ as distinct from the initial $x^1, ..x^{p-1}$. Two sub-models detailing distinct component variables of the upper layer model but involving some common input variables may of course generate competing effects, threatening chained monotony in spite of the separate monotony of each. Note that is does not mean that the model inputs of the two sub-models must be independent in a probabilistic sense: at this stage dependence of model inputs has no impact on the monotony property.

Secondly, the monotonous property of the failure function remains if the model input variables are changed through injective monotonous transformations component per component. Suppose that

$u^i = T^i(x^i)$ (for $i = 1$ to p) denote a set of monotonous functions gathered altogether in the vector function $T(\boldsymbol{x}) = (T^1(x^1),..,T^i(x^i),..T^p(x^p))'$ mapping the domain D_x into an image subset D_u of R^p. Suppose that all T^i are injective, which enables defining the inverse $T^{-1}(\boldsymbol{u})$ over the image subset D_u: it is straightforward to see that $T^{-1}(\boldsymbol{u})$ and $GoT^{-1}(\boldsymbol{u})$ follow the property of Equation (10.74) as functions of $\boldsymbol{u}$ over the image subset D_u.

This is particularly the case for the classical structural reliability transformation into standard Gaussian space for independent variables: in that case the transformation simply goes $T^i(u^i) = \Phi^{-1}oF_{xi}(x^i)$ where $F_{xi}(.)$ stands for the i-th marginal cumulative distribution function (cdf) of $\boldsymbol{X}$ and Φ stands for the cdf of a standard Gaussian random variable. It is a composition of cdf (or their inverse) that are monotonous (increasing) simply because of their probabilistic definition: $F(x) = P(X < x)$. It will also be the case when considering an alternative transformation to the standard one, which will prove quite useful hereafter: the *transformation into the uniform space* defined by $v^i = F_{xi}\ (x^i)$ (supposing all are injective) generates a set of p random variables V^i over [0,1], the marginal distributions of which are uniformly-distributed. If the X^i are not independent, neither are the V^i and their joint pdf defines the associated copula function (see Chapter 5). $GoT^{-1}(\boldsymbol{v})$ still retains the global monotony property, however. It may not be the case when applying a transformation on a dependent $\boldsymbol{X}$ into the standard *independent Gaussian* space (such as Rosenblatt or Nataf transformation (Madsen *et al.*, 1986): in the Nataf case, linear transformations involving opposite-signed coefficients may be necessary to de-correlate the uncertain vector.

Conversely, considering the associated inverse transformations, one ends up with the fact that the global monotony of the failure function is equivalent whether considering the physical space or any space obtained through monotonous injective transformations; in particular physical space monotony and Uniform-transformed space are equivalent. It is also the case with the Gaussian-transformed standard space, but only in the independent case.

10.5.5 Proofs on level-2 quantiles of monotonous system models

We now consider a p-dimensional system model that is at least partially monotonous with respect to p-th component.

Note that EZ increases when any of the expectations of input uncertain components $\theta^p = E(X^p)$ increases. If composed with a monotonous function, the increase of the expectation of a random variable (with unchanged distribution otherwise) is preserved. In other words, the output expectation is monotonous with the input expectations. Consider U and V, two random variables with identical distributions, although shifted by a positive increase of the expectation:

$$E(U) = E(V) + a \tag{10.78}$$

$$\begin{aligned} U &\sim f_U(U = u) = k(u - E(U)) \\ V &\sim f_V(V = v) = k(v - E(V)) \end{aligned} \tag{10.79}$$

Suppose that $G(.)$ is 1-dimensional monotonously increasing system model, then:

$$\begin{aligned} E[G(U)] &= \int_u G(u) f_U(U = u)du = \int_u G(u)k(u - E(U))du \\ &= \int_u G(w + E(U))k(w)dw \geq \int_u G(w + E(V))k(w)dw = E[G(V)] \end{aligned} \tag{10.80}$$

This is easily generalised into the case p-dimensional case. Consider now that G is partially monotonous (increasing) with respect to x^p with $U = X^p$ and $V = Y^p$:

$$\begin{aligned} E[G(X^1,..X^p)] &= E_{X^1,..X^{p-1}}[E_{X^p}[G(X^1,..X^p)|X^1,..X^{p-1}]] \\ &\quad \forall x^1,..x^{p-1} \quad E_{X^p}[G(X^1,..X^p)|x^1,..x^{p-1}] = E[H(U)] \\ \geq E[H(V)] &= E_{Y^p}[G(X^1,..X^{p-1},Y^p)|x^1,..x^{p-1}] \end{aligned} \tag{10.81}$$

$$\begin{aligned} &\Rightarrow E_{X^1,..X^{p-1}}[E_{X^p}[G(X^1,..X^p)|X^1,..X^{p-1}]] \\ &\geq E_{X^1,..X^{p-1}}[E_{Y^p}[G(X^1,..X^{p-1},Y^p)|x^1,..x^{p-1}]] \\ &\Rightarrow E[G(X^1,..X^{p-1},X^p)] \geq E[G(X^1,..X^{p-1},Y^p)] \end{aligned} \tag{10.82}$$

10.5.6 Proofs on the estimator of adaptive Monte-Carlo under monotony (section 7.4.3)

Consider a globally monotonous system model (or failure function) $G(.)$. We wish to prove in the following the unbiasedness and variance properties of the S-MRM estimator, that is a Monte-Carlo adaptive sampling estimator under monotony defined as follows:

$$\hat{P}_f^{mis} = \frac{1}{N}\sum_{j=1..N}\left[P(E_{j-1}^+) + 1_{G(x_j)\leq 0}\left[1 - P(E_{j-1}^+) - P(E_{j-1}^-)\right]\right] \tag{10.83}$$

with an adaptive sampling density defined as follows:

$$\tilde{f}_{N+1}(\boldsymbol{x}_{N+1}) = f_x(\boldsymbol{x}_{N+1})\frac{1_{x\in E_N}(\boldsymbol{x}_{N+1})}{P(E_N)} \tag{10.84}$$

For commodity of notation, we will denote $p_N(.)$ as the estimator above for a N-sample, considered as a function of non-*i.i.d.* random variables $\boldsymbol{X}_1...\boldsymbol{X}_N$:

$$\hat{P}_f^{mis} = \frac{1}{N}\sum_{j=1..N}\left[P(E_{j-1}^+) + 1_{G(\boldsymbol{X}_j)\leq 0}\left[1 - P(E_{j-1}^+) - P(E_{j-1}^-)\right]\right] = p_N(\boldsymbol{X}_1,..,\boldsymbol{X}_N) \tag{10.85}$$

Unbiasedness

We wish to prove that:

$$E[p_N(\boldsymbol{X}_1,..,\boldsymbol{X}_N)] = P_f \tag{10.86}$$

Proof will be built recursively. Note first that for $j = 1$, adaptive sampling density is just the original density f_X. Thus, the estimator is identical to that of MCS:

$$p_1(\boldsymbol{X}_1) = P(E_0^+) + 1_{G(\boldsymbol{X}_1)\leq 0}\left[1 - P(E_0^+) - P(E_0^-)\right] = 1_{G(\boldsymbol{X}_1)\leq 0} \tag{10.87}$$

So that:

$$E(p_1(\mathbf{X}_1)) = E(1_{G(\mathbf{X}_1)\leq 0}) = p_f \tag{10.88}$$

Supposing Equation (10.86) is true at step N, consider then step $N+1$. We will decompose the expectation and variance through the classical probabilistic results on conditional expectation and variance, conditioning successively on $(\mathbf{X}_1\ldots\mathbf{X}_N)$ and then on $\mathbf{X}_{N+1}$:

$$E[p_{N+1}(\mathbf{X}_1,..,\mathbf{X}_{N+1})] = E_{\mathbf{X}_1,..,\mathbf{X}_N}\left[E_{\mathbf{X}_{N+1}}[p_{N+1}(\mathbf{X}_1,..,\mathbf{X}_{N+1})|\mathbf{X}_1,..,\mathbf{X}_N]\right] \tag{10.89}$$

Thus:

$$\begin{aligned}E_{\mathbf{X}_{N+1}}[p_{N+1}(\mathbf{X}_1,..,\mathbf{X}_{N+1})|\mathbf{X}_1,..,\mathbf{X}_N] &= (N+1)^{-1}\left(E_{\mathbf{X}_{N+1}}\left[N.p_N(\mathbf{X}_1,..,\mathbf{X}_N)+P(E_N^+)+1_{G(\mathbf{X}_{N+1})\leq 0}P(E_N)|\mathbf{X}_1,..,\mathbf{X}_N\right]\right)\\ &=\frac{N}{N+1}p_N(\mathbf{X}_1,..,\mathbf{X}_N)+\frac{1}{N+1}\left(P(E_N^+)+P(E_N)E_{\mathbf{X}_{N+1}}\left[1_{G(\mathbf{X}_{N+1})\leq 0}\right]\right)\end{aligned} \tag{10.90}$$

Note that:

$$E_{\mathbf{X}_{N+1}}\left[1_{G(\mathbf{X}_{N+1})\leq 0}\right] = E_{\mathbf{X}_{N+1}}[P(G(\mathbf{X}_{N+1})\leq 0|\mathbf{X}_{N+1}\in E_N)] = \frac{p_f - P(E_N^+)}{[P(E_N)]} \tag{10.91}$$

Thus:

$$\begin{aligned}&E_{\mathbf{X}_{N+1}}[p_{N+1}(\mathbf{X}_1,..,\mathbf{X}_{N+1})|\mathbf{X}_1,..,\mathbf{X}_N]=\frac{N}{N+1}p_N(\mathbf{X}_1,..,\mathbf{X}_N)+\frac{1}{N+1}p_f\\ &E[p_{N+1}(\mathbf{X}_1,..,\mathbf{X}_{N+1})]=E_{\mathbf{X}_1,..,\mathbf{X}_N}\left[\frac{N}{N+1}p_N(\mathbf{X}_1,..,\mathbf{X}_N)+\frac{1}{N+1}p_f\right]=\frac{N}{N+1}E_{\mathbf{X}_1,..,\mathbf{X}_N}[p_N(\mathbf{X}_1,..,\mathbf{X}_N)]+\frac{1}{N+1}p_f\end{aligned} \tag{10.92}$$

So that Equation (10.86) is true at step $N+1$.

Proof of variance reduction of the S-MRM estimator
We wish to prove that:

$$\begin{aligned}&\mathrm{var}[p_N(\mathbf{X}_1,..,\mathbf{X}_N)]\\ &=\frac{p_f(1-p_f)}{N}\left(1-N^{-1}\sum_{j=1..N-1}E_{\mathbf{X}_1,..,\mathbf{X}_j}\left[\frac{P(E_j^-)}{1-p_f}+\frac{P(E_j^+)}{p_f}-\frac{P(E_j^-)P(E_j^+)}{(1-p_f)p_f}\right]\right)\\ &=N^{-1}p_f(1-p_f)\left(1-N^{-1}\sum_{j=1..N-1}c_j\right)\end{aligned} \tag{10.93}$$

At step $N=1$, sampling density is the original one, so that:

$$\mathrm{var}[p_1(\boldsymbol{X}_1)] = p_f(1-p_f) \tag{10.94}$$

Again, it will be built recursively, by a successive conditioning on $(\boldsymbol{X}_1 \ldots \boldsymbol{X}_N)$ and then on $\boldsymbol{X}_{N+1}$:

$$\begin{aligned}&\mathrm{var}[p_{N+1}(\boldsymbol{X}_1, .., \boldsymbol{X}_{N+1})]\\ &= \mathrm{var}_{\boldsymbol{X}_1,..,\boldsymbol{X}_N}\left[E_{\boldsymbol{X}_{N+1}}[p_{N+1}(\boldsymbol{X}_1, .., \boldsymbol{X}_{N+1})|\boldsymbol{X}_1, .., \boldsymbol{X}_N]\right] + E_{\boldsymbol{X}_1,..,\boldsymbol{X}_N}\left[\mathrm{var}_{\boldsymbol{X}_{N+1}}[p_{N+1}(\boldsymbol{X}_1, .., \boldsymbol{X}_{N+1})|\boldsymbol{X}_1, .., \boldsymbol{X}_N]\right]\end{aligned} \tag{10.95}$$

The first term is:

$$\begin{aligned}&\mathrm{var}_{\boldsymbol{X}_1,..,\boldsymbol{X}_N}\left[E_{\boldsymbol{X}_{N+1}}[p_{N+1}(\boldsymbol{X}_1, .., \boldsymbol{X}_{N+1})|\boldsymbol{X}_1, .., \boldsymbol{X}_N]\right]\\ &= (N+1)^{-2}\mathrm{var}_{\boldsymbol{X}_1,..,\boldsymbol{X}_N}\left[N.p_N(\boldsymbol{X}_1, .., \boldsymbol{X}_N) + E_{\boldsymbol{X}_{N+1}}\left[P(E_N^+) + 1_{G(\boldsymbol{X}_{N+1})\leq 0}P(E_N)|\boldsymbol{X}_1, .., \boldsymbol{X}_N\right]\right]\\ &= (N+1)^{-2}\mathrm{var}_{\boldsymbol{X}_1,..,\boldsymbol{X}_N}[N.p_N(\boldsymbol{X}_1, .., \boldsymbol{X}_N) + p_f] = N^2(N+1)^{-2}\mathrm{var}_{\boldsymbol{X}_1,..,\boldsymbol{X}_N}[p_N(\boldsymbol{X}_1, .., \boldsymbol{X}_N)]\end{aligned} \tag{10.96}$$

And the second term is:

$$\begin{aligned}&E_{\boldsymbol{X}_1,..,\boldsymbol{X}_N}\left[\mathrm{var}_{\boldsymbol{X}_{N+1}}[p_{N+1}(\boldsymbol{X}_1, .., \boldsymbol{X}_{N+1})|\boldsymbol{X}_1, .., \boldsymbol{X}_N]\right]\\ &= (N+1)^{-2}E_{\boldsymbol{X}_1,..,\boldsymbol{X}_N}\left[\mathrm{var}_{\boldsymbol{X}_{N+1}}\left[P(E_N^+) + 1_{G(\boldsymbol{X}_{N+1})\leq 0}P(E_N)|\boldsymbol{X}_1, .., \boldsymbol{X}_N\right]\right]\\ &= (N+1)^{-2}E_{\boldsymbol{X}_1,..,\boldsymbol{X}_N}\left[p_f - P(E_N^+)\right]\left[1 - p_f - P(E_N^-)\right]\\ &= \frac{p_f(1-p_f)}{(N+1)^2}\left(1 - E_{\boldsymbol{X}_1,..,\boldsymbol{X}_N}\left[\frac{P(E_N^-)}{1-p_f} + \frac{P(E_N^+)}{p_f} - \frac{P(E_N^-)P(E_N^+)}{(1-p_f)p_f}\right]\right)\\ &= (N+1)^{-2}p_f(1-p_f)(1-c_N)\end{aligned} \tag{10.97}$$

So that:

$$\mathrm{var}p_{N+1}(.) = (N+1)^{-2}[N^2\,\mathrm{var}\,p_N(.) + p_f(1-p_f)(1-c_N)] \tag{10.98}$$

Supposing that Equation (10.93) is true at step N, easy algebra shows that it is true at step $N+1$. The quantity c_N can be interpreted as the average reduction of the failure and safe domain by the multiple-result hypercorners after N steps. It is not easy to compute in general. It is, however, certainly positive since $P(E_N^+)$ and $P(E_N^-)$ are lower than p_f and than 1-p_f respectively so that variance of the adaptive estimator is certainly lower than the variance of classical MCS:

$$\begin{aligned}&\mathrm{var}[p_N(\boldsymbol{X}_1, .., \boldsymbol{X}_N)] = N^{-1}p_f(1-p_f)\left(1 - N^{-1}\sum_{j=1..N-1} c_j\right)\\ &< N^{-1}p_f(1-p_f) \quad \mathit{if} \quad N > 1\end{aligned} \tag{10.99}$$

References

de Rocquigny E., Devictor N. and Tarantola S. (eds) (2008) *Uncertainty in Industrial Practice, A Guide Q42 to Quantitative Uncertainty Management*, John Wiley & Sons, Ltd.

Helton, J.C. (1993) Uncertainty and sensitivity analysis techniques for use in performance assessment for radioactive waste disposal. *Reliability Engineering and System Safety*, **42**, 327–367.

Kaplan, S., Garrick, B.J. (1981) On the quantitative definition of risk. *Risk Analysis*, **1**(1), 11–27.

Knight, F.H. (1921) *Risk, Uncertainty and Profit*, Hart. Schaffner & Marx.

Madsen, H.O., Kenk, S. and Lind, N.C. (1986) *Methods of Structural Safety.* Prentice-Hall Inc.

Epilogue

Luke 13: 4-5

Or those eighteen on whom the tower at Siloam fell, do you suppose they had failed in their duty more than all the rest of the people who live in Jerusalem?

Gospels enjoy a variety of interpretations in association to any of the rich theological, historical, sociological or poetic dimensions surrounding the material; probably not the least of reasons being the full linguistic spectrum of modern translations that are being given to the original texts, themselves written in a Greek influenced by a different native, cultural or liturgical language of the authors. Without any form of exegetical ambition, the passage of *Luke 13* attracted the author as an intriguing metaphor of some of the ideas brought forward in the present book.

The fall of the Siloam tower can be partially traced back into archaeological evidence, as visitors to modern Jerusalem are guided into the rests of the Siloam district of the 1st century AD-Jerusalem. It must have been a building of certain importance at the time, the failure of which brought casualties and made an *adverse event* significant enough to be referred to in the news and the topics addressed by Jesus and to mark his contemporaries and disciples. Yet, even if the text itself could probably be supplemented by additional evidence coming from archaeology, a number of sources and types of uncertainty surround the evidence.

It is ambiguous for instance whether the 18 casualties involved builders or workers associated to the tower erection or to the tower activity, actively and deliberately present; or whether those were merely passing by – say 'at random' – at the time of the event. In that sense, the peculiar translation that was retained for the quote in the present book involves the two words 'fail' and 'duty' that get a particular meaning in the context of engineering design, risk and reliability. One interpretation could be for instance that some of the men were actively involved in building or operating the tower and 'failed in their duty' as they did not comply with the prescribed safety regulations and margins involved – be it by lack of knowledge, by laxism or by opportunism as full compliance with those regulations would constrain the optimisation of other variables of interest such as the load flexibility or cost of infrastructure. Either, it could be that those men were mere visitors passing by without following the safety procedures – again either by ignorance, lack of knowledge, absence or shortcomings of the public risk information or even mischievous instructions by the tower operators.

Alternatively, it could be that unknown or non-anticipated internal phenomena affected the behaviour of building material so that ageing or phenomenological degradation ruined the structure well beyond what could reasonably be anticipated even when following the state-of-the-art or best practices of

---

Modelling Under Risk and Uncertainty: An Introduction to Statistical, Phenomenological and Computational Methods, First Edition. Etienne de Rocquigny.

engineering, building or operation at that time, hence *irreducible uncertainty.* Or when following practices of engineering or building at the level of sophistication that could be afforded within the reasonable budget dedicated to the tower erection (as compared to *all the rest* of the building projects *in Jerusalem*), hence amounting to *reducible* lack of knowledge, but *non economical*. Moreover, it could be that unexpected *external events* affected the integrity of the structure, such as local quakes, landslide or geological instabilities, or a storm, heavy rain, flood of the nearby Kidron brook, corrosive seepage from the Siloam spring etc.: no information is given as to whether the building professionals had any prior experience or knowledge about the occurrence of natural events in the area and would therefore have 'failed in their duty' of taking into account those events inside the design or whether such external aggressions were completely outside of likelihood.

Facing such wide amount of uncertainty (or ambiguity, lack of knowledge . . .), it is tempting to indulge in causal thoughts and infer that those who died *deserved* it to some extent; as tempting it was to 1st-century dwellers of Jerusalem as it is to modern democracies or time-pressured journalists in identifying the culprit for most major detrimental events such as the eruption of a disease, the association of earthquake, tsunami and nuclear plant failure, flooded houses or industrial explosion. Culprits could be: the improper building or protection regulations – either experts of industry, research or governments arguing of their irrelevance because there are made of objectionable deterministic or non-controllable probabilistic targets; the improper application of regulations – either experts arguing that those regulations are over-costly, too complex, or that the operators are ignorant, laxist or profit-driven; the uncontrollable behaviour of public goods because no one can be held as responsible, or alternatively the uncontrollable behaviour of private operators that do not care for the outside public; the lack of budget for sound research and code of practice to develop and increase the reliability of new technologies, or for control authorities to enforce the codes of practice; industry's, agriculture's or the society's impact onto the natural environment or the climate that increase the frequency of external aggressions and hence etc.

Moreover, at a spiritual level, responsibility or culpability for casualty can temptingly be connected to past blameworthy behaviour or sin. Alternative translations of the passage would translate the word *όφειλέται* attached to the dead men rather than 'failed in their duty' into 'debtors', 'guilty', 'sinners', 'offenders', 'culprits' etc. thus explicitly setting the question in the moral domain. Scholars have commented some of the religious views combated by Jesus referring to the causal link made between fatalities and deeds; it is classically exemplified elsewhere in the Bible, e.g. in the book of Job where the misfortunate Just fights against the dominating view that he had to commit some blameworthy act for his suffering to happen.

Yet, significantly enough for analysts, scientists or risk professionals, one may understand Jesus as deliberately escaping any causal interpretation of the adverse event. Putting the assertion into a *question* directed to his audience, he opens up widely for the complexity, if not mystery, of reality and associated uncertainty. A risk and uncertainty expert could also appreciate that this answer does not discuss (or diminish) the interest or importance of either of the following: research and development of robust building or protection standards; enforcement of professional practice through regulation or courts; increased monitoring of internal or external events; etc. Somehow, the quest for *external* responsibility is converted into *internal* responsibility as the question requests each auditor to make a personal judgment and decision in the face of uncertainty. Each of them is urged not to simplify too quickly the extent of reality: rather keep humble or possibly make a responsible use of the mysterious though consubstantial extent of freedom opened up by an uncertain reality?

Index

Modelling Under Risk and Uncertainty: An Introduction to Statistical, Phenomenological and Computational Methods, First Edition. Etienne de Rocquigny.

WILEY SERIES IN PROBABILITY AND STATISTICS
ESTABLISHED BY WALTER A. SHEWHART AND SAMUEL S. WILKS

Editors: *David. J. Balding, Noel A.C. Cressie, Garrett M. Fitzmaurice, Harvey Goldstein, Iain M. Johnstone, Geert Molenberghs, David W. Scott, Adrian F. M. Smith, Ruey S. Tsay, Sanford Weisberg*

Editors Emeriti: *Vic Barnett, Ralph A. Bradley, J. Stuart Hunter, J.B. Kadane, David G. Kendall, Jozef L. Teugels*

The ***Wiley Series in Probability and Statistics*** is well established and authoritative. It covers many topics of current research interest in both pure and applied statistics and probability theory. Written by leading statisticians and institutions, the titles span both state-of-the-art developments in the field and classical methods.

Reflecting the wide range of current research in statistics, the series encompasses applied, methodological and theoretical statistics, ranging from applications and new techniques made possible by advances in computerized practice to rigorous treatment of theoretical approaches.

This series provides essential and invaluable reading for all statisticians, whether in academia, industry, government, or research.

† ABRAHAM and LEDOLTER · Statistical Methods for Forecasting
AGRESTI · Analysis of Ordinal Categorical Data, *Second Edition*
AGRESTI · An Introduction to Categorical Data Analysis, *Second Edition*
AGRESTI · Categorical Data Analysis, *Second Edition*
ALTMAN, GILL, and McDONALD · Numerical Issues in Statistical Computing for the Social Scientist
AMARATUNGA and CABRERA · Exploration and Analysis of DNA Microarray and Protein Array Data
ANDĚL · Mathematics of Chance
ANDERSON · An Introduction to Multivariate Statistical Analysis, *Third Edition*
* ANDERSON · The Statistical Analysis of Time Series
ANDERSON, AUQUIER, HAUCK, OAKES, VANDAELE, and WEISBERG · Statistical Methods for Comparative Studies
ANDERSON and LOYNES · The Teaching of Practical Statistics
ARMITAGE and DAVID (editors) · Advances in Biometry
ARNOLD, BALAKRISHNAN, and NAGARAJA · Records
* ARTHANARI and DODGE · Mathematical Programming in Statistics
* BAILEY · The Elements of Stochastic Processes with Applications to the Natural Sciences
BAJORSKI · Statistics for Imaging, Optics, and Photonics
BALAKRISHNAN and KOUTRAS · Runs and Scans with Applications
BALAKRISHNAN and NG · Precedence-Type Tests and Applications
BARNETT · Comparative Statistical Inference, *Third Edition*
BARNETT · Environmental Statistics
BARNETT and LEWIS · Outliers in Statistical Data, *Third Edition*
BARTHOLOMEW, KNOTT, and MOUSTAKI · Latent Variable Models and Factor Analysis: A Unified Approach, *Third Edition*
BARTOSZYNSKI and NIEWIADOMSKA-BUGAJ · Probability and Statistical Inference, *Second Edition*

*Now available in a lower priced paperback edition in the Wiley Classics Library.
† Now available in a lower priced paperback edition in the Wiley-Interscience Paperback Series.

BASILEVSKY · Statistical Factor Analysis and Related Methods: Theory and Applications
BATES and WATTS · Nonlinear Regression Analysis and Its Applications
BECHHOFER, SANTNER, and GOLDSMAN · Design and Analysis of Experiments for Statistical Selection, Screening, and Multiple Comparisons
BEIRLANT, GOEGEBEUR, SEGERS, TEUGELS, and DE WAAL · Statistics of Extremes: Theory and Applications
BELSLEY · Conditioning Diagnostics: Collinearity and Weak Data in Regression
† BELSLEY, KUH, and WELSCH · Regression Diagnostics: Identifying Influential Data and Sources of Collinearity
BENDAT and PIERSOL · Random Data: Analysis and Measurement Procedures, *Fourth Edition*
BERNARDO and SMITH · Bayesian Theory
BHAT and MILLER · Elements of Applied Stochastic Processes, *Third Edition*
BHATTACHARYA and WAYMIRE · Stochastic Processes with Applications
BIEMER, GROVES, LYBERG, MATHIOWETZ, and SUDMAN · Measurement Errors in Surveys
BILLINGSLEY · Convergence of Probability Measures, *Second Edition*
BILLINGSLEY · Probability and Measure, *Anniversary Edition*
BIRKES and DODGE · Alternative Methods of Regression
BISGAARD and KULAHCI · Time Series Analysis and Forecasting by Example
BISWAS, DATTA, FINE, and SEGAL · Statistical Advances in the Biomedical Sciences: Clinical Trials, Epidemiology, Survival Analysis, and Bioinformatics
BLISCHKE and MURTHY (editors) · Case Studies in Reliability and Maintenance
BLISCHKE and MURTHY · Reliability: Modeling, Prediction, and Optimization
BLOOMFIELD · Fourier Analysis of Time Series: An Introduction, *Second Edition*
BOLLEN · Structural Equations with Latent Variables
BOLLEN and CURRAN · Latent Curve Models: A Structural Equation Perspective
BOROVKOV · Ergodicity and Stability of Stochastic Processes
BOSQ and BLANKE · Inference and Prediction in Large Dimensions
BOULEAU · Numerical Methods for Stochastic Processes
* BOX · Bayesian Inference in Statistical Analysis
BOX · Improving Almost Anything, *Revised Edition*
* BOX and DRAPER · Evolutionary Operation: A Statistical Method for Process Improvement
BOX and DRAPER · Response Surfaces, Mixtures, and Ridge Analyses, *Second Edition*
BOX, HUNTER, and HUNTER · Statistics for Experimenters: Design, Innovation, and Discovery, *Second Editon*
BOX, JENKINS, and REINSEL · Time Series Analysis: Forcasting and Control, *Fourth Edition*
BOX, LUCEÑO, and PANIAGUA-QUIÑONES · Statistical Control by Monitoring and Adjustment, *Second Edition*
* BROWN and HOLLANDER · Statistics: A Biomedical Introduction
CAIROLI and DALANG · Sequential Stochastic Optimization
CASTILLO, HADI, BALAKRISHNAN, and SARABIA · Extreme Value and Related Models with Applications in Engineering and Science
CHAN · Time Series: Applications to Finance with R and S-Plus®, *Second Edition*
CHARALAMBIDES · Combinatorial Methods in Discrete Distributions
CHATTERJEE and HADI · Regression Analysis by Example, *Fourth Edition*
CHATTERJEE and HADI · Sensitivity Analysis in Linear Regression

*Now available in a lower priced paperback edition in the Wiley Classics Library.
† Now available in a lower priced paperback edition in the Wiley-Interscience Paperback Series.

CHERNICK · Bootstrap Methods: A Guide for Practitioners and Researchers, *Second Edition*
CHERNICK and FRIIS · Introductory Biostatistics for the Health Sciences
CHILES and DELFINER · Geostatistics: Modeling Spatial Uncertainty, *Second Edition*
CHOW and LIU · Design and Analysis of Clinical Trials: Concepts and Methodologies, *Second Edition*
CLARKE · Linear Models: The Theory and Application of Analysis of Variance
CLARKE and DISNEY · Probability and Random Processes: A First Course with Applications, *Second Edition*
* COCHRAN and COX · Experimental Designs, *Second Edition*
COLLINS and LANZA · Latent Class and Latent Transition Analysis: With Applications in the Social, Behavioral, and Health Sciences
CONGDON · Applied Bayesian Modelling
CONGDON · Bayesian Models for Categorical Data
CONGDON · Bayesian Statistical Modelling, *Second Edition*
CONOVER · Practical Nonparametric Statistics, *Third Edition* COOK · Regression Graphics
COOK and WEISBERG · An Introduction to Regression Graphics
COOK and WEISBERG · Applied Regression Including Computing and Graphics
CORNELL · A Primer on Experiments with Mixtures
CORNELL · Experiments with Mixtures, Designs, Models, and the Analysis of Mixture Data, *Third Edition*
COX · A Handbook of Introductory Statistical Methods
CRESSIE · Statistics for Spatial Data, *Revised Edition*
CRESSIE and WIKLE · Statistics for Spatio-Temporal Data
CSÖRGŐ and HORVÁTH · Limit Theorems in Change Point Analysis
DAGPUNAR · Simulation and Monte Carlo: With Applications in Finance and MCMC
DANIEL · Applications of Statistics to Industrial Experimentation
DANIEL · Biostatistics: A Foundation for Analysis in the Health Sciences, *Eighth Edition*
* DANIEL · Fitting Equations to Data: Computer Analysis of Multifactor Data, *Second Edition*
DASU and JOHNSON · Exploratory Data Mining and Data Cleaning
DAVID and NAGARAJA · Order Statistics, *Third Edition*
* DEGROOT, FIENBERG, and KADANE · Statistics and the Law
DEL CASTILLO · Statistical Process Adjustment for Quality Control
DEMARIS · Regression with Social Data: Modeling Continuous and Limited Response Variables
DEMIDENKO · Mixed Models: Theory and Applications
DENISON, HOLMES, MALLICK and SMITH · Bayesian Methods for Nonlinear Classification and Regression
DETTE and STUDDEN · The Theory of Canonical Moments with Applications in Statistics, Probability, and Analysis
DEY and MUKERJEE · Fractional Factorial Plans
DE ROCQUIGNY · Modelling Under Risk and Uncertainty: An Introduction to Statistical, Phenomenological and Computational Models
DILLON and GOLDSTEIN · Multivariate Analysis: Methods and Applications
* DODGE and ROMIG · Sampling Inspection Tables, *Second Edition*
* DOOB · Stochastic Processes
DOWDY, WEARDEN, and CHILKO · Statistics for Research, *Third Edition*
DRAPER and SMITH · Applied Regression Analysis, *Third Edition*
DRYDEN and MARDIA · Statistical Shape Analysis

*Now available in a lower priced paperback edition in the Wiley Classics Library.
† Now available in a lower priced paperback edition in the Wiley-Interscience Paperback Series.

DUDEWICZ and MISHRA · Modern Mathematical Statistics
DUNN and CLARK · Basic Statistics: A Primer for the Biomedical Sciences, *Fourth Edition*
DUPUIS and ELLIS · A Weak Convergence Approach to the Theory of Large
Deviations EDLER and KITSOS · Recent Advances in Quantitative Methods in Cancer and Human Health Risk Assessment
* ELANDT-JOHNSON and JOHNSON · Survival Models and Data Analysis ENDERS · Applied Econometric Time Series, *Third Edition*
† ETHIER and KURTZ · Markov Processes: Characterization and Convergence
EVANS, HASTINGS, and PEACOCK · Statistical Distributions, *Third Edition*
EVERITT, LANDAU, LEESE, and STAHL · Cluster Analysis, *Fifth Edition*
FEDERER and KING · Variations on Split Plot and Split Block Experiment Designs
FELLER · An Introduction to Probability Theory and Its Applications, Volume I, *Third Edition,* Revised; Volume II, *Second Edition*
FITZMAURICE, LAIRD, and WARE · Applied Longitudinal Analysis, *Second Edition*
* FLEISS · The Design and Analysis of Clinical Experiments
FLEISS · Statistical Methods for Rates and Proportions, *Third Edition*
† FLEMING and HARRINGTON · Counting Processes and Survival Analysis
FUJIKOSHI, ULYANOV, and SHIMIZU · Multivariate Statistics: High-Dimensional and Large-Sample Approximations
FULLER · Introduction to Statistical Time Series, *Second Edition*
† FULLER · Measurement Error Models
GALLANT · Nonlinear Statistical Models
GEISSER · Modes of Parametric Statistical Inference
GELMAN and MENG · Applied Bayesian Modeling and Causal Inference from Incomplete-Data Perspectives
GEWEKE · Contemporary Bayesian Econometrics and Statistics
GHOSH, MUKHOPADHYAY, and SEN · Sequential Estimation
GIESBRECHT and GUMPERTZ · Planning, Construction, and Statistical Analysis of Comparative Experiments
GIFI · Nonlinear Multivariate Analysis
GIVENS and HOETING · Computational Statistics
GLASSERMAN and YAO · Monotone Structure in Discrete-Event Systems
GNANADESIKAN · Methods for Statistical Data Analysis of Multivariate Observations, *Second Edition*
GOLDSTEIN · Multilevel Statistical Models, *Fourth Edition*
GOLDSTEIN and LEWIS · Assessment: Problems, Development, and Statistical Issues
GOLDSTEIN and WOOFF · Bayes Linear Statistics
GREENWOOD and NIKULIN · A Guide to Chi-Squared Testing
GROSS, SHORTLE, THOMPSON, and HARRIS · Fundamentals of Queueing Theory, *Fourth Edition*
GROSS, SHORTLE, THOMPSON, and HARRIS · Solutions Manual to Accompany Fundamentals of Queueing Theory, *Fourth Edition*
* HAHN and SHAPIRO · Statistical Models in Engineering
HAHN and MEEKER · Statistical Intervals: A Guide for Practitioners
HALD · A History of Probability and Statistics and their Applications Before 1750
† HAMPEL · Robust Statistics: The Approach Based on Influence Functions
HARTUNG, KNAPP, and SINHA · Statistical Meta-Analysis with Applications

*Now available in a lower priced paperback edition in the Wiley Classics Library.
† Now available in a lower priced paperback edition in the Wiley-Interscience Paperback Series.

HEIBERGER · Computation for the Analysis of Designed Experiments
HEDAYAT and SINHA · Design and Inference in Finite Population Sampling
HEDEKER and GIBBONS · Longitudinal Data Analysis
HELLER · MACSYMA for Statisticians
HERITIER, CANTONI, COPT, and VICTORIA-FESER · Robust Methods in Biostatistics
HINKELMANN and KEMPTHORNE · Design and Analysis of Experiments, Volume 1: Introduction to Experimental Design, *Second Edition*
HINKELMANN and KEMPTHORNE · Design and Analysis of Experiments, Volume 2: Advanced Experimental Design
HINKELMANN (editor) · Design and Analysis of Experiments, Volume 3: Special Designs and Applications
* HOAGLIN, MOSTELLER, and TUKEY · Fundamentals of Exploratory Analysis of Variance
* HOAGLIN, MOSTELLER, and TUKEY · Exploring Data Tables, Trends and Shapes
* HOAGLIN, MOSTELLER, and TUKEY · Understanding Robust and Exploratory Data Analysis
HOCHBERG and TAMHANE · Multiple Comparison Procedures
HOCKING · Methods and Applications of Linear Models: Regression and the Analysis of Variance, *Second Edition*
HOEL · Introduction to Mathematical Statistics, *Fifth Edition*
HOGG and KLUGMAN · Loss Distributions
HOLLANDER and WOLFE · Nonparametric Statistical Methods, *Second Edition*
HOSMER and LEMESHOW · Applied Logistic Regression, *Second Edition*
HOSMER, LEMESHOW, and MAY · Applied Survival Analysis: Regression Modeling of Time-to-Event Data, *Second Edition*
HUBER · Data Analysis: What Can Be Learned From the Past 50 Years
HUBER · Robust Statistics
† HUBER and RONCHETTI · Robust Statistics, *Second Edition*
HUBERTY · Applied Discriminant Analysis, *Second Edition*
HUBERTY and OLEJNIK · Applied MANOVA and Discriminant Analysis, *Second Edition*
HUITEMA · The Analysis of Covariance and Alternatives: Statistical Methods for Experiments, Quasi-Experiments, and Single-Case Studies, *Second Edition*
HUNT and KENNEDY · Financial Derivatives in Theory and Practice, *Revised Edition*
HURD and MIAMEE · Periodically Correlated Random Sequences: Spectral Theory and Practice
HUSKOVA, BERAN, and DUPAC · Collected Works of Jaroslav Hajek—with Commentary
HUZURBAZAR · Flowgraph Models for Multistate Time-to-Event Data
INSUA, RUGGERI and WIPER · Bayesian Analysis of Stochastic Process Models
JACKMAN · Bayesian Analysis for the Social Sciences
† JACKSON · A User's Guide to Principle Components
JOHN · Statistical Methods in Engineering and Quality Assurance
JOHNSON · Multivariate Statistical Simulation
JOHNSON and BALAKRISHNAN · Advances in the Theory and Practice of Statistics: A Volume in Honor of Samuel Kotz
JOHNSON, KEMP, and KOTZ · Univariate Discrete Distributions, *Third Edition*
JOHNSON and KOTZ (editors) · Leading Personalities in Statistical Sciences: From the Seventeenth Century to the Present

*Now available in a lower priced paperback edition in the Wiley Classics Library.
† Now available in a lower priced paperback edition in the Wiley-Interscience Paperback Series.

JOHNSON, KOTZ, and BALAKRISHNAN · Continuous Univariate Distributions, Volume 1, *Second Edition*
JOHNSON, KOTZ, and BALAKRISHNAN · Continuous Univariate Distributions, Volume 2, *Second Edition*
JOHNSON, KOTZ, and BALAKRISHNAN · Discrete Multivariate Distributions
JUDGE, GRIFFITHS, HILL, LÜTKEPOHL, and LEE · The Theory and Practice of Econometrics, *Second Edition*
JUREK and MASON · Operator-Limit Distributions in Probability Theory
KADANE · Bayesian Methods and Ethics in a Clinical Trial Design
KADANE AND SCHUM · A Probabilistic Analysis of the Sacco and Vanzetti Evidence
KALBFLEISCH and PRENTICE · The Statistical Analysis of Failure Time Data, *Second Edition*
KARIYA and KURATA · Generalized Least Squares
KASS and VOS · Geometrical Foundations of Asymptotic Inference
† KAUFMAN and ROUSSEEUW · Finding Groups in Data: An Introduction to Cluster Analysis
KEDEM and FOKIANOS · Regression Models for Time Series Analysis
KENDALL, BARDEN, CARNE, and LE · Shape and Shape Theory
KHURI · Advanced Calculus with Applications in Statistics, *Second Edition*
KHURI, MATHEW, and SINHA · Statistical Tests for Mixed Linear Models
* KISH · Statistical Design for Research
KLEIBER and KOTZ · Statistical Size Distributions in Economics and Actuarial Sciences
KLEMELÄ · Smoothing of Multivariate Data: Density Estimation and Visualization
KLUGMAN, PANJER, and WILLMOT · Loss Models: From Data to Decisions, *Third Edition*
KLUGMAN, PANJER, and WILLMOT · Solutions Manual to Accompany Loss Models: From Data to Decisions, *Third Edition*
KOSKI and NOBLE · Bayesian Networks: An Introduction
KOTZ, BALAKRISHNAN, and JOHNSON · Continuous Multivariate Distributions, Volume 1, *Second Edition*
KOTZ and JOHNSON (editors) · Encyclopedia of Statistical Sciences: Volumes 1 to 9 with Index
KOTZ and JOHNSON (editors) · Encyclopedia of Statistical Sciences: Supplement Volume
KOTZ, READ, and BANKS (editors) · Encyclopedia of Statistical Sciences: Update Volume 1
KOTZ, READ, and BANKS (editors) · Encyclopedia of Statistical Sciences: Update Volume 2
KOWALSKI and TU · Modern Applied U-Statistics
KRISHNAMOORTHY and MATHEW · Statistical Tolerance Regions: Theory, Applications, and Computation
KROESE, TAIMRE, and BOTEV · Handbook of Monte Carlo Methods
KROONENBERG · Applied Multiway Data Analysis
KULINSKAYA, MORGENTHALER, and STAUDTE · Meta Analysis: A Guide to Calibrating and Combining Statistical Evidence
KULKARNI and HARMAN · An Elementary Introduction to Statistical Learning Theory
KUROWICKA and COOKE · Uncertainty Analysis with High Dimensional Dependence Modelling
KVAM and VIDAKOVIC · Nonparametric Statistics with Applications to Science and Engineering
LACHIN · Biostatistical Methods: The Assessment of Relative Risks, *Second Edition*
LAD · Operational Subjective Statistical Methods: A Mathematical, Philosophical, and Historical Introduction
LAMPERTI · Probability: A Survey of the Mathematical Theory, *Second Edition*

*Now available in a lower priced paperback edition in the Wiley Classics Library.
† Now available in a lower priced paperback edition in the Wiley-Interscience Paperback Series.

LAWLESS · Statistical Models and Methods for Lifetime Data, *Second Edition*
LAWSON · Statistical Methods in Spatial Epidemiology, *Second Edition*
LE · Applied Categorical Data Analysis, *Second Edition*
LE · Applied Survival Analysis
LEE · Structural Equation Modeling: A Bayesian Approach
LEE and WANG · Statistical Methods for Survival Data Analysis, *Third Edition*
LEPAGE and BILLARD · Exploring the Limits of Bootstrap
LESSLER and KALSBEEK · Nonsampling Errors in Surveys
LEYLAND and GOLDSTEIN (editors) · Multilevel Modelling of Health Statistics
LIAO · Statistical Group Comparison
LIN · Introductory Stochastic Analysis for Finance and Insurance
LITTLE and RUBIN · Statistical Analysis with Missing Data, *Second Edition*
LLOYD · The Statistical Analysis of Categorical Data
LOWEN and TEICH · Fractal-Based Point Processes
MAGNUS and NEUDECKER · Matrix Differential Calculus with Applications in Statistics and Econometrics, *Revised Edition*
MALLER and ZHOU · Survival Analysis with Long Term Survivors
MARCHETTE · Random Graphs for Statistical Pattern Recognition
MARDIA and JUPP · Directional Statistics
MARKOVICH · Nonparametric Analysis of Univariate Heavy-Tailed Data: Research and Practice
MARONNA, MARTIN and YOHAI · Robust Statistics: Theory and Methods
MASON, GUNST, and HESS · Statistical Design and Analysis of Experiments with Applications to Engineering and Science, *Second Edition*
McCULLOCH, SEARLE, and NEUHAUS · Generalized, Linear, and Mixed Models, *Second Edition*
McFADDEN · Management of Data in Clinical Trials, *Second Edition*
* McLACHLAN · Discriminant Analysis and Statistical Pattern Recognition
McLACHLAN, DO, and AMBROISE · Analyzing Microarray Gene Expression Data
McLACHLAN and KRISHNAN · The EM Algorithm and Extensions, *Second Edition*
McLACHLAN and PEEL · Finite Mixture Models
McNEIL · Epidemiological Research Methods
MEEKER and ESCOBAR · Statistical Methods for Reliability Data
MEERSCHAERT and SCHEFFLER · Limit Distributions for Sums of Independent Random Vectors: Heavy Tails in Theory and Practice
MENGERSEN, ROBERT, and TITTERINGTON · Mixtures: Estimation and Applications
MICKEY, DUNN, and CLARK · Applied Statistics: Analysis of Variance and Regression, *Third Edition*
* MILLER · Survival Analysis, *Second Edition*
MONTGOMERY, JENNINGS, and KULAHCI · Introduction to Time Series Analysis and Forecasting
MONTGOMERY, PECK, and VINING · Introduction to Linear Regression Analysis, *Fifth Edition*
MORGENTHALER and TUKEY · Configural Polysampling: A Route to Practical Robustness
MUIRHEAD · Aspects of Multivariate Statistical Theory
MULLER and STOYAN · Comparison Methods for Stochastic Models and Risks
MURTHY, XIE, and JIANG · Weibull Models
MYERS, MONTGOMERY, and ANDERSON-COOK · Response Surface Methodology: Process and Product Optimization Using Designed Experiments, *Third Edition*

*Now available in a lower priced paperback edition in the Wiley Classics Library.
† Now available in a lower priced paperback edition in the Wiley-Interscience Paperback Series.

MYERS, MONTGOMERY, VINING, and ROBINSON · Generalized Linear Models. With Applications in Engineering and the Sciences, *Second Edition*
NATVIG · Multistate Systems Reliability Theory With Applications
† NELSON · Accelerated Testing, Statistical Models, Test Plans, and Data Analyses
† NELSON · Applied Life Data Analysis
NEWMAN · Biostatistical Methods in Epidemiology
NG, TAIN, and TANG · Dirichlet Theory: Theory, Methods and Applications
OKABE, BOOTS, SUGIHARA, and CHIU · Spatial Tesselations: Concepts and Applications of Voronoi Diagrams, *Second Edition*
OLIVER and SMITH · Influence Diagrams, Belief Nets and Decision Analysis
PALTA · Quantitative Methods in Population Health: Extensions of Ordinary Regressions
PANJER · Operational Risk: Modeling and Analytics
PANKRATZ · Forecasting with Dynamic Regression Models
PANKRATZ · Forecasting with Univariate Box-Jenkins Models: Concepts and Cases
PARDOUX · Markov Processes and Applications: Algorithms, Networks, Genome and Finance
PARMIGIANI and INOUE · Decision Theory: Principles and Approaches
* PARZEN · Modern Probability Theory and Its Applications
PEÑA, TIAO, and TSAY · A Course in Time Series Analysis
PESARIN and SALMASO · Permutation Tests for Complex Data: Applications and Software
PIANTADOSI · Clinical Trials: A Methodologic Perspective, *Second Edition*
POURAHMADI · Foundations of Time Series Analysis and Prediction Theory
POWELL · Approximate Dynamic Programming: Solving the Curses of Dimensionality, *Second Edition*
POWELL and RYZHOV · Optimal Learning
PRESS · Subjective and Objective Bayesian Statistics, *Second Edition*
PRESS and TANUR · The Subjectivity of Scientists and the Bayesian Approach
PURI, VILAPLANA, and WERTZ · New Perspectives in Theoretical and Applied Statistics
† PUTERMAN · Markov Decision Processes: Discrete Stochastic Dynamic Programming
QIU · Image Processing and Jump Regression Analysis
* RAO · Linear Statistical Inference and Its Applications, *Second Edition*
RAO · Statistical Inference for Fractional Diffusion Processes
RAUSAND and HØYLAND · System Reliability Theory: Models, Statistical Methods, and Applications, *Second Edition*
RAYNER, THAS, and BEST · Smooth Tests of Goodnes of Fit: Using R, *Second Edition*
RENCHER · Linear Models in Statistics, *Second Edition*
RENCHER · Methods of Multivariate Analysis, *Second Edition*
RENCHER · Multivariate Statistical Inference with Applications
RIGDON and BASU · Statistical Methods for the Reliability of Repairable Systems
* RIPLEY · Spatial Statistics
* RIPLEY · Stochastic Simulation
ROHATGI and SALEH · An Introduction to Probability and Statistics, *Second Edition*
ROLSKI, SCHMIDLI, SCHMIDT, and TEUGELS · Stochastic Processes for Insurance and Finance
ROSENBERGER and LACHIN · Randomization in Clinical Trials: Theory and Practice
ROSSI, ALLENBY, and McCULLOCH · Bayesian Statistics and Marketing
† ROUSSEEUW and LEROY · Robust Regression and Outlier Detection

*Now available in a lower priced paperback edition in the Wiley Classics Library.
† Now available in a lower priced paperback edition in the Wiley-Interscience Paperback Series.

ROYSTON and SAUERBREI · Multivariate Model Building: A Pragmatic Approach to Regression Analysis Based on Fractional Polynomials for Modeling Continuous Variables
* RUBIN · Multiple Imputation for Nonresponse in Surveys
RUBINSTEIN and KROESE · Simulation and the Monte Carlo Method, *Second Edition*
RUBINSTEIN and MELAMED · Modern Simulation and Modeling
RYAN · Modern Engineering Statistics
RYAN · Modern Experimental Design
RYAN · Modern Regression Methods, *Second Edition*
RYAN · Statistical Methods for Quality Improvement, *Third Edition*
SALEH · Theory of Preliminary Test and Stein-Type Estimation with Applications
SALTELLI, CHAN, and SCOTT (editors) · Sensitivity Analysis
SCHERER · Batch Effects and Noise in Microarray Experiments: Sources and Solutions
* SCHEFFE · The Analysis of Variance
SCHIMEK · Smoothing and Regression: Approaches, Computation, and Application
SCHOTT · Matrix Analysis for Statistics, *Second Edition*
SCHOUTENS · Levy Processes in Finance: Pricing Financial Derivatives
SCOTT · Multivariate Density Estimation: Theory, Practice, and Visualization
* SEARLE · Linear Models
† SEARLE · Linear Models for Unbalanced Data
† SEARLE · Matrix Algebra Useful for Statistics
† SEARLE, CASELLA, and McCULLOCH · Variance Components
SEARLE and WILLETT · Matrix Algebra for Applied Economics
SEBER · A Matrix Handbook For Statisticians
† SEBER · Multivariate Observations
SEBER and LEE · Linear Regression Analysis, *Second Edition*
† SEBER and WILD · Nonlinear Regression
SENNOTT · Stochastic Dynamic Programming and the Control of Queueing Systems
* SERFLING · Approximation Theorems of Mathematical Statistics
SHAFER and VOVK · Probability and Finance: It's Only a Game!
SHERMAN · Spatial Statistics and Spatio-Temporal Data: Covariance Functions and Directional Properties
SILVAPULLE and SEN · Constrained Statistical Inference: Inequality, Order, and Shape Restrictions
SINGPURWALLA · Reliability and Risk: A Bayesian Perspective
SMALL and McLEISH · Hilbert Space Methods in Probability and Statistical Inference
SRIVASTAVA · Methods of Multivariate Statistics
STAPLETON · Linear Statistical Models, *Second Edition*
STAPLETON · Models for Probability and Statistical Inference: Theory and Applications
STAUDTE and SHEATHER · Robust Estimation and Testing
STOYAN · Counterexamples in Probability, *Second Edition*
STOYAN, KENDALL, and MECKE · Stochastic Geometry and Its Applications, *Second Edition*
STOYAN and STOYAN · Fractals, Random Shapes and Point Fields: Methods of Geometrical Statistics
STREET and BURGESS · The Construction of Optimal Stated Choice Experiments: Theory and Methods
STYAN · The Collected Papers of T. W. Anderson: 1943-1985

*Now available in a lower priced paperback edition in the Wiley Classics Library.
† Now available in a lower priced paperback edition in the Wiley-Interscience Paperback Series.

SUTTON, ABRAMS, JONES, SHELDON, and SONG · Methods for Meta-Analysis in Medical Research
TAKEZAWA · Introduction to Nonparametric Regression
TAMHANE · Statistical Analysis of Designed Experiments: Theory and Applications
TANAKA · Time Series Analysis: Nonstationary and Noninvertible Distribution Theory
THOMPSON · Empirical Model Building: Data, Models, and Reality, *Second Edition*
THOMPSON · Sampling, *Third Edition*
THOMPSON · Simulation: A Modeler's Approach
THOMPSON and SEBER · Adaptive Sampling
THOMPSON, WILLIAMS, and FINDLAY · Models for Investors in Real World Markets
TIERNEY · LISP-STAT: An Object-Oriented Environment for Statistical Computing and Dynamic Graphics
TSAY · Analysis of Financial Time Series, *Third Edition*
UPTON and FINGLETON · Spatial Data Analysis by Example, Volume II: Categorical and Directional Data
† VAN BELLE · Statistical Rules of Thumb, *Second Edition*
VAN BELLE, FISHER, HEAGERTY, and LUMLEY · Biostatistics: A Methodology for the Health Sciences, *Second Edition*
VESTRUP · The Theory of Measures and Integration
VIDAKOVIC · Statistical Modeling by Wavelets
VIERTL · Statistical Methods for Fuzzy Data
VINOD and REAGLE · Preparing for the Worst: Incorporating Downside Risk in Stock Market Investments
WALLER and GOTWAY · Applied Spatial Statistics for Public Health Data
WEISBERG · Applied Linear Regression, *Third Edition*
WEISBERG · Bias and Causation: Models and Judgment for Valid Comparisons
WELSH · Aspects of Statistical Inference
WESTFALL and YOUNG · Resampling-Based Multiple Testing: Examples and Methods for *p*-Value Adjustment
* WHITTAKER · Graphical Models in Applied Multivariate Statistics
WINKER · Optimization Heuristics in Economics: Applications of Threshold Accepting
WOOD WORTH · Biostatistics: A Bayesian Introduction
WOOLSON and CLARKE · Statistical Methods for the Analysis of Biomedical Data, *Second Edition*
WU and HAMADA · Experiments: Planning, Analysis, and Parameter Design Optimization, *Second Edition*
WU and ZHANG · Nonparametric Regression Methods for Longitudinal Data Analysis
YIN · Clinical Trial Design: Bayesian and Frequentist Adaptive Methods
YOUNG, VALERO-MORA, and FRIENDLY · Visual Statistics: Seeing Data with Dynamic Interactive Graphics
ZACKS · Stage-Wise Adaptive Designs
* ZELLNER · An Introduction to Bayesian Inference in Econometrics
ZELTERMAN · Discrete Distributions—Applications in the Health Sciences
ZHOU, OBUCHOWSKI, and McCLISH · Statistical Methods in Diagnostic Medicine, *Second Edition*

*Now available in a lower priced paperback edition in the Wiley Classics Library.
† Now available in a lower priced paperback edition in the Wiley-Interscience Paperback Series.